The Waltham Book
of Clinical Nutrition of the Dog & Cat

The Waltham Book
of Clinical Nutrition of the Dog & Cat

Edited by

JOSEPHINE M. WILLS

Waltham Centre for Pet Nutrition
Melton Mowbray, Leicestershire

and

KENNETH W. SIMPSON

Royal Veterinary College, Hertfordshire

PERGAMON

U.K.   Elsevier Science Ltd, The Boulevard, Langford Lane, Kidlington,
            Oxford OX5 1GB, U.K.

U.S.A.   Elsevier Science Inc, 660 White Plains Road,
            Tarrytown, New York 10591-5153, U.S.A.

JAPAN   Elsevier Science Japan, Tsunashima Building Annex,
            3-20-12 Yushima, Bunkyo-ku, Tokyo 113, Japan

First edition 1994

Library of Congress Cataloging in Publication Data

The Waltham Book of Clinical Nutrition of the Dog and Cat
edited by Josephine Wills, Kenneth Simpson.
p.   cm. Includes index.
1. Dogs–Diseases–Diet therapy. 2. Cats–Diseases–Diet therapy.
3. Dogs–Nutrition. 4. Cats–Nutrition. I. Wills, Josephine.
II. Simpson, Kenneth. III. Waltham Centre for Pet Nutrition.
SF991.W34   1993   636.7'0895854–dc20   93-27493

British Library Cataloguing in Publication Data

A catalogue record for this book is available from the British Library

ISBN 0 08 042294 2 Hardcover (US Edition)
ISBN 0 08 040839 7 Hardcover (Rest of the World)

Printed in the United States of America

Contents

Foreword

The understanding of the basic nutrition of the dog and cat has evolved over many years. The data currently available on this subject emphasize the need for companion animal species to be studied as separate entities with respect to their individual needs and idiosyncrasies in metabolism of specific nutritional elements. Basic nutrition has been reviewed comprehensively in the recently published *Waltham Book of Companion Animal Nutrition*. The discipline and scientific investigation applied to the basic nutrition of the dog and cat has only recently been applied to the clinical aspects of nutrition and as a result scientific data are replacing long established "folklore."

The purpose of this book is to present the currently available data and the impact that these have on clincial nutrition of the dog and cat. The editors have divided the book into two sections with the first part dedicated to the understanding of the basic principles of clinical nutrition and the second to the practice of clinical nutrition. Each chapter presents information that reflects the current knowledge and views. There will undoubtedly be significant progress in both technology and our understanding of disease processes, so the opinion on certain issues will continue to be modified.

It was a long-held view that clinical nutrition was simply a discipline where nutritionists corrected individual deficiencies. Clinical nutrition is of course much more than this and is of major importance in veterinary practice. Our understanding shows quite clearly that appropriate nutrition has a significant impact on the moderation of disease processes in many organs. The impact of a growing body of expertise on our management of these cases is significant, and as a result our ability to manage rationally the nutritional requirements of these patients has been transformed. The scientific validation of such relationships has not always been forthcoming but such data are now emerging and are reviewed in the clinical chapters in part two of this text.

The text reaffirms the accepted view that clinical nutrition has to be considered as an integral part of the management of many cases in clinical veterinary practice. The editors are to be congratulated in assembling a distinguished panel of authors and ensuring that the material is presented in a clear and logical manner.

Neil T. Gorman BVSc PhD FRCVS DipACVIM
Head of Research, WALTHAM Centre for Pet Nutrition

List of Contributors

ROGER M. BATT MSc PhD MRCVS:

Qualified from Bristol University in 1972 and following a year as intern at the University of Pennsylvania, he joined the Department of Medicine at the Royal Postgraduate Medical School in London, where he received an MSc in Biochemistry and a PhD for studies on the effects of glucocorticoids on the small intestine. He then began investigations on naturally occurring gastrointestinal diseases in dogs, and established a Comparative Gastroenterology Research Group at the University of Liverpool where he was Reader in Veterinary Pathology. In 1990 he was appointed Professor of Veterinary Medicine and head of the Department of Small Animal Medicine and Surgery at the Royal Veterinary College, University of London. He has published more than 200 papers, book chapters and research abstracts, and for his research received a 1989 Ralston Purina Award from the AVMA, the 1990 Walter-Frei Prize from the University of Zurich and the 1991 Woodrow Award from the BSAVA.
Address: Department of Small Animal Medicine, Royal Veterinary College, University of London, Hawkshead Lane, North Mymms, AL9 7TA, U.K.

JOHN E. BAUER DVM PhD DipACVN:

Dr. John E. Bauer earned the DVM and PhD (Nutritional Biochemistry) degrees at the University of Illinois at Urbana–Champaign, USA. He is a Charter Diplomate and Immediate Past-Chairman of the American College of Veterinary Nutrition. Currently he is the Mark L. Morris Professor of Veterinary Clinical Nutrition at Texas A&M University, College of Veterinary Medicine and Director of the Comparative Nutrition Research Laboratory and Clinical Nutrition Support Services in the Veterinary Teaching Hospital. Previously a Professor of Physiological Sciences at the University of Florida, College of Veterinary Medicine, Dr. Bauer directed the Clinical Chemistry and Nutrition Laboratories. He is a member of the American Oil Chemists' Society, the AVMA, and recipient of the Cornelius Research Award. Dr. Bauer's research interests are in the area of molecular and biochemical nutrition specifically relating to dietary effects on lipid and lipoprotein metabolism. His studies

have included lipoprotein metabolism of domestic and exotic animals as well as animal models of hypercholesterolemia and atherogenesis in man.
Address: Department of Physiology, Texas Veterinary Medical Center, Texas A&M University, College Station, TX 77843–4474, U.S.A.

SCOTT A. BROWN VMD PhD DipACVIM:

Scott was awarded a VMD from the University of Pennsylvania School of Veterinary Medicine in 1982. He completed a postdoctoral fellowship at the Nephrology Research and Training Center at the School of Medicine of the University of Alabama-at-Birmingham in 1987 where he studied single nephron function in dogs with renal dysfunction. He was awarded a PhD in Renal Physiology in 1989 by the University of Georgia for studies of the role of phosphate in the progression of canine chronic renal disease. Scott is currently an Associate Professor at the College of Veterinary Medicine of the University of Georgia where he studies the role of nutrition in the course of chronic renal disease in dogs and cats.
Address: Department of Physiology, University of Georgia, College of Veterinary Medicine, Athens, GA 30602, U.S.A.

TONY C. BUFFINGTON DVM BS MS PhD DipACVN:

Tony received the DVM, BS, MS and PhD degrees in Nutrition from the University of California at Davis, and is board certified in Veterinary Nutrition. He joined the faculty of veterinary clinical sciences at The Ohio State University College of Veterinary Medicine in 1987, where he currently is an Associate Professor and Chief of the hospital's Nutrition Support Service. His professional interests include feline lower urinary tract diseases, nutrition and urolithiasis and nutritional support of hospitalized patients.
Address: The Ohio State University, 601 Vernon L. Tharp Street, Columbus, OH 43210–1089, U.S.A.

IVAN H. BURGER BSc PhD:

Ivan graduated in Physiology and Biochemistry from Southampton University in 1968. After graduation he joined the Leatherhead Food Research Association to carry out research in the biochemistry of meat curing, with particular reference to the involvement of the mitochondrial electron transport chain. This work was presented as a PhD thesis (awarded 1972) on a collaborative basis with the Biochemistry Department of Surrey University. In 1973 Ivan moved to WCPN. His first job was Nutritional Biochemist and he conducted work on the nutritional requirements of the cat, in particular protein and amino acids. Following this, he assumed responsibility for food safety in relation to additives and contaminants while retaining an interest in the nutritional work. In 1987 he started a 2-year assignment into the interactions between people and companion animals with particular reference to the effects of pet animals on the health of their owners. In 1989 Ivan rejoined the nutrition group as Senior Nutritionist in charge of studies in the dog, while retaining responsibilities for food safety. In 1992 he further extended his job to include responsibility for equine nutrition.
Address: WALTHAM Centre for Pet Nutrition, Waltham-on-the-Wolds, Melton Mowbray, LE14 4RT, U.K.

COLIN F. BURROWS BVetMed PhD MRCVS DipACVIM:

After graduating from the Royal Veterinary College, University of London in 1969, Dr. Burrows spent a year as house physician at the Beaumont Animal Hospital and in 1970 joined the faculty of the University of Pennsylvania School of Veterinary Medicine. While at the University of Pennsylvania, Dr. Burrows developed interests in critical care, gastroenterology, nutrition and gastrointestinal physiology and pathophysiology, became board certified in internal medicine and in 1980 earned a PhD in gastrointestinal physiology. Dr. Burrows has ongoing research interests in canine nutrition and in the motility of the gastrointestinal tract in health and disease, with particular emphasis on the disruptions that occur in gastric dilatation–volvulus and acute diarrhea. Dr. Burrows is a past-President of the Comparative Gastroenterology Society and the Eastern States Veterinary Association. He has been a member of the faculty at the University of Florida since 1980.

Address: University of Florida, College of Veterinary Medicine, Small Animal Clinical Sciences, PO Box 100126 HSC, Gainesville, FL 32610–0126, U.S.A.

RICHARD F. BUTTERWICK BSc PhD:

Richard graduated from the Department of Agricultural Biochemistry and Nutrition, Newcastle University in 1985. He was awarded a PhD in 1989 by Newcastle University for studies on the long-term effects of bovine somatotrophin (BST) in lactating dairy cows. During 1989 Richard joined the Samora Machel School of Veterinary Medicine, Zambia as a visiting lecturer in Biochemistry. He then rejoined the Department of Agricultural Biochemistry and Nutrition, Newcastle University as a postdoctoral research fellow. In 1990 he joined the Department of Paediatric Endocrinology at the Medical College of Saint Bartholomew's Hospital as a postdoctoral research fellow and was involved in clinical and research studies in short-statured children. Richard joined WCPN in 1991 and is currently involved in the research and development of clinical diets, with an emphasis on the management of obesity and postoperative nutritional support.

Address: WALTHAM Centre for Pet Nutrition, Waltham-on-the-Wolds, Melton Mowbray, LE14 4RT, U.K.

W. JEAN DODDS DVM:

Jean graduated from the Ontario Veterinary College, University of Toronto in 1964 with a Doctor of Veterinary Medicine degree. She then did graduate studies in Pathology at the Albany Medical College and embarked on a 25-year career as a research scientist with the State Department of Health in Albany, New York. Her research expertise is primarily in the areas of comparative hematology and immunology with emphasis on bleeding disorders. More recently this has expanded into the study of autoimmune diseases and endocrinology, and the role of nutrition in maintaining the integrity of the immune system. The long-standing research program she has directed is funded by the National Institutes of Health and supports a unique colony of dogs with inherited bleeding disorders. In 1987 Jean relocated to Southern California where she developed the first national not-for-profit animal blood bank that provides blood components for clinical veterinary transfusions throughout the United States and Canada. Jean maintains an advisory role with the research program in New York and frequently travels nationally and internationally to teach professionals and the public. She has been an active dog fancier and exhibitor for many years.

Address: 938 Standard Street, Santa Monica, CA 90403, U.S.A.

SUSAN DONOGHUE MS VMD DipACVN:

Susan is president of Nutrition Support Services, Inc., Pembroke, Virginia. She trained in Animal Nutrition, University of Connecticut, and in Veterinary Medicine and Clinical Nutrition, University of Pennsylvania. She is board certified in nutrition. She has studied vitamin A nutriture in rabbits, sheep and horses. Dr. Donoghue also developed the clinical specialty of nutrition support for sick animals, especially cats and dogs, for captive mammals, birds and reptiles and for oil-spill damaged mammals and birds. She has published over 50 scientific papers and presented over 50 invited talks in six countries. When not farming her tortoises and draught horses, she writes for fanciers of dogs, cats and other companion animals, reptiles and amphibians and consults on clinical nutrition for veterinarians, wildlife rehabilitators and curators.
Address: Nutrition Support Services, Inc., Box 189, Pembroke, VA 24136, U.S.A.

CLIVE M. ELWOOD MA VetMB CertSAC MRCVS:

Clive qualified as a veterinarian from Cambridge University in 1989. After a short period in general practice he was appointed Demonstrator in the Department of Small Animal Studies, University of Liverpool. From 1990 to 1993 he was a Resident in Small Animal Internal Medicine at the Royal Veterinary College, gaining his Cardiology Certificate in September 1991. He is currently undertaking a PhD project on cell-mediated immune responses in the canine gut, funded by the Wellcome Trust.
Address: Department of Small Animal Medicine, Royal Veterinary College, University of London, Hawkshead Lane, North Mymms, AL9 7TA, U.K.

PETER A. GRAHAM BVMS CertVR MRCVS:

Peter Graham qualified from Glasgow University in 1989. He spent the following two years as a Small Animal House Physician in the Department of Veterinary Medicine at Glasgow University and was awarded the RCVS Certificate in Veterinary Radiology in 1991. Since then, he has remained in the Department of Veterinary Medicine at Glasgow and is now researching diabetes mellitus in companion animals.
Address: Department of Veterinary Medicine, Glasgow University Veterinary School, Bearsden Road, Glasgow, GO1 1QH, U.K.

W. GRANT GUILFORD BVSc PhD:

Grant graduated from Massey University, New Zealand in 1982. Prior to entering academic life he spent 4 years in private practice in New Zealand and England. He undertook a residency in small animal medicine at the University of Missouri. Grant then accepted an appointment as a lecturer at the University of California at Davis. He then completed a nutrition PhD program at Davis, before heading home to his native New Zealand where he is currently a senior lecturer in small animal medicine at Massey. Grant is a Diplomate of the American College of Veterinary Internal Medicine and a Fellow of the Australian College of Veterinary Scientists.
Address: Department of Veterinary Clinical Sciences, Palmerston North 5301, Massey University, New Zealand.

DUNCAN K. HALL BVSc MRCVS GradDipMktg GradDipBus(Acc):

Duncan graduated from the University of Melbourne (Australia) with an honors degree in Veterinary Medicine in 1982. After 18 months in practice in both Australia and England Duncan joined the pharmaceutical industry working in product marketing for 6 years and during this time he completed Graduate Diplomas in both Marketing and Accounting. Duncan joined WALTHAM in Australia as a Veterinary Adviser in 1990 and moved to England in mid-1992 following his appointment as Communications Manager at WCPN.
Address: WALTHAM Centre for Pet Nutrition, Waltham-on-the-Wolds, Melton Mowbray, LE14 4RT, U.K.

RICHARD E. W. HALLIWELL MA VetMB PhD MRCVS:

Richard Halliwell received his veterinary degree from Cambridge in 1961. After a period in practice, he returned to Cambridge and was awarded his PhD in 1970 for studies on canine IgE. He spent 5 years at the University of Pennsylvania and then 10 years at the University of Florida where he was Chairman of the Department of Medical Sciences in the College of Veterinary Medicine. In 1987, he was appointed William Dick Professor of Veterinary Clinical Studies at The University of Edinburgh. He has published extensively in the area of allergic diseases of domestic animals.
Address: Royal (Dick) Veterinary School, Summerhall, Edinburgh, U.K.

ALAN S. HAMMER DVM DipACVIM-Oncology:

Alan graduated from The Ohio State University, College of Veterinary Medicine in 1985 and completed a residency in oncology and internal medicine in 1990. He is currently an Assistant Professor at The Ohio State University, Department of Veterinary Clinical Sciences where he practices clinical oncology. His areas of interest include chemotherapy, immunotherapy and the detection and imaging of cancer in animals.
Address: Veterinary Teaching Hospital, 601 Vernon L. Tharp Street, The Ohio State University, Columbus, OH 43210, U.S.A.

JOHN G. HARTE MVB MRCVS:

John graduated from University College Dublin in 1989, with an honors degree in Veterinary Medicine. He returned to practice in the British Isles after completing an internship in small animal medicine and surgery at the University of Minnesota. He joined WCPN in 1991 and is involved in the research development of clinical diets.
Address: WALTHAM Centre for Pet Nutrition, Waltham-on-the-Wolds, Melton Mowbray, LE14 4RT, U.K.

RICHARD G. HARVEY BVSc CertSAD MIBiol MRCVS:

Richard graduated from the University of Bristol in 1978 and is a partner in a small animal practice in Coventry. He obtained his Certificate in Small Animal Dermatology in 1988 and has a number of clinics in which he sees cases referred by colleagues. He has a general interest in the relationship between nutrition and the skin and, in particular, the role of essential fatty acids. He has published a series of three papers exploring the role of essential fatty acids in the management of feline dermatoses.

Address: Quinten Veterinary Centre, 207 Daventry Road, Cheylesmore, Coventry, CV3 5HH, U.K.

HERMAN A. W. HAZEWINKEL DVM PhD DiplEurCollVetSurgery:

Dr. Hazewinkel graduated as veterinarian in 1976 from Utrecht University and after working in private practice he joined the university's department of veterinary sciences in companion animals. Responsible for referred orthopedic patients, he became interested in nutritional and hormonal relations in skeletal diseases. Dr. Hazewinkel is currently involved in research focused on metabolic influences on skeletal development as well as in other aspects of small animal orthopedics, including his membership of the national hip dysplasia committee.
Address: University of Utrecht, Faculty of Veterinary Medicine, Department of Clinical Science of Companion Animals, Yalelaan 8, PO Box 80.154, 3508 TD Utrecht, The Netherlands.

RICHARD C. HILL MA VetMB PhD:

Richard completed his veterinary training at Cambridge University in 1980 and spent 5 years as a Small Animal Practitioner in a large practice north of London. Subsequently, he has completed a 2-year residency in Small Animal Medicine at the University of Pennsylvania and has just been awarded a PhD in Veterinary Medicine by the University of Florida for his work on nutrition and gastrointestinal physiology in dogs. While studying for his PhD he has been running a Clinical Nutrition Service at the University of Florida Small Animal Hospital and has just joined the Faculty as the Waltham Assistant Professor of Clinical Nutrition. His current responsibilities involve teaching both Nutrition and Small Animal Medicine, but he is continuing his research into colonic physiology and assisting with a study into the effect of diet in pancreatic disease.
Address: University of Florida, College of Veterinary Medicine, Small Animal Clinical Sciences, PO Box 100126, HSC, Gainesville, FL 32610–0126, U.S.A.

ASTRID E. HOPPE DVM PhD:

In 1976 Astrid graduated from The Royal Veterinary School in Stockholm, Sweden. Since 1983 she has been working as an Assistant Professor at the Department of Medicine and Surgery, Swedish University of Agricultural Sciences, Uppsala, teaching internal medicine. She was awarded a PhD in 1992 at the Swedish University of Agricultural Sciences for studies on cystinuria in the dog. For her postdoctoral research, she studies long-term treatment of cystinuria, progressive nephropathy in the dog and bacteriuria in dogs and cats. Astrid is now currently involved in teaching nephrology and urology as well as seeing referred cases at the small animal clinic at the university.
Address: Morby, Lagga, S-755 90, Uppsala, Sweden.

ELLEN KIENZLE DMU DMVA:

Ellen studied Veterinary Medicine at Tierärztlichen Hochschule Hannover from 1976–1981. She then completed her dissertation at the Institute for Animal Nutrition of Tierärztlichen Hochschule Hannover and was promoted to DrMedVet in 1983. Dr. Kienzle then became the assistant of Professor Meyer at the above mentioned institute. She became Veterinary

Specialist for nutrition and dietetics in 1988 and Dr. MedVetHabit in 1989. She has recently been appointed head of the department of Veterinary Nutrition in the Veterinary Faculty of Ludwig-Maximilians-Universität München. Dr. Kienzle's scientific work includes nutrition studies in dogs, cats, horses and pigs as well as nutrition advice in all species of domestic animals.
Address: Institut für Physiologie, Physiologische Chemie und Ernährungsphysiologie der Tierärztlichen Fakultät der Ludwig-Maximilians-Universität München, Germany.

DAVID S. KRONFELD MA PhD DSc MVSc MRCVS DipACVN DipACVIM:

David is the Paul Mellon Distinguished Professor of Agriculture and Veterinary Medicine, Virginia Polytechnic University. He trained in Veterinary Science and Biochemistry, University of Queensland, and in Comparative Physiology, University of California at Davis. He is board certified in two clinical specialties: nutrition and veterinary internal medicine. He has studied metabolic responses to pregnancy, parturition, lactation, growth, old age and strenuous exercise, mainly in dairy cows, sheep, dogs and horses. He has published over 400 scientific papers and presented over 200 invited talks in 17 countries. Address: Box 189, Pembroke, VA 24136, U.S.A.

PETER J. MARKWELL BSc, BVetMed, MRCVS:

Peter graduated from the Royal Veterinary College, University of London, in 1981, having previously obtained a degree in basic medical sciences with anatomy from King's College, University of London. He then worked in small animal practice for 3 years, before joining the staff of the Royal Veterinary College as a lecturer in companion animal husbandry in the Department of Animal Health. He joined the WALTHAM Centre for Pet Nutrition at the end of 1985 as Veterinary Advisor and subsequently became Senior Nutritionist. He is currently Senior Clinical Nutritionist with responsibility for the research and development of clinical diets.
Address: WALTHAM Centre for Pet Nutrition, Waltham-on-the-Wolds, Melton Mowbray, LE14 4RT, U.K.

IAN E. MASKELL BSc PhD:

Ian graduated from the department of Animal Physiology and Nutrition, Leeds University in 1986. He was awarded a PhD in 1990 by Newcastle University for studies of nutritional and toxicological aspects of rapeseed meal in porcine diets. Ian joined WCPN in 1990 and has worked on the research development of clinical diets, with an emphasis on gastro-intestinal-related illness.
Address: WALTHAM Centre for Pet Nutrition, Waltham-on-the-Wolds, Melton Mowbray, LE14 4RT, U.K.

MATTHEW W. MILLER DVM DipACVIM:

In 1984, Matthew graduated from The Ohio State University with a Doctor of Veterinary Medicine degree. He completed a small animal medicine and surgery internship at the West Los Angeles Veterinary Medical Group from 1984 to 1985. In 1985, he returned to The Ohio State University College of Veterinary Medicine to start a cardiology residency that he

completed in 1988. He joined the faculty of the Department of Small Animal Medicine and Surgery at Texas A&M University in 1988 and attained diplomate status in the American College of Veterinary Internal Medicine, subspecialty Cardiology, in the same year. He is currently the staff cardiologist at the Texas Veterinary Medical Center.
Address: Texas Veterinary Medical Center, Texas A&M University, College Station, TX 77843–4474, U.S.A.

ROBERT MORAILLON DVM:

Robert graduated from the Ecole Nationale Vétérinaire d'Alfort in 1961 and is a holder of two license certificates from the Faculty of Sciences of Paris. Robert Moraillon was assistant at the Ecole Nationale Vétérinaire d'Alfort in 1963, Assistant Professor in 1964 and Professor in 1970. Since 1973 he has been the manager of the Department of Medicine for Horses and Carnivores and was responsible for the Department of Companion Animal Medicine at the Ecole Nationale Vétérinaire d'Alfort from 1987 to 1991. He is also vice-president of the French Small Animal Veterinary Association (CNVSPA).
Address: Ecole Nationale Vétérinaire, Service de Médecine, 7 Avenue du Général de Gaulle, 94704 Maisons-Alfort Cedex, France.

H. CAROLIEN RUTGERS DVM MS MRCVS DipACVIM:

Dr. Rutgers qualified from Utrecht State University in 1978. Between 1979 and 1980, she followed an internship in small animal medicine and surgery at the University of Pennsylvania. After a year in small animal practice in The Netherlands, she returned to the U.S.A. in 1982 for a residency in small animal internal medicine at The Ohio State University, where she also gained an MSc for studies on canine liver disease. She was at the University of Liverpool from 1985 to 1990 as a lecturer in small animal medicine, and became a diplomate of the American College of Veterinary Internal Medicine in 1988. Since 1990 she has been lecturer in small animal internal medicine at The Royal Veterinary College in London, where she specializes in gastroenterology, liver disease and feline medicine. She has published several papers and book chapters and has lectured at national and international meetings on these subjects.
Address: Department of Small Animal Medicine, Royal Veterinary College, University of London, Hawkshead Lane, North Mymms, AL9 7TA, U.K.

KENNETH W. SIMPSON BVM&S PhD MRCVS DipACVIM:

Dr. Simpson received his veterinary degree (BVM&S) from the University of Edinburgh in 1984. Postgraduate studies of canine pancreatic and intestinal function at the University of Leicester (PhD, 1988) were followed by an internship in small animal medicine and surgery at the University of Pennsylvania and a residency in small animal internal medicine at The Ohio State University. Dr. Simpson is a Diplomate of the American College of Veterinary Internal Medicine and a recipient of the National Phi Zeta Award. He is currently a lecturer at the Royal Veterinary College, London with clinical and research interests in small animal internal medicine and gastroenterology.
Address: Department of Small Animal Medicine, Royal Veterinary College, University of London, Hawkshead Lane, North Mymms, AL9 7TA, U.K.

REBECCA L. STEPIEN BS DVM MS DipACVIM:

Rebecca Stepien was a graduate in 1987 of the University of Wisconsin School of Veterinary Medicine. She completed an internship at The Animal Medical Center in New York, NY in 1988, and completed a cardiology residency at The Ohio State University in 1991. She received her MSc from The Ohio State University and became board certified in Cardiology through the American College of Veterinary Internal Medicine in 1991. She is currently a Lecturer in Cardiology and Small Animal Medicine at the Royal Veterinary College in London. Dr. Stepien's research interests include electrolyte imbalances in the therapy of refractory cardiovascular disease and cardiovascular drug interactions.
Address: Department of Small Animal Medicine, Royal Veterinary College, University of London, Hawkshead Lane, North Mymms, AL9 7TA, U.K.

ANGELE THOMPSON BS PhD:

Angele graduated from the University of California at Los Angeles in Biochemistry in 1975. She was awarded a PhD in Nutritional Biochemistry from the University of Arizona in 1980 for her studies on the effects of dietary fibers on trace minerals. Angele then joined Heinz Pet Products as a Nutritionist. While at Heinz, she gained experience in nutrition, product development and regulatory affairs. She was ultimately promoted to Manager Laboratory Technical Services. In 1991, Angele joined Kal Kan Foods as a Product Development Manager. In the same year, Angele was selected as a member of the AAFCO Feline Nutrition Expert Subcommittee, which developed the AAFCO Cat Food Nutrient Profile, AAFCO Lifestage Feeding Protocols and the AAFCO Digestibility Protocol. Angele is currently the Chairperson of the Nutrition Task Force for the Pet Food Institute.
Address: Kal Kan Foods, Inc., 3250 East 44th Street, Vernon, CA 90068–0853, U.S.A.

VICTORIA L. VOITH DVM MA PhD:

Victoria is a veterinarian (The Ohio State University, 1968) with advanced degrees in Psychology (OSU, 1975) and Neuroanatomy/Animal Behavior (UC Davis, 1982). She has been engaged in clinical animal behavior since 1970 and developed the Animal Behavior Clinic of the Hospital of the University of Pennsylvania. Her research interests center around human–companion animal interactions and attachment as well as the treatment and diagnoses of behavior problems in animals. She has written many articles concerning the aforementioned subjects, is a charter diplomate in the American College of Veterinary Behavior and received the State of Pennsylvania Leo Bustad Companion Animal Veterinarian of the Year Award in 1986. She currently serves as an animal behavior consultant based at a ranch in Spring Branch, Texas.
Address: 1015 Flying R Ranch Road East, Spring Branch, TX 78070, U.S.A.

JOSEPHINE M. WILLS BVetMed PhD MRCVS:

Jo qualified from the Royal Veterinary College, London, in 1981, then spent 2 years in small animal practice, where an interest in feline medicine developed. She then joined the department of veterinary medicine at Bristol University, and researched into *Chlamydia psittaci* infection in cats, and other feline infectious diseases. She was awarded a PhD for her work on *Chlamydia* in cats in 1986. After a period of postdoctoral research at the University of Manchester Medical School, she joined WCPN in 1988. Jo was Scientific Editor for the

Bulletin of the Feline Advisory Bureau for 5 years. She is now Scientific Affairs Manager at WCPN.

Address: The WALTHAM Centre for Pet Nutrition, Waltham-on-the-Wolds, Melton Mowbray, LE14 4RT, U.K.

ROGER WOLTER DVM PhD:

Dr. Wolter studied at the Ecole Nationale Vétérinaire d'Alfort (1956–1960), where he later became Assistant Professor. In 1968 he received his doctorate. He moved to Ecole Nationale Vétérinaire of Lyon where he was appointed as Professor in 1971. Since 1980 he has been the Professor of Nutrition and Food Supply at the ENV of Alfort. Research interests have included nutrition of horses in relation to sport, food additives and metabolic nutrition diseases and nutrition of the cow. Dr. Wolter has been the author of about 400 publications on topics including bovine, equine, feline and canine nutrition and the relationship between diet and sporting ability. He is also an expert at the court of appeals, Paris, a member of the interministerial commission of animal feeding and an expert at the European Commission for regulations in Animal Nutrition.

Address: Ecole Nationale Vétérinaire, Service d'Alimentation, 7 Avenue du Général de Gaule, 94704 Maisons-Alfort Cedex, France.

PART 1

Principles of
Clinical Nutrition

CHAPTER 1

Inappropriate Feeding: The Importance of a Balanced Diet

ELLEN KIENZLE and DUNCAN K. HALL

Introduction

Inappropriate feeding practices can result in the development of a wide variety of nutritional imbalances in cats and dogs that may lead to clinical disease. Such nutritional disorders represent a significant diagnostic challenge for veterinarians due to the often complex pathogenesis and subsequent diversity of presenting clinical signs. There are many reasons why pets are often fed inappropriate diets. In some cases it is due to an inappropriate choice of food by the pet owner, such as when cat owners attempt to feed their pets on an entirely vegetarian diet. In other cases inappropriate pet feeding practices reflect a more complex problem that results from a genuine lack of awareness among new pet owners as to how to care for a dog or cat. This is commonly encountered in countries where pet ownership has not previously been widespread but with improved economic circumstances is now increasing rapidly. This presents veterinarians with a rewarding educational challenge to provide advice regarding all aspects of pet care to their clients.

Veterinarians are familiar with the well-documented problems such as nutritional secondary hyperparathyroidism in young dogs and cats fed predominantly meat diets and consequently practitioners endeavor to provide sound nutritional advice to pet owners to avoid this disease state. Veterinarians should, however, consider nutrition-related disorders in the differential diagnosis when presented with medical cases because many of these conditions can have subtle presenting signs in the early stages of disease and may take a considerable time to develop. Examples of such diseases in companion animals include the gradual depletion of body stores of nutrients, such as taurine deficiency or the recently reported episodic hypokalemic polymyopathy in cats fed a high protein, low potassium vegetarian diet.[44] Similarly, clinical signs associated with excessive intake of nutrients can also develop over a protracted period of time, as is seen where cat owners feed a diet consisting predominantly of liver with the subsequent development of clinical signs associated with hypervitaminosis A. Fortunately, with the dramatic increase in our knowledge regarding the nutritional requirements of companion animals and the widespread feeding of well-formulated petfoods,

1

some conditions such as niacin deficiency in dogs, historically associated with the feeding of home-prepared diets based on flaked maize, are now rarely seen.

This chapter reviews diseases of cats and dogs that can develop as a result of inappropriate nutrient supply. It does not deal with other nutrition-related problems that might develop as a consequence of poor food hygiene or an inadequate feeding technique.

Potential Problems Associated with Nutrient Deficiencies

Nutrient deficiencies are most likely to occur in growing animals. In adults in a maintenance state, deficiencies of most nutrients are less likely to develop due to body stores and comparatively low requirements. The development of nutrient deficiencies in adults usually takes several weeks or even months.

Protein

Isolated protein deficiency is unusual in pet dogs and cats because low protein diets are rather unpalatable. When fed a diet with 2.5 g digestible protein/100 kcal digestible energy (DE), cats did not eat enough food to meet their maintenance energy requirement.[8] In dogs, a high content of sugar or other highly palatable compounds may increase acceptance of low protein diets. Therefore by feeding sweetmeats an isolated protein deficiency could be induced.[52] Signs of deficiency are dull coats, skin lesions, apathy and hypoproteinemia. In growing puppies, the cartilage in the joints is thinned and in severe cases mineralization of bones may be impaired. Newborn puppies may be underdeveloped after severe protein deficiency in pregnant bitches. Protein deficiency during lactation leads to exhaustion of the dam as well as to insufficient milk yield.

Taurine

Taurine is an amino sulfonic acid that is an end product of the metabolism of the sulfur-containing amino acids, methionine and cystine. It appears to have an important role in regulating calcium fluxes across membranes, osmoregulation and neuromodulation. Most mammals are able to synthesize sufficient taurine to meet their metabolic needs. Cats, however, have only a very limited ability to synthesize taurine and consequently must consume sufficient preformed taurine in their diet.[9] This requirement for dietary taurine is increased due to the cats' obligatory loss of taurine in conjugating bile salts. Unlike other placental mammals, cats cannot substitute glycine when taurine is limited.[57] The bioavailability of taurine varies in relation to the protein source and preparation (intensity and time of heating).[29] Current recommendations for the inclusion in commercial cat foods are 1200 mg taurine/kg dry matter (DM) for a dry diet and 2500 mg/kg DM for a moist food.[57]

Taurine deficiency in the cat can result in abnormalities in a number of body systems. Conditions that have been associated with taurine deficiency in the cat include central retinal degeneration, dilated cardiomyopathy (see Chapter 20), reproductive failure and developmental abnormalities in kittens.

Taurine deficiency is most likely to develop where the cat is fed inappropriately, such as when the cat is fed commercial dry dog foods (which contain very limited taurine)[1] or where the owner has attempted to feed the cat a vegetarian diet, which contains little or no taurine, or a diet with large amounts of low quality protein (offals) that may impair bioavailability.

Diagnosis can be made from plasma or urine taurine levels. Plasma taurine levels below 40 μmol/l are marginal.[57] A taurine-deficient state can be assumed if taurine is not detectable in urine samples.[58]

Calcium and Phosphorus

The absolute requirement for calcium and phosphorus depends on the stage of life. Growing puppies, especially of fast growing large and giant breeds have a higher requirement than adults of the same breed. The situation is further complicated by different published requirement figures. Disagreements between sources are mainly due to different figures on availability of calcium and phosphorus; the net requirement (inevitable losses in feces, urine and skin as well as calcium accretion during pregnancy or growth or calcium losses via milk) can be calculated more accurately. Calcium and phosphorus availability can be impaired by diets high in phytate or low in vitamin D.[62]

For dogs, Meyer suggests a calcium content in the food of 9–16 g/kg DM and a phosphorus content of 7–8 g/kg DM (lower range for maintenance, higher range for growth) if the energy content of the food amounts to about 3824 kcal DE/kg DM.[52] This is the case for many dry foods. Moist foods often have a higher energy content (about 4780 kcal DE/kg DM) and consequently must contain more calcium (11–19 g/kg DM) and phosphorus (8–10 g/kg DM). For growing cats the National Research Council recommends a calcium content of 8 g/kg DM and a phosphorus content of 6 g/kg DM[63] (energy content in food 5020 kcal/kg of metabolizable energy (ME)).* The ratio calcium to phosphorus (by weight) should not be lower than 1:1 for dogs and 0.9:1 for cats or exceed 2:1 for either species. Calcium deficiency is well described, but inadequate supply of phosphorus has so far only been described in connection with calcium excess.[26]

Nutritional Secondary Hyperparathyroidism

Nutritional secondary hyperparathyroidism arises when there is an absolute or relative deficiency of calcium in the diet. An imbalance of calcium to phosphorus in diets consisting of predominantly meat can lead to hypocalcemia and subsequent hyperparathyroidism, which stimulates bone resorption to maintain blood calcium levels within their normal

---

* Metabolizable energy (ME) and digestible energy (DE) are different approaches to energy evaluation. DE depends on experimental data on nutrient digestibility. ME is calculated by multiplying the content of nutrients with factors reflecting their digestibility and utilization. In practice, differences between both systems are small as long as diets do not show a very unusual composition (e.g. diets high in fiber).

range. Young animals are especially susceptible to this condition. Nutritional secondary hyperparathyroidism is the most frequently observed clinical result when young animals are fed an all-meat diet, such as the recently reported case of nutritional secondary hyperparathyroidism in a 6-month-old kitten that had been fed exclusively on a meat food formulated for human babies.[13]

All meats, including offal (except bones) as well as cereals contain very limited amounts of calcium and have adverse calcium to phosphorus ratios for optimum skeletal development. Therefore home-made diets based on these compounds must be supplemented with mineral mixtures containing substantial percentages of calcium (>8% on an "as is" basis if the amount of mineral mix is to remain within a reasonable range); commercial pet-foods containing meat and cereals are usually balanced with respect to the calcium and phosphorus ratio, and do not require additional mineral supplementation.

Clinically, affected animals present in a good body condition with variable degrees of locomotor problems. Typically there is pain upon palpation of the bones and affected animals may show signs from a reluctance to use one limb, to complete inability to stand. Radiographically, bones appear poorly mineralized with thin cortices and there is poor contrast between bones and soft tissues. Pathological fractures are often visible.[35] Plasma calcium level is not suitable for diagnostic purposes, because it will not reflect the total body calcium status.

Eclampsia

The etiology of eclampsia (hypocalcemia) in lactating bitches is not fully understood. There is a predisposition in toy breeds.[2] This may be due to the feeding practice of toy breed owners, who give their dogs a diet containing substantial quantities of meat, which fails to provide sufficient calcium. Alternatively, it has been proposed that oversupplementation of calcium can contribute to inducing hypocalcemia due to decreased parathyroid hormone release, as in dairy cows.

FIG. 1.1 Typical posture exhibited by cats with hypokalemia. (Photograph courtesy of Professor J. Morris, University of California at Davis.)

Magnesium, Sodium and Potassium

Deficiencies of these minerals are uncommon in cats and dogs. Magnesium deficiency may occur on poorly formulated low magnesium "struvite prevention" diet, sodium deficiency on a low sodium "heart" diet, especially if diuretics are used. Potassium deficiency has been reported in healthy adult cats on a low potassium vegetarian diet (850 mg K/kg DM).[44] Polyuric renal failure has also been associated with potassium deficiency due to increased renal losses.[17] Increased fecal potassium losses have been observed in connection with diarrhea.[37] Clinical signs of deficiency are muscular weakness, ventroflexion of the head and a stiff gait. A diagnosis can be made from the plasma potassium level (Fig. 1.1).

Iron

In suckling puppies mild iron deficiency develops at the end of the suckling period if additional feeding starts too late (>5 weeks). This condition, however, must be carefully discriminated from a physiological anemia that has been observed both in puppies and kittens during the first 4 weeks of life[11,13,36,39,40,46] and that cannot completely be prevented by iron supplementation.[46] Blood hematocrit and hemoglobin levels should not fall below 30% and 10 g/100 ml, respectively, in suckling puppies. Severe iron deficiency (probably associated with copper deficiency) and fatal anemia have been observed in 3-week-old hand-reared puppies, who were fed on a milk substitute containing limited amounts of iron and copper, amounting to only 30% and 10%, respectively, of the content in bitch's milk.[39] It has also been suggested that the iron requirement may increase considerably in bitches during late pregnancy, where a marginal supply might occur on milk- or cereal-based diets.[36] During lactation and maintenance, iron deficiency is unlikely even on a low iron diet (i.e. milk- or cereal-based diet without meat or offals).[59] Nonnutritive iron deficiency can occur after acute or chronic blood losses and in parasitism during all stages of life.

Copper

Experimentally induced copper deficiency has been described in growing puppies.[4,80]

FIG. 1.2 Silver discoloration of coat (left) typical of copper deficiency compared with normal coat coloring in tabby domestic short haired kittens.

Signs of deficiency are anemia, depigmentation of the hair and disorders of skeletal development (Fig. 1.2). The latter were more severe in the older study, probably because the supply of other nutrients (calcium, phosphorus, vitamins) was not in the optimal range. The copper content of the diet amounted to about 1 mg/kg DM in both studies. Cats appear to be less susceptible to copper deficiency. Doong *et al.*[16] did not succeed in inducing clinical signs of deficiency in kittens fed diets with less than 1 mg Cu/kg DM. Under practical conditions copper deficiency might occur in rapidly growing puppies on diets based on milk, milk products or eggs. A secondary deficiency could be induced by excessive use of zinc supplements that might improve coat condition. Chelating agents (for prophylaxis of cystine uroliths) may also reduce copper availability in a balanced diet.

Zinc

Zinc deficiency in most cases is associated with reduced bioavailability of dietary zinc due to high levels of calcium or phytate, found in plant materials such as soybean products and cereals. Zinc deficiency has been experimentally induced in dogs on a dry diet marginal in zinc (33 mg/kg) and high in calcium (10 g/kg)[68] and on a low zinc diet (8.5 mg/kg DM) without calcium excess.[31] Onset of signs (parakeratosis, depigmentation of hair) took several weeks. In growing puppies, the zinc requirement in relation to food intake and body weight is highest at the age of 4–6 months.[36] At this age, zinc-related skin diseases have been observed in practice several times.[7,43,64,79] The condition has been reproduced in growing puppies, with a high calcium diet (30 g/kg DM) with marginal zinc supply (35–40 mg/kg DM).[70] Suckling puppies have also been reported to suffer from zinc deficiency.[64] Cats appear to be less susceptible. Signs of zinc deficiency were not produced in growing kittens on diets with 15 mg Zn/kg, other than degenerated testicular tubules, even with the addition of 2% dicalcium phosphate to an already calcium-supplemented diet. Signs of zinc deficiency (retinal dysfunction, decreased food intake and weight gain, dull coat) were reported in kittens (age 130–290 days) fed a diet with less than 7.5 mg Zn/kg for 4 months.[33] Diagnosis can be made from plasma zinc levels (<50 μg/100 ml are regarded as marginal).

A non-nutritive hereditary zinc deficiency due to zinc malabsorption has been described in Alaskan Malamutes.[6] An insufficient availability of dietary zinc is also likely to occur in animals with exocrine pancreatic insufficiency.

In Siberian Huskies a zinc-responsive dermatosis has been recorded.[43] In an affected bitch, Meyer *et al.* could not demonstrate any relationship to zinc supply.[55] Plasma zinc levels in several dogs with this disease were within the normal range. Dermatological conditions associated with zinc deficiency are described in Chapter 23.

Iodine

Iodine deficiency is possible under various conditions. It is one of the features of the all-meat disease in carnivorous species. It occurs in iodine deficient areas, usually inland, if meat and cereals produced in that area are fed. Further losses of iodine occur during cooking of the food. Therefore it is not surprising that iodine deficiency is observed from time to time in practice. Recently a case

FIG. 1.3 Iodine deficient poodle puppy showing stunted limb development, hyperplasia of the thyroid gland, edema and loss of hair. The hyperplastic thyroid gland is shown as an inset. (Photograph courtesy of Dr. M. Kölling, Germany.)

has been reported in Germany,[42] (Fig. 1.3) where home-made diets from offals and cereals were quite common until very recently. Signs of deficiency are an enlarged thyroid gland, lethargy, alopecia, disorders of growth or fertility, weight loss and edema.[52] Diagnosis can be made from plasma levels of T_3 and T_4, where combined T_3 and T_4 levels below 0.5 µg/100 ml and T_3 levels below 70 µg/100 ml in dogs indicates a deficiency in dogs without other thyroid disorders.[5]

Selenium

Selenium deficiency is a very rare condition in dogs and cats, which has been observed in New Zealand. It developed as a result of dogs being fed offals from sheep in a selenium deficient area. Clinical signs were muscular weakness resembling white muscle disease in other species, and decreased viability of puppies.[50]

Other Minerals

There are no reports of deficiencies of manganese, cobalt (provided vitamin B_{12} supply is adequate), fluorine or molybdenum in dogs and cats under practical conditions.

B Group Vitamins

Due to the water-soluble nature of these vitamins, there is minimal storage in the body and therefore clinical signs of deficiency develop comparatively quickly.

Thiamin

Thiamin deficiency has been described in both the cat and the dog. Thiamin is a sulfur-containing water-soluble vitamin that, in its pyrophosphate form is involved in carbohydrate metabolism. Because of its water soluble nature and the body's limited capacity to store thiamin, clinical signs due to thiamin deficiency may be observed after a relatively short period of ingestion of a thiamin-deficient diet. Read and Harrington induced clinical signs of thiamin deficiency in young Beagles by feeding a diet containing 20–30 µg thiamin per kg of diet.[66] They reported three phases of disease: an initial phase where dogs appeared healthy but grew suboptimally, lasting 18 ± 8

days, followed by an intermediate stage of variable duration (59 ± 37 days) of anorexia, loss of body weight and coprophagy, followed by either a short period of neurological illness or sudden death.[66]

A similar pattern of clinical signs has been described in thiamin deficient cats.[20,34] Anorexia and sometimes emesis occurs within 2 weeks of ingestion of a thiamin-deficient diet and is followed by the sudden development of neurological disorders including abnormal posture, ataxia and seizures culminating in progressive weakness and death. Affected cats often show ventroflexion of the head.

Thiamin is heat-labile and is progressively destroyed by cooking.[49] Reputable manufacturers of petfoods supplement prepared foods with thiamin to compensate for degradation during the cooking process. Some fish flesh contains the enzyme thiaminase and thiamin deficiency can develop when such fish is fed raw to cats. Thiaminase is readily inactivated by heating,[19] so the feeding of cooked fish to cats is unlikely to result in a thiamin-deficient state. Sulfites cleave thiamin into pyrimidine and thiazole compounds, rendering it inactive. Clinical signs of thiamin deficiency have been reported in both cats and dogs fed uncooked meat that contained sulfur dioxide as a preservative.[19] This destruction of thiamin has important implications in the feeding of cats and dogs, as sulfur dioxide-treated meat can induce a thiamin deficient state when fed alone or when fed mixed with thiamin-replete commercial petfoods. Although legislation in many countries prohibits the use of sulfur dioxide in human foods, veterinarians should be aware of the health risks that these treated meats pose to cats and dogs.

Niacin/Tryptophan

Niacin deficiency in dogs provided a valuable insight into the etiology of pellagra.[10] It has occurred when dogs have been fed on corn-based diets.[62] The condition is called black tongue disease because of the severe ulceration of the oral and pharyngeal mucosa resulting in drooling of bloody saliva and halitosis. Affected dogs usually exhibit anorexia and weight loss and other signs associated

with the condition may include bloody diarrhea due to the resultant hemorrhagic necrosis of duodenal and jejeunal epithelium. If uncorrected, niacin deficiency can result in dehydration and death.

Biotin

Biotin is required for a range of reactions involving the metabolism of carboxy groups. A deficiency of biotin results in impaired protein synthesis, thought to be associated with a reduction in the synthesis of dicarboxylic acids.

Biotin deficiency has been recorded in both the cat and dog, although it is not often identified as most, if not all, of the animal's requirement for biotin is met by synthesis by gut microflora. A deficient state can be induced by including significant quantities of raw egg white in the diet of dogs and cats. Egg white contains avidin, a relatively heat stable glycoprotein, which forms a stable complex with biotin, rendering the biotin unavailable to the animal.[65] Evidence of biotin deficiency in dogs was produced by feeding diets containing dried egg whites and sulfaguanidine.[62] It appears that the administration of sulfa drugs and some intestinally active antibiotics results in a reduction of available biotin due to their effect on gut microflora biotin synthesis.

Biotin deficiency in the dog has been associated with hyperkeratosis of epithelia, resulting in a dry scurfy skin. Affected cats develop signs of alopecia, dried secretions around the nose, eyes, mouth and feet, focal dermatitis of the lips and a brown discoloration of the skin.[65]

Vitamin A

Vitamin A deficiency is rarely seen in practice. It could be induced by diets composed of offals or cereals without liver or milk products. In dogs, vegetables could serve as a source of provitamin A (carotene); however, cats lack the metabolic pathway necessary to convert carotene to the active form of vitamin A. Clinical signs of deficiency include skin and eye lesions, increased susceptibility to infections, problems with bone development and reproduction. Plasma levels below 300 IU/100 ml are regarded as marginal.[52]

Vitamin D

Vitamin D plays an important role in the metabolism of calcium and phosphorus. The evidence on the dietary requirement is rather confusing because of interactions with calcium and phosphorus as well as the possibility of synthesis of vitamin D in the skin following exposure to UV light.[35] The ability of dogs to synthesise vitamin D under the influence of UV light has been doubted by Hazewinkel *et al*.[27] Clinical signs of deficiency are ricketts or osteomalacia. The condition is rarely seen in practice. If an animal is presented with signs resembling vitamin D deficiency, it is most likely that this is due to inappropriate calcium intake or simply to overfeeding of growing puppies. Vitamin D supplementation might even be harmful in such a case.

Vitamin E

Vitamin E functions as an antioxidant and, together with selenium, is important in maintaining the integrity of cell membranes by preventing peroxidation and the adverse effects of free radicals. The dietary requirement for vitamin E is closely associated with the dietary intake of polyunsaturated fatty acids as this can affect the composition of cell membranes.[62]

Experimentally induced vitamin E deficiency has resulted in a wide range of effects involving muscle, reproductive, nervous and vascular systems. In dogs, vitamin E deficiency has been associated with skeletal muscle dystrophy, impaired immune response and reproductive problems including degeneration of testicular tissue, failure of spermatogenesis and the birth of weak and dead pups.

Vitamin E deficiency in the cat has been associated with steatitis, with a number of reports proposing an association between steatitis and the consumption of predominantly fish diets, particularly tuna.[12,21,24,60] It appears that the cat's need for dietary vitamin E is influenced by the composition of its diet. It has been recommended to include 0.7 mg vitamin E per g of unsaturated fatty acids in a dog's diet.[52] This requirement is higher if the dietary fat is rancid.[24]

Hagiwara *et al*. described a clinical case of pansteatitis in a cat fed exclusively on fried sardines.[24] Signs included nonspecific signs of listlessness, fever and anorexia together with intense abdominal pain and palpable nodular masses subcutaneously. The adipose tissue of affected cats is typically discolored yellow to orange–brown. Diagnosis can be confirmed by plasma tocopherol levels (<300 µg/100 ml) or erythrocyte membrane stability testing.

The prognosis is variable depending on the severity of clinical signs and the timeliness of vitamin E supplementation.

Essential Fatty Acids

A requirement for unsaturated fatty acids of the linoleic acid family (n6) was shown for many mammals including the dog and cat.[25,67] Dogs need linoleic acid only; arachidonic acid is also essential for the cat.[48,67]

Signs associated with linoleic and arachidonic acid deficiency in dogs and cats include hair loss, fatty liver degeneration, anemia and reduced fertility in females and males. A diagnosis can be confirmed by history and the fatty acid pattern in plasma lipids. A deficiency of essential fatty acids (EFAs) is most unlikely in adult healthy animals because of the stores in adipose tissue. In cats with the opportunity to hunt, a deficiency is also not probable. Growing or lactating animals (especially bitches who have a high content of linoleic acid in their milk) are at risk if their diet does not contain sufficient essential fatty acids (in dog food 20 g linoleic acid/kg DM, cat food 5 g linoleic and 0.2 g arachidonic acid/kg DM[53]). Home-made diets composed mainly of beef and rice may be deficient in EFAs, as may low fat dry type foods, and in cats a vegetarian diet is certain to induce arachidonic acid deficiency sooner or later. A deficiency can also be due to the intake of rancid fats, where most unsaturated fatty acids are destroyed. In both species, essential fatty acid deficiency may occur as a consequence of maldigestion or malabsorption.[3] In human patients, fatty acid deficiency has been reported during total parenteral nutrition.[32]

Positive effects on coat quality reported by owners after the addition of unsaturated fatty acids (vegetable oil, egg yolk, linseed oil) to the diet of their animals are probably not due to repletion of a deficiency but to pharmacodynamic effects of these fatty acids. Such effects have been postulated for several non-essential polyunsaturated fatty acids (C18:3 n6, C20:5 n3) for several species including man, dogs and cats (see Chapter 23).[47]

Potential Problems Associated with Nutrient Excesses

Carbohydrates

Excessive intake of carbohydrate can lead to digestive disorders in dogs and cats. Lactose intolerance is the best known carbohydrate intolerance, an effect due to low lactase levels in the small intestinal mucosa of adult mammals.[22] Other carbohydrates may also cause problems, especially in carnivorous species. The pathogenesis of intestinal carbohydrate intolerance in dogs and cats is essentially similar for all carbohydrates concerned. When the capacity of carbohydrate digestion of the small intestine is exceeded, carbohydrates reach the large bowel where they are fermented by microorganisms. Metabolites of microbial carbohydrate fermentation, such as short chain fatty acids and lactate, reduce the pH and increase the osmolality of the chyme. The end result may be acidic osmotic diarrhea. Fecal pH usually is below 6.3 in such cases. The severity of the disorder depends on the amount of carbohydrates escaping digestion in the small intestine and on the degradability of the carbohydrates by microorganisms. Small amounts of sugars or decomposed starch, for example, can have more dramatic effects than large amounts of raw starch of low digestibility.[37,71]

Dogs and cats differ considerably in their capacity to digest carbohydrates. Activity of pancreatic amylase and of the disaccharidases in the small intestinal mucosa (measured by identical methods) are much lower in the cat.[37,38] The activity of small intestinal lactase is lowest in adults of both species. Nevertheless there is little lactase activity left in adults, varying considerably between individuals and

allowing the digestion of limited amounts of lactose. This variation accounts for higher (or lower) individual tolerance of milk than average maximum lactose tolerance (50 ml milk/kg BW in dogs and 25 ml/kg BW in cats correspond to the average maximum tolerance of lactose; Fig. 1.4).

Meyer and Kienzle reviewed carbohydrate tolerance in adult cats and dogs (Fig. 1.4) and calculated the maximal tolerable content in the food.[54] The latter depends on the amount of food that is eaten, based on the energy density of the food and the energy requirement of the animal.[54] The lower the energy density of the food and the higher the energy requirement of the animal, the higher the food intake and the lower the tolerable level of carbohydrates in the food. Otherwise, small intestinal capacity of carbohydrate digestion will be exceeded. Therefore lactating bitches, queens or working sled dogs who need large amounts of food will not tolerate high carbohydrate diets as well as animals on maintenance feeding. A higher food intake relative to body weight is also necessary for growth. In puppies and kittens the picture is further complicated by a different enzyme activity. Amylase activity is low until approximately the third month of life, while lactase is high in suckling kittens and puppies.[37,38] Therefore at weaning age, the starch tolerance of puppies and kittens is considerably lower than in adults, while they eat approximately double the amount of food per kilogram body weight. Thus, the starch content in the food should amount only to a quarter of the amount recommended for adults.

Besides intestinal disorders induced by high sugar diets in cats, potential adverse effects in intermediary metabolism (hyperglycemia, glucosuria) have been shown.[37] Therefore, the intake of sugar should be limited to even smaller amounts than intestinal tolerance allows.

Calcium and Phosphorus

Excesses of calcium alone or together with phosphorus can contribute to the development of skeletal problems in growing puppies. Hedhammar *et al.* demonstrated that Great

Ellen Kienzle and Duncan K. Hall

CARBOHYDRATE TOLERANCE IN DOGS AND CATS				
	Intake (g/kg BW)		Content (% DM) in food	
	Dogs	Cats	Dogs	Cats
Starch				
decomposed	10*	5	40*	24
raw potato	8[†]	8[†]	35[†]	35[†]
Sugars				
lactose	2–3	1[‡]	10	5[‡]
sucrose	6–8	1[‡]	35	5[‡]
stachyose &				
raffinose	1	1	5	5
glucose	10[§]	1[‡]	40[§]	5[‡]

* Higher intake not tested experimentally.
[†] For dietary purposes.
[‡] Limited tolerance in intermediary metabolism.
[§] Estimated from starch data; not tested experimentally.

FIG. 1.4 Carbohydrate tolerance in adult dogs and cats and maximum content for maintenance in food with average energy density.

Dane puppies having free access to a diet rich in energy, protein and calcium developed more severe skeletal problems than a control group where food intake was restricted.[28] Because several parameters were varied it was not possible to separate the effect of calcium oversupply from that of general overnutrition. Hazewinkel *et al.* demonstrated that a high intake of calcium (33 g Ca/kg DM, either 9 or 30 g P/kg DM) may lead to disturbed enchondral ossification that may result in clinical problems such as radius curve syndrome and osteochondrosis.[26] The problems were less severe when the phosphorus content of the diet was 30 g/kg DM (Ca:P = 1.1:1) than when it amounted to 9 g/kg DM (Ca:P = 3.7:1). This shows once again the importance of a balanced calcium to phosphorus ratio.

The point should be stressed, however, that in large growing dogs, especially Great Danes, skeletal problems have been induced by feeding a complete and balanced diet *ad libitum*, thus inducing rapid growth.[56] Therefore, overnutrition should be considered as a differential diagnosis in cases of disorders of skeletal development. This area is reviewed in Chapter 22. In cats a high intake of calcium (as carbonate or organic calcium salts) is considered to enhance the risk of struvite urolithiasis because of an alkalinizing effect on urine pH.[41]

Others

Excessive intake of magnesium is not recommended because it might contribute to the risk of struvite urolithiasis. It has been postulated that chronic excessive sodium supply may lead to hypertension, an analogy to human nutrition that is indeed possible; but so far a convincing long-term study on sodium overnutrition in healthy dogs or cats is not available.[45]

Trace Elements

Oversupplementation of trace elements is usually done with the intention of enhancing their physiological effects. It goes without saying that such excesses will not produce the desired results. Because trace elements will usually be added as supplements, the maximum tolerable levels are given per kilogram of body weight wherever possible.

Iron

Oversupplementation of iron is unusual in dogs. It might occur in sled and racing dogs, in order to increase hematocrit levels. Under practical conditions it is most likely that the dogs or cats will vomit before severe metabolic damage is done and the owner will discontinue supplementation because of obvious gastrointestinal signs. A chronic toxicity has never been reported in dogs or cats, but in other species excess iron is known to impair the availability of other trace elements.[51]

Copper

Acute copper intoxication in dogs has been described.[23] The intake of a single dose of 165 mg copper sulfate/kg BW led to vomitus and death within 4 hours. Chronic copper toxicity has not been hitherto reported in healthy dogs and cats. Nevertheless the addition of copper salts or other copper mineral supplements to a diet that is not deficient in copper in order to improve collagen synthesis and hair pigmentation may be harmful. In Bedlington Terriers with hereditary copper toxicosis a normal copper intake leads to copper accumulation in the liver and liver degeneration.[78]

Zinc

Oversupplementation with zinc is possible, although dogs and cats appear to be rather tolerant to zinc.[18] On the other hand, these toxicity studies were carried out with a high calcium diet and this may have influenced the level at which zinc toxicity was induced. Vomiting, weight loss and histological changes of the pancreas have been reported in cats after excessive zinc intake of 50–100 mg/kg BW.[73] In dogs, intoxication has been observed after the ingestion of pennies. Symptoms were anemia, anorexia and vomiting.[69] A diagnosis can be made from plasma zinc levels. Judging from experience in other species, plasma levels may increase to values above 150 μg/100 ml without ill effects.[30]

Others

Iodine can be supplied in excess by seaweed products. On the other hand, dogs appear to be fairly tolerant to iodine compared to other species. Cases of intoxication have not been observed under practical conditions. Selenium toxicity might occur when pet owners oversupplement the diet in the belief that additional selenium may improve the condition of the muscles. A concentration of 2 mg/kg DM in the food (0.04–0.12 mg/kg BW) is suggested as a maximum tolerable level for all species.[61]

Vitamin A

Chronic excessive consumption of vitamin A, either through oversupplementation of diets (e.g. with cod liver oil or commercial supplements) or through feeding cats on a diet consisting predominantly of liver, can result in clinical signs of hypervitaminosis A.

Such signs have been induced within 14 weeks of feeding a diet containing excessive levels of vitamin A in cats.[63] Skeletal lesions associated with hypervitaminosis A have been induced when the dietary vitamin A levels were between 57,000 and 117,000 IU vitamin A/kg BW; a dose of 100,000 IU/kg BW resulted in lethargy after 10 weeks and spondylosis after 24 weeks.[74] Dogs appear to be less sensitive to the effects of excessive intake of vitamin A[72] but an intake of 300,000 IU/kg BW leads to signs of intoxication.[62,63]

Donoghue *et al.* described a case of combined chronic vitamin A and D toxicity in a miniature Dachshund (home-made diet based on liver and cod liver oil).[15] Clinical signs included weakness, anorexia, heat seeking, alopecia, lameness, shivering and pain upon handling. After the removal of vitamin A from the diet, the dog recovered completely within 9 months.

In the cat, the primary lesion associated with hypervitaminosis A is osteocartilaginous hyperplasia, primarily affecting the cervical vertebrae. This is usually exhibited as difficulty and pain upon flexion of the neck. Exostoses and soft tissue calcification can also occur. Although some clinical improvement can be expected in affected animals once the source of excessive vitamin A is removed from the diet, complete resolution of pathological changes is unlikely if extensive bony alterations are evident. This area is reviewed in Chapter 22.

Vitamin D

Toxic effects of hypervitaminosis D include soft tissue mineralization and resultant multiple organ dysfunction (including nephropathy) associated with the development of hypercalcemia and hyperphosphatemia.

Chronic intake of 10,000 IU/kg BW/day or a single dose of 200,000 IU/kg BW in dogs leads to clinical signs.[77] Toxicity can be produced relatively easily in both the dog and cat with oversupplementation of concentrated sources of vitamin D. Fatal iatrogenic hypervitaminosis has been recorded following oral supplementation with 5 million IU vitamin D_3 and 2.5 million IU vitamin A over a 6-month period as treatment for skin disease.[76] Diagnosis is based on history, clinical signs and plasma calcium and vitamin D levels.

References

1. Aguirre, G. D. (1978) Retinal degeneration associated with the feeding of dog foods to cats. *Journal of the American Veterinary Medical Association,* **172,** 791–796.
2. Austad, S. and Bjerkas, E. (1976) Eclampsia in the bitch. *Journal of Small Animal Practice,* **17,** 793–798.
3. Batt, R. (1988) Beziehungen zwischen Futter und Malabsorption beim Hund. In *Ernährung, Fehlernährung und Diätetik bei Hund und Katze.* Eds. H. Meyer and E. Kienzle. (Int. Symp. Hannover, 1987) pp. 154–158.
4. Baxter, J. H. and van Wyk, J. J. (1953) A bone disorder associated with copper deficiency. *Bulletin of John Hopkins Hospital,* **93,** 1–13.
5. Belshaw, B. E., Cooper, T. B. and Becker, D. V. (1975) The iodine requirement and influence of iodine intake on iodine metabolism and thyroid function in the adult Beagle. *Endocrinology,* **96,** 1280–1284.
6. Brown, R. G., Hoag, N. H., Smart, M. E. and Mitchell, L. H. (1978) Alaskan malamute chondrodysplasia V: Decreased gut zinc absorption. *Growth,* **42,** 1–6.
7. Burger, I. H. (1984) The zinc story. *Pedigree Digest,* **11(2),** 6–7.
8. Burger, I. H., Blaza, S. E., Kendall, P. T. and Smith, P. M. (1984) The protein requirement of adult cats for maintenance. *Feline Practice,* **14,** 8–14.
9. Burger, I. and Earle, K. (1992) Understanding taurine. *Waltham International Focus,* **2(2),** 9–13.
10. Carpenter, K. J. (1991) Contribution of the dog to the science of nutrition. *Journal of Nutrition,* **121,** S1–S7.
11. Chausow, D. G. and Czarnecki–Maulden, G. L.

12. Cordy, D. R. (1954) Experimental production of steatitis ("yellow fat disease") in kittens fed a commercial canned cat food and prevention of the condition by vitamin E. *Cornell Veterinarian,* **44,** 310.
13. Crabo, B., Kjellgren, E. and Bäckgren, A. W. (1970) Jörninnehall och röd blodbild hos hundvalpar under diperioden. *Svensk Veterinaertidning,* **22,** 857–861.
14. Crager, C. S. and Nachreiner, R. F. (1993) Increased parathyroid hormone concentration in a Siamese kitten with nutritional secondary hyperparathyroidism. *Journal of the American Animal Hospital Association,* **29,** 331–336.
15. Donoghue, J., Szanto, J. and Kronfeld, D. S. (1988) Vitamin A—Überdosierung beim Hund: Ein Beispiel für klinische Diätetik. In *Ernährung, Fehlernährung und Diätetik bei Hund und Katze.* Eds. H. Meyer and E. Kienzle. (Int. Symp. Hannover, 1987) pp. 171–174.
16. Doong, G., Keen, C. L., Rogers, Q. R., Morris, J. G. and Rucker, R. B. (1983) Selected features of copper metabolism in the cat. *Journal of Nutrition,* **113,** 1963–1971.
17. Dow, S. W., Fettman, M. J. and LeCouteur, R. A. (1988) Muscle weakness syndrome of cats: results of an owner and veterinarian questionnaire. *Companion Animal Practice,* **2(10),** 11–14.
18. Drinker, K. R., Thompson, R. K. and Marsh, M. (1927) An investigation of the effect of long continued ingestion of zinc in the form of zinc oxide by cats and dogs, together with observations upon the excretion and the storage of zinc. *American Journal of Physiology,* **80,** 31–64.
19. Edney, A. T. B. (Ed.) (1988) *The Waltham Book of Dog and Cat Nutrition.* 2nd edn. Oxford: Pergamon.
20. Everett, G. M. (1944) Observations on the behaviour and neurophysiology of acute thiamin deficient cats. *American Journal of Physiology,* **141,** 439–448.
21. Griffiths, R. C., Thornton, G. W. and Wilson, J. E. (1960) Eight additional cases of pansteatitis (yellow fat) in cats fed canned red tuna. *Journal of the American Veterinary Medical Association,* **137,** 126–128.
22. de Groot, A. P. and Hoogendorn, P. (1957) The detrimental effect of lactose. II. Quantitative lactase determinations in various mammals. *Nederlands Melk Zuivelttijdschrift,* **11,** 290–303.
23. Gubler, C. J., Lahey, M. E., Cartwright, G. E. and Wintrobe, M. M. (1953) Studies on copper metabolism. IX. The transportation of copper in blood. *Journal of Clinical Investigation,* **32,** 405–414.
24. Hagiwara, M. K., Guerra, J. L. and Maeoka,

(1987) Estimation of the dietary iron requirement for the weanling puppy and kitten. *Journal of Nutrition,* **117,** 928–932.

M. R. M. (1986) Pansteatitis (yellow fat disease) in a cat. *Feline Practice,* **16(5),** 25–27.

25. Hansen, A. E. and Wiese, H. F. (1943) Studies with dogs maintained on diets low in fat. *Proceedings of the Society of Experimental Biology and Medicine,* **52,** 205–208.

26. Hazewinkel, H. A. W., van't Klooster, A. T., Voorhout, G. and Goedegebuure, S. A. (1988) Skelettentwicklung bei erhöhter Ca- und P-Aufnahme. In *Ernährung, Fehlernährung und Diätetik bei Hund und Katze.* Eds. H. Meyer and E. Kienzle. (Int. Symp. Hannover, 1987) pp. 184–188.

27. Hazewinkel, H. A. W., How, K. L., Bosch, R., Goodegebuure, S. A. and Voorhout, G. (1988) Ungenügende Photosynthese von Vitamin D bei Hunden. In *Ernährung, Fehlernährung und Diätetik bei Hund und Katze.* Eds. H. Meyer and E. Kienzle. (Int. Symp. Hannover, 1987) pp. 125–129.

28. Hedhammar, A., Wu, F., Krook, L., Schriver, H. F., de Lahunta, A., Whalen, J. P., Kallfelz, F., Nunez, E. A., Hintz, H. F., Sheffy, B. E. and Ryan, A. D. (1974) Overnutrition and skeletal disease: an experimental study in growing Great Dane dogs. *Cornell Veterinary Supplement,* **5,** 11–16.

29. Hickman, M. A., Morris, J. G. and Rogers, Q. R. (1992) Intestinal taurine and the enterohepatic circulation of taurocholic acid in the cat. In *Taurine. Nutritional Values and Mechanisms of Action.* Eds. J. B. Lombardini, S. W. Schaffer and J. Azuma. pp. 45–54. New York: Plenum.

30. Hill, G. M. and Miller, E. R. (1981) Effect of dietary zinc levels on the growth and development of the gilt. *Journal of Animal Science,* **57,** 106–113.

31. Hommerich, G. (1983) Untersuchungen über die Auswirkungen eines chronischen Zinkmangels beim ausgewachsenen Hund. Tierärztliche Hochschule, Hannover, Diss.

32. Horrobin, D. F. (1989) Essential fatty acids in clinical dermatology. *Journal of the American Academy of Dermatology,* **20,** 1045–1053.

33. Jacobsen, S. G., Meadows, N. J., Keeling, P. W. N., Mitchell, W. D. and Thompson, R. P. H. (1986) Rod mediated retinal dysfunction in cats with zinc depletion: comparison with taurine depletion. *Clinical Science,* **71,** 559–564.

34. Jubb, K. V., Saunders, L. Z. and Coates, H. V. (1956) Thiamine deficiency encephalopathy in cats. *Journal of Comparative Pathology,* **66,** 217–227.

35. Kealy, J. K. and McAllister, H. (1991) Metabolic bone disease. *Waltham International Focus,* **1(2),** 21–27.

36. Kienzle, E. (1988) Spurenelementbedarf des Hundes. *Übersichten zur Tierernährungg,* **16,** 153–212.

37. Kienzle, E. (1989) Untersuchungen zum Intestinal- und Intermediärstoffwechsel von Kohlenhydraten (Stärke verschiedener Herkunft und Aufbereitung, Mono- und Disaccharide) bei der Hauskatze (*Felis catus*). Hannover, Tierärztliche Hochschule, Habilschrift.

38. Kienzle, E. (1988) Enzymaktivität in Pancreas, Darmwand und Chymus des Hundes in Abhängigkeit von Alter und Futterart. *Journal of Animal Physiology and Animal Nutrition,* **60,** 276–288.

39. Kienzle, E. (1991) Praxis der mutterlosen Aufzucht von Hunde- und Katzenwelpen. *Proceedings XVI. WSAVA Congress,* Vienna. pp. 266–268.

40. Kienzle, E., Meyer, H., Dammers, C. and Lohrie, H. (1985) Milchaufnahme, Gewichtsentwicklung, Futterverdaulichkeit sowie Energie- und Nährstoffretention bei Saugwelpen. *Advances in Animal Physiology and Animal Nutrition,* **16,** 26–50.

41. Kienzle, E. and Schuknecht, A. (1993) Untersuchungen zur Struvitsteindiätetik: 1. Einfluszlig; verschiedener Futterrationen auf den Harn-pH-Wert der Katze. *Deutsche Tierärztliche Wochenschrift,* **100,** 198–203.

42. Kölling, M. personal communication.

43. Kunkle, G. A. (1980) Zinc responsive dermatoses in dogs. In *Current Veterinary Therapy VII. Small Animal Practice.* Ed. R. W. Kirk. pp. 472–476. Philadelphia: W. B. Saunders.

44. Leon, A., Bain, S. A. F. and Levick, W. R. (1992) Hypokalaemic episodic polymyopathy in cats fed a vegetarian diet. *Australian Veterinary Journal,* **69,** 249–254.

45. Lewis, L. D., Morris, M. L. and Hand, M. S. (1990) *Klinische Diätetik Für Hund und Katze.* Mark Morris Associates.

46. Lintzel, W. and Radeff, R. (1931) Über den Eisengehalt und Eisenansatz neugeborener und saugender Tiere (nach Versuchen an Kaninchen, Meerschweinchen, Ratte, Hund, Katze, Ziege, Rind). *Archiv fur Tierernährung und Tierzucht,* **6,** 314–356.

47. Lloyd, D. H. (1989) Essential fatty acids and skin disease. *Journal of Small Animal Practice,* **30,** 207–212.

48. MacDonald, M. L., Rogers, Q. R., Morris, J. G. and Cupps, P. T. (1984) Effects of linoleate and arachidonate deficiencies on reproduction and spermatogenesis in the cat. *Journal of Nutrition,* **114,** 719–726.

49. McDowell, L. R. (1989) *Vitamins in Animal Nutrition.* San Diego: Academic Press.

50. Manktelow, B. W. (1963) Myopathy in dogs resembling white muscle disease of sheep. *New Zealand Veterinary Journal,* **11,** 52–55.

51. Mertz, W. (1987) *Trace Elements in Human and Animal Nutrition.* San Diego: Academic Press.

52. Meyer, H. (1990) *Ernährung des Hundes.* 2nd edn. Stuttgart: Eugen Ulmer.

53. Meyer, H., Kienzle, E. and Dammers, C. (1985) Milchmenge und Milchzusammensetzung bei der Hündin sowie Futteraufnahme und Gewichtsentwicklung ante und post partum. *Advances in Animal Physiology and Animal Nutrition,* **16**, 51–72.

54. Meyer, H. and Kienzle, E. (1991) Dietary protein and carbohydrates: relationship to clinical disease. Purina International nutrition symposium in association with the Eastern States Veterinary Conference, Orlando, Florida. pp. 12–28.

55. Meyer, H., Rodenbeck, H., Zentek, J., Grübler, B., Kienzle, E. and Carlos, G. M. (1989) Ein Beitrag zum klinischen Bild und zur Ätiologie einer Dermatose bei Siberian Huskies. *Effem-Report,* **29**, 1–11.

56. Meyer, H. and Zentek, J. (1992) Über den Einfluszlig; einer unterschiedlichen Energieversorgung wachsender Doggen auf Körpermasse und Skelettentwicklung. 1. Mitteilung: Körpermasseentwicklung und Energiebedarf. *Journal of Veterinary Medicine, A,* **39**, 130–141.

57. Morris, J. H. and Rogers, Q. R. (1992) The metabolic basis for the taurine requirement of cats. In *Taurine. Nutritional Values and Mechanisms of Action.* Eds. J. B. Lombardini, S. W. Schaffer and J. Azuma. pp. 33–44. New York: Plenum.

58. Mühlum, A. and Meyer, H. (1991) Influence of taurine intake on plasma values an renal taurine excretion of cats. *Journal of Nutrition,* **121**, S175–S176.

59. Müller, V. R. (1988) Untersuchungen über die endogene faecale und renale Eisen-, Kupfer- und Zinkausscheidung beim Hund. Tierärztliche Hochschule Hannover, Diss.

60. Munson, R. O., Holzworth, J. and Small, E. (1958) Steatitis (yellow fat) in cats fed canned red tuna. *Journal of the American Veterinary Medical Association,* **133**, 563–568.

61. National Research Council (1980) *Mineral Tolerance of Domestic Animals.* Washington, DC: National Academy of Sciences.

62. National Research Council (1985) *Nutrient Requirements of Dogs.* Washington, DC: National Academy of Sciences.

63. National Research Council (1986) *Nutrient Requirements of Cats.* Washington, DC: National Academy of Sciences.

64. Öhlen, B. (1985) Zinkreaktive Dermatitis beim Hund. *Kleintier-Praxis,* **30(4)**, 185–188.

65. Pastoor, F. J. H., van Herck, H., van't Klooster, A. and Beynen, A. C. (1991) Biotin deficiency in cats induced by feeding a purified diet containing egg white. *Journal of Nutrition,* **121**, S73–S74.

66. Read, D. H. and Harrington, D. D. (1981) Experimentally induced thiamine deficiency in Beagle dogs: clinical observations. *American Journal of Veterinary Research,* **42**, 984–991.

67. Rivers, J. P. (1982) Essential fatty acids in cats. *Journal of Small Animal Practice,* **23**, 563–576.

68. Robertson, B. T. and Burns, M. J. (1963) Zinc metabolism and the zinc deficiency syndrome in the dog. *American Journal of Veterinary Research,* **24**, 997–1001.

69. Robinson, F. R., Mason, R. M., Fulton, R. M., Martinez, M. and Everson, R. J. (1991) Zinc toxicosis in a dog. *Canine Practice,* **16**, 27–31.

70. Sanecki, R. K., Corbin, J. E. and Forbes, R. M. (1982) Tissue changes in dogs fed a zinc deficient ration. *American Journal of Veterinary Research,* **43**, 1642–1646.

71. Schünemann, C., Mühlum, A., Junker, S., Wilfarth, H. and Meyer, H. (1989) Praecaecale und postileale Verdaulichkeit verschiedener Stärken sowie pH-Werte und Gehalte an organischen Säuren in Darmchymus und Faeces. *Advances in Animal Physiology and Animal Nutrition,* **19**, 44–58.

72. Schweigert, F. J. (1988) Insensitivity of dogs to the effects of nonspecific bound vitamin A in plasma. *International Journal of Vitamin Nutrition Research,* **58**, 22–25.

73. Scott, D. A. and Fisher, A. M. (1938) Studies on the pancreas and liver of normal and zinc fed cats. *American Journal of Physiology,* **121**, 253–260.

74. Seawright, A. A., English, P. B. and Gartner, R. J. W. (1967) Hypervitaminosis A and deforming cervical spondylosis of the cat. *Journal of Comparative Pathology,* **77**, 29–39.

75. Studdert, V. P. and Labuc, R. H. (1991) Thiamin deficiency in cats and dogs associated with feeding meat preserved with sulphur dioxide. *Australian Veterinary Journal,* **68**, 54–57.

76. Suter, P. (1957) Zur Gefahr der Überdosierung von Vitamin-D-Präparaten. *Schweizer Archiv Tierheilkunde,* **99**, 421–433.

77. Spangler, W. L., Gribble, D. H. and Lee, T. C. (1979) Vitamin D intoxication and the pathogenesis of vitamin D nephropathy in the dog. *American Journal of Veterinary Research,* **40**, 73–83.

78. Twedt, D. C., Sternlieb, I. and Gilbertson, S. R. (1979) Clinical morphology, and chemical studies on copper toxicosis of Bedlington terriers. *Journal of the American Veterinary Medical Association,* **175**, 269–275.

79. van den Broek, A. H. M. and Thoday, K. L. (1986) Skin disease in dogs associated with zinc deficiency: a report of five cases. *Journal of Small Animal Practice,* **27(5)**, 313–323.

80. Zentek, J., Dämmrich, K. and Meyer, H. (1991) Untersuchungen zum Cu–Mangel beim wachsenden Hund. *Journal of Veterinary Medicine, A,* **38**, 561–570.

CHAPTER 2

Reading a Petfood Label

IVAN H. BURGER and ANGELE THOMPSON

Introduction

The petfood label is a complex, highly regulated communication device that must be carried on every can, box and bag of petfood. In relation to its size, the label must represent an extremely concentrated information package. Within its confines, a manufacturer must make certain declarations that are strictly defined and that provide useful data to the pet owner. This, of course, is in addition to the brand name and characteristic design and image of the product.

Regulation

In Europe, petfood labels are largely controlled by legislation, originating in European Community (EC) Directives, which is then implemented through national regulations. For example, in the U.K., these Directives are promulgated through the Feeding Stuffs Regulations that have recently been updated.[5]

Many agencies regulate petfood labels in the U.S.A. The Federal Government regulates via the Food and Drug Administration (FDA) and the Department of Agriculture. Each of the 50 states also regulates the labels via its own state laws. There has been an attempt to promote more uniform regulations by a non-regulating group called the Association of American Feed Control Officials (AAFCO). AAFCO is composed of state officials responsible for feed regulation. Approximately 27 of the 50 states have adopted AAFCO's Uniform Feed Bill and Pet Food Regulations.

The regulations can change on a yearly basis. Often the changes are minor, but they can be extensive. The changes usually take about 6 months to phase in, owing to backlogs of label inventories. Currently in the U.S.A., the Federal Government is revamping the labelling for human foods. As a result of Federal preemption, however, some of these new regulations may flow over into petfoods. Thus, it is possible that the labels will change within the next 2–3 years.

Label Design

Every label has two basic parts (Fig. 2.1):
(1) **The Principal Display Panel** (PDP);
(2) **The Information Panel** (IP) or **Statutory Statement**.

The PDP is the first line of communication. It is the primary means of attracting the

15

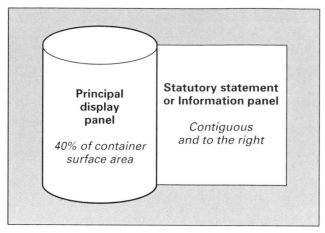

FIG. 2.1 Label design.

consumer's attention to a product and must immediately communicate the product identity. The IP (statutory statement in the U.K.) provides the second line of communication and its contents are strictly regulated. It supplies critical data about the product, which may influence the purchase decision.

Regulations concerning the required declarations from the manufacturers vary between countries. In the U.K., these declarations are encapsulated in the statutory statement that, in addition to being visible, legible and indelible, must be separate from all other information on the label. Other important information that may be shown outside the statutory statement includes the

"best before" date, batch number, net weight and the name and address of the company responsible for the product. This is not necessarily the manufacturer but could be a packer, importer or distributor. If the date, number or weight is shown outside the statement, then there must be an indication within the statutory statement of where to find the information.

In the U.S.A., however, net weight must be displayed on the PDP, together with a statement of identity for the product. Other information may appear on the PDP provided that it does not obscure the required information and is not misleading in any way.

Principal Display Panel

The PDP is, simply, the front of the container, i.e. that part of the label that faces the purchaser (Figs 2.2 and 2.3). The function of the PDP is to provide visual impact, which is designed to grab attention and readily communicate the product identity. The PDP's functions are achieved via marketing designs that are executed over a strict, tightly regulated framework. Once the baseline essentials are in place, the dramatic flair makes the difference between dull reading and impressive communication.

FIG. 2.2 The statement of intent.

FIG. 2.3 Bursts and flags on petfood labels.

FDA regulations in the U.S.A. are that the area of the PDP for cylindrical containers is 40% of the height of the container times its circumference. The area of the PDP determines the size of the type face. The rule of thumb is, the larger the area of the PDP, the larger the print.

Brand Name

The brand name provides the overall image and company affiliation of the product.

Product Name

The product name provides the information about the individual identity of the particular product within the brand. The name can be a general name or a more specific name that includes an ingredient. An ingredient used in the product name is known as a "namer." The way the namer is used in the product name indicates the amount of namer in the total product as shown in the following example.

- *Beef*: 70–100% of total product is beef.
- *Beef Dinner* (platter, entrée, etc.): 10–70% of the total product is beef.

- *Beef Flavor*: Enough beef is added to the product to give a discernible, characteristic beef taste.

When two or more namers are used in the dinner-type product, the total of both namers must be 10–70% of the product and all namers must be at least 3% of the product.

- *Beef and Liver Dinner*: 10–70% beef plus liver. Beef must be more than liver and both must exceed 3%.

Statement of Intent

This component identifies the target animal for which the product is intended (Fig. 2.2). In the U.K. it is *normal*, but not essential, for this information to be conveyed on the PDP, but it must also be included in the statutory statement. U.S. regulations, however, *require* that this is displayed on the PDP. There are two ways to accomplish this communication: (1) with words only and (2) with words plus a "vignette" (a picture of the target animal). For example, the words must say either "Dog Food" or "Food for Dogs." The size of these words is governed by whether or not the words are accompanied by the vignette. Without the vignette, the words must be twice the size of the product name; with the vignette they may be one-half the size of the name. There is no regulation specifying the size of the vignette.

Net Weight

This communicates the amount of product inside the container (Figs. 2.2 and 2.3). This small piece of information is highly regulated and stringently inspected. U.S. regulations dictate the type size of the statement (reasonable size is related to the largest type on the PDP), the location of the statement (bottom 30% of the label, perpendicular to bottom) and the units of measurement (which varies by package size and will change in the U.S.A. in 1994). The Packaged Goods Regulations govern the manner in which the statement may appear in the U.K. The net weight of the product is easily verified by government officials, who vigorously enforce the laws in order to assure that the consumer receives the full amount purchased.

Bursts and Flags

Bursts and flags are set-off areas on the PDP that communicate specific information of impact, which is not included in the items listed above (Fig. 2.3). The time allowed for the burst to be on the label varies with the type of burst. "New" or "New and Improved"

can only be on the label for 6 months, while a comparison such as "Preferred 3 to 1" can remain on the label for 1 year.

Product Picture and Slogan

These components are not specifically regulated. They form part of the optional marketing design for grasping attention and for communication.

There is clearly a large amount of information available on the PDP. To be effective, however, the PDP must be clear, readable and designed to be understood at a glance.

Statutory Statement or IP

At this point in the purchasing decision, the consumer has picked up the container and read the PDP; the intention is now to find out more about the product. As previously mentioned, the IP information may be critical to swaying the actual purchase.

The statutory statement carries the required declarations for petfoods marketed in the U.K. or, in the case of the best before date, batch number and net weight, must include an indication of where to find the information. In addition to being visible, legible and indelible, the statutory statement must be separate from all other information on the label.

The IP on U.S. petfood labels carries the required information that is not included on

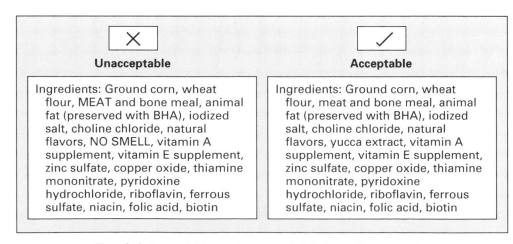

Fig. 2.4 Acceptable and unacceptable ingredient panels.

the PDP. FDA regulations[4] define the IP by location. The IP is immediately contiguous and to the right of the PDP as observed by an individual facing the PDP. If that panel is too small, the next panel may also be used. However, any area outside of the PDP is used for this part of the communication. Although the content and size of items of information carried on the IP are regulated, their order and exact placement on the label are not.

Ingredient Statement

This is a listing of the product's ingredients, which must be made in descending order of predominance by weight (Fig. 2.4). The requirements for the naming of the ingredients vary tremendously from country to country.

In the U.K., the type of ingredient can be stated by an individual name or may be grouped under various categories as stipulated in the regulations. For example, the category name "meat and animal derivatives" is used to describe the fleshy parts of slaughtered warm-blooded land animals, fresh or preserved by appropriate treatment and all products and derivatives of the processing of the carcass or parts of the carcass of such animals. There are similar categories for derivatives of vegetable origin, milk and milk derivatives, fish and fish derivatives and egg and egg derivatives. These categories are designed to provide the consumer with an indication of the source of raw materials while allowing the manufacturer some flexibility in the selection of the ingredients within a

specified category. Nevertheless, the manufacturer must be aware of the effects of any changes on nutritional content through the analytical declaration.

In the U.S.A., every ingredient must be listed separately either by its official name or its common name (no brand names are allowed). No single ingredient can be highlighted or given undue emphasis. Often, in reading the statement, the purchaser may be concerned about the "chemical" sounding ingredients (such as pyridoxine hydrochloride or ferrous sulfate) that are listed. Most of these, however, are vitamins and minerals (often taken every day by the purchaser in their own multivitamin).

Typical (U.K.) or Guaranteed (U.S.A.) Analysis

The U.S.A. requires that a guaranteed analysis shows the guaranteed amounts of specified nutrients in the product (Fig. 2.5). The order and format of this statement is dictated by the regulations. Certain nutrients (crude protein, crude fat, crude fiber, moisture) are required by U.S. law, but other nutrients (calcium, phosphorus, taurine) are optional. The values listed in the guarantee are given as maxima or minima (by law) and they are not the *actual* values of the nutrients in the product. Because of natural variation of ingredients, the actual values vary within a small range close to either the maximum or the minimum, depending on the nutrient. Government officials randomly analyze the products (sometimes just a single can) for compliance

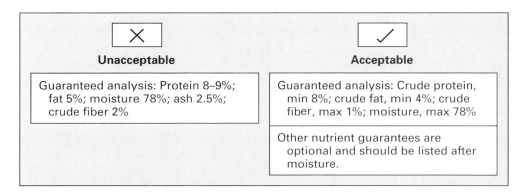

FIG. 2.5 Acceptable and unacceptable guaranteed analysis statement (U.S.A.).

TYPICAL PRODUCT ANALYSES

	Canned petfood	Dry petfood	Dressed tripe*	Lean chicken meat*
Protein %	8.0	30.0	9.4	21.8
Moisture %	78.0	8.0	88.1	74.4
Dry matter %	22.0	92.0	11.9	25.6
Protein % (in dry matter)	36.4	32.6	79.0	85.1

* Analyses from Holland *et al.* (1991).[6]

FIG. 2.6 Typical product analyses.

with these guarantees. Noncompliance can result in the product being removed from sale.

EC regulations dictate that the typical concentrations of the following nutrients must be declared as percentages in the product: protein, oil, fiber and ash. The 1991 Feeding Stuffs regulations[5] also require a statement of the percentage moisture in the product if this is over 14%. These values can be used to compare nutrient values between different products but it is important always to compare similar products. If the protein and moisture contents of typical petfoods and raw materials are compared (Fig. 2.6), then the values all seem very different. Nevertheless, the important aspect is the level of nutrient that is ingested by the animal. This depends on energy and the energy content is proportional to the dry matter content. If the percentage protein in *dry matter* of the products in Fig. 2.6 are compared, the two petfoods show similar values of around 30–35%. This is comfortably adequate to support all life stages of the dog or cat. The tripe and chicken meat show much higher contents that, although meeting the nutritional requirement for protein, indicate that other nutrients in the product may not be in balance.

This discussion raises the question as to whether the declaration of energy content on the label is the next logical step. This may offer some advantages but is a very complex issue; in the U.S.A. energy content claims have been discussed for over 10 years. The energy content depends on the protein, fat and carbohydrate contents of the product (i.e. the energy-containing nutrients). The main challenge is to find an acceptable way to calculate the available or metabolizable energy (ME) of the food from these constituents. The ME content depends both upon the composition of the food and the animal eating it. For example, the digestive system of the dog seems to be more efficient than that of the cat. Thus, the *same food* fed to dogs or cats will yield different ME values in the two species. Until an agreed method has been finalized that gives an ME value for petfoods that is neither misleading nor over-complicated, the EC has ruled that energy declarations on petfoods are illegal. This reinforces the importance of reliable feeding recommendations. U.S. regulations are currently being modified to permit voluntary energy content claims on the label.

Product Description (U.K.) and Nutritional Adequacy Statement (U.S.A.)

The label must state whether the food is nutritionally complete or complementary. A complete food can satisfy the particular nutritional demands of the animal without any additional ration. A complementary food must be fed in conjunction with another product in order to complete the ration, in which case, the additional food must be stated. This description must be considered in relation to the intended purpose of the food or the particular life stage for which it is defined. For example, the food might be designed for adult maintenance or growth or for all life stages. The species or category of animal is an integral part of this description and this aspect is particularly important for cats. The cat is an obligate nutritional carnivore and

Meets the minimum nutritional levels established by the AAFCO Cat Food Nutrient Profile for the maintenance of cats	Complete and balanced nutrition for all stages of a dog's life as substantiated by testing in accordance with AAFCO procedures
Animal feeding tests using AAFCO procedures substantiate that (Brand) provides complete and balanced nutrition for the growth of kittens	

FIG. 2.7 Nutritional adequacy statement.

has an absolute requirement for some animal tissue in its diet.[2] It is therefore crucial that cats are fed only cat foods: some dog foods may not contain a satisfactory nutritional profile for cats. The overall nutritional description of the food often incorporates these three important factors in phrases such as "a complete food for growing cats" or "a complementary food for adult dogs."

The nutritional adequacy statement (Fig. 2.7) required by U.S. law indicates the nutritional delivery of the product. It shows the level of nutrition provided (i.e. whether it is complete or complementary and for which life stage it is designed), the method of substantiation of the statement and the target animal (species or other category). Complementary foods in the U.S.A. must carry the statement "Intended for limited or supplemental feeding only." There are several options for establishing the level of nutritional adequacy. These options can be placed in a hierarchy by the stringency of the tests.

Profile Comparison Via Calculation

The least stringent option is to calculate the nutrient levels in the product and compare those levels with an accepted authoritative nutrient profile (such as *AAFCO Dog/Cat Nutrient Profiles*). This option is based on standard tables and does not reflect the actual product values. The values obtained do not account for ingredient variation from "standard," postharvest changes, processing losses or bioavailability of the nutrients.

Profile Comparison Via Analysis

The product may be analyzed chemically and those values compared to the nutrient profile. This option takes into account the actual ingredients and the possible nutrient changes from processing. It does not account for bioavailability of the nutrients.

Animal Feeding Trials

The best test options are those using animal feeding tests (also known as feeding protocols) because they show that the product actually supports the animals during various life stages. The minimum requirements for these tests have been published.[1] The maintenance protocol using adult nonreproducing animals is the least stringent feeding test. These animals are in a steady state, that is no change in nutrient requirements. The adult animals are fed the test diet for 6 months. The next tier of feeding trials is the growth protocol. This test is performed with "just weaned" animals that are fed the diet throughout the most active growth period. The animals are fed the test diet from 8–18 weeks-of-age. The most rigorous protocol is called an "all stages" or "unqualified" protocol. In this test, reproducing females are mated, go through gestation, give birth and go through lactation on the same diet. Their offspring are then tested in a growth protocol while still on the same diet. The product must be able to supply adequate nutrients in enough quantity to support animals through the difficult times of lactation and growth. Although other life stage diets (such as senior) or other purpose products (reducing diets or clinical diets) exist, there are no standard protocols to show their efficacy. Each manufacturer must "adequately" test the product to show that it can meet its intended purpose.

Feeding Guidelines

Petfood labels must provide the consumer with directions for use. In addition to defining the intended purpose of the food (and what, if any, other products must be fed to complete the daily ration), the directions for use must also incorporate the feeding recommendations or guidelines. These may be fairly simple (for adult maintenance in cats) or encompass a wide range (for growing puppies). The amount of food required is dependent on the energy requirement of the animal and it is a particular challenge to provide this information for the dog. The range of body weight in dogs is uniquely wide (from around 1 to 100 kg for normal adult nonobese animals) and this means that the derivation of a single equation for energy requirement is somewhat daunting.[3] Other variations in activity, body conformation, type of hair covering and so on mean that the values given by manufacturers can never be more than a good average guideline. This reinforces the message that the pet owner's judgement as to whether the animal is too fat or too thin (and making corresponding adjustments to the food offered) is just as important a factor as the label information.

Feeding guidelines are required by U.S. regulations, but are not strictly dictated. The requirement is that the guideline be accurate and communicated in commonly understood units. A simple statement on a 3-oz can of cat food may be "feed one can per meal per 6-lb cat." For large bags of dog food, the guidelines may be in the form of a chart showing various breeds, weights and ages. These statements are all, at best, guidelines. Quite commonly, labels will carry the *most* accurate feeding guideline: "Feed your animal according to its age and activity level to maintain optimal weight. *Adjust as needed.*" Every animal is different and the proper adjustments should be made according to the individual animal's requirements. The best way for clients to monitor their animal's feed intake is to use a consistent measure and to weigh the animal.

Nutrient Declaration

Declaration is obligatory in EC countries if the fat-soluble vitamins A, D and E or the trace element copper have been added to the product. The units used must be mg/kg for copper and International Units (IU), mg or µg (per kg) for the vitamins as appropriate. The stated concentration must include the *total* content, that is, that which is naturally present plus that which is added. For vitamins, a guaranteed stability and shelf-life must be stated and this is usually linked to the use before date. The declaration of other nutrient concentrations is optional but becomes obligatory if a claim is made within the statutory statement, for example "high in zinc."

Additives Declaration

Under EC regulations, three groups of additives must be declared if they have been

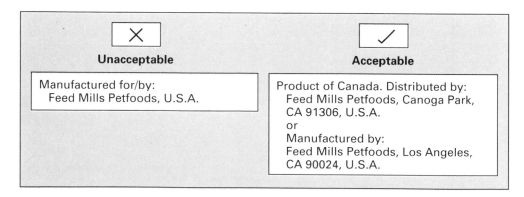

FIG. 2.8 Acceptable and unacceptable display of manufacturers and distributors.

added to the product: preservatives, anti-oxidants and colors. The manufacturer may use categories (e.g. "contains permitted anti-oxidant") or chemical names (or EC reference numbers). In addition, the manufacturer must keep a record of which batches of products contain particular additives and relate this information to a reference number on the label (this is often the batch number). In this way, a consumer can find out which additives were used in a product by writing to the relevant company and quoting the reference or batch number.

Manufacturer or Distributor

Under U.S. regulations, the party responsible for either manufacture or distribution of the product must be shown on the label (Fig. 2.8). The address must be complete enough to allow identification and contact via the telephone information listing. If the product is manufactured in a country other than where it is sold, this statement will be accompanied by "Product of ———" (stating country of origin). "Distributed by" indicates that the product has been manufactured by a company other than the one selling the product. This is called co-manufacture or co-packing.

Petfoods marketed in the U.K. must bear the name and address of the company responsible for the product. This could be the manufacturer, packer, importer, seller or distributor. This information need not appear within the statutory statement.

The IP thus provides a great deal of information about the product, but it requires a lingering and more intense attention to detail than does the information on the PDP.

General Information

Optional data such as batch numbering and date marking have been stated on the label for many years. The new Feeding Stuffs Regulations (1991)[5] have now incorporated these as obligatory declarations for petfoods marketed in the U.K. Although not a legal requirement, most manufacturers also include

a bar code, or Universal Product Code (UPC), on the label. This is used in many countries to identify the product in a machine readable form, which allows information about sales, stock holdings and physical distribution of products to be retrieved through scanning at the point of purchase.

Any additional information to be communicated to the purchaser may be provided as "label copy." Feeding stuffs regulations in both Europe and the U.S.A. require that general information must not mislead the purchaser. For example, the label must not suggest that the product possesses properties that it does not, nor must it imply that it is special when similar properties are found in other products. Any information outside the statutory statement is controlled in the U.K. by the Trade Descriptions Act.

Furthermore, the label must not claim that the product will prevent or cure disease; in the U.K. such products will normally be controlled by the Medicines Act. Nevertheless, some petfoods are designed to aid dietary management of disease and make claims to that effect. In the EC, discussions between the Commission and Member States, in consultation with the industry, have been in progress since June 1991 to ratify the categorization of this type of product. There is, as yet, no agreed document but it is likely that any future legislation will relate to the specific claims made for a particular product on the label. In the U.S.A. such claims are regulated by the FDA and are currently not allowed. There are certain "grey areas" concerning physiological changes (such as urine pH control) that are assessed by the FDA on a case-by-case basis and approval granted if considered appropriate.

References

1. AAFCO Official Publication (1993) Association of American Feed Control Officials Inc., Atlanta, GA, U.S.A.
2. Burger, I., Edney, A. and Horrocks, D. (1991) Basics of feline nutrition. In *Feline Practice*. Ed. E. Boden. pp. 101–115. London: Ballière Tindall.

3. Burger, I. H. and Johnson, J. V. (1992) Dogs large and small: the allometry of energy requirements within a single species. *Journal of Nutrition,* **121,** S18–S21.

4. FDA publication. U.S.A. Code of Federal Regulations, Chapter 21, April 1, 1992.

5. HMSO (1991) The Feeding Stuffs Regulations 1991. Statutory Instrument No. 2840. London: Her Majesty's Stationery Office.

6. Holland, B., Welch, A. A., Unwin, I. D., Buss, D. H., Paul, A. A. and Southgate, D. A. T. (1991) In *McCance and Widdowson's The Composition of Foods.* 5th edn. Cambridge: Royal Society of Chemistry and Ministry of Agriculture, Fisheries and Food.

7. PFMA (1991) PFMA profile. Pet Food Manufacturers' Association, London.

CHAPTER 3

Feeding Hospitalized Dogs and Cats

SUSAN DONOGHUE and DAVID S. KRONFELD

Introduction

Good feeding management requires a knowledge of the animal's nutritional requirements for specific purposes, the design of diets that meet these requirements and the best ways of delivering these diets. It is helped by an appreciation of the animal's instincts and emotions, and it is more than simply sound nutritional science. Good feeding management is the mainstay of responsible care of animals.

For hospitalized dogs and cats, feeding management influences their progress to varying degrees. Each case needs at least one crucial nutritional decision: whether it is to be fed routinely or specially. Most will fare well on a sound routine feeding program. Some will be identified for individual attention. Hospital nutrition tends to become preoccupied with the nutritional support of these few selected patients. Those that need tube feeding or intravenous glucose attract attention. Hospital nutrition, however, should never overlook the general feeding program in clinic wards.

General Feeding Plans

The routine feeding program in a veterinary clinic may be regarded as a treatment adjunct. To be effective, however, feeding programs require attention to several points, including the animal at presentation to the clinic and the ambiance of the wards, as well as diets and nutritional requirements.

Patient Presentation

The medical record of each dog and cat at presentation should include body weight and body condition, together with feeding history. Some breeds, such as Yorkshire Terriers, have relatively wide distributions of body weight, so body condition assumes greater importance than comparisons to an "ideal" body weight. Cats lose or gain significant body condition with only a few pounds lost or added. Clinic scales are not always accurate enough to note significant changes in cats' weights, but physical examination usually picks up loss of condition.

We use a simple five-point system for body condition scoring of dogs and cats, where "1"

STANDARDIZED SCORING SYSTEM FOR BODY CONDITION	
Score	**Condition**
1	Cachectic; no obvious body fat
2	Thin; limited body fat evident
3	Optimal; ribs palpated easily but not observed readily
4	Overweight; ribs not visible and difficult to palpate
5	Obese; large amounts of subcutaneous fat, obvious incapacity

FIG. 3.1 A standardized system for body condition scoring of hospitalized dogs and cats. Patients are palpated over ribs and spine. Distributions of body condition scores from patients in veterinary hospitals vary with breed (for dogs) and age (for dogs and cats).[8]

is cachectic, "3" is optimal, and "5" is obese (Fig. 3.1).[8] The subpopulation of hospitalized dogs or cats that requires special feeding tended to be relatively more extreme in weight (thin or fat) than the general hospital population.[8]

Also at presentation, the pet's usual and recent diets (these may differ) should be inquired about and recorded. It may help in diagnosis and always aids management of finicky eaters. For example, sick dogs and cats that were fed canned food and table foods at home may not eat dry petfood in the hospital. Finicky eaters with unknown food preferences may be offered initially a choice of canned and dry diets.

When the dog or cat is first admitted, abrupt changes in diet should be avoided if possible, because such changes often aggravate inappetence in sick animals or may lead to digestive upsets, such as flatus and diarrhea. For example, we worked with a 10-year-old German Shepherd that had been hit by a car and suffered several bone fractures, abrasions and lacerations. The dog had been fed a low calorie low protein "senior diet." Because of the dog's injuries, he now needed high energy and high protein. In this type of case, a total diet change is necessary in order to increase energy and protein, necessitating a switch from a low fat to a high fat diet, which might induce diarrhea. For such cases, a feeding plan is designed with gradual introduction of high energy food over about 3 days to avoid digestive upsets. After about 12 weeks, when recovery is almost complete, a plan is written for the owners gradually to change their pet's diet back to the petfood of their choice.

Wards

Stress can be caused by unfamiliar people and surroundings as well as by disease and injury. Most hospitalized animals are sick or injured, and even healthy boarders may become unsettled in unusual confinement. Healthy animals often enjoy company, but sick dogs and cats may prefer to be left alone. Privacy is offered to shy patients, especially when anorexic. A towel over part of the cage may suffice, although adequate ventilation should be assured. A cardboard box is better for hiding. The box is placed in the cage with its open side facing the back wall. Extra effort is made to ensure adequate observation of patients in hide boxes.

Food intake is encouraged when wards are well ventilated, cages and bowls are clean and personnel are observant and nonthreatening. Feed intake of healthy dogs in balance experiments can change with personnel and, generally, sick dogs and cats are even more sensitive to the feeder's attitude. Less is known about the impact of background sounds. A quiet radio masks sudden startling noise.

Wards should take notice of the shrinking thermoneutral zone of the sick, but the clinic population is complex. There may be a mix of indoor or outdoor pets, and short- or long-haired dogs and cats. Some may be pyrectic and others chilly. Further, ambient conditions have an impact on energy needs. Dogs and cats need up to 15% more food energy for cooling in hot environments and up to perhaps 50% more for heating in a cold ward. All ambient conditions in the wards should be examined if food intake and maintenance of body condition are problems in patients.

Diet Selection

A good clinic feeding program requires a selection of commercial products and also foods suitable for home-made diets. The number and nature of diets is affected by cost and convenience as well as nutritional and medical considerations. The system should comprise mixtures of a few carefully chosen commercial products or recipes of a few selected ingredients used in varying proportions to meet the nutritional goals of the wards in general and of particular individuals.

Hospitalized animals may be assumed to be stressed to some degree. Thus the optimal routine hospital diet should have a stress profile, that is, contain certain nutrients (those affected by stress) at higher levels relative to energy: protein, vitamin A, vitamin E, zinc, copper, iron, perhaps ascorbic acid, carnitine and choline. At present, however, we know of no commercial stress diet, containing extra nutrients but restricted energy, that is suited for certain hospitalized dogs and cats. Diets currently marketed for sick dogs and cats are energy-dense and suited for hypermetabolic patients (see below). Mixing different foods, life-stage products as well as diets for hypermetabolism, achieves nutrient and energy profiles suited for hospitalized pets.[7]

For years, when the better canned products contained mainly meat (protein and fat) and the dry foods contained mainly plants (carbohydrate), a simple and effective approach to a hospital diet was to mix canned with dry. Meat and meat by-products used to make up 95% of the meat-type canned products; now only 25% meat is found in some products. The chunks in a can may look meaty but may well be soy-based. On the other hand, the dense-dry super-premiums are made with meat and poultry ingredients. The ingredient list on the label must be read more carefully than before. For hospital use, the majority of ingredients above salt should be predominantly meat-based, at least three of five, or four of six.

Healthy dogs and, to a lesser extent, cats, adapt well to a wide range of proportions of protein, fat and carbohydrate (the primary fuel sources), including much more carbohydrate than was common during their evolution. During stress, however, carnivores revert to using more protein and fat than carbohydrate. Stress reduces adaptability, and most hospitalized dogs and cats draw closer to their carnivorous heritage. Within a type of pet-food (dry or canned) look for the highest levels of protein and fat, and, in most instances, the lowest levels of fiber, for feeding hospitalized pets.

Palatability and digestibility are generally (though not invariably) higher in meat-based than in plant-based products. Foods of animal origin are digested hydrolytically in the small

MIXING FOODS TO ACHIEVE A RANGE OF ENERGY DENSITIES AND FUEL SOURCES				
	% Metabolizable energy (kcal(kJ)/g)			
	Protein	Fat	Carbohydrate	Energy
Dense–dry product	32	35	33	4.5 (19)
Low–energy product	24	10	66	2.5 (11)
Dense–dry:low–energy				
75:25	30	29	41	4.0 (17)
50:50	28	22	50	3.5 (15)
25:75	26	16	58	2.9 (12)

Fig. 3.2 Mixing dense-dry and low-energy dog foods in varying proportions achieves a wide range of energy density and fuel sources. In this example, a dense-dry product contains (on an as-fed basis, from the label) a minimum of 34% protein and 16% fat. In contrast, the carbohydrate is estimated to be about 35%. A low-energy dry dog food contained (on an as-fed basis, from the label) 14% protein, 2.5% fat and 39% carbohydrate. Energy values were calculated using 3.8 kcal/g (16 kJ/g) protein and carbohydrate, and 8.8 kcal/g (37 kJ/g) fat and 10% moisture.

intestine. In contrast, the galactosides in soybeans and the fibers (plant cell walls) in fillers are left to undergo fermentative digestion in the large intestine. The by-products of fermentation are gases, short-chain fatty acids and lactic acid. Acetic acid is rapidly absorbed. In contrast, lactic acid accumulates during rapid fermentation and attracts water; this process may lead to bowel distension, looser stools and, if more severe, osmotic diarrhea.

Some ingredients in hospital liquid enterals derive from plant sources, such as isolated soy protein and maltodextrins. Enteral ingredients are highly processed and purified, with very high digestibility, no fermentation from soy protein and with minimal residue. Enterals with plant-source ingredients are indicated for sick animals, whereas commercial petfoods with plant-source ingredients are contraindicated for most sick dogs and cats.

For dogs in wards, a good working system is to mix a dense-dry super-premium (relatively high in protein and fat) with a mild weight-control product (relatively low in fat) to approximate a stress mixture for inactive dogs (Fig. 3.2). Those requiring more energy receive more of the high fat product; those requiring less energy receive more of the low-energy food (Fig. 3.2).

For cats in wards, the feeding program is simpler. The variation between cat foods is not especially large, partly because the nutrient requirements for cats are strict, and partly because cats are so finicky. One key point to remember is that the cat needs high protein throughout its life. Its liver does not adapt to low protein diets. Also, cats cannot be forced to eat a food they reject. Simply "waiting out" the cat does not work. For voluntary intake, the cat must be given a food it likes. Good quality canned cat foods should be offered to all cats; dry foods are recommended for the rare cat refusing canned foods.

Puppies and kittens should be fed products containing label statements that the food meets requirements for growth or all stages of the life cycle. Likewise, patients that are pregnant or lactating should be fed products meeting requirements for reproduction or all life stages.

The food value of the product depends not only on its formula and ingredients but also on processing, packaging and shelf-life. Here we must trust and choose between manufacturers. In the end, one's actual experiences in feeding the product and observing the animal's responses—its body condition and stool quality, appetite and attitude, activity and responsiveness and coat—determine a product's appropriateness for hospitalized animals. Knowledge of metabolic changes with disease and injury, and the effects on nutritional needs of dogs and cats, permit informed selection of appropriate commercial products.

Nutritional Needs of Patients

The keys to feeding hospitalized dogs and cats are the identification of individual patients in need of specialized nutritional support and the determination of appropriate nutritional goals. This requires a working knowledge of metabolic aspects of starvation, trauma, surgery, sepsis, stress and specific diseases. These metabolic alterations affect energy expenditure, proportions of protein, fat and carbohydrate used and fluxes of minerals and vitamins.

Nutritional goals are reached through a global assessment of the nutritional status of the patient.[3,4] The patient's overall condition and the need for patient comfort contribute to the selection of specific diets and to the timing and feeding schedule.

Nutritional Status

This is assessed through a combination of dietary history, physical examination and laboratory data. It is one of several factors in the selection of dogs and cats for nutritional support.[2,4]

Diet History

The pet's prior diet may have led to malnutrition, explaining why a particular patient has a poor recovery from illness or surgery. Maintenance-type, plant-based, low-cost petfoods may chronically deplete tissue proteins and other nutrients lost during stress, and hence disadvantage an animal prior to hospitalization.

Body Weight

A pet's weight serves as the initial basis for estimating energy goals. It provides limited quantitative information on lean body mass. However, all weight loss, even in healthy patients, is accompanied by loss of lean tissue as well as adipose tissue. In sickness and after surgery, the loss of tissue protein is often accelerated and always involves negative nitrogen balance. Protein catabolism during illness results in cumulative losses of substance (e.g. skeletal muscle mass) and eventually function (e.g. enzyme systems). Thus, loss of body weight signals loss of vital protein, even in overweight patients.

Tissue proteins usually continue to be depleted during initial recovery from illness or surgery. Weight gain immediately following illness or surgery represents more water and fat than protein. Tissue protein is restored later during convalescence.

Body Condition

Standardized scores provide a subjective means of assessing nutritional status (Fig. 3.1). Patients are observed and palpated to assess degrees of subcutaneous fat and skeletal muscle. Severe protein losses, irrespective of body weight loss, may be grossly evident in the size of the muscle mass in the pet's lumbar region and limbs, and over the frontal and parietal bones of the skull. Protein loss from muscle is more evident, but protein loss from tissues such as heart and intestine are more life threatening. Animals exhibiting muscle atrophy should be considered to be in serious negative nitrogen balance and in need of immediate nutritional attention.

Laboratory Analysis

Blood and urine measurements reflect metabolic changes and help to assess the degree of change. Low serum albumin may indicate protein deficiency, although the response may be too slow for useful evaluation of acutely ill patients. Proteins with more rapid turnover, such as transthyretin, are better indicators of protein status but are not measured in most veterinary hospitals.

Our best current approach is to examine the hospitalized dog or cat with all of the above in mind and to rely on our clinical experience and judgement, the global approach.[3,4]

Energy Needs: Hypo- and Hypermetabolism

Much is known about the effects of trauma and illnesses on metabolism, and hence nutritional needs, in humans.[6,13] Related experimental studies used laboratory animals, frequently dogs but rarely cats. Many of these have been reviewed.[4] This knowledge is combined with information about the nutrition of healthy dogs and cats, and integrated with clinical veterinary experiences feeding hospitalized animals.

Energy expenditure of any hospitalized animal will be below, equal to or above the average number of calories required for a healthy individual of the same species, breed, sex, age and weight. Hence, hospitalized patients may be characterized as hypo-, normo- or hypermetabolic.

Hypometabolism

Hypometabolism occurs during starvation and occasionally with disease. Animals that are hypothyroid, flaccid, comatose, moribund or in the terminal stages of a disease may be hypometabolic. These animals often have lower than normal body temperatures and heart rates.

Inactivity lowers energy expenditure. Because the total energy needs represent a sum of all the factors that decrease as well as increase energy expenditure, an injured animal that is less active than usual, perhaps because of pain, leg splints or cage confinement, may require less energy than if it was mobile.

Food deprivation also lowers energy expenditure. Carnivores are especially tolerant of short fasts. Previously healthy dogs and cats deprived of food for a few days, because of anorexia or intentional fasting for diagnostic tests or surgery, do not suffer the metabolic disturbances (such as ketosis) typical of fasting in most species. Responses to fasting may be thought of as exaggerations of the usual feeding cycle.

Feeding Cycle

The normal feeding cycle is a complex set of hormonal and enzymatic adaptations that regulate the utilization, storage and mobilization of nutrients. During and immediately after a meal containing carbohydrates, glucose is used for synthesis of glycogen in liver and triglycerides in adipose cells. After a few hours, glucose utilization decreases, and fatty acids are released from stores. In time, amino acids are mobilized and used for gluconeogenesis.

Most cells can use a variety of substrates for energy, but those of the nervous system, renal medulla, bone marrow and circulating blood are obligate glucose users. Repairing tissues and certain neoplasms are also dependent on glucose. Thus gluconeogenesis in the short term sustains life by providing glucose to neurons and nephrons, but in the long run progressively diminishes vital functions dependent on tissue proteins.

Cats are strict carnivores that have high rates of protein utilization for gluconeogenesis when in a fed state.[10] When fasted, their blood glucose levels remain constant and the high rates of gluconeogenesis continue.

Review of many studies indicates that fatty acid utilization accounts for 70–85% of energy expenditure, ketones up to 15%, protein up to 25% and carbohydrate less than 10% during starvation in dogs and cats.[4]

Hypermetabolism

Hypermetabolism occurs during stress associated with injury, sepsis and many diseases. It is characterized by increased needs for energy and certain nutrients, notably protein. Even relatively simple surgical procedures increase energy expenditure, often 5 or 10% above normal. Energy needs follow an increasing gradient with the seriousness of the injuries. Thus two broken bones result in higher energy needs than one broken bone. A combination of broken bones and infection result in higher energy needs than either disorder occurring singly.

Some of the greatest increases in energy requirements, perhaps twice normal, are seen following severe head trauma, a common injury in dogs and cats hit by cars. The brain is one of the biggest energy users, even when healthy. Certain therapeutic drugs, such as high doses of steroids, exacerbate energy needs. Other patients with extreme hypermetabolism are those suffering extensive burns.

Hypermetabolism is also characterized by peripheral insulin resistance, prolonged, marked protein catabolism and a negative nitrogen balance. Fat and protein are the primary fuel sources. Significant amounts of carbohydrate are contraindicated during stress, especially when hyperglycemia indicates insulin resistance.

Catabolic Stress

Catabolic stress tends to override the protein-sparing adaptations of starvation and is seen especially in pets that are septic or have certain cancers. In contrast to simple fasting, the metabolically stressed dog or cat continues to exhibit a negative nitrogen balance, accelerated gluconeogenesis and insulin resistance when fed. Catabolic patients may be hypo-, normo- or, typically, hypermetabolic.

These expensive metabolic responses support the healing of wounds and resistance to infection for days. The cumulative drain on tissues may continue for weeks, necessitating special feeding management well beyond the hospital stay.

Energy Intakes

In hospital practice, despite the dearth of hard data, optimal daily intakes of energy are

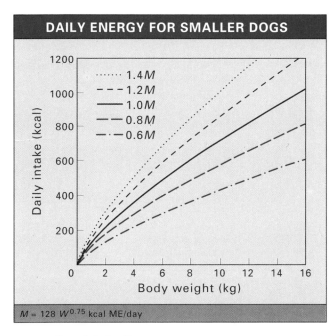

FIG. 3.3 Daily energy for smaller dogs.

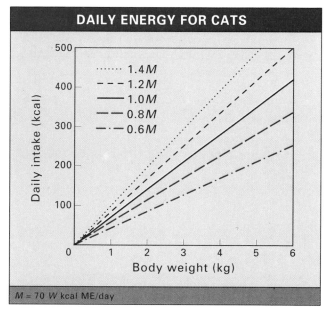

FIG. 3.5 Daily energy for cats.

estimated for sick dogs and cats.[4,7] For the general ward population, the daily energy intakes are based on cage-rest maintenance (Figs. 3.3–3.5). Further estimates are provided for patients that may be hypo- or hypermetabolic (Figs. 3.6 and 3.7).

Because variation is large between individuals, some dogs and cats require more energy than that calculated and some require less. For long-term hospitalization and for feeding at home, we recommend that personnel first follow the guides presented here, then adjust intakes gradually to meet desired body condition.

Protein Needs

In patients, protein is used for anabolism (such as antibody production and wound healing) and energy. Except for a few specific disorders (such as hepatic encephalopathy), dietary protein is not spared in diets for hospitalized dogs and cats. Our recommendations for protein are expressed on a metabolizable energy (ME) basis (Fig. 3.8).

Levels of protein above 30% ME can be achieved with mixtures of dense-dry premium petfoods, select meat-based canned petfoods, commercial diets for hypermetabolism and by additions of protein modules, such as dehydrated cottage cheese.

Food Amounts and Feeding Schedules

From our observations on routine ward feeding, cats tend to be fed at suboptimal frequencies, small dogs tend to be overfed and large dogs underfed. The latter is especially

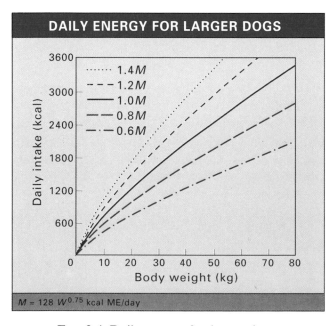

FIG. 3.4 Daily energy for larger dogs.

AVERAGE DAILY ENERGY INTAKES RECOMMENDED FOR HOSPITALIZED DOGS*					
Body weight		Multiples of cage–rest maintenance (kcal (kJ) ME/day)				
lb	kg	0.7	0.85	1	1.15	1.3
2.5	1.1	99 (416)	119 (505)	141 (594)	162 (683)	183 (772)
5	2.3	166 (700)	201 (849)	237 (999)	272 (1149)	308 (1299)
7.5	3.4	225 (948)	273 (1151)	321 (1354)	369 (1558)	417 (1761)
10	4.5	279 (1176)	339 (1429)	398 (1681)	458 (1933)	518 (2185)
15	6.8	378 (1595)	459 (1936)	540 (2278)	621 (2620)	702 (2961)
20	9.1	469 (1979)	570 (2403)	670 (2827)	771 (3251)	871 (3675)
25	11.4	554 (2339)	673 (2840)	792 (3342)	911 (3843)	1030 (4344)
30	13.6	636 (2682)	772 (3256)	908 (3831)	1044 (4406)	1181 (4980)
40	18.2	789 (3328)	958 (4041)	1127 (4754)	1296 (5467)	1465 (6180)
50	22.7	933 (3934)	1132 (4777)	1332 (5620)	1532 (6463)	1732 (7306)
60	27.3	1069 (4510)	1298 (5477)	1528 (6443)	1757 (7410)	1986 (8376)
70	31.8	1200 (5063)	1458 (6148)	1715 (7233)	1972 (8318)	2229 (9403)
80	36.4	1327 (5596)	1611 (6796)	1895 (7995)	2180 (9194)	2464 (10393)
90	40.9	1449 (6113)	1760 (7423)	2070 (8733)	2381 (10043)	2692 (11353)
100	45.4	1568 (6616)	1905 (8034)	2241 (9451)	2577 (10869)	2913 (12287)
120	54.5	1798 (7585)	2184 (9211)	2569 (10836)	2954 (12462)	3340 (14087)
140	63.6	2019 (8515)	2451 (10340)	2884 (12164)	3316 (13989)	3749 (15814)
160	72.7	2231 (9412)	2710 (11429)	3188 (13446)	3316 (13989)	3749 (15814)

*$M = 128 (BW_{kg}^{0.75})$ where M = cage – rest maintenance energy requirement of metabolizable energy, kcal/day and BW = body weight in kg.[7] It is likely that one dog in seven needs 20% more, and another from the same seven needs 20% less energy than the average M for a given weight.

FIG. 3.6 Average daily energy intakes recommended for hospitalized dogs. Calories (kcal) were calculated as multiples of cage-rest maintenance energy (M).

prevalent if only one size of feeding bowl is used for all dogs. Three changes in feeding management solve these problems.

How Much to Feed

First calculate the energy contents of ¼ bowl, ½ bowl, ¾ bowl and so on of the usual ward diets. Estimating the energy content of ward diets may be difficult if the only information comes from labels. Generally, one 8-fl oz cup of dense-dry premium dog and cat foods contains 75–100 g and 350–450 kcal (1475–1900 kJ) and an equal volume of plant-based dry food contains about 250–350 kcal (1050–1475 kJ). A 13-oz (368 g) can of petfood contains about 400 kcal (1690 kJ), and a 6-oz (170 g) can contains about 200 kcal (840 kJ). These are approximations. More precise numbers can be calculated from the typical analysis, if it is available. For specific nutritional information, manufacturers should be contacted.

Second, charts for daily ME intakes of dogs and cats should be posted where food bowls are prepared. Charts of daily intakes may be in the form of tables (Figs. 3.6 and 3.7) or graphs (Figs. 3.3–3.5). Because only a few people are responsible for daily feedings in most clinics, personnel quickly develop a sense about the correct volumes to feed. Large dogs may need to be provided with more than one bowl of food.

When to Feed

Healthy dogs are commonly offered food once daily, cats twice daily (that is, the total daily intake divided into two meals). Hospitalized dogs and cats may handle smaller meals better than larger ones, for digestion is facilitated by increasing the number of meals without changing the total daily intake. Cats respond especially well to many small meals offered throughout the day. Their natural inclination is to eat 10–16 meals daily; this instinct should be considered when faced with finicky cats.

Routine meal times are recommended: two meals daily for dogs, and more meals for cats, at about the same times each day, at least 6 hr apart. Meal times are an opportunity for social interactions and for careful observation of the animal's demeanor and responses to the feeders as well as the food.

The total daily food intake offered initially to each patient should supply the ME needs for cage-rest maintenance (Figs. 3.3–3.7). If less than about 85% is consumed, then the patient usually qualifies for a specific plan of nutritional support and perhaps a special diet.

Moderately ill patients should be fed smaller amounts of food, 25–75% of their usual intake, for the first 1 or 2 days. Most injured patients reach 105–120% of their usual ME intake by 4–10 days posttrauma; increases in food intake must be gradual and monitored. Those patients with multiple fractures, extensive burns or serious head injuries may need to be eventually built up to 130–200% of cage-rest maintenance.

For patients with poor food intake, ensure that the food is fresh and wholesome, and utensils are clean. Food intake may be improved by warming food, and by offering many small meals. Consider behavioral characteristics too (Fig. 3.9). Offer different flavors or brands, top-dress with tasty table foods and invite the patient's family in to offer food to their pet.

Amounts of food offered and then actually consumed should be written in the case record. Failure to reach a specified goal, usually 85% consumption in 2 or 3 days, requires further attention; intake of only 70% requires action.

Adverse Reactions

Anecdotal assertions in the veterinary literature suggest that food intakes for sick dogs and cats based on maintenance (rather than basal) requirements of healthy animals lead to digestive and metabolic upsets. This assertion appears to be without foundation in

AVERAGE DAILY ENERGY REQUIREMENT FOR SICK CATS*						
Body weight		Multiples of maintenance (kcal (kJ) ME/day)				
lb	kg	0.7	0.85	1	1.15	1.3
4.0	1.8	89 (376)	108 (376)	127 (537)	146 (617)	165 (698)
4.5	2.0	100 (423)	121 (513)	143 (604)	165 (694)	186 (785)
5.0	2.3	111 (470)	135 (570)	159 (671)	183 (772)	207 (872)
5.5	2.5	122 (517)	149 (627)	175 (738)	201 (849)	228 (960)
6.0	2.7	134 (564)	162 (684)	191 (805)	220 (926)	248 (1047)
6.5	3.0	145 (611)	176 (742)	207 (872)	238 (1003)	269 (1134)
7.0	3.2	156 (658)	189 (798)	223 (939)	256 (1080)	290 (1221)
7.5	3.4	167 (704)	203 (856)	239 (1006)	274 (1158)	310 (1308)
8.0	3.6	178 (752)	216 (913)	254 (1074)	293 (1235)	331 (1396)
8.5	3.9	189 (798)	230 (967)	270 (1141)	311 (1312)	352 (1483)
9.0	4.1	200 (846)	243 (1027)	286 (1208)	329 (1389)	372 (1570)
9.5	4.3	212 (892)	257 (1084)	302 (1275)	348 (1466)	393 (1657)
10.0	4.5	223 (939)	270 (1141)	318 (1342)	366 (1543)	414 (1745)
10.5	4.8	234 (986)	284 (1198)	334 (1409)	384 (1621)	434 (1832)
11.0	5.0	245 (1033)	298 (1255)	350 (1476)	402 (1698)	455 (1919)
11.5	5.2	256 (1080)	311 (1312)	366 (1543)	421 (1775)	476 (2006)
12	5.4	267 (1127)	324 (1369)	382 (1611)	439 (1852)	496 (2094)
13	5.9	290 (1221)	352 (1483)	414 (1745)	476 (2006)	538 (2268)
14	6.4	312 (1315)	379 (1597)	445 (1879)	512 (2161)	579 (2443)
15	6.8	334 (1409)	406 (1711)	477 (2013)	549 (2315)	620 (2617)
16	7.3	356 (1503)	433 (1825)	509 (2147)	585 (2469)	662 (2792)
18	8.2	401 (1691)	487 (2053)	573 (2416)	659 (2778)	744 (3140)
20	9.1	445 (1879)	541 (2282)	636 (2684)	732 (3087)	827 (3489)

*$M = 70 (BW_{kg})$ where M = maintenance energy requirement of metabolizable energy, kcal/day and BW = body weight in kg.[12] It is likely that one cat in seven needs 20% more, and another from the same seven needs 20% less energy than the average M for a given body weight.

FIG. 3.7 Average daily energy requirement for sick cats. Calories were calculated as multiples of maintenance energy (M).

ESTIMATED RANGES OF DIETARY PROTEIN FOR HOSPITALIZED DOGS AND CATS		Equivalent as fed	
Species and condition	**% ME**	**Can***	**Dry†**
Healthy young, mature dogs at *M*	>20	>5.5	>23
Sick young dogs, depending on stress	20 – 48	5.5 – 13.2	23 – 55
Healthy cats	>24	>6.6	>27
Old dogs at *M*	24 – 38	6.6 – 10.5	27 – 43
Sick cats and old dogs, depending on stress	24 – 48	6.6 – 13.2	27 – 55
Growth and, presumably, tissue repair	28 – 43	7.7 – 11.8	32 – 49

*1.1 kcal (4.6 kJ)/g as fed.
†3.5 kcal (14.8 kJ)/g as fed.
ME = metabolizable energy, *M* = maintenance.

FIG. 3.8 Estimated ranges of dietary protein for hospitalized dogs and cats (ME = metabolizable energy, *M* = maintenance).

experimental science or human clinical experience,[7] and it is contrary to our substantial clinical experience.[4,5]

For all patients, feeding is started gradually. Although the full goals for water are offered every day, the full measures of food are reached progressively in 2–6 days. Special care is needed, however, when rehabilitating (refeeding) starved dogs and cats.

Refeeding

This term is used for nutritional repletion of a previously starved animal. Upon refeeding, restoration of nutritional balance encourages a shift from the conserving adaptations of starvation to repletion and anabolism. During refeeding, initial diets are predominantly fat and protein, and change progressively from simpler to more complex ingredients.

Provision of too much carbohydrate must be avoided. When excess carbohydrate is given to starved patients, it initiates a set of metabolic derangements, termed the refeeding syndrome, that can be fatal. Excessive carbohydrate leads to insulin-induced transport of phosphorus and potassium into cells and subsequent life-threatening hypophosphatemia and hypokalemia. The syndrome is well described experimentally in fasted dogs.[14] It will probably be observed clinically in starved dogs and cats.

For refeeding, small amounts of food are offered in several daily meals. The exact amount and frequency depends on the animal's condition, health and appetite. Most dogs do well with three small meals for the first day of refeeding. Cats do better with four or more daily meals. Food consumption is recorded as well as the animal's attitude, patterns of urination and defecation and the condition of urine and feces.

Drug–Nutrient Interactions

The nutritional status of sick patients affects drug metabolism. For example, enzyme activities decreased in the hepatic cytochrome P-450-dependent mixed function oxidase detoxification system due to fasting, decreased dietary protein quantity or quality, increased dietary sugars, decreased dietary fat, deficiencies of several vitamins and deficiencies of trace minerals such as copper and magnesium.[1] For hepatic metabolism of drugs, diets should contain optimal levels of fat and high quality protein, minimal carbohydrate, little sugar and optimal amounts of vitamins and minerals.

Drug therapies can affect nutritional status, too, through their action on absorption and specific organs. Drugs that alter hepatic metabolism, for example, also affect nutrients such as vitamin A and protein. Penicillamine binds copper and tends to prevent copper accumulation in Terriers predisposed to copper

IMPROVING FOOD INTAKE OF HOSPITALIZED ANIMALS BY UTILIZING BEHAVIORAL CHARACTERISTICS	
Species	**Characteristics**
Dogs and cats	Prefer animal protein
	Prefer animal fat
	Prefer several small meals
	Prefer warmed food
Dogs	Social carnivores, eating with a pack (for domestic dogs, its family)
Cats	Solitary carnivores, eating alone
	Sensitive to off-odors from utensils (plastics, chlorine, molds)
	Increase food intake when flavors varied
	Habituate to odors, so increase food intake when offered food as multiple meals instead of free choice

FIG. 3.9 Food intake in hospitalized animals may be improved by making use of behavioral characteristics.

intoxication (see Chapter 8). Other nutrients affected by drugs include folacin (by cholestyramine, bicarbonate, sulfasalazine, aspirin and phenytoin), calcium (by tetracycline, neomycin and methotrexate), phosphate (by aluminum hydroxide), vitamins A and K (by cholestyramine), vitamin B_{12} (by cimetidine), vitamin D (by isoniazid, phenytoin and phenobarbital), vitamin K (by cephalosporin) and vitamin B_6 (by isoniazid).[11] Addition of aluminum antacids that bind phosphorus for the management of chronic renal failure is an example of a desired drug–nutrient interaction.

Last, nutritional management of hospitalized patients also affects drug therapies. For example, humans on enteral tube feedings exhibit reduced blood concentrations of phenytoin and warfarin, and additions of potassium or aluminum to enterals adversely affects the physical characteristics of the diets.[9]

Transitions

Nutritional support in the hospital usually requires a discharge session with the client, because it is continued in the home. This counselling is often the responsibility of the veterinary nurse or technician who has been feeding the patient in the hospital.

As a general guide, special feeding needs to be continued for 6 weeks for hypometabolic cases such as neglect or starvation, 2 weeks for uncomplicated major surgeries, 2–4 weeks for uncomplicated trauma cases and 4–12 weeks for severe trauma. Cases of chronic disease, such as neoplasia, may need special attention for months.

Diet transitions should be made slowly, from chemically simple to more complex ingredients, and from individualized to more general feeding programs. Each change of diet should be made in a series of partial replacement steps and take about 2–4 days for simple cases, 4–8 days for cases of serious illness and as long as 10–14 days for dogs with severe gastrointestinal disease.

Clients and others responsible for feeding are counselled about diets and amounts to feed, and the dangers of overfeeding. Charts or bar graphs of increasing daily energy needs or food volumes are useful visual aids for clients. The animal's body weight can be expressed on a graph alongside the diet changes.

Feeding management in convalescent pets is best combined with physical training, such as progressively longer walks, to return body condition to an optimal state. Once optimal physical condition has returned, plan to stabilize it, using diets carefully chosen to suit the lifestyle of the pet and its owners.

Special Diets

Supplements

The most common supplements contain micronutrients (vitamins and trace minerals).

High quality petfoods aim to supply micro-nutrients in optimal ranges. Adding just one or two micronutrients may lead to imbalances, a risk that is minimized by using broad-spectrum supplements. Only one such supplement should be used at a time, and then only according to its maker's recommendations.

Supplements for hospital use often contain just one ingredient or nutrient and are termed *modules*. Carbohydrate for hospital feeding is usually rice because other sources are relatively indigestible, but rice is rarely added, except for home-made diets. Fiber modules, usually powdered cellulose, are added to alter intestinal motility.

Protein is the module added most often in nutritional support of dogs and cats. It is used to raise the dietary protein content to more than 30% of energy. This level is used for all cats, and for dogs that are catabolic, such as those with fractures, infections or cancer. Protein supplements may also be used to enhance palatability. One might use fresh lean meat, poultry or fish. Liver provides most vitamins and trace elements as well as protein. Dehydrated cottage cheese is marketed commercially for dogs and cats.

Fat is added to increase energy density. The most common is vegetable oil. Second are medium-chain triglycerides, providing 8 kcal/g (34 kJ/g), that require no lymphatics for absorption. They are useful when feeding dogs and cats with chylothorax or lymphangectasia.

Fat modules are used in hospitalized animals to increase energy density of diets for very small dogs and cats who have high energy needs but sometimes are finicky eaters. In addition, dietary fat may be used to reduce the work-load of a compromized respiratory system because it produces less carbon dioxide than carbohydrate produces when used as a fuel.

Commercial products marketed as energy sources may contain only 1% protein and over 60% fat. Energy sources diminish nutrient density on an ME basis and tend to induce multiple deficiencies, so they should be used only in judicious amounts.

Enterals

Enterals are sterile liquid nutritionally complete dietary products manufactured primarily for hospitalized humans. Used commonly in critically ill patients, enterals also have a place in the wards. Enterals may be blended with canned petfoods, forming energy- and nutrient-dense palatable slurries suitable for voluntary intake by sick dogs and cats. A mix of one 6-oz (170 g) can of meat-based petfood and about 170 ml enteral provides a mix containing 1–2 kcal/ml (4.2–8.4 kJ/ml). For further information, see Chapters 4 and 5.

Home-Made Diets

Often containing special nutrient modifications, such as low protein or low fat, home-made diets may be used when patients refuse, or owners cannot afford, commercial pre-scribed diets.[5] If well-balanced and whole-some, home-made diets may be fed as the sole diet to sick dogs and cats (see Appendix for recipes).

References

1. Bidlack, W. R., Brown, R. C. and Mohan C. (1986) Nutritional parameters that alter hepatic drug metabolism, conjugation, and toxicity. *Federation Proceedings*, **45**, 142–148.
2. Carnevale, J. M. *et al.* (1991) Nutritional assessment: guidelines to selecting patients for nutritional support. *Compendium of Continuing Education for Practicing Veterinarians*, **13**, 255–261.
3. Detsky, A. S. *et al.* (1987) What is global assessment of nutritional status? *Journal of Parenteral and Enteral Nutrition*, **11**, 8–11.
4. Donoghue, S. (1989) Nutritional support of hospitalized patients. *Veterinary Clinics of North America: Small Animal Practice*, **19(3)**, 475–495.
5. Donoghue, S. (1991) A quantitative summary of nutrition support services in a veterinary teaching hospital. *Cornell Veterinarian*, **81**, 109–128.
6. Kinney, J. M. *et al.* (1988) *Nutrition and Metabolism in Patient Care*. Philadelphia: W. B. Saunders.
7. Kronfeld, D. S. (1991) Protein and energy estimates for hospitalized dogs and cats. *Proceedings of the Purina International Nutrition Symposium*. pp. 5–12. St Louis: Ralston Purina.

8. Kronfeld, D. S., Donoghue, S. and Glickman, L. T. (1991) Body condition and energy intakes of dogs in a referral teaching hospital. *Journal of Nutrition,* **121**, S157–S158.

9. Melnick G. (1990) Pharmacologic aspects of enteral nutrition. In *Enteral and Tube Feeding.* Eds. J. L. Rombeau and M. D. Caldwell. 2nd edn. pp. 472–509. Philadelphia: W. B. Saunders.

10. Morris, J. G. and Rogers, Q. R. (1989) Comparative aspects of nutrition and metabolism of dogs and cats. In *Nutrition of the Dog and Cat,* Waltham Symposium Number 7. Eds. I. H. Burger and J. P. W. Rivers. pp. 35–66. Cambridge: Cambridge University Press.

11. Munro, H. N. (1988) Aging. In *Nutrition and Metabolism in Patient Care.* Eds. J. M. Kinney, K. N. Jeejeebhoy, G. L. Hill and O. E. Owen. pp. 145–166. Philadelphia: W. B. Saunders.

12. National Research Council (1986) *Nutrient Requirements of Cats.* Washington, DC: National Academy Press.

13. Rombeau J. L. and Caldwell, M. D. (1990) *Clinical Nutrition: Enteral and Tube Feeding.* 2nd edn. Philadelphia: W. B. Saunders.

14. Solomon, S. M. and Kirby, D. F. (1990) The refeeding syndrome: a review. *Journal of Parenteral and Enteral Nutrition,* **14**, 90–97.

CHAPTER 4

Critical Care Nutrition

RICHARD C. HILL

Introduction

Most critical patients are catabolic. Hypovolemia and shock may reduce metabolic rate initially, but catabolism predominates after volume repletion. This includes patients after surgery or trauma, patients with sepsis, diabetes mellitus or multiple organ failure, and patients treated with corticosteroids or anticancer drugs.[106]

The nutrient requirements for catabolic dogs and cats have yet to be determined so it has been necessary to extrapolate from the nutrient requirements of normal dogs and cats, assuming a response to injury similar to that in humans (Fig. 4.1). It is important to bear in mind, however, that metabolic differences exist between humans and small animals, for example, the normal rate of free fatty acid utilization in dogs is twice that in humans,[26] so caution is necessary when adapting data from humans and rats to small animal patients.

Nutritional support (NS) in critical patients frequently involves feeding by tube or intravenous parenteral infusion. This chapter focuses on enteral NS because it is the safest, least expensive and most practical method of involuntary feeding in most small animal patients. Nevertheless, some space is devoted to parenteral NS because it is a complex and hazardous technique and may be the only method available. Emphasis is placed on the provision of calories and protein, but the composition of the protein, fat and carbohydrate (CHO) is also important, so nonstandard nutrients, particularly glutamine and fiber, will also be discussed.

Incidence

The use of NS has increased with increased understanding of the principles and techniques involved and with the appearance of NS services at major referral centers. One of these institutions has provided the only survey reporting the use of NS in small animals.[32] NS was requested for 2.1% of canine and 3.7% of feline admissions. Almost half of these requests were for sick or anorectic patients. Involuntary supplementation of food intake was necessary in 32% of canine and 16% of feline consultations. Supplementation was mostly by syringe or by nasogastric tubes so that only a quarter were fed by gastrostomy or enterostomy, and less than 2% were fed parenterally.

Nasogastric tubes do not require anesthesia but are easily dislodged, carry a risk of aspiration and need careful management. Gastrostomy, on the other hand, requires

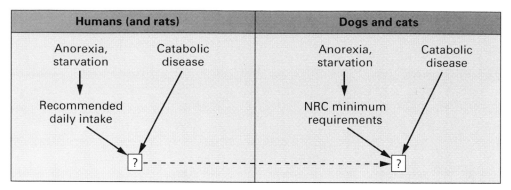

FIG. 4.1 Derivation of small animal requirements in critical care nutrition
extrapolated from human/rat data.

anesthesia but is easier to manage, has a low complication rate and allows an earlier return home. The clinicians at the University of Florida Veterinary Clinic, therefore, use percutaneous endoscopic gastrostomy with a greater frequency than the report above indicates.

Pathophysiology

Humans respond to injury with increased secretion of corticosteroids, adrenalin, glucagon and growth hormone.[23,106] Body weight may increase initially because increased secretion of aldosterone and ADH increase sodium retention and extracellular fluid volume; however, metabolic rate, gluconeogenesis and lipolysis increase despite reduced food intake and there is a rapid loss of muscle protein and body fat.[106]

If the injury resolves, for example, after elective surgery, this systemic response to injury is transient, but if the injury persists or is accompanied by malnutrition there may be immunosuppression, increased migration of bacteria from the intestine, increased risk of sepsis, delayed wound healing and prolonged hospitalization.[106] Eventually, metabolic rate declines as the body attempts to conserve any remaining protein and death may ensue.[23] NS is designed to ameliorate these effects, but careful studies in humans show that NS alone does not completely eliminate protein wasting.[106] The best treatment remains resolution of the inciting disease.

The stress response in dogs appears to be similar to that in humans. Corticosteroids and adrenalin are released in response to immobilization, hemorrhage and burn injury.[84] Metabolic rate increases 25% during sepsis[88] and 35% after burns.[107] Glucagon increases 5- to 10-fold during sepsis but insulin increases only slightly, so most of the increase in metabolic rate is fueled by increased oxidation of fat and there is very little change in glucose oxidation.[88] In normal cats fed high protein diets, gluconeogenesis is active at all times[71] so gluconeogenesis may not increase when cats are exposed to stress. Nevertheless, hyperglycemia is common in stressed cats, which implies insulin resistance from increased circulating glucagon and corticosteroids.

Assessment

Nutritional assessment is important because the risks and expense of involuntary feeding often outweigh any physiological benefits. Controlled prospective studies in human medicine have often failed to establish an overall benefit when all patients receiving NS are considered, but have done so when patients are stratified according to degree of malnutrition.[101] For example, in normally nourished human surgical patients, there was no overall benefit from parenteral nutrition because an increased incidence of sepsis outweighed a decrease in the incidence of major complications.[96] In severely malnourished patients, however, an overall benefit was observed because the frequency of major complications decreased more than the frequency of sepsis increased. Similar studies are needed in small animals, but, in the interim, parenteral NS in surgical patients at the University of Florida Veterinary Clinic is confined to severely malnourished animals.

The nutritional status of a patient may be assessed either objectively or subjectively. Objective measures include body weight, plasma albumin concentration and lymphocyte count. All of these parameters decrease during nutritional deprivation, but their utility is limited because body weight is greatly affected by fluid shifts, and hypoalbuminemia and lymphopenia have many causes unrelated to nutritional status. Albumin also has a long half-life (8 days in the normal dog) and is slow to respond to changes in nutritional status. Various other circulating proteins with shorter half-lives have been used in human medicine including prealbumin, transferrin and retinol-binding protein, but none of these tests has been validated in dogs and cats. Total iron-binding capacity provides a measure of transferrin concentration and has been found to decline more than 30% in 15 days in malnourished dogs[98] but this test is not easily available and has not been widely used.

Subjective assessment is derived from human medicine.[27] It is based primarily on body weight, appetite and loss of fat or muscle mass. Patients are categorized as severely malnourished if there is a history of >10% weight loss, anorexia and muscle wasting. A patient that has recently regained some weight is regarded as better nourished than one that has a similar amount of weight loss but continues to lose weight.[27] Subjective assessment is readily adapted to small animals but weight loss >20% may be more representative of severe malnutrition in dogs because they lose weight with relative impunity.[26]

NS should also be considered in patients where food is withheld or who do not eat for more than a week. Fat cats especially tolerate starvation poorly.[13] Normal dogs tolerate starvation well because starvation causes only a slight increase in circulating glucagon so hypoglycemia and ketosis are rare.[26] Nevertheless, markers of immune function (complement and neutrophil chemotaxis) decline in normal dogs after only one week of malnutrition[29]; thymus, spleen, lymph nodes and Peyer's patches become markedly atrophic.[30] NS returns these indicators of immune function to normal or above normal levels and supports repopulation of lymphoid tissue.[29,30]

Route of Administration

The gut should be used whenever possible because the intestine regulates the absorption of nutrients. Also, the intestinal mucosa acts as a barrier to bacteria only when nutrients are present in the lumen. Bacteria translocate out of the intestine during parenteral NS[5] so even patients requiring parenteral NS may benefit from enteral supplementation. Critically ill malnourished patients previously thought to be too sick for surgical enterostomy or gastrostomy may, in fact, be too sick *not* to have a feeding tube implanted.

Rate of Initiation

The rule of thumb is to start slowly! NS is *not* an emergency procedure and should be instituted only after careful assessment of the potential risks and benefits. Rapid initiation of nutrition is possible in animals that have only recently been deprived of food but most candidates for NS have become malnourished over a prolonged period of time. In these patients, food should be reintroduced only gradually.

To introduce NS slowly, the caloric requirement should be estimated based on *current* body weight. Half this amount is then administered on day 1, three quarters on day 2 and the full amount on day 3, subject to patient tolerance. No attempt should be made in the unstable critical patient to supply the extra calories necessary to replace lost weight. Extra calories (~6 kcal/g of new tissue) should be supplied only after the patient has stabilized. In practice, the patient often starts to put on weight as the disease resolves because the extra calories supplied for increased metabolism are used for anabolism as the metabolic rate declines.

Parenteral glucose must *always* be introduced gradually. If lipid is providing some non-protein calories so that glucose provides no more than 50% of total calories (% kcal), then glucose is infused at half rate for the first 12 hr before being increased. If glucose is the only source of nonprotein calories, then glucose infusion should be increased very cautiously in increments over 2–3 days. Parenteral glucose should also be withdrawn gradually by reversing this procedure to prevent hypoglycemia.

Hyperosmolar enteral solutions should also be introduced gradually. Intravascular fluid equilibrates with hyperosmolar solutions in the proximal small intestine so rapid infusion can cause hypotension from fluid leaving the blood too quickly. This is called the "dumping syndrome."[89] Initially, therefore, hyperosmolar solutions should be diluted to 300–400 mOsm/l with water (1:1 dilution for a 600 mOsm/l formula). The osmolality of the solution may then be increased gradually over several days. Hyperosmolar solutions do not need to be diluted for infusion into the stomach because the stomach regulates the rate at which nutrients are submitted to the small intestine.[42] The volume should be restricted and only increased slowly.

Energy Requirements

The first step in formulating NS is calculating the patient's daily caloric requirement. Very few clinicians have access to equipment that measures energy consumption directly, so most authors have recommended estimating basal (B) or maintenance (M) energy requirements using regression equations based on body weight in kilograms (W). This estimate can then be multiplied by arbitrary factors that allow for activity, stress or severity of underlying disease. Energy intakes for hospitalized dogs and cats can be found in Figs. 3.3–3.7.

Many authors have used the interspecies estimate of B for dogs and cats[61,83,105]:

$$B = 70W^{0.75} \text{ kcal/day} \qquad (1)^{60}$$

or its linear approximation:

$$B = 70 + 30W \text{ kcal/day}. \qquad (2)^{57}$$

Equation (2) is easy to calculate but inaccurate[58] so others have used the National Research Council recommendations for estimating M[18,31]:

$$M \text{ (dog)} = 132W^{0.75} \text{ kcal/day} \qquad (3)^{73}$$
$$M \text{ (cat)} = 70W \text{ kcal/day}. \qquad (4)^{75}$$

This author prefers to use the recently published regression equations for M in caged animals:

$$M \text{ (dog)} = 162W^{0.64} \sim 1500 \times BSA$$
$$\text{kcal/day} \qquad (5)$$
$$M \text{ (cat)} = 136W^{0.404} \qquad (6)$$

They are represented graphically in Figs. 4.2 and 4.3.

Equation (2) derives metabolizable energy whereas Eqn (1) derives digestible energy. BSA represents the body surface area in m^2. M in kcal should be multiplied by 4.2 to obtain the value in kJ.

Equations (5) and (6) appear to be the best estimates of energy expenditure in normal dogs and cats because they were obtained from animals with a wide range of body weights.

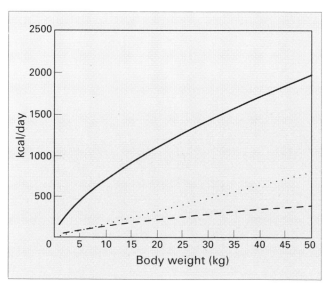

FIG. 4.2 Daily energy requirement (—) and protein requirement (20% kcal, - - -; 4 g/kg, ...) for normal dogs at cage rest.

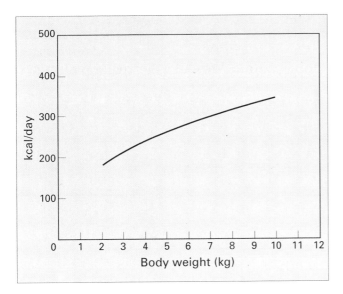

FIG. 4.3 Daily energy requirement for normal cats at cage rest.

Estimates of *B* have to be increased for *both* activity and stress whereas estimates of *M* have to be *decreased* for reduced activity or *increased* for stress. In practice, the final estimate of energy requirement is not greatly different from *M* because *M* has to be reduced in most critical patients by ~25% for reduced activity and increased by ~25% for catabolic illness. The metabolic rate may increase more than 25% with more severe injury. Adjustments of energy requirement for severity of injury in human patients are shown in Fig. 4.4.

The estimate of energy requirement only represents a mean for the population of dogs. The coefficient of variation for Eqn (5) was 5%,[19] but there was greater variation in dogs of large breeds possibly because of differences in breed, conformation and coat length: Newfoundlands required less calories, Great Danes more.[56] Variation may be greater still, however, in catabolic patients because individual variation in energy consumption was 12% in stable but only 46% in stable postoperative human patients.[104] Careful monitoring is, therefore, necessary to prevent under- or overnutrition.

Monitoring

Parameters used to assess nutritional status can also be used to monitor patient performance. A suggested schedule for patient monitoring is shown in Fig. 4.5 but an unstable patient may need more frequent examination. Care is particularly necessary to prevent overfeeding because body weight responds slowly. Physical examination may be helpful, however, because normal cats given excess calories parenterally vomited and developed lingual ulcers.[62] Puppies given excess calories parenterally developed hepatomegaly and increased serum ALT, AST and alkaline phosphatase due to an increased deposition of water and glycogen in the liver.[65]

Enteral Tube Feeding

Amount and Type of Food to Feed

Once the estimate of caloric requirement has been obtained, a formula or petfood must be chosen that provides calories in the

HUMAN ENERGY REQUIREMENTS IN ACTIVITY AND STRESS		
Condition	**Factor x *M***	**Examples**
Physical inactivity, starvation	0.5–0.9	
Catabolic disease: Mild	1.0–1.2	Surgery, trauma, sepsis, cancer
Moderate	1.2–1.5	
Severe	1.5–2.0	Extensive burns, head injuries, cancer
M = maintenance energy requirement.		

FIG. 4.4 Factors by which estimates of energy requirements are changed in humans to make allowance for activity and stress. *M* = maintenance energy requirement.

PATIENT MONITORING PROTOCOL	
Parameter	**Minimum frequency**
Physical exam	2 times daily
Body weight	Daily
PCV/total solids*	1–2 times daily
Blood glucose	Daily. 2 times daily during parenteral NS and 2 hr after a change of dextrose infusion rate
CBC/chemistries	2 per week. 24 hr after parenteral NS begun

* Hyperlipidemia may be evident from blood centrifuged for PCV determination.

FIG. 4.5 Patient monitoring protocol.

optimum proportions of protein, fat and CHO. The proportions of calories from protein, fat and CHO for various enteral formulae are shown in Fig. 4.6 together with their caloric densities. The daily food requirement can be obtained by dividing the desired number of calories by the caloric density of the food or formula.

The number of calories attributable to protein, fat and CHO depends on the digestibility of the nutrient. The factors used here are shown in Fig. 4.7. The recommendation of the Canine Nutrition Expert Committee of the American Association of Feed Control Officials was followed for petfoods to allow for the lower digestibilities of their ingredients.[35] Atwater factors were used for human enteral products and parenteral solutions because they contain more bioavailable ingredients.[6,32] Protein is used catabolically rather than anabolically so a caloric value of 4.0 kcal/g (17 kJ/g) was used rather than 5.25 kcal/g (22 kJ/g).

Protein

The protein requirement of the stressed small animal patient has not been determined. There has been much confusion because some have made recommendations in g/kg[61,83,105] and others have recommended percentages of total calories (% of kcal).[18,31] M is related

CHARACTERISTICS OF DIFFERENT TYPES OF ENTERAL DIETS					
Type of diet	**Energy density in kcal/ml (kJ/ml)**	**Osmolality (mOsm/ml)**	**Macronutrient content (% kcal)**		
			Protein	**Fat**	**CHO**
Petfood gruels Feline pediatric/growth diet					
+ Water	0.75 (3.1)		35	55	10
+ High fat formula	1.5 (6.3)		26	55	19
Small animal polymeric diets					
Canine	0.9–0.92 (3.8–4.1)	265–340	20–25	50–55	25–26
Feline	0.9–1.2 (3.8–4.9)	265–380	25–37	41–50	22–25
Mixed (Hydrolyzed protein)					
Canine	1.5	700	38	34	27
Human polymeric liquid diets					
Standard ± fiber	1.0–1.1 (4.2–4.6)	300–450	13–14	30–40	45–60
High calories/high N	1.5–2.0 (6.3–8.4)	400–800	16–20	30–40	45–53
High fat	1.5 (6.3)	500	17	55	28
100 ml + 20 g protein module			36	42	22
Critical care	1.0–1.5 (4.2–6.3)	350–500	22–25	25–40	38–53
Elemental diets					
Amino acids	0.8–1.1 (3.4–4.6)	450–630	10–15	1–14	76–91
Peptides	0.9–1.1 (3.8–4.6)	300–650	12–18	3–36	47–83
Protein modules	3–4.2/g (≈16/g)		75–97	9–19	1–10

FIG. 4.6 Characteristics of different types of enteral diets.

ENERGY FROM MAJOR NUTRIENTS IN kcal/g (kJ/g)			
Type of food	**Protein**	**Fat**	**CHO**
Petfoods	3.5 (15)	8.5 (35)	3.5 (15)
Human enteral products and parenteral solutions	4.0 (17)	9.0 (37)	4.0 (17)

FIG. 4.7 Calories (ME) from major nutrients in kcal/g (kJ/g).

to W exponentially whereas g/kg is linear, so % of kcal from protein changes with W if the g/kg method is used (Fig. 4.2). Protein recommendations will be expressed here as % of kcal to avoid this complication. Most experimental studies are performed on 15–25 kg dogs and an energy requirement of 80 kcal/kg/day (335 kJ/kg/day) is assumed. Results from experimental studies have been converted to % of kcal on this basis.

The minimum maintenance protein requirement for normal dogs is 6–7% of kcal (ME), as determined by nitrogen balance experiments that disregard prior depletion of tissue protein. Studies of repletion indicate minimum casein requirements of 13% of kcal for young mature dogs and 19% of kcal for old dogs.[102] Assuming protein quality to be lower in petfoods than in casein, protein contents of 16 and 24% of kcal have been recommended as minima for young mature dogs and old dogs, respectively.[58] A wide range of protein, 20–48% of kcal, has been used for hospitalized dogs and cats (see Fig. 3.8). These starting points are based on experimental data on dogs, clinical experience with dogs and clinical data on humans.[31] Further evaluation is needed in the animal hospital.

Fat

High fat diets have been recommended because triglyceride (TG) rather than glucose provides the principal fuel for increased metabolism in the catabolic patient. High fat diets also tend to be more palatable, digestible and calorie dense. One report suggests a more rapid progression of infectious hepatitis in Beagle pups fed a diet with increased fat

(25 vs. 9% as fed).[39] The two diets contained similar amounts of protein (27%) on an as-fed basis, but closer inspection reveals that the high fat diet contained a lower protein content (26 vs. 33% kcal) on an energy basis and dogs fed the low protein diet had lower albumin levels. This report, therefore, lends support to the high protein requirement in infected dogs and suggests that protein should be at least 33% of kcal in pups.

Vitamins and Minerals

It is generally assumed that normal petfoods contain more than sufficient minerals and vitamins so that provision of adequate calories will also result in adequate provision of these noncaloric ingredients. Not all petfoods contain minerals in a form that is readily absorbed, however, and some specialized diets are deliberately designed to contain minimal amounts of important electrolytes and minerals. Starvation and catabolic breakdown of tissue result in loss of potassium, phosphorus and magnesium so careful monitoring is advisable when renal failure diets with restricted phosphorus, or feline urological diets with restricted magnesium, are fed. These diets contain sufficient magnesium and phosphorus for normal animals but may not contain enough for tissue repletion.

Petfoods

"Pediatric" or "growth" petfoods designed for feeding puppies are commonly used because they are highly digestible and contain increased calories from protein (≥35%), calories from fat (≥40%), minerals and vitamins. Canned rather than dry diets are used because they contain more fat and because the higher water and fat content improves palatability.

Baby foods are commonly fed because they are very palatable and dogs and cats will often eat them voluntarily. Meat-based baby foods contain 30–70% of kcal as protein and 20–60% of kcal as fat but are deficient in calcium, vitamin A and thiamine so should not be relied on as a sole source of nutrients.

Pureed Petfoods

The simplest and cheapest enteral diets are obtained by blending commercial canned petfoods (1.3–1.5 kcal/g, ~6 kJ/g) with a similar amount of water to produce a gruel with half the caloric density of the parent petfood (0.75 kcal/ml, ~3 kJ/ml). Petfoods have the advantage that they are specifically designed to supply the dietary needs of the cat or dog and also contain ingredients not usually present in liquid diets such as glutamine, nucleic acids and carnitine. The caloric density can be increased to 1.5 kcal/ml (6 kJ/ml) by mixing the petfood with an equal quantity of liquid human enteral diet (1.5 kcal/ml, ~6 kJ/ml). This results in a gruel with a nutrient composition half-way between the petfood and the formula.

A gruel made from a low fiber diet containing low (≤20% kcal) or moderate fat (≤30% kcal) may be preferred in patients with esophageal disease or reduced gastric emptying. Cholecystokinin (CCK) secretion is stimulated by fat and fiber in the duodenum. This slows gastric emptying, reduces lower esophageal sphincter (LES) pressure and stimulates pancreatic secretion. Diets low in fat and fiber tend to increase LES pressure and gastric emptying.

Liquid Enteral Formulae

Gruels are viscous so they can only be infused through wide-bore gastrostomy or pharyngostomy tubes. Extra fluid may even have to be added to low fat diets to produce the required consistency for tube feeding. Narrow-bore nasogastric and enterostomy tube feeding requires the use of liquid enteral diets.

There has been a proliferation of liquid diets in human medicine and a few have been developed specifically for small animals. The characteristics of the common types of formulae are summarized in Fig. 4.6. It is important, however, to ascertain the precise composition of any formula before it is used because there are many small differences.

The principal difference between human liquid enteral formulae lies in the degree of hydrolysis of the ingredients. At one extreme, polymeric diets contain intact protein isolates and maltodextrins whereas, at the other, elemental diets contain amino acids and glucose. Polymeric diets tend to be isoosmotic (300 mOsm/l) but elemental diets contain smaller molecules and tend to be hyperosmotic (600 mOsm/l). Diets also vary in energy density with high fat diets (1.5 kcal/ml, ~6.3 kJ/ml) tending to be more energy dense than low fat diets (1 kcal/ml, ~4.2 kJ/ml).

Elemental amino acid formulations were founded on the belief that proteins have to be hydrolyzed to amino acids for absorption. Some amino acids are absorbed as di- or possibly tripeptides.[89] Formulations containing protein hydrolyzed to small oligopeptides (<50 amino acids in length) have been developed to take advantage of this mechanism but it has been difficult to establish an advantage in absorption over amino acids or intact protein in humans.[89] Polymeric diets may be used for most purposes but amino acid or peptide-based diets may still be absorbed better in patients with pancreatic insufficiency or malabsorption. Amino acid based diets may also reduce the requirement for corticosteroid in patients with inflammatory bowel disease from allergy to intact protein.

Most standard human enteral diets contain 14–17% kcal as protein or amino acids. This leaves little room for the increased requirements of the stressed patient. Arginine and methionine also tend to be low, especially for cats, so special protein modules have been mixed with human liquid formulae to produce a diet with adequate protein and arginine (20 g Promod to 100 ml Pulmocare). These mixtures may cause hyperammonemia in cats, however, and do not support serum arginine[13,28] so they may require further supplementation with citrulline (see hepatic lipidosis below).

A number of critical care diets have been developed that contain increased protein and arginine and show some promise for use in dogs. These diets contain medium chain triglycerides (MCTs) so should be used in cats with caution. Some of these formulae also contain increased ω3 fatty acids, branched-chain amino acids (BCAA), glutamine, nucleic acids and fiber. These formulae are still being evaluated but the principles behind them are discussed below. Caution is necessary in their use because each ingredient tends to have many diverse roles in the body.

Special Considerations

Albumin

Hypoalbuminemia is common in critical patients because of malnourishment, hepatic disease, protein-losing nephropathy or poor intestinal function. Protein is also rapidly lost from the circulation during surgical manipulation of the bowel (~10 g/hr).[12] Normal plasma osmotic pressure is necessary, however, if the gut is to perform normally.[72] Hypoproteinemia in the dog inhibits gastric emptying and intestinal absorption.[8,44,67] Net absorption is replaced by secretion when the plasma protein concentration decreases below 4.5 g/l.[44] In patients with hypoproteinemia, therefore, it is advisable to administer fresh frozen plasma or plasma substitutes before enteral feeding.

Glutamine and Bacterial Translocation

Glutamine is a major substrate for the increased gluconeogenesis observed in stressed dogs. Glutamine is the principal amino acid in plasma and muscle tissue but is released from muscle into the circulation after enterectomy or corticosteroid administration. Its concentration in both plasma and muscle declines, however, because intestinal glutamine uptake and alanine synthesis increase, the alanine being converted to glucose by the liver.[92] Glutamine is also used by many other tissues important in the stress response, including the kidneys, white blood cells and fibroblasts. Thus, although glutamine can be synthesized in adequate quantities by normal animals and has been regarded as a nonessential amino

acid, it appears not to be synthesized in adequate quantities during stress.[92]

The intestine contains large numbers of bacteria but the intestinal mucosa, secretory IgA and gut-associated lymphoid tissue (GALT) present a barrier to bacterial invasion in normal animals.[5] Increased bacterial translocation (BT) to mesenteric lymph nodes was observed in rats after stresses such as burn injury, hemorrhage, administration of endotoxin and radiation injury or following disruption of the intestinal flora with oral antibiotics.[5] Such a breakdown of the intestinal mucosal barrier represents a major risk for infection in critical patients.

NS, glutamine and the route by which they are administered appear to influence this mucosal barrier. The systemic response to injury was attenuated and survival improved when rats were fed enterally rather than parenterally.[5] Starvation and parenteral NS both result in marked mucosal atrophy, reduced s-IgA and IgA-producing cells in the GALT and increased BT when compared to enteral NS.[5] Addition of a 2% solution of glutamine to parenteral NS increased IgA and IgA-producing cells in the GALT and decreased BT to levels similar to those in rats fed enterally. Survival was still less, however, when compared to a similar solution given enterally.[5]

It remains to be seen, however, whether an increase in glutamine content above that in normal diets is necessary for maximum survival in catabolic animals. In rats, intestinal mucosal repair after radiation injury improved with a simple 3% solution of glutamine,[58] but extremely large doses (50% of nonessential amino acids, 25% of total amino acids) are necessary both parenterally and enterally for a maximal trophic effect on the mucosa.[92] These large doses were also necessary for a maximal weight gain rate after intestinal resection.[92]

Glutamine is present as 4–10% of the amino acids in normal petfoods and protein based enteral formulae[92] although this may not be readily apparent because glutamine is extremely labile and is converted to glutamate by the acid hydrolysis used in most analyses.

The situation is different for parenteral solutions because glutamine is labile and has not been included. Stable dipeptides of glutamine may be added to parenteral NS in the future, but enteral NS must be preferred over parenteral NS in the interim.

Arginine

Arginine is an essential amino acid for dogs and cats but not for humans and rats. The NRC recommendation is for <137 and <200 mg arginine/100 kcal for growth in dogs and cats, respectively.[74,75] Most petfoods contain adequate arginine for normal animals but arginine deficiency may be exacerbated by increased urea synthesis during stress and the optimum dose in this situation has yet to be determined. Arginine also stimulates the immune system, increases secretion of growth hormone, prolactin, insulin and glucagon and is a major source for nitric oxide. Nitric oxide has numerous effects including vascular dilation and as a neurotransmitter.[22] Increased arginine has improved survival in septic rats and reduced hospital stay in humans.[22]

Highly digestible petfoods probably contain enough excess arginine for safety but standard human enteral diets are deficient in arginine when compared to the NRC recommendation. Protein modules can be added to liquid diets to increase arginine, or critical care formulae with increased arginine may be used. Serum arginine decreased and also hyperammonemia developed in cats fed a feline liquid enteral diet, so there appears to be inadequate arginine in these diets.[28]

Nucleic Acids

Purines and pyrimidines have not been considered essential nutrients because adequate amounts are synthesized by normal animals or salvaged during cell breakdown. Dietary nucleotides, particularly uracil, however, appear to support normal T-cell responses in rodents.[22] Nucleotides are present in petfoods and some human critical care formulae, but not standard human liquid formulae.

MCTs

MCTs, containing fatty acids with chains of 8–12 carbons, are absorbed directly into the portal vein rather than through the lymphatic system. They also do not require hydrolysis or bile acids for absorption and are rapidly oxidized to ketones in the liver.[89] They increase energy expenditure (30%), however, and are therefore less effective at preventing protein and weight loss in rats when compared to an equivalent amount of long chain triglycerides.[89]

MCTs cause anorexia and vomiting in dogs when given at >3 g/kg (~33% kcal) and have a significant dose related central nervous system toxicity when given at ~50% kcal.[68,69] Human critical care formulae contain ≤20% kcal as MCT. They are safe to use in dogs but MCT may exacerbate encephalopathy so should be avoided in decompensated liver failure. Cats would not eat and developed fatty liver when fed diets containing MCTs.[64] Lipidosis was probably due to anorexia, however, so human enteral diets may still be safe for tube feeding.

ω3 Fatty Acids

The systemic response to injury is mediated by cytokines (tumor necrosis factor, interleukins-1, -2 and -6) and lipid mediators (prostaglandins, thromboxanes and leukotrienes).[106] Hence, catabolism and nitrogen loss may be moderated by using cyclooxygenase inhibitors such as ibuprofen, and by increasing the dietary ratio of ω3 to ω6 fatty acids.[3,106] ω3 Fatty acids have other effects, however, including the inhibition of clotting and the reduction of glomerular hypertension.[17]

Petfoods contain primarily saturated and ω6 unsaturated fat. Menhaden oil can be added as a source of ω3 fatty acids but considerable quantities of ω3 fatty acids are required to change the ω3:ω6 ratio significantly. This dilutes other important nutrients. A human enteral formula is available with increased ω3 fatty acids but it is expensive and should be used with caution in cats because it also contains MCT.

Fiber

Fiber consists of lignin and polysaccharide resistant to digestion in the small intestine. Fiber may be insoluble like cellulose, or soluble like pectin. Soluble fiber, starch or simple CHOs that pass unabsorbed through the small intestine are fermented to short chain fatty acids (SCFA), acetate, propionate and butyrate by bacteria in the large intestine. SCFA are rapidly absorbed along with sodium and water in the large intestine. Butyrate is used as a fuel by the colonocyte and propionate and acetate provide energy for the periphery.[63] SCFA reduce mucosal atrophy associated with parenteral NS, enhance healing of colonic anastomoses in rats and reduce colitis caused by diverting the nutrient stream in humans.[63] Fiber also moderates intestinal transit and slows gastric emptying, but nutrient absorption may be decreased because increased viscosity reduces mixing.[63]

The amount of indigestible CHO in petfoods is difficult to assess. Some petfood diets have cellulose added deliberately as a source of insoluble fiber but the indigestible CHO in most petfoods is composed primarily of soluble fiber and starch resistant to digestion in the small intestine. These are not detected by the crude fiber method used in proximate analysis and it is necessary to examine the ingredient list to assess the amount of indigestible CHO in the diet.

Most highly digestible canned diets contain very little soluble or insoluble fiber but some starch passes undigested through the small intestine and supports fermentation in the large intestine. Standard human liquid enteral diets contain no fiber but 4–14 g/l of soy polysaccharide has been added to some formulae. Normal dogs fed diets containing soy CHO were able to ferment 14 g CHO/1000 kcal in their large intestine so this level of inclusion appears to be within the fermentative capacity of the dog's large intestine.[47] Nevertheless, adaptation to the diet was slow and stool tended to be sloppy, so lower levels of inclusion may be preferable and infusion rates must be increased slowly to allow the bacterial flora to adapt. Liquid diets

are also viscous so require an infusion pump or large-bore tube (>10 French) for delivery.

See Chapter 5 for more information on enteral tube feeding.

Parenteral Nutrition

Parenteral NS should be used only in patients that cannot be fed enterally and it is difficult to justify for periods of less than 5 days. Amino acid and glucose solutions are hyperosmotic and must be delivered via a central catheter. Sepsis is a major concern so catheter insertion should be treated as a surgical operation with antiseptic scrub and use of gloves, etc. Each break in the administration line represents a potential contamination so the catheter should not be used for purposes such as drawing blood or giving medication. Even changing bags should involve surgical standards of sterility.

Energy

Normal puppies grow when given 125–140 kcal/kg.[34,95] Normal 8-kg dogs maintained weight with parenteral NS of 86 kcal/kg/day, which closely approximates the energy requirement obtained with Eqn (5).[20] Cats maintained weight when given calories equivalent to 1.4 times B determined using Eqn (2).[62] This is slightly more than the requirement obtained from Eqn (6). Equations (5) and (6) can, therefore, be used to calculate the daily caloric requirement. Calories are then provided using a combination of amino acids, glucose and lipid. The number of calories

ENERGY DENSITY OF SELECTED PARENTERAL SOLUTIONS	
Solution	Energy density in kcal/ml (kJ/ml)
8.5% Amino acids	0.34 (1.4)
10% Amino acids	0.4 (1.7)
50% Dextrose	1.71 (7.2)
10% Lipid emulsion	1.1 (4.6)
20% Lipid emulsion	2.0 (8.4)

FIG. 4.8 Caloric density of selected parenteral solutions.

EXAMPLES OF PARENTERAL MIXTURES			
	No lipid		With lipid
Recipe (ml)	Moderate protein	High protein	High protein
10% Amino acids	500	500	500
50% Dextrose	500	400	250
20% Lipid emulsion			125*
Ratio of volumes	1:1	5:4	4:2:1
Caloric density in kcal/ml	1.05	0.98	1.0
(kJ/ml)	(4.4)	(4.2)	(4.2)
Protein % kcal	19	23	23
CHO % kcal	81	77	48
Fat % kcal	0	0	29
* 250 ml of 10% lipid emulsion.			

FIG. 4.9 Examples of parenteral mixtures.

from each ingredient are obtained by multiplying the volume by the caloric density. The caloric densities of these solutions are shown in Fig. 4.8. Some standard parenteral mixtures are shown in Fig. 4.9.

Fluids

The fluid provided by parenteral solutions is substantial. The normal maintenance requirement for fluids (~2 ml/kg/hr) is increased in catabolic patients, but parenteral NS may deliver more than twice maintenance levels, for example, when 20% kcal are provided as amino acids with an 8.5% solution, and 10% lipids, 50% dextrose provide the nonprotein calories. Other intravenous fluids must be reduced proportionately or discontinued.

Fluid in excess of maintenance levels can be excreted by patients with good renal function but critical patients may retain fluid because of increased ADH secretion. This can be a serious problem for patients with poor cardiac or renal function. Amino acid solutions contribute proportionately the most fluid because they are the least calorie-dense of all the major ingredients. Using more concentrated solutions, however, reduces the amount of fluid administered. Amino acid solutions are available at concentrations ranging from 5.5 to 15%. Most veterinary literature has advocated 8.5% solutions, 10% solutions are

inexpensive and reduce fluid administration because they are more calorie-dense; 15% solutions tend to be too expensive; 20% emulsions are more calorie-dense and provide less fluid than 10% solutions.

Amino Acids

Most amino acid solutions for parenteral NS are designed to optimize the plasma amino acid profile in humans. They contain the eight amino acids essential to humans plus arginine and histidine. Glycine, alanine and proline are provided as nonessential amino acids. One solution ("Vamin") is based on the amino acid profile found in egg and contains less arginine and methionine. Although this suggests that Vamin may not be suitable for small animals, hyperammonemia did not develop in normal dogs infused intravenously with Vamin at a high rate.[9] Alternative mixtures with increased BCAA, lysine and histidine have been developed but these solutions are still being evaluated and are expensive.

The concentration of amino acids in parenteral NS necessary to maximize nitrogen balance has yet to be determined. Recommendations in dogs have been for 4–6 g/kg (~13–26% kcal) as amino acids.[18,83] Recommendations for cats have been for 6 g/kg

(~20–26% kcal) as amino acids.[18,83] Parenteral NS containing 13% kcal from amino acids maintained adult dogs and supported growth in pups.[20,34] Nitrogen balance was slightly positive when 20% kcal were provided as amino acids.[46] Nevertheless, nitrogen balance did not reach a plateau in dogs fed 20% kcal (4 g/kg) as amino acids parenterally.[7] This last

---

ADVANTAGES AND DISADVANTAGES OF PROVIDING CALORIES AS GLUCOSE ALONE OR AS LIPID/GLUCOSE MIXTURES

Glucose alone

Advantages
- Inexpensive.
- Do not support the growth of bacteria.[4]
- *Candida* spp. grow relatively slowly at room temperature.[4]
- 0.22 μm filters can be used to remove bacteria and contaminants.[4]
- Positive inotropic effect on heart at very high infusion rates.[1]

Disadvantages
- Hyperglycemia: Catabolic patients are insulin resistant so regular insulin (0.25 IU/kg IM) may be required to prevent hyperglycemia.[18] (Advocates of this method say that insulin is rarely required in dogs[18] but this has not been this author's experience.)
- Increased risk of hypoglycemia during withdrawal of glucose infusion if insulin has previously been administered.[18]
- Hepatic lipidosis may be exacerbated because glucose, not used peripherally, is converted to lipid in the liver.[76]
- Increased risk of hypophosphatemia.[59]
- Hyperosmolar solution: The final concentration of glucose (20–25%) after dilution with amino acids is still enough to cause endothelial injury in the dog even when infused with a central catheter into the right atrium.[41,52] When there is a breakdown of the blood–brain barrier, e.g. following trauma, hyperosmolar solutions potentiate vasogenic edema in the cat.[102]
- Fluid retention is increased because insulin has significant anti-natriuretic activity.[33]
- Fine 0.22 μm filters have a tendency toward air-locks and blockages. This increases the risk of contamination below the filter because each intervention represents a break in the line.[4]
- Essential fatty acid (EFA) deficiency developed within 1 week in pups.[95] EFA deficiency increased the risk of BT in rodents.[10]
- Increased respiratory quotient (RQ) and CO_2 excretion may compromise patients with respiratory disease.[33]

Lipid emulsions

Advantages
- Isosmolar so reduces hyperosmolarity when mixed with glucose.[41]
- Insulin is rarely necessary and withdrawal of glucose is easier.
- EFA deficiency prevented when >4% kcal (=2% linoleic acid) are provided as lipid emulsion.[95]
- Reduced RQ improves recovery in patients with respiratory compromise.[33]
- Fat and glucose appear to be equally efficacious at sparing nitrogen in humans[108] but one report suggests that lipid and glucose together are better at maintaining nitrogen balance than glucose alone in dogs with abdominal infections. Glucose alone performed better in normal dogs.[50]
- Lipid particles are cleared by lipoprotein lipase in many tissues other than the liver so lipid may not exacerbate hepatic lipidosis.[108]

Disadvantages
- Expensive.
- Both bacteria and fungi proliferate rapidly in lipid emulsions at room temperature so stringent cleanliness is necessary and unmixed lipids should be kept for a maximum of 12 hr. Bacteria grow slower when lipids are mixed with amino acids and glucose so all-in-one mixtures are licensed by the FDA to be kept for 24 hr, but should not be stored at 4°C for more than 24 hr.[4,33]
- Emulsion particles have an average size of 0.5 μm that prevents the use of in-line bacterial filters. 1.2 μm filters can be used to remove *Candida* and oversized lipid particles.[33]
- RE function may be compromised in septic patients.[88]
- Lipid infusions have a negative inotropic effect on the left ventricle and decrease systemic vascular resistance.[43]
- Soybean emulsions reduce hematocrit, hemoglobin and plasma proteins and increase serum alkaline phosphatase when infused chronically at >40% kcal.[80,81]

FIG. 4.10 Advantages and disadvantages of providing calories as glucose alone or as lipid/glucose mixtures.

experiment had a flaw, in that total calories increased slightly as protein increased, but the implication is that >20% kcal as amino acids may be necessary for optimum nitrogen balance in dogs. A further increase may be necessary to provide the extra protein required by a catabolic patient. Normal cats were in positive nitrogen balance and maintained weight and plasma proteins when fed parenterally for 2 weeks with 35% kcal as amino acids.[62] Whether lesser amounts are adequate has not been determined.

Nonprotein Calories

Nonprotein calories can be provided by glucose alone or a combination of glucose and lipid. The advantages and disadvantages of glucose alone vs. lipid and glucose are listed in Fig. 4.10. Much debate has centered on the safety of lipid emulsions. What follows pertains to lipid emulsions composed of soybean TG with glycerol and egg phospholipid as emulsifying agents ("Intralipid") because most of the safety studies have been performed on dogs using Intralipid. Other emulsions may not behave similarly because the behavior of lipid emulsions is greatly affected by the composition of the TG and emulsifying agent.[33]

Normal dogs tolerate long-term infusion from Intralipid at high rates (60% kcal infused over 3–4 hr/day).[37] Exogenous TG particles incorporate lipoproteins and are cleared by lipoprotein lipase. Clearance rates are similar to those of endogenous chylomicrons[97,99] but can be affected by factors that affect lipoprotein lipase.[97] Thus, clearance is increased by insulin and heparin, and decreased by glucagon.[97] Clearance is also decreased in diabetes mellitus, nephrotic syndrome, splenectomy and terminal renal failure, but may be increased after hemorrhage, hepatitis and cirrhosis.[97] It is important to monitor for evidence of lipidemia as an indicator of slow lipid clearance. It is also important to monitor for evidence of hyperlipidemia as an indicator of slow lipid clearance.

Some lipid particles accumulated in the reticuloendothelial (RE) cells of the liver and spleen during infusion and seem to compromise the RE system. Dogs with sepsis did not survive when infused with Intralipid at rates that provided ~50% kcal/day as fat.[88] Puppies with severe septic shock also did not metabolize "Lyposyn" (a safflower oil emulsion) well.[25] Whether RE compromise is a problem at lower rates of infusion has not been established.

Veterinary authors advocating the use of lipid have suggested dividing nonprotein calories equally between fat and glucose (~40% kcal as lipid infused over 24 hr).[61,82] This should perhaps represent a maximum for lipid infusion, however, until more is known about RE compromise, that is, lipids should be limited to ≤40% kcal over 24 hr or ≤20% kcal over 12 hr. Further restriction may be necessary in conditions where RE function plays a crucial role in survival such as sepsis and pancreatitis.

A second concern has been the stability of lipid emulsions when mixed with glucose and electrolytes.[33] Amino acids increase stability but glucose and electrolytes, particularly divalent cations, decrease stability.[33] Emulsions also become less stable toward the end of their shelf-life.[33] Standard mixtures similar to those described here are well within concentrations that have been shown to be stable for periods of a week when stored at 4°C and have been infused in normal dogs without ill effects.[87] The order of mixing is important, however, and is listed in Fig. 4.11.

Electrolytes and Trace Minerals

Phosphorus, magnesium and potassium are lost during tissue breakdown from starvation and may be further depleted by cellular uptake when calories are reintroduced. Glucose

---

ORDER OF MIXING OF SOLUTIONS FOR PARENTERAL NS

1. Amino acids, dextrose, electrolytes and minerals
2. Lipid emulsion
3. Vitamins

FIG. 4.11 The order of mixing the solutions for parenteral NS.

stimulates insulin secretion and increases use of phosphorus for the phosphorylated intermediates of glycolysis.[109] Hypophosphatemia caused by too rapid administration of calories in the form of glucose occurs more rapidly in starved dogs than in normal animals.[90] As phosphorus decreases in the serum, ATP decreases in red cells, platelets and nervous tissue causing increased red cell fragility, anemia, hemorrhage and neurological signs (ataxia, nystagmus, coma).[90,109]

Electrolytes are important, therefore, but the parenteral electrolyte requirements for catabolic small animals have not been determined. The only study of note showed that fluids containing 45 mEq/l Na^+, 45 mEq/l Cl^-, 36 mEq/l K^+, ~18 mmol/l PO_4^{2-}, 7 mEq/l Mg^{2+}, 23 mEq/l Ca^{2+} maintained serum electrolyte concentrations in starved and normal dogs hyperalimented with glucose but did not prevent the demise of starved dogs.[91]

Most veterinarians have relied on amino acid solutions designed for humans with electrolytes already included.[18,82] The electrolyte concentrations of one such amino acid mixture before and after a standard 1:1 dilution are shown in Fig. 4.12. The concentration of electrolytes varies, however, with the amino acid concentration in the final mixture.

Amino acid solutions with electrolytes do not contain calcium or trace minerals so some authors have added calcium gluconate 10% (1 ml = 0.45 mEq Ca/l) and trace minerals (2 mg Zn, 0.8 mg Cu, 0.2 mg Mn, 0.008 mg Cr/l).[18] Other authors have suggested that trace minerals are not necessary when parenteral NS is used for short periods.[82] This must be questioned, however, because diarrhea and intestinal dysfunction can result in deficiencies of zinc and copper. Zinc and copper are important in protein synthesis because they are essential components of metalloenzymes, and serum zinc was halved within 1 week in dogs given zinc-free parenteral NS.[49] In these zinc-depleted animals, collagen hydroxyproline within the wound and serum albumin were reduced after enterotomy compared to a

ELECTROLYTE SOLUTIONS FOR PARENTERAL INFUSIONS					
Product	8.5% Travasol + electrolytes	10% Calcium gluconate	M.T.E. 4	Tracelyte II	Potassium phosphate
Dilution	500 ml/l	10 ml/l	3 ml/l	20 ml/l	5–10 ml/l
Sodium	70 mEq/l (35 mEq/l)			1.75 mEq/l (35 mEq/l)	
Potassium	66 mEq/l (33 mEq/l)			1 mEq/l (20 mEq/l)	4.4 mEq/ml (22–44 mEq/l)
Chloride	96 mEq/l (48 mEq/l)			1.75 mEq/l (35 mEq/l)	
Magnesium	10 mEq/l (5 mEq/l)			0.25 mEq/l (5 mEq/l)	
Phosphate	30 mEq/l (15 mEq/l)				3 mmol/ml (15–30 mmol/l)
Calcium		0.45 mEq/ml (4.5 mEq/l)		0.225 mEq/l (4.5 mEq/l)	
Zinc			1 mg/ml (3 µg/ml)	0.15 mg/ml (3 µg/ml)	
Copper			0.4 mg/ml (1.2 µg/ml)	0.06 mg/ml (1.2 µg/ml)	
Manganese			0.1 mg/ml (0.3 µg/ml)	0.015 mg/ml (0.3 µg/ml)	
Chromium			4 mg/ml (12 µg/ml)	0.6 mg mg/ml (12 µg/ml)	

FIG. 4.12 Electrolyte solutions for parenteral infusions. Concentrations after standard dilutions are shown in parentheses.

control group. The control group maintained plasma zinc when given 70 µg/kg/day zinc IV, which suggests that at least 1 µg/kcal (~1 mg/l) should be added to canine parenteral NS.

An alternative method of supplying electrolytes involves the use of amino acid solutions without electrolytes. Mixtures of trace elements and electrolytes are available that can then be added (20 ml/l) to give consistent concentrations of electrolytes (Fig. 4.12). These mixtures are deficient in phosphate so 5–10 ml/l (15–30 mmol/l) potassium phosphate should

also be added. Excess phosphate should be avoided because a high concentration product of calcium and phosphorus will result in precipitation.

A list of potential complications and their treatment is shown in Fig. 4.13.

Vitamins should also be added to parenteral solutions. Many patients present after prolonged anorexia with multiple vitamin deficiencies so that even fat-soluble vitamins may become depleted. Vitamins A and E, as well as zinc and selenium, stimulate immunity.[24] Folate and vitamin B_6 are essential for protein

COMPLICATIONS OF PARENTERAL NS: THEIR CAUSE AND TREATMENT		
Complication	**Possible explanations**	**Treatment**
Sepsis	Inadequate sterility of catheter or solution Endogenous nidus of infection	Check for endogenous nidus R_x antibiotics Remove and culture catheter Replace using sterile technique Use enteral NS
Phlebitis	Hyperosmolar solutions	Add lipid. Remove catheter R_x heparin, corticosteroids
Hyperglycemia	Rapid dextrose infusion Glucose intolerance	Reduce rate of dextrose infusion and increase slowly R_x insulin (0.25 IU/kg)
Hypoglycemia	Sepsis; insulin active during abrupt dextrose withdrawal	As for sepsis Withdraw dextrose slowly
Hypertriglyceridemia	Diminished clearance	Reduce rate of lipid infusion
Fatty liver	Overfeeding High CHO diet	Reduce calories Reduce CHO and increase fat
Hyperammonemia/ Encephalopathy	Liver dysfunction Insufficient arginine Old blood transfusion GI hemorrhage	Reduce rate of amino acid infusion Increase nonprotein calories Increase arginine or BCAA R_x H–2 blocker, lactulose
Osmotic diuresis	Hyperglycemia	Reduce dextrose infusion rate Correct dehydration
Azotemia	Dehydration; low kcal/nitrogen ratio; GI bleeding, renal failure	Increase fluids. As for osmotic diuresis Increase nonprotein calories R_x H–2 blocker
Hypokalemia	Insufficient K+ Increased GI or renal loss Rapid glucose infusion	Increase K+ in solution R_x for vomiting, diarrhea
Hyperkalemia	Acidosis; renal failure Sepsis; tissue necrosis	Decrease K+ in fluids As for acidosis
Hypophosphatemia	Rapid glucose infusion	Reduce rate of glucose Increase P in solution May have to decrease Ca
Hypomagnesemia	Inadequate Mg or P Excess GI or renal loss	Increase Mg
Metabolic acidosis	Renal or GI loss of base Inadequate acetate to neutralize acid products of amino acid metabolism	Increase acetate in solution Decrease Cl if hyperchloremic
Metabolic alkalosis	Excessive base administration	NaCl or KCl. If restrictions to these, 0.1 N (100 mEq) HCl/l in 5% dextrose via central vein

FIG. 4.13 Complications caused by parenteral NS, their cause and treatment.

turnover and serum folate has been shown to be below normal in dogs with small intestinal disease.[11] Most authors have recommended adding simple B-complex vitamins and giving fat-soluble vitamins intramuscularly.[18,82] Simple B-complex vitamins are deficient in folate, however, so more expensive vitamin formulae designed for human parenteral solutions, which include folate and fat-soluble vitamins, are preferred. Folate is susceptible to light, and vitamin B_{12} interacts chemically with other ingredients, so these vitamins should be mixed and added just before administration.

Selected Diseases

This section gives a brief synopsis of the role of enteral and parenteral nutritional support in selected common diseases in critical patients. The reader is referred to specific chapters for further details of pathophysiology and treatment.

Surgery

Early enteral NS is indicated in any patient where oral food intake is likely to be delayed after surgery. In the absence of food intake, there is a lag between the resorption of mature collagen and the deposition of new collagen in surgical wounds so the wound is at its weakest after 4–5 days.[72] Early enteral nutrition in Beagles curtailed the lag phase of wound healing after bowel resection, increased wound protein and DNA synthesis, increased plasma insulin and consumption of glucose.[72] Supplemental tube feeding resulted in a marked reduction in hospital stay for thin elderly women hospitalized for fracture repair.[101]

Abdominal Trauma

In human patients fed by enterostomy with a high-nitrogen amino acid-based liquid diet ("Vivonex") immediately after celiotomy for major abdominal trauma, there was no reduction in the major complication rate nor a reduction in hospital stay, but the incidence

of sepsis was greatly reduced and nitrogen balance and overall cost per patient were improved.[70] Patients with the most severe injury, however, did not tolerate full scale enteral feeding so they had to be supplemented parenterally.

Hemorrhagic Shock

A consistent sequence of events follows sublethal hemorrhagic shock. The intestinal villi become rapidly denuded of epithelial cells then extraintestinal lesions develop over 2–3 days including cardiac infarction, renal congestion, tubular necrosis and hepatic centrilobular vacuolation.[15] Two interesting reports suggest that an elemental diet fed to dogs before and after the induction of hemorrhagic shock reduces epithelial necrosis and completely prevents the extraintestinal lesions seen when regular diet is fed.[16] The same authors also report clinical studies in humans where elemental diets moderated the injury caused by chemotherapy and radiation injury.[14]

Pancreatitis

See also Chapter 13.

NS is not required in most cases of acute pancreatitis because inflammation is mild and resolves if nothing is given orally for 3–5 days. The more severe hemorrhagic, necrotic form of the disease, however, causes massive multiple organ damage and treatment usually involves prolonged starvation. NS in these patients is, therefore, designed to provide increased protein and calories while minimizing pancreatic stimulation.

Pancreatic secretion is stimulated if food is given orally or if acid, fat or amino acids are infused into the stomach or duodenum.[48,55,93,94] Intrajejunal or intravenous infusion appear to be the best methods of infusing nutrients but there has been some debate as to the degree of pancreatic stimulation induced. Infusion of elemental liquid diet ("Vivonex"; pH 5.4) into the proximal jejunum stimulates pancreatic secretion but the enzyme response is relatively poor.[21,86] Elemental diets appear to cause less stimulation than puréed or polymeric diets, although this is probably due to their lower fat content.[48,55] One report found less protein secretion when Vivonex was infused at neutral pH[79] and another suggested that secretion was least when infusion was beyond the first meter of jejunum.[86]

Parenteral NS without lipid does not affect or decrease pancreatic secretion.[2,55,85,93] When lipid was infused intravenously, one report suggests a slight increase in pancreatic secretion but all other reports found no change.[40,66,94,97,100] Parenteral NS appears, therefore, to be the method of choice but, as NS may have to be prolonged, enteral infusion into the mid-jejunum provides an alternative.

Hepatic Encephalopathy

See also Chapter 15.

Human patients with chronic liver disease are catabolic and have an increased protein requirement for nitrogen balance because of portosystemic shunting, impaired hormonal degradation and peripheral insulin resistance. Nevertheless, most NS in hepatic encephalopathy has been directed at limiting the amino acid supply or adapting the amino acid pattern by increasing BCAA at the expense of aromatic amino acids. Fischer *et al.*, for example, report that all dogs with artificially created portosystemic shunts died when given 4 g/kg of a standard amino acid solution but not when given a solution with increased BCAA.[38]

Unfortunately, most of the enteral and parenteral formulae with increased BCAA are expensive and have not been widely used in veterinary medicine. A recent report by Kearns *et al.* is therefore encouraging.[54] In contrast to most studies that have compared increased BCAA formulae with formulae containing restricted protein, these investigators examined the effect of supplementing a regular diet with a standard casein-based enteral supplement in humans with chronic alcoholic liver disease. The supplemented group received twice the protein and calories of the control group because the control group had a limited appetite. Hepatic encepha-

RECIPE FOR CATS WITH HEPATIC LIPIDOSIS	
Ingredient	Amount
High fat liquid diet	100 ml
Protein supplement	15–20 g
KCl	2.5 g
Citrulline	500 mg
or arginine	250 mg
Threonine	250 mg
Methionine	250 mg
Taurine	125 mg
Inositol	10 mg
Choline chloride	50–250 mg

FIG. 4.14 Recipe for cats with hepatic lipidosis.

lopathy and serum bilirubin improved in the supplemented group. This suggests that patients with chronic liver disease benefit from additional NS using standard enteral formulae.

Hepatic Lipidosis

See also Chapter 15.

Tube feeding is the treatment of choice for hepatic lipidosis in cats but the optimum mix of nutrients has yet to be determined. Eleven out of 15 cats survived when they were fed via gastrostomy with a gruel plus added L-carnitine.[51] Biourge *et al.* observed hyperammonemia and a slow recovery in cats fed a high fat enteral formula fortified with protein. These authors increased the rate of recovery and avoided hyperammonemia by adding citrulline, choline and potassium chloride.[13] Hayes has recommended adding arginine rather than citrulline, the amino acids threonine and methionine, and choline and inositol to improve lipoprotein mobilization.[45] The suggested amounts are listed in Fig. 4.14.

Renal Failure

See also Chapter 19.

Patients with acute renal failure are frequently catabolic and require NS to minimize azotemia from tissue breakdown and to promote renal regeneration. Glucose or glucose with amino acids doubled survival time in anephric dogs.[78] High protein diets have also been advocated to compensate for protein loss in nephrotic syndrome. Nevertheless, high protein diets increase glomerular hypertension so proteinuria can best be minimized by feeding a low protein diet (14% kcal).[78] Diets containing increased ω3 fatty acids also reduce glomerular hypertension and so may be beneficial.[17]

Cancer

See also Chapter 6.

Human patients with upper gastrointestinal cancer have reduced morbidity and mortality if given parenteral NS preoperatively. This is one example where parenteral NS was associated with fewer complications than enteral NS.[101] Tumor tissue metabolizes glucose using anaerobic glycolysis but has difficulty using lipid as an energy source.[77] Lactate generated by glycolysis is converted back to glucose by the liver at considerable energetic cost to the patient. Dogs with lymphoma have normal glucose tolerances but elevated serum lactate and insulin.[77] Solutions containing lactate or excess glucose should probably be avoided whereas increased lipid is desirable.[77]

Neurological Injury

Head injury in humans causes a doubling of the metabolic rate.[31] Metabolic rate decreases 10% following spinal cord injury but nitrogen and calcium excretion in the urine is increased; body weight decreased 10% and metabolic rate was still greater than similarly immobilized patients.[53] Overall metabolism decreased because of immobility and loss of muscle activity due to denervation but these patients were also catabolic. Poor nutrition is one of the risk factors for decubitus ulcers so these patients are good candidates for NS but can usually be fed orally.

References

1. Abel, R. A., Subramanian, V. A. and Gay, W. A. (1977) Effects of an intravenous amino acid nutrient solution on left ventricular contractility in dogs. *Journal of Surgery Research*, **23**, 201–206.
2. Adler, M., Pieroni, P. L., Takeshima, T., Nacchiero, M., Dreiling, D. A. and Rudick, J. (1975) Effects of parenteral hyperalimentation on pancreatic and biliary secretion. *Surgery Forum*, **26**, 445–446.
3. Alexander, J. W., Saito, H., Ogle, C. K. and Trocki, O. (1986) The importance of lipid type in the diet after burn injury. *Annals of Surgery*, **204(1)**, 1–8.
4. Allen, J. R. (1978) The incidence of nosocomial infection in patients receiving total parenteral nutrition. In *Advances in Parenteral Nutrition: Proceedings of an International Symposium.* Ed. I. D. A. Johnston. pp. 339–377. Baltimore, MD: University Park Press.
5. Alverdy, J. C. (1990) Effects of glutamine supplemented diets on immunology of the gut. *Journal of Parenteral and Enteral Nutrition*, **14**, 109S–113S.
6. Atwater, W. O. (1910) Principles of nutrition and the nutritive value of food. Farmers bulletin No. 142. Washington, DC: U.S. Department of Agriculture.
7. Ausman, R. K. and Meng, H. C. (1976) The effect of nitrogen balance and other parameters of varying doses of intravenous protein and calories. *Acta Chirurgica Scandinavica, Supplement*, **466**, 26–27.
8. Barden, R. P., Thompson, W. D., Ravdin, I. S. and Frank, I. L. (1939) The influence of the serum protein on the motility of the small intestine. *Surgery of Gynecology and Obstetrics*, **66**, 819–821.
9. Bark, S. (1976) Amino acid concentration after gastrointestinal, intraportal and intravenous administration of crystalline amino acids. Preliminary report. *Acta Chirurgica Scandinavica*, **142(4)**, 79–284.
10. Barton, R. G., Cerra, F. B. and Wells, C. L. (1992) Effect of a diet deficient in essential fatty acids on the translocation of intestinal bacteria. *Journal of Parenteral and Enteral Nutrition*, **16(2)**, 22–128.
11. Batt, R. M. and Hall, E. J. (1989) Chronic enteropathies in the dog. *Journal of Small Animal Practice*, **30**, 3–12.
12. Beard, J. W. and Blalock, A. (1932) Intravenous injections: A study of the composition of the blood during continuous trauma to the intestines when no fluid is injected and when fluid is injected continuously. *Journal of Clinical Investigations*, **111**, 249–265.
13. Biourge, V., Pion, P., Lewis, J., Morris, J. G. and Rogers, Q. R. (1991) Dietary management of idiopathic feline hepatic lipidosis with a liquid diet supplemented with citrulline and choline. *Journal of Nutrition*, **121**, S155–S156.
14. Bounos, G. (1972) Enteral hyperalimentation with

elemental diet. *Canadian Medical Association Journal*, **107(7)**, 607–608.

15. Bounos, G., Cronin, R. F. P. and Gurd, F. N. (1967) Dietary prevention of experimental shock lesions. *Archives of Surgery*, **94**, 46–60.

16. Bounos, G., Sutherland, N. G., McArdle, A. H. and Gurd, F. N. (1967) The prophylactic use of an "Elemental" diet in experimental hemorrhagic shock and intestinal ischemia. *Annals of Surgery*, **166(3)**, 12–342.

17. Brown, S. A. (1992) Role of dietary lipids in renal disease in the dog. In *Proceedings of the Tenth Annual Veterinary Medical Forum*. Ed. W. B. Morrison. pp. 568–570. Madison, WI: Omnipress.

18. Buffington, C. A. T. (1991) Nutritional management of critical care patients. In *14th Kal Kan Waltham Symposium: Emergency Medicine and Critical Care Medicine*. Ed. W. W. Campfield. pp. 133–139. Vernon, CA: Kal Kan Foods.

19. Burger, I. H. and Johnson, J. V. (1991) Dogs large and small: the allometry of energy requirements within a single species. *Journal of Nutrition*, **121**, S18–S21.

20. Carter, J. M. and Freedman, A. B. (1977) Total intravenous feeding in the dog. *Journal of the American Veterinary Medical Association*, **171**, 71–76.

21. Cassim, M. M. and Allardyce, D. B. (1974) Pancreatic secretion in response to jejunal feeding of elemental diet. *Annals of Surgery*, **180(2)**, 228–231.

22. Cerra, F. B. (1991) Nutrient modulation of inflammatory and immune function. *American Journal of Surgery*, **161**, 230–234.

23. Chandler, M. L., Greco, D. S. and Fettman, M. J. (1992) Hypermetabolism in illness and injury. *Compendium of Continuing Education*, **14(10)**, 1284–1290.

24. Chandra, R. K. (1991) Nutrition and immunity: lessons from the past and new insights into the future. *American Journal of Clinical Nutrition*, **53**, 1087–1101.

25. Coran, A. G., Dr.ongowski, R. A., Lee, G. S., Klein, M. D. and Wesley, J. R. (1984) The metabolism of an exogenous lipid source during septic shock in the puppy. *Journal of Parenteral and Enteral Nutrition*, **8(6)**, 652–656.

26. de Bruijne, J. J. (1981) *Ketone Body Metabolism in Fasting Dogs*. Utrecht: Dr.ukkerij Elinkwijk BV.

27. Detsky, A. S., McLaughlin, J. R., Baker, J. P., Johnston, N., Whittaker, S., Mendelson, R. A. and Jeejeebhoy, K. N. (1987) What is subjective global assessment of nutritional status? *Journal of Parenteral and Enteral Nutrition*, **11**, 8–13.

28. Diehl, K. J. and Wheeler, S. L. (1992) Evaluation of three enteral feeding formulas in cats. In *Proceedings of the Tenth Annual Veterinary Medical Forum*. Ed. W. B. Morrison. p. 813. Madison, WI: Omnipress.

29. Dionigi, R., Zonta, A., Dominioni, L., Gnes, F. and Ballabio, A. (1977) The effects of total parenteral nutrition on immunodepression due to malnutrition. *Annals of Surgery*, **185(4)**, 467–474.

30. Dominioni, L., Gnes, F., Dionigi, R., Zonta, A. and Prati, A. (1976) Histopathological studies on dog lymphoid structures during malnutrition and total parenteral nutrition. *Bolletinodi Istituto Sieroter Milan*, **55(4)**, 311–316.

31. Donoghue, S. (1989) Nutritional support of hospitalized patients. *Veterinary Clinics of North America (Small Animal Practice)*, **19(3)**, 475–495.

32. Donoghue, S. (1991) A quantitative summary of nutrition support services in a veterinary teaching hospital. *Cornell Veterinarian*, **81(2)**, 109–128.

33. Dr.iscoll, D. F. (1990) Clinical issues regarding the use of total nutrient admixtures. *DICP Annals of Pharmacotherapy*, **24**, 296–303.

34. Dudrick, S. J., Wilmore, D. W., Vars, H. M. and Rhoads, J. E. (1968) Longterm total parenteral nutrition with growth, development, and positive nitrogen balance. *Surgery*, **64(1)**, 134–142.

35. Dzanzis, D. A., Corbin, J. E., CzarneckiMaulden, G. L., Hirakawa, D. A., Kallfelz, F. A., Morris, M. L. and Sheffy, B. E. (1993) *AAFCO Nutrient Profiles for Dog Foods*. American Association of Feed Control Officials.

36. Earle, K. E. and Smith, P. M. (1991) Digestible energy requirements of adult cats at maintenance. *Journal of Nutrition*, **121**, S45–S46.

37. Edgren, B., Hallberg, D., Hakansson, I., Meng, H. C. and Wretland, A. (1964) Longterm tolerance study of two fat emulsions for intravenous nutrition in dogs. *American Journal of Clinical Nutrition*, **14**, 28–36.

38. Fischer, J. E., Funovics, J. M., Aguirre, A., James, J. H., Keane, J. M., Wesdorp, R. I., Yoshimura, H., Mashima, Y. and Iwasaki, I. (1975) The role of plasma amino acids in hepatic encephalopathy. *Surgery*, **78(3)**, 276–290.

39. Fiser, R. H., Rollins, J. B. and Beisel, W. R. (1972) Decreased resistance against infectious canine hepatitis in dogs fed a highfat ration. *American Journal of Veterinary Research*, **33**, 713–719.

40. Fried, G. M., Ogden, W. D., Rhea, A., Greeley, G. and Thompson, J. C. (1982) Pancreatic protein secretion and gastrointestinal hormone release in response to parenteral amino acids and lipid in dogs. *Surgery*, **92(5)**, 902–905.

41. Fujiwara, T., Kawarasaki, H. and Fonkalsrud, E. W. (1984) Reduction of postinfusion venous endothelial

injury with intralipid. *Surgery of Gynecology Obstetrics*, **158(1)**, 57–65.

42. Goldblum, O. M., Holzman, I. R. and Fisher, S. E. (1981) Intragastric feeding in the neonatal dog. Its effect on intestinal osmolality. *American Journal of Disease in Children*, **135(7)**, 631–633.

43. Grimes, J. B. and Abel, R. M. (1979) Acute hemodynamic effects of intravenous fat emulsion in dogs. *Journal of Parenteral and Enteral Nutrition*, **3(2)**, 40–44.

44. Hakim, A. A. and Lifson, N. (1969) Effects of pressure on water and solute transport by dog intestinal mucosa. *American Journal of Physiology*, **216(2)**, 276–284.

45. Hayes, K. C. (1992) Inositol as part of therapy for feline hepatic lipidosis. *Viewpoints in Veterinary Medicine*, **2(2)**.

46. Heller, L. (1969) Clinical and experimental studies on complete parenteral nutrition. *Scandinavian Journal of Gastroenterology, Supplement*, **3**, 7–16.

47. Hill, R. C., Ellison, G. W., Bauer, J. E. and Burrows, C. F. (1992) The effect of texturized vegetable protein on apparent digestibility in the dog. In *Proceedings of the Tenth Annual Veterinary Medical Forum*. Ed. W. B. Morrison. p. 814. Madison, WI: Omnipress.

48. Hisano, S., Yokota, T., Nara, M., Nakase, A. and Tobe, T. (1980) Effects of long term enteral hyperalimentation on pancreatic secretion in dogs. *Gastroenterology, Japan*, **15(5)**, 500–509.

49. Iriyama, K., Mori, T., Takenaka, T., Teranishi, T. and Mori, H. (1982) Effect of serum zinc level on amount of collagenhydroxyproline in the healing gut during total parenteral nutrition: an experimental study. *Journal of Parenteral and Enteral Nutrition*, **6(5)**, 416–420.

50. Iriyama, K., Nishiwaki, H., Kusaka, N., Teranishi, T., Mori, H. and Suzuki, H. (1985) Nitrogen sparing effect of lipid emulsion in septic dogs. *Japanese Journal of Surgery*, **15(4)**, 321–323.

51. Jacobs, G., Cornelius, L., Allen, S. and Greene, C. (1989) Treatment of idiopathic hepatic lipidosis in cats: 11 cases (1986–1987). *Journal of the American Veterinary Medical Association*, **195(5)**, 635–638.

52. Kawarasaki, H., Fujiwara, T. and Fonkalsrud, E. W. (1985) The effects of administering hyperalimentation solutions into the atrium and pulmonary artery. *Journal of Paediatric Surgery*, **20(3)**, 205–210.

53. Kearns, P. J., Thompson, J. D., Werner, P. C., Pipp, T. L. and Wilmot, C. B. (1992) Nutritional and metabolic response to acute spinal cord injury. *Journal of Parenteral and Enteral Nutrition*, **16**, 11–15.

54. Kearns, P. J., Young, H. and Garcia, G. (1992) Accelerated improvement of alcoholic liver disease with enteral nutrition. *Gastroenterology*, **102**, 200–205.

55. Kelly, G. A. and Nahrwold, D. L. (1976) Pancreatic secretion in response to an elemental diet and intravenous hyperalimentation. *Surgery of Gynecology and Obstetrics*, **143(1)**, 87–91.

56. Kienzle, E. and Rainbird, A. (1991) Maintenance energy requirement of dogs: what is the correct value for the calculation of metabolic body weight in dogs? *Journal of Nutrition*, **121**, S39–S40.

57. Kleiber, M. (1975) *The Fire of Life: An Introduction to Animal Energetics*. New York: Krieger.

58. Kronfeld, D. S. (1991) Protein and energy estimates for hospitalized dogs and cats. In *Proceedings of Purina International Nutrition Symposium*. pp. 5–12. St Louis: Ralston Purina.

59. Langaniere, S. and Maestracci, D. (1985) A technique for enteral feeding in unrestrained chronic dogs. *Journal of Parenteral and Enteral Nutrition*, **9**, 370–374.

60. Lewis, L. D., Morris, M. L. and Hand, M. S. (1987) *Small Animal Clinical Nutrition*. 3rd edn. Topeka: Mark Morris Associates.

61. Lippert, A. C. and Armstrong, P. J. (1989) Parenteral nutritional support. In *Current Veterinary Therapy*. Ed. R. W. Kirk. pp. 25–30. Philadelphia: W. B. Saunders.

62. Lippert, A. C., Faulkner, J. E., Evans, A. T. and Mullaney, T. P. (1989) Total parenteral nutrition in clinically normal cats. *Journal of the American Veterinary Medical Association*, **194(5)**, 669–675.

63. MacBurney, M. M., Russell, C. and Young, L. S. (1990) Formulas. In *Clinical Nutrition. Enteral and Tube Feeding*. Eds. J. L. Rombeau and M. D. Caldwell. pp. 149–173. Philadelphia: W. B. Saunders.

64. MacDonald, M. L., Anderson, B. C., Rogers, Q. R., Buffington, C. A. and Morris, J. G. (1984) Essential fatty acid requirements of cats: pathology of essential fatty acid deficiency. *American Journal of Veterinary Research*, **45(7)**, 1310–1317.

65. Mashima, Y. (1979) Effect of caloric overload on puppy livers during parenteral nutrition. *Journal of Parenteral and Enteral Nutrition*, **3(3)**, 139–145.

66. Matsuno, S., Miyashita, E., Sasaki, K. and Sato, T. (1981) Effects of intravenous fat emulsion administration on exocrine and endocrine pancreatic function. *Japanese Journal of Surgery*, **11(5)**, 323–329.

67. Mecray, P. M., Barden, R. P. and Ravdin, I. S. (1937) Nutritional edema: its effect on the gastric emptying before and after gastric operations. *Surgery*, **1**, 53–64.

68. Meyer, H. and Kienzle, E. (1991) Dietary protein and carbohydrates: relationship to clinical disease. In *Proceedings of the Purina International Symposium.* pp. 13–26. St Louis: Ralston Purina.

69. Miles, J. M., Cattalani, M., Sharbrough, F. W., Wold, L. E., Wharen, R. E., Gerich, J. E. and Haymond, M. W. (1991) Metabolic and neurologic effects of an intravenous medium chain triglyceride emulsion. *Journal of Parenteral and Enteral Nutrition,* **15(1)**, 37–41.

70. Moore, E. E. and Jones, T. N. (1986) Benefits of immediate jejunostomy feeding after major abdominal trauma—a prospective, randomized study. *Journal of Trauma,* **26(10)**, 874–881.

71. Morris, J. G. and Rogers, Q. R. (1989) Comparative aspects of nutrition and metabolism of dogs and cats. In *Nutrition of the Dog and Cat. Waltham Symposium 7.* Eds. I. H. Burger and J. P. W. Rivers. pp. 35–66. Cambridge: Cambridge University Press.

72. Moss, G. (1985) Early enteral feeding after abdominal surgery. In *Nutrition in Clinical Surgery.* Ed. M. Deitel. pp. 220–231. Baltimore, MD: Williams & Wilkins.

73. National Research Council (1974) *Nutrient Requirement of Dogs.* Washington, DC: National Academy Press.

74. National Research Council (1985) *Nutrient Requirements of Dogs.* Washington, DC: National Academy Press.

75. National Research Council (1986) *Nutrient Requirements of Cats.* Washington, DC: National Academy Press.

76. Nuusbaum, M. S., Li, S., Bower, R. H., McFadden, D. W., Dayal, R. and Fischer, J. E. (1992) Addition of lipid to total parenteral nutrition prevents hepatic steatosis in rats by lowering the portal venous insulin/glucagon ratio. *Journal of Parenteral and Enteral Nutrition,* **16(2)**, 106–109.

77. Ogilvie, G. K. and Vail, D. M. (1990) Nutrition and cancer. *Veterinary Clinics of North America (Small Animal Practice),* **20(4)**, 969–985.

78. Polzin, D., Osborne, C. and O'Brien, T. (1989) Diseases of the kidneys and ureters. In *Textbook of Veterinary Internal Medicine.* Ed. S. J. Ettinger. pp. 1962–2046. Philadelphia: W. B. Saunders.

79. Ragins, H., Levenson, S. M., Signer, R., Stamford, W. and Seifter, E. (1973) Intrajejunal administration of an elemental diet at neutral pH avoids pancreatic stimulation. Studies in dog and man. *American Journal of Surgery,* **126(5)**, 606–614.

80. Reimold, E. W. (1979) Studies of the toxicity of an intravenous fat emulsion I: Hematologic changes and survival after administration of a soybean oil (FES15) in beagles. *Journal of Parenteral and Enteral Nutrition,* **3(5)**, 328–334.

81. Reimold, E. W. (1979) Studies of the toxicity of an intravenous fat emulsion. II Blood chemical changes after administration of a soybean oil (FES15) in Beagles. *Journal of Parenteral and Enteral Nutrition,* **3(5)**, 335–339.

82. Remillard, R. L. and Martin, R. A. (1990) Nutritional support in the surgical patient. *Seminars in Veterinary Medicine and Surgery (Small Animals),* **5(3)**, 197–207.

83. Remillard, R. L. and Thatcher, C. D. (1989) Parenteral nutritional support in the small animal patient. *Veterinary Clinics of North America (Small Animal Practice),* **19(6)**, 1287–1306.

84. Rosin, E. (1981) The systemic response to injury. In *Pathophysiology in Small Animal Surgery.* Ed. M. J. Bojrab. pp. 3–11. Philadelphia: Lea and Febiger.

85. Salto, Y., Tokutake, K., Matsuno, S., Noto, N., Honda, T. and Sato, T. (1978) Effects of hypertonic glucose and amino acid infusions on pancreatic exocrine function. *Tohoku Journal of Experimental Medicine,* **124(2)**, 99–115.

86. Sapy, P., Furka, I., Fabian, E., Miko, I. and Balaza, G. (1990) Experimental study of parenteral nutrition and of the exocrine function of the pancreas. *Acta Chirurgia Hungary,* **31(2)**, 145–150.

87. Sayeed, F. A., Johnson, H. W., Sukumaran, K. B., Raihle, J. A., Mowles, D. L., Stelmach, H. A. and Majors, K. R. (1986) Stability of Liposyn II fat emulsion in total nutrient admixtures. *American Journal of Hospital Pharmacy,* **43**, 1230–1235.

88. Shaw, J. H. F. and Wolfe, R. R. (1984) A conscious septic dog model with hemodynamic and metabolic responses similar to responses of humans. *Surgery,* **95(5)**, 553–560.

89. Shils, M. E. (1988) Enteral (tube) and parenteral nutrition support. In *Modern Nutrition in Health and Disease.* Ed. M. E. Shils and V. R. Young. pp. 1023–1066. Philadelphia: Lea and Febiger.

90. Silvis, S., DiBartolomeo, A. G. and Aaker, H. M. (1980) Hypophosphatemia and neurological changes secondary to oral caloric intake. *American Journal of Gastroenterology,* **73(3)**, 215–222.

91. Silvis, S. E. and Paragas, P. V. (1971) Fatal hyperalimentation syndrome. Animal studies. *Journal of Laboratory and Clinical Medicine,* **78**, 918–930.

92. Smith, R. J. and Wilmore, D. W. (1990) Glutamine nutrition and requirements. *Journal of Parenteral and Enteral Nutrition,* **14(4)**, 94S–99S.

93. Stabile, B. E. Borzatta, M. and Stubbs, R. S. (1984) Pancreatic secretory responses to intravenous hyperalimentation and intraduodenal elemental and full liquid diets. *Journal of Parenteral and Enteral Nutrition,* **8(4)**, 77–380.

94. Stabile, B. E., Borzatta, M., Stubbs, R. S. and Debas, H. T. (1984) Intravenous mixed amino acids and fats do not stimulate exocrine pancreatic secretion. *American Journal of Physiology,* **246,** G274–G280.

95. Tashiro, T., Ogata, H., Yokoyama, H., Mashima, Y. and Iwasaki, I. (1975) The effect of fat emulsion on essential fatty acid deficiency during intravenous hyperalimentation in pediatric patients. *Journal of Paediatric Surgery,* **10(2),** 203–213.

96. The Veterans Affairs Total Parenteral Nutrition Cooperative Study Group (1991) Perioperative total parenteral nutrition in surgical patients. *New England Journal of Medicine,* **325,** 525–532.

97. Thompson, S. W. (1974) *The Pathology of Parenteral Nutrition with Lipids.* Springfield, IL: Charles C. Thomas.

98. Toledo–Pereyra, L. H., Zammit, M. and Mittal, V. K. (1983) Total parenteral nutrition in renal transplantation. Experimental observations. *American Surgery,* **49(7),** 396–399.

99. Tonouchi, H., Iriyama, K. and Carpentier, Y. A. (1990) Transfer of apolipoproteins between plasma lipoproteins and exogenous lipid particles after repeated bolus injections or during a continuous infusion of fat emulsion. *Journal of Parenteral and Enteral Nutrition,* **14(4),** 381–385.

100. Traverso, L. W., Abou Zamzam, A. M., Maxwell, D. S., Lacy, S. M. and Tomkins, R. K. (1981) The effect of total parenteral nutrition or elemental diet on pancreatic activity and ultrastructure. *Journal of Parenteral and Enteral Nutrition,* **5(6),** 496–500.

101. Twomey, P. (1990) Cost and effectiveness of enteral nutrition. In *Clinical Nutrition. Enteral and Tube Feeding.* Eds. J. L. Rombeau and M. D. Caldwell. pp. 532–539. Philadelphia: W. B. Saunders.

102. Wannemacher, R. W. and McCoy, J. R. (1966) Determination of optimal dietary protein requirements of young and old dogs. *Journal of Nutrition,* **88,** 66–74.

103. Waters, D. C., Hoff, J. T. and Black, K. L. (1986) Effect of parenteral nutrition on cold induced vasogenic edema in cats. *Journal of Neurosurgery,* **64,** 460–465.

104. Weissman, C., Kemper, M. and Hyman, A. I. (1989) Variation in the resting metabolic rate of mechanically ventilated critically ill patients. *Anesthetics and Analgesics,* **68(4),** 457–461.

105. Wheeler, S. L. and McGuire, B. H. (1989) Enteral nutritional support. In *Current Veterinary Therapy.* Ed. R. W. Kirk. pp. 30–37. Philadelphia: W. B. Saunders.

106. Wilmore, D. W. (1991) Catabolic illness: strategies for enhancing recovery. *New England Journal of Medicine,* **325,** 695–702.

107. Wolfe, R. R., Durkot, M. J. and Wolfe, M. H. (1982) Effect of thermal injury on energy metabolism, substrate kinetics, and hormonal concentrations. *Circulatory Shock,* **9,** 383.

108. Wretlind, A. (1981) Development of fat emulsions. *Journal of Parenteral and Enteral Nutrition,* **5(3),** 230–235.

109. Yawata, Y., Hebbel, R. P., Silvis, S., Howe, R. and Jacob, H. (1974) Blood cell abnormalities complicating the hypophosphatemia of hyperalimentation: erythrocyte and platelet ATP deficiency associated with hemolytic anemia and bleeding in hyperalimented dogs. *Journal of Laboratoy and Clinical Medicine,* **84(5),** 643–653.

CHAPTER 5

Techniques for Enteral Nutritional Support

KENNETH W. SIMPSON and CLIVE M. ELWOOD

Introduction

Many sick animals are at risk of becoming nutritionally compromised. Stress and disease can induce neuroendocrine changes that cause catabolism, which is often compounded by inadequate nutrient intake because the animal does not want to eat, has difficulty eating (e.g. oral trauma, oral neoplasia) or cannot be fed orally (e.g. after esophageal surgery). The net effect of these changes is protein calorie malnutrition and the associated problems of weight loss, poor wound healing, abnormal gastrointestinal function and immune incompetence. To combat or prevent the adverse effects of protein calorie malnutrition, nutritional support has become an integral part of the medical management of sick animals.

Animals with inadequate nutrient intake (see Chapter 3) who have not responded to short-term appetite stimulation (alternative foods or drugs), assisted feeding or orogastric tube feeding, require more aggressive nutritional support using the enteral or parenteral route. The enteral route is more physiological and economical than the parenteral route, is generally easier to access and is the route of choice in animals with normal gastrointestinal

function. Where gastrointestinal function is compromised it may still be possible to feed enterally by an appropriate choice of route (e.g. jejunostomy tube feeding in animals with gastric disorders) or diet (e.g. elemental diets in animals with intestinal malabsorption).

The aim of this chapter is to give a practical overview of the techniques used to provide enteral nutritional support in dogs and cats. Details of the nutritional assessment of hospitalized animals, calculation of nutrient requirements and specific dietary recommendations are given in Chapters 3 and 4.

Techniques for Enteral Nutritional Support

Enteral nutritional support is provided by administering nutrients directly into the esophagus, stomach or jejunum. Common techniques used for enteral nutritional support require nasoesophageal, pharyngostomy, gastrostomy or jejunostomy tube placement.

Choice of Technique

Once the decision to institute enteral feeding has been made the most appropriate

63

POLYVINYLCHLORIDE OR SILICONE ELASTOMER FEEDING TUBES		
Route	**Size**	**Tube type**
Nasoesophageal	6–16 Fr	PVC or Silastic feeding tube
Pharyngostomy	10–24 Fr	Latex Foley catheter
Percutaneous endoscopic gastrostomy	16–24 Fr	Depezzer urinary catheter
Surgical gastrostomy	10–24 Fr	Latex Foley catheter
	16–24 Fr	Depezzer urinary catheter
Jejunostomy	5 Fr	PVC or Silastic feeding tube

FIG. 5.1 Feeding tubes available for the provision of enteral nutritional support.

technique is selected. The choice of technique is based on the nature of the animal's disease, the length of time support is required and the capability of the practitioner to perform the technique. In order to optimize the assimilation of nutrients it is important to make use of as much of the gastrointestinal tract as possible.

Nasoesophageal Intubation

Nasoesophageal tubes are appropriate for the short- to medium-term support of anorectic hospitalized animals that have a normal nasal cavity, pharynx and esophagus. Termination of the tube in the thoracic esophagus minimizes the risk of reflux esophagitis.[5,15] Contraindications to the use of a nasoesophageal tube include disease and dysfunction of the nares, pharynx, larynx, swallowing reflex, esophagus and stomach. Nasoesophageal tubes should not be placed in comatose animals.

Nasoesophageal tubes can be placed in the conscious animal with minimal restraint and topical anesthesia of the nasal chamber.

EQUIPMENT
Topical anesthetic
Long PVC or silicone infant feeding tube (5–10 Fr)
Lubricating jelly
Adhesive tape
Suture material or fast-acting adhesive
Sterile water or saline
Elizabethan collar

Some animals may require sedation for successful placement. Polyvinylchloride or silicone elastomer infant feeding tubes are available (Fig. 5.1) that are flexible, have an injection cap and range in size from 5 French, small enough for small cats, to 16 French, for large dogs. Silicone elastomer tubes have been recommended because they are more pliable and less irritating, but they may be more expensive.[3] Tubes that are too flexible may be chilled before placement to increase stiffness.[3]

Two to three drops of the topical anesthetic are placed into the selected nostril. The tube is measured from the nares to the ninth rib in the cat, to the seventh intercostal space in the dog (Fig. 5.2) and the length marked with a tape.[3] The tube tip is lubricated and passed into the ventral meatus of the nasal chamber. In cats the ventral meatus is most

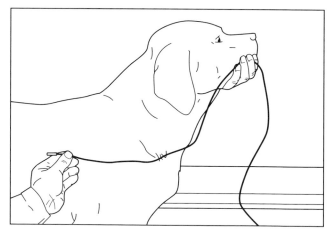

FIG. 5.2 The nasoesophageal tube is measured to the seventh intercostal space in the dog so that it will terminate in the thoracic esophagus.

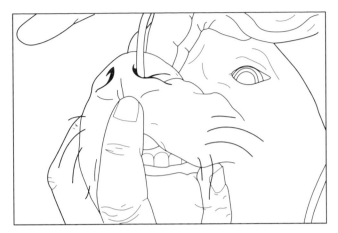

FIG. 5.3 Passage of the nasoesophageal tube into the ventral meatus of the dog is made easier by simultaneous dorsal deviation of the external nares.

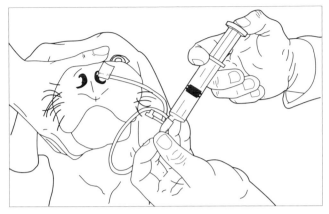

FIG. 5.5 The position of the nasoesophageal tube is checked by instilling a small volume of sterile saline, which should evoke a cough if the tube has been placed in the airway.

easily entered by directing the tube ventrally and medially.[12] In dogs, entry into the ventral meatus is facilitated by directing the tube ventrally and medially, to the point of contact with the medial septum, then pushing the external nares dorsally to open the ventral meatus (Fig. 5.3).[1] As the tube is gently advanced through the nasal cavity some sneezing, head shaking and discomfort may be seen, but rarely persists. If firm resistance is felt the tube should be withdrawn and redirected, as it might be impinging upon the ethmoturbinates within the middle meatus.[8] To encourage the tube to enter the esophagus rather than the airway the head and neck should be in a normal resting position as the tube reaches the pharynx. The tube is advanced

into the esophagus, often with a swallowing action from the animal, until the marker tape is positioned against the external nares (Fig. 5.4). The tube position is checked by injecting 5–10 ml of air while auscultating for borborygmi, or by injecting 2 ml of sterile water and observing for a cough response (Fig. 5.5).[8,19] In some cats no cough reflex is elicited so care should be taken. If doubt about the tube position remains, a radiograph should be obtained.[3] Once tube position has been checked, it is secured with the marker tape as close to the nostril as possible, either by suturing the tape to the skin, or by gluing it with fast acting adhesive to surrounding hair. Fixing the tape as close to the external nares

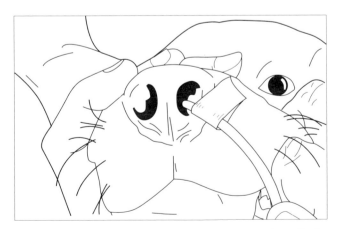

FIG. 5.4 The nasoesophageal tube is passed until the previously placed marker tape reaches the level of the nostril.

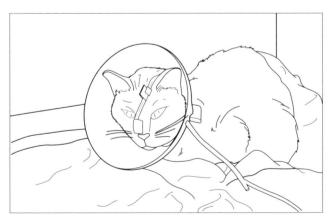

FIG. 5.6 The nasoesophageal tube is secured in place by suturing or gluing tapes at the level of the nares and between the eyes. An Elizabethan collar may or may not be necessary.

as possible minimizes the chances of tube withdrawal. The free end of the tube is then directed along the side of the maxilla, or between the eyes and sutured or glued in place using a second "butterfly" tape (Fig. 5.6).[3] The patient is less likely to interfere with the tube if it does not enter its field of vision. An Elizabethan collar prevents interference, but is not always necessary. The tube should be capped when not in use and its position checked before each feed, or at least once daily if drip feeding. The tube is removed by cutting the sutures or glued hair and gently withdrawing.

For smaller bore tubes (8 French or less) liquid enteral diets are preferred because prepared canned foods may block the tube. Diets can be administered in boluses or drip-fed, although in the latter case the animal should be under continuous observation while being fed. Old, clean fluid bags and giving sets can be reused for gravity feeding, but should be clearly marked "not for intravenous use."[10]

Minor complications associated with naso-esophageal tubes include epistaxis, rhinitis, dacryocystitis and esophageal reflux or vomiting.[8] Major complications include aspiration of the tube and/or diet into the airway, esophagitis and stricture formation and pneumothorax due to nasopleural intubation.[2,3] Should complications occur appropriate therapy should be instituted and the tube withdrawn as necessary.

Pharyngostomy Tube Placement

Pharyngostomy tubes are now less popular in institutions where percutaneous endoscopic gastrostomy (PEG) tubes can be placed, because of reported complications, but they still have a role in general practice.[11,15] They are appropriate for short- to long-term support of animals with a functioning pharynx, larynx, esophagus and stomach. They may be placed in animals with mandibular, maxillary or nasal disease, but otherwise the contraindications are those for nasoesophageal tubes and general anesthesia.

The animal is anesthetized and placed in

EQUIPMENT

14–18 Fr soft feeding tube or Foley catheter
Long curved hemostats
Scalpel blade
Suture material
Selected general anesthetic
Adhesive tape

lateral recumbency. An area of skin over the ventrolateral head and neck, centered on the hyoid apparatus, is aseptically prepared.[11] The tube is measured from the cranial hyoid apparatus to the ninth rib in the cat and the seventh intercostal space in the dog and marked with tape, so that it will terminate in the thoracic esophagus. A mouth gag is placed and an index finger is used to palpate the intended stoma site from the inside, dorsally and caudal to the stylohyoid and epihyoid bones, as close as possible to the rostral esophageal opening (Fig. 5.7).[11] The pulsating carotid artery is palpated and avoided as a 1 cm skin incision is made over the finger which is pushing the pharyngeal wall outwards.[11] Curved hemostats are used to bluntly dissect soft tissue, to create the stoma.[11,14] The tube is then grasped with the hemostats, drawn into the mouth and fed into the esophagus up to the level of the pre-placed marker tape. To avoid interference with the glottis or epiglottis the tube should run straight into the esophagus without kinking.[11] The tube is secured with sutures through butterfly tapes. It should not be sharply kinked externally because this may cause internal deviation and risk of pharyngeal obstruction.[11] Because of the risk of respiratory embarrassment, animals should be monitored during recovery from anesthesia.[3] The exit site should be treated with antiseptic, an encircling neck bandage applied and the wound inspected and cleaned regularly while the tube is *in situ*.

Before removal the tube should be flushed or aspirated to remove any contents. A tube may be removed by withdrawing it through the stoma, but this can contaminate the wound site. Therefore, sedation, manual withdrawal of the esophageal portion, division

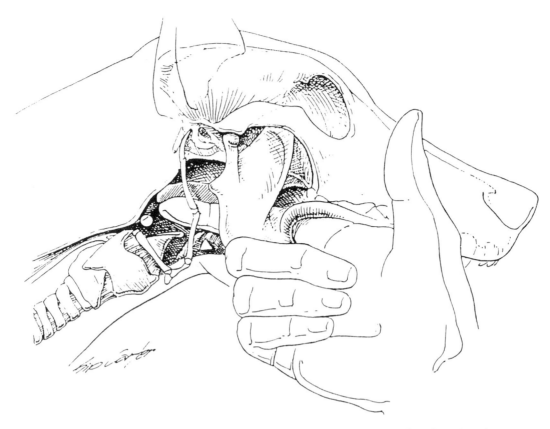

FIG. 5.7 The preferred site of entry for a pharyngostomy tube, dorsal and caudal to the stylohyoid and epihyoid bones. Reprinted with permission from Crowe and Downs.[11]

of the tube at the stoma and removal of the esophageal portion orally has been recommended.[14] The wound is left to heal by second intention.

Early reports of pharyngostomy placement recommended the creation of a stoma site rostral to the hyoid apparatus; in this position, however, the tube curved over the epiglottis and could cause obstruction of the glottis or interference with epiglottic function, leading to the current recommendations for placement.[11,14] Large-bore tubes are also avoided for similar reasons. Complications of pharyngostomy tubes include airway obstruction, aspiration of food, vomiting or reflux, tube displacement, tongue manipulation of the tube, chewing by the patient, esophagitis and infection of the wound site.[3,11,15] If the tube is displaced and undamaged it may be replaced in the conscious animal by opening the mouth and redirecting the tube down the esophagus.[11] If an obvious

stoma has formed and the tube requires complete replacement, the new tube may be passed through the old stoma.[11]

Gastrostomy Tube Placement

Where inadequate food intake is caused by oronasal trauma or neoplasia, esophageal disease, or anorexia and nutritional support is required for more than 5 days, enteral nutritional support can be provided via a gastrostomy tube. Contraindications to gastrostomy tube placement include neurological dysfunction of the esophagus, persistent vomiting, gastric and intestinal disorders and ascites. The development of new techniques enables gastrostomy tubes to be placed nonsurgically using an endoscope[6,16] or a blindly inserted semirigid stomach tube.[13] Endoscopic and blind tube placement are more rapid and less traumatic than surgical gastrostomy.

Surgical gastrostomy tubes are, however, often placed in animals undergoing celiotomy.

PEG

PEG requires a flexible endoscope and a specialized feeding tube. The high cost of gastrostomy tubes manufactured for humans often precludes their use in animals, so a cheaper but effective technique using a urological catheter with a "mushroom tip" (the feeding tube) (Fig. 5.8) is generally used.[6,16]

The urologic catheter is modified by removing the flare end and the small nipple on the mushroom tip. A further portion is then cut off and divided into two pieces (approximately 3 cm long) each of which has a slit made through it with a scalpel (Fig. 5.8). Two operators are necessary, the first to guide the endoscope and the second to manipulate the body wall aseptically. The PEG tube is inserted under general anesthesia with the animal in right lateral recumbency. An area of skin caudal to the left costal arch is shaved and aseptically prepared. The endoscope is passed into the stomach and the stomach is inflated. The left body wall is transilluminated with the endoscope to ensure that the spleen is not between the stomach and body wall (Fig. 5.9) and an appropriate site for tube insertion is selected by endoscopically monitored digital palpation (Fig. 5.10). A small incision is made in the skin with a scalpel and an intravenous catheter is stabbed through the body wall into the lumen of the stomach. The catheter stylet is removed and a long suture is inserted into the stomach via the catheter and grasped with the

EQUIPMENT
General anesthetic
Endoscope
Mushroom-tipped feeding tube (16–20 Fr)
16 gauge over the needle catheter
Pipette tip (250 µl) or catheter
Monofilament nylon suture
Scalpel blade
Sterile surgical gloves
Stockinette bandage

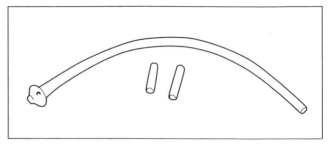

FIG. 5.8 Mushroom-tipped catheter and stents suitable for use as a percutaneous endoscopic gastrostomy tube.

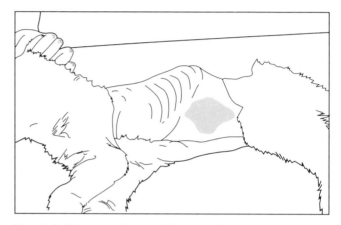

FIG. 5.9 Endoscopic transillumination ensures that the spleen is not between the stomach and the body wall.

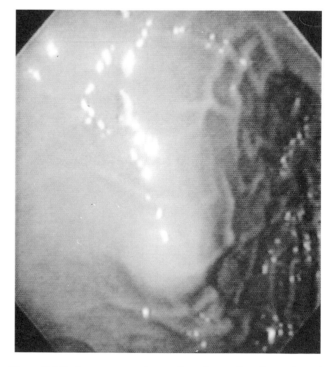

FIG. 5.10 An appropriate site for tube insertion is selected by endoscopically monitored digital palpation.

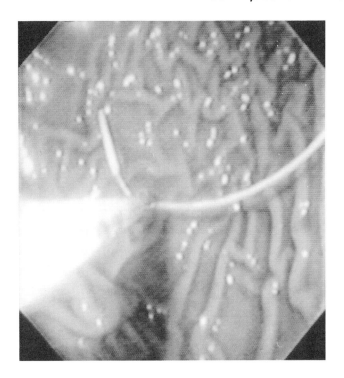

FIG. 5.11 A long nylon suture introduced into the stomach through an intravenous catheter is grasped with endoscopic biopsy forceps.

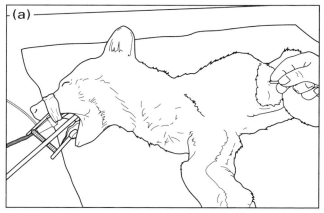

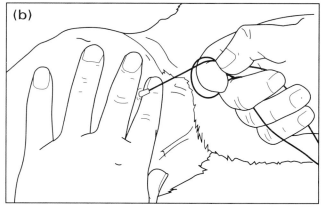

(c)

endoscopic biopsy forceps (Fig. 5.11). The endoscope and forceps are then withdrawn and the oral end of the suture is secured to the feeding tube (Fig. 5.12). Gentle traction is applied to the suture where it exits the body wall to draw the tube assembly into the stomach and the passage of the tube is monitored with the endoscope (Fig. 5.13a and b). The feeding tube is pulled out through the body wall until the mushroom end fits snugly against the gastric mucosa (Fig. 5.13c) and secured with the second stent (Fig. 5.14).

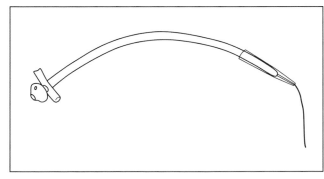

FIG. 5.12 The feeding tube is secured to the oral end of the suture.

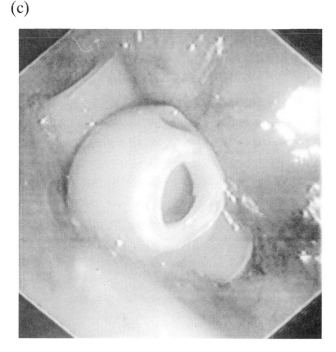

FIG. 5.13 The feeding tube is introduced into the stomach and pulled out through the body wall using gentle traction (a and b) until the mushroom end fits snugly against the gastric mucosa (c).

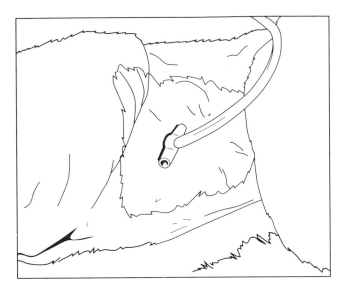

FIG. 5.14 The tube is secured by a stent placed next to the body wall. The tube should not be secured too firmly to the body wall or pressure necrosis will occur.

A stockinette jacket can be fitted to prevent patient access and the tube catching on things. Once experience is gained the whole procedure can be completed in 10–15 min.

A detailed account of diet selection and feeding regimens has been published by Armstrong *et al.*[3] The authors do not start feeding until the day after tube placement and introduce food gradually, feeding one third of caloric requirements on day one, two thirds on day two and the full amount on day three in divided portions qid. To avoid blockage the tube should be flushed with water before and after feeding. Should the tube become blocked, flushing with water or a solution of pancreatic enzyme extract usually clears it. The tube should remain in place for at least 5 days before removal to allow a peritoneal seal to form and is removed when oral food intake is sufficient to meet the animal's caloric requirements.

Complications related to PEG tubes include procedure-related problems such as splenic laceration, mild gastric hemorrhage and pneumoperitoneum and delayed complications such as vomiting, aspiration pneumonia, inadequate gastric emptying, tube extraction, tube migration and infection around the tube.[4] Splenic laceration is fatal and highlights the need for transillumination before insert-

ing the catheter. Vomiting can usually be resolved by decreasing then gradually increasing the amount of food given per meal. Animals with esophageal dysfunction appear to have a high incidence of aspiration pneumonia associated with PEG tube feeding.[4]

Blind Gastrostomy Tube Placement

Fulton and Dennis have recently described a technique that enables a gastrostomy tube to be placed without celiotomy or endoscopy.[13] The technique is very similar to endoscopic tube placement, except that a lubricated vinyl tube, rather than an endoscope, is passed through the mouth and into the stomach. The vinyl tube is inserted until the gastric end is seen to displace the lateral abdominal wall. The tube is then grasped and transabdominally manipulated until it lies 2–3 cm caudal to the end of the left 13th rib. A small skin incision is made over the lumen of the tube and a 14 gauge over-the-needle catheter inserted into the tube through the stomach and body walls. Correct positioning is confirmed by wiggling the needle about and feeling the side of the tube. A guide wire made from a Banjo string is attached to a suture 60 cm longer than the stomach tube. The wire is introduced through the catheter into the tube, grasped when it reaches the oral end and the attached suture pulled through. The suture is cut from the wire and the stomach tube is removed. A mushroom tipped catheter is attached to the suture and secured in an identical fashion to a PEG tube.

The authors stated a similar range and rate of complications to PEG tube placement.[3] Potential limitations are stated as obesity and esophageal disease. This technique appears to be relatively straightforward and suited to general clinical practice. As the clinical application of this technique was restricted to 6 dogs and 14 cats further studies are therefore necessary before firm statements on its efficacy are made.

Surgical Gastrostomy Tube Placement

Surgical gastrostomy tube placement has generally been superceded by PEG tube

EQUIPMENT
Surgical pack suitable for laparotomy Allis or Babcock forceps Mushroom-tipped feeding tube (16–20 Fr) Monofilament sutures Stockinette

placement. Surgical gastrostomy is still performed when an endoscope is not available or when the patient is undergoing celiotomy.

The following technique has been used extensively in small animals.[7,9] Tube placement is performed under general anesthesia with the animal in right lateral recumbency. An area of skin caudal to the left costal arch is shaved and aseptically prepared. A 3–5 cm incision is made just caudal to and parallel to the last rib. The underlying fascia is incized and a tunnel is made through the abdominal oblique muscles by dissecting them parallel to their fiber length. The transverse abdominal oblique muscle, the transverse fascia and the peritoneum are incized. The stomach wall is located digitally and retracted through the incision using Allis or Babcock forceps. In deep-chested dogs the stomach can be difficult to locate and the incision may need to be enlarged or a stomach tube may be passed by an assistant and 10–15 ml/kg of air insufflated.[19] The stomach is carefully examined to verify its identity and a site on the left lateral aspect of the gastric body or caudal aspect of the stomach is selected for tube insertion.[9] The area between the exteriorized stomach and the body wall is packed off with gauze swabs. Two full-thickness purse-string sutures are placed in the stomach wall around the selected tube insertion site (Fig. 5.15).[7,9] A stab incision is made in the center of the purse string and a balloon-tipped Foley catheter or a mushroom-tipped urologic catheter (Fig. 5.8) is introduced into the stomach. A stylet is inserted into the mushroom-tipped catheter to "flatten" the mushroom tip prior to introduction. The stylet is then removed (or the balloon is inflated with saline if using a Foley) and the inner, then outer purse strings

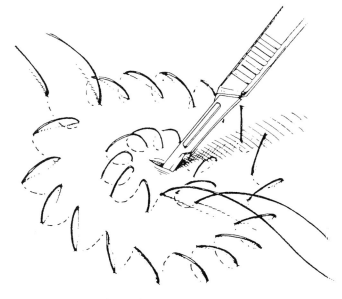

FIG. 5.15 Two full-thickness purse-string sutures are placed in the stomach wall around the tube insertion site. A stab incision is made in the center and a catheter introduced. Reprinted with permission from Crowe.[9]

secured. The tube can exit the body wall via the initial grid incision or via a separate stab incision.[7,9] Two layers of omentum are wrapped around the tube and interposed between the stomach and the body wall to help prevent leakage of the gastric contents. The stomach is sutured to the body wall using a simple interrupted pattern (Fig. 5.16). The

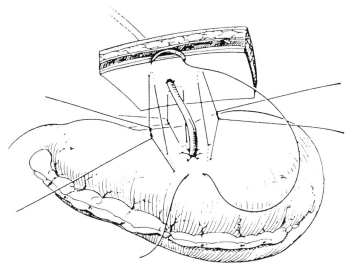

FIG. 5.16 The stomach is sutured to the body wall using a simple interrupted pattern. Reprinted with permission from Crowe.[9]

tube is secured to the skin of the outer body wall using a Chinese finger trap friction suture. The tube is capped and a stockinette vest placed around the animals trunk to protect it.

The balloon tips of Foley catheters are not very durable in the stomach and often burst within a short time. This can lead to tube displacement. Peritonitis may ensue if an adequate seal between the stomach and body wall has not formed. For this reason the authors prefer to use mushroom-tipped catheters. Despite the widespread use of surgical gastrostomy tubes there are no reports that document the incidence of complications associated with the technique.

Enterostomy Tubes

Enterostomy tubes are placed when intragastric or intraesophageal feeding is inappropriate, for example following gastric surgery. They are appropriate for medium-term support of the hospitalized patient and extrapolation from human data suggests that normal intestinal function is not necessary. Contraindications include ascites, peritonitis, profound immunosuppression, distal small bowel obstruction and coagulopathies.[3] Percutaneous duodenostomy and jejunostomy tube feeding may enable complications such as vomiting and aspiration pneumonia associated with PEG tubes to be overcome. Percutaneous duodenostomy and jejunostomy tubes can be placed by passing them through the gastrostomy tube into the stomach and then feeding the tube into the jejunum endoscopically. Preliminary experimental studies indicate that percutaneous duodenostomy is a safe and

feasible technique in the dog but specific recommendations based on clinical application have not yet been reported.[17]

Surgical placement of a jejunostomy tube during laparotomy should add only 10–15 min to the surgery time. Tube placement is achieved through a large bore needle, therefore needle and tube compatibility should be checked before use. Alternatively a "through-the-needle" catheter may be used. The tube gauge should be sufficient to allow the intended liquid food to flow freely and the tube should be long enough to allow passage through the abdominal wall and for 20–30 cm of the tube to lie within the bowel lumen.[18] The following method is suitable when the intended feeding tube has a permanent Luer connector.[3]

A loop of distal duodenum or proximal jejunum is selected and a small purse-string suture placed on the antimesenteric border (Fig. 5.17). This piece of bowel is then opposed to the right ventrolateral body wall to determine a suitable exit site for the tube and a needle is passed through the body wall from inside to out. The feeding tube is then passed through the needle from outside to inside and the needle withdrawn.

The bowel loop is retrieved and the needle is passed directly into the bowel lumen, on the antimesenteric border, 2–3 cm opposite from the purse-string suture. The needle is then directed orad and tunnelled within the bowel wall, to emerge through the purse string suture (Fig. 5.18). The feeding tube is now passed down the needle and the needle

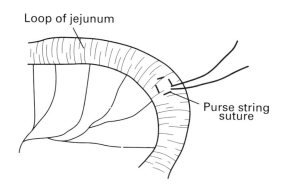

FIG. 5.17 Jejunosotomy tube placement. A loop of jejunum is selected and a purse-string suture is placed on the antimesenteric border.

EQUIPMENT
Long feeding tube
Large bore hypodermic needle (to pass the feeding tube through) (Alternatively, a large through-the-needle catheter unit, disassembled)
Surgical set
General anesthetic

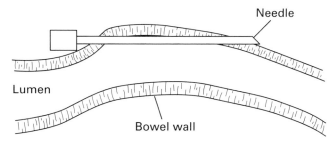

FIG. 5.18 Jejunostomy tube placement. The needle passes into the bowel lumen, runs within the wall then exits through the purse-string suture.

carefully withdrawn so that the tube passes through the purse string, tunnels 2–3 cm in the bowel wall and emerges into the lumen of the bowel (Fig. 5.19). Tunnelling the tube submucosally helps provide a seal, preventing leakage of bowel contents.

The tube is then advanced so that 20–30 cm is within the bowel lumen. The purse string suture is tied and the needle entry site closed with a simple interrupted suture. The enterostomy site and body wall are opposed and sutured with interrupted or continuous sutures passing through intestinal submucosa and abdominal fascia.[19]

The site is then omentalized. The tube is fixed to the exterior body wall using a butterfly tape and sutures, labelled to avoid

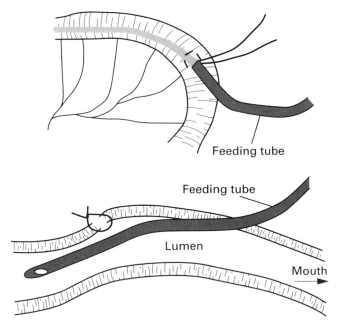

FIG. 5.19 Final position of the jejunostomy tube within the bowel. The needle entry site is sutured closed.

infusion errors and incorporated into an abdominal wrap.

Tube removal is accomplished by cutting the skin sutures and withdrawing the tube, leaving the small wound to heal by second intention.

Complications of enterostomy tube placement have included diarrhea, peritonitis secondary to tube displacement and peritonitis following catheter-induced bowel rupture.[3,18] To minimize the risk of tube displacement patients should be immobilized as much as possible and the tube carefully wrapped. Passing 20–30 cm of tube into the bowel lumen helps minimize the risk of displacement and stiff tubes are avoided to reduce the risk of bowel rupture.[3] Continuous feeding of a liquid diet formulation, building up to full energy requirements over 2–3 days, minimizes the risk of diarrhea due to osmolar load on the intestine.[9] Should diarrhea be a problem, diluting the diet to half-strength may help.

References

1. Abood, S. K. and Buffington, C. A. (1991) Improved nasogastric intubation technique for administration of nutritional support in dogs. *Journal of the American Veterinary Medical Association,* **199**, 577–579.
2. Abood, S. K. and Buffington, C. A. (1992) Enteral feeding of dogs and cats: 51 cases (1989–1991). *Journal of the American Veterinary Medical Association,* **201**, 619–622.
3. Armstrong, P. J., Hand, M. S. and Frederick, G. S. (1990) Enteral nutrition by tube. *Veterinary Clinics of North America: Small Animal Practice,* **20**, 237–275.
4. Armstrong, P. J. and Hardie, E. M. (1990) Percutaneous endoscopic gastrostomy. A retrospective study of 54 clinical cases in dogs and cats. *Journal of Veterinary Internal Medicine,* **4**, 202–206.
5. Balkany, T. J., Baker, B. B., Bloustein, P. A. and Jafek, B. M. (1977) Cervical esophagostomy in dogs. Endoscopic, radiographic and histopathologic evaluation of esophagitis induced by feeding tubes. *Annals of Otology, Rhinology and Laryngology,* **86**, 588–593.
6. Bright, R. M. and Burrows, C. F. (1988) Percutaneous endoscopic tube gastrostomy in dogs. *American Journal of Veterinary Research,* **49**, 629–633.
7. Crane, S. W. (1980) Placement and maintenance of a temporary feeding tube gastrostomy in the dog and cat. *Compendium on Continuing Education for the Practising Veterinarian,* **11**, 770–776.

8. Crowe, D. T. (1986) Clinical use of an indwelling nasogastric tube for enteral nutrition and fluid therapy in the dog and cat. *Journal of the American Animal Hospital Association,* **22**, 675–682.

9. Crowe, D. T. (1986) Enteral nutrition for critically ill or injured patients—Part II. *Compendium on Continuing Education for the Practising Veterinarian,* **8**, 719–732.

10. Crowe, D. T. (1989) Nutrition in critical patients: administering support therapies. *Veterinary Medicine,* **84**, 152–180.

11. Crowe, D. T. and Downs, M. O. (1986) Pharyngostomy complications in dogs and cats and recommended technical modifications: experimental and clinical investigations. *Journal of the American Animal Hospital Association,* **22**, 493–503.

12. Forenbacher, S. (1950) Passing the stomach tube through the nose in the cat. *Veterinary Medicine,* **45**, 407–410.

13. Fulton, R. B. and Dennis, J. S. (1992) Blind percutaneous placement of a gastrostomy tube for nutritional support in dogs and cats. *Journal of the American Veterinary Medical Association,* **201**, 697–700.

14. Lantz, G. C. (1981) Pharyngostomy tube installation for the administration of nutritional and fluid requirements. *Compendium on Continuing Education for the Practising Veterinarian,* **3**, 135–142.

15. Lantz, G. C., Cantwell, H. D., VanVleet, J. F., Blakemore, J. C. and Newman, S. (1983) Pharyngostomy tube induced esophagitis in the dog: an experimental study. *Journal of the American Animal Hospital Association,* **19**, 207–212.

16. Mathews, K. A. and Binnington, A. G. (1986) Percutaneous incisionless placement of a gastrostomy tube utilizing a gastroscope: preliminary observations. *Journal of the American Animal Hospitals Association,* **22**, 601–610.

17. McCrackin, M. A., Bright, R. M., DeNovo, R. C. and Toal, R. L. (1992) Endoscopic placement of a percutaneous gastroduodenostomy feeding tube in the dog. *Journal of Veterinary Internal Medicine,* **6** (Abstr.), 130.

18. Orton, E. C. (1986) Enteral hyperalimentation administered via needle catheter-jejunostoma as an adjunct to cranial abdominal surgery in dogs and cats. *Journal of the American Veterinary Medical Association,* **188**, 1406–1411.

19. Wheeler, S. L. and McGuire, B. H. (1989) Enteral Nutritional Support. In *Current Veterinary Therapy X.* Ed. R. W. Kirk. pp. 30–37. Philadelphia: W. B. Saunders.

CHAPTER 6

Nutrition and Cancer

ALAN S. HAMMER

Introduction

Nutrition and cancer is a multifaceted subject and includes such areas as dietary carcinogens, prevention of cancer through dietary manipulations and cachexia and weight loss in the cancer patient. This discussion will focus on weight loss and cachexia (general physical wasting and malnutrition) in dogs or cats with cancer, the advances in the understanding of metabolic abnormalities associated with cachexia and supportive therapeutic measures that have resulted in improved quality of life for cancer patients. Prevention of cancer by the use of dietary carcinogen inhibitors will also be briefly discussed as that promises to be an active area of future investigation.

Cachexia in the cancer patient arises because tumors are metabolically active and capable of uncoupling metabolism, resulting in accelerated wasting and failure of adequate nutrient utilization. Although cachexia is a common disorder in human cancer patients the exact incidence in veterinary patients is not known. One report indicated that 50% of human cancer patients experienced weight loss and that between 15 and 24% of patients have greater than 10% weight loss.[11,56] Interestingly, subclinical metabolic abnormalities have been detected in both human and veterinary cancer patients not experiencing weight loss.[5,64] This indicates that the syndrome of cachexia is more common than is detected clinically and that the spectrum of involvement ranges from subclinical to overt cachexia.

Cancer cachexia is clinically important because of its prevalence and its poor prognosis. Patients with cancer and no weight loss have longer survival times than those with weight loss.[10,11] Humans and animals with poor performance status, including weight loss, have more complications and higher morbidity associated with chemotherapy, radiotherapy and surgery.[10,11,17] Improvement in nutritional status has decreased surgical morbidity in cancer patients, but increases in survival time have not yet been reported.[32,33] Improved immunological function has also been demonstrated in cancer patients through nutritional supplementation.[8,9] It appears that correcting the underlying nutritional problems of cancer patients may be the foundation upon which more specific anticancer therapy may then be added. Sole reliance on chemotherapy, surgery or radiotherapy and not addressing clinical weight loss and cachexia in a cancer patient may deprive the patient of important supportive therapy.

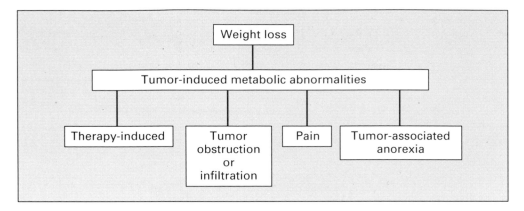

FIG. 6.1 The multifactorial etiology of weight loss in the cancer patient. Tumor-induced metabolic abnormalities are superimposed on other etiologies in most patients.

Overview of the Cancer Patient with Weight Loss

In approaching the veterinary cancer patient with anorexia and weight loss, it is important to be aware of the multifactorial nature of cachexia (see Fig. 6.1). Iatrogenic etiologies should be considered initially. Weight loss may be a consequence of chemotherapy or radiotherapy or due to daily testing or anesthetic procedures that result in withholding food for several days.

Another etiology associated with weight loss is directly related to the physical presence of the tumor. The tumor may cause obstruc-

tion of the gastrointestinal tract or malabsorption due to infiltration of the bowel (e.g. intestinal lymphoma). Oral tumors frequently cause cats to become anorectic.

Pain associated with a tumor may result in decreased food intake and weight loss. Cancer-associated pain is a nebulous entity in veterinary medicine and difficult to quantify. Dogs with osteosarcoma may refuse to eat because of tumor-associated pain and difficulty or reluctance to move to the food bowl. In humans, appropriate management of pain is considered to be important to improve the quality of life of cancer patients and to maintain food intake.

COMMON METABOLIC ABNORMALITIES IN CANCER PATIENTS	
Carbohydrate	Glucose intolerance
	Hyperinsulinemia
	Insulin resistance
	Hyperlactatemia
	Increased glucose turnover
	Increased gluconeogenesis
	Increased Cori cycle activity
Protein	Increased muscle protein catabolism
	Increased hepatic protein synthesis
	Increased plasma branched chain amino acids
Fat	Increased lipolysis
	Increased lipogenesis
	Decreased serum lipoprotein lipase activity
	Failure of glucose to suppress lipolysis
	Increased VLDL and free fatty acids
	Decreased HDL and cholesterol

FIG. 6.2 Common metabolic abnormalities in cancer patients.

Anorexia associated with cancer is apparently mediated through humoral effects on the central nervous system.[4,27] The mechanisms underlying this phenomenon have not been well characterized, but it is interesting to speculate on the role this particular etiology may have in causing anorexia in cats. Additionally, as depicted in Fig. 6.1, tumor-induced abnormalities in patient metabolism are frequently present although they may range from subclinical to overtly cachectic and will act in combination with therapy-induced anorexia, tumor obstruction or pain and centrally mediated anorexia.

Pathophysiology

The types of abnormalities of metabolism in the patient with cancer have been reasonably well characterized and involve carbohydrate, protein and fat metabolism. Figure 6.2 lists some of the common abnormalities of metabolism in humans and animals with cancer.

Carbohydrate Metabolism

Alterations in carbohydrate metabolism have been repeatedly recognized in humans, experimental animals and domestic animals with cancer.[26,64] These abnormalities affect numerous metabolic pathways and some investigators believe that altered carbohydrate metabolism alone may account for much of the cachexia seen in cancer patients.[14] There is increased glucose turnover due to anaerobic metabolism of glucose by the tumor, resulting in increased lactate production. Lactate is converted back to glucose by the Cori cycle in the patient's liver (Fig. 6.3). The tumor gains 2 ATP and the patient has a potential loss of 42 ATP and 2 GTP per glucose molecule metabolized. Increased hepatic gluconeogenesis is frequently observed in cancer patients and certain plasma amino acids may decrease in concentration secondary to utilization in glucose synthesis.[68] Gluconeogenesis is not entirely a substrate-driven reaction and humoral factors produced by the tumor or the patient in response to the

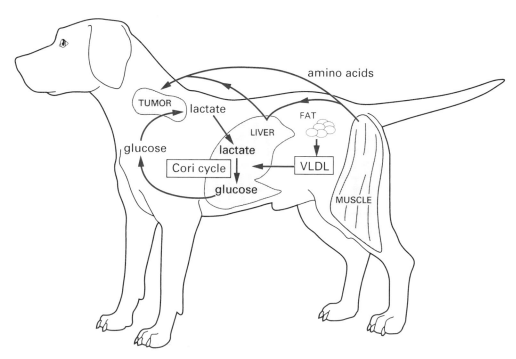

FIG. 6.3 A simplified diagram of the abnormal metabolism induced by the tumor that results in supplying the tumor with nutrients and energy and causing cachexia in the patient.

tumor may account for much of the increased gluconeogenesis.[15,18] This feeding of the tumor by the liver represents a tremendous energy drain on the patient.

A series of experiments performed on noncachectic dogs with lymphoma demonstrated glucose tolerance curves similar to control dogs.[64] This is different from humans, where glucose intolerance is commonly observed in cancer patients.[21,46] Dogs with lymphoma, however, do have hyperlactatemia and hyperinsulinemia similar to human patients.[64] The hyperlactatemia can be exacerbated by lactated Ringer's solution in dogs with lymphoma (serum bicarbonate concentrations are decreased before and after lactated Ringer's solution infusion compared to control dogs).[63] Induction of remission with doxorubicin in dogs with lymphoma does not correct the hyperlactatemia or hyperinsulinemia.[48] This indicates that even subclinical neoplasia can significantly alter patient metabolism. The hyperinsulinemia and relative insulin resistance may result from beta cell hypersecretion, or a receptor, or postreceptor defect. Research indicates postreceptor non responsiveness is possibly due to increased serum concentrations of free fatty acids, lipoproteins or other metabolic alterations.[2,22,36,51] This metabolic state in the cancer patient resembles a noninsulin dependent diabetes mellitus-like state in some respects. This correlation may aid in correcting or manipulating the underlying carbohydrate metabolic abnormalites in cancer patients.

Protein Metabolism

Clinicians are familiar with abnormal protein metabolism in cancer patients in the form of severe muscle atrophy. As shown in Fig. 6.3, the tumor acts as a nitrogen sink and will induce increased hepatic synthesis of proteins.[66,67] Hepatic protein synthesis has been found almost to double in an attempt to meet the protein requirements of the patient *and* the tumor. There is decreased muscle protein synthesis in the face of continued muscle protein catabolism.[35] The amino acids supplied to the tumor are utilized for synthesis

and gluconeogenesis. Certain amino acids, for example threonine, glutamine, glycine, valine, cystine and arginine, appear to be preferentially utilized and are found to be decreased in dogs with cancer when compared to control dogs, whereas others such as isoleucine and phenylalanine are increased in the plasma of dogs with cancer.[47]

The abnormal protein metabolism has clinical implications beyond muscle atrophy. Decreased immune response, gastrointestinal function and healing have been identified in animals with cancer. Currently, there is significant data that correcting the amino acid abnormalities by supplementing the diet may improve the immune function.[9] Whether this will result in decreased morbidity and increased survival in clinical patients remains to be seen.

Fat Metabolism

Hyperlipidemia and depletion of fat stores are frequently observed in humans with cancer. Although loss of body fat is frequently seen in veterinary patients, hyperlipidemia is not observed.[13,47] Hypocholesterolemia, often seen in humans with cancer, has not been found in dogs with cancer.[13,47] The cholesterol content of high-density lipoproteins (HDL), however, is decreased when compared to control dogs and the triglyceride content of very low-density lipoproteins (VLDL) and low-density lipoproteins has been reported to be greater in dogs with cancer than control dogs.[13] As with the abnormalities in carbohydrate metabolism in dogs with lymphoma, there was no correction of lipid metabolism (apart from increased cholesterol) after doxorubicin-induced remission.[13]

A more detailed evaluation of fat metabolism in cancer patients has found increased lipolysis that was not suppressed by glucose.[53] This results in increased lipid mobilization. The resultant hyperlipidemia may be immunosuppressive and may indirectly decrease survival.[57] Concurrently, serum lipoprotein lipase activity is decreased and lipogenesis also declines.[65] Fat metabolism in the cancer patient differs from starvation in that increased lipolysis and decreased lipogenesis are not accompanied by concurrent declines in plasma insulin concentrations. This is a maladaptive response because insulin normally promotes lipid storage. Tumors do not appear to utilize fatty acids directly as an energy source, however, the liver may convert the glycerol into glucose, a more readily available energy source (Fig. 6.3).

Humoral Abnormalities

The alterations of metabolism associated with cancer are known, but the mechanism by which these abnormalities are produced is only beginning to be identified. Humoral factors tentatively identified as causing cancer cachexia include tumor necrosis factor-α (TNF-α) and interleukin-1 (IL-1).[37,41,55] Discovery of TNF as a mediator of cachexia occurred because of parallel investigations. Rabbits with chronic trypanosomiasis develop cachexia, hyperlipidemia and decreased lipoprotein lipase activity.[50] A humoral factor that could induce similar changes in normal rabbits was identified and called cachectin. At the same time, TNF was identified as the agent inducing hemorrhagic necrosis in tumors and was elaborated by the macrophage. In time, the amino acid sequences of cachectin and TNF were shown to be identical.[1] Other mediators of cachexia have been proposed but are not as well characterized.

TNF suppresses lipogenesis and lipoprotein lipase activity in fat cells, thus preventing the storage and synthesis of lipids.[25,62] It is also believed to increase skeletal muscle catabolism and may be the factor that stimulates hepatic gluconeogenesis.[26] TNF and IL-1 may be the agents that act centrally to cause anorexia in cancer patients.[26,37,41] This may occur through decreased taste and smell or through a more central neural network.[12,19] Serotonin (5-HT) may also play a role in this centrally mediated anorexia in that humans with serotonin-producing carcinoid tumors experience anorexia.[29]

These observations have led to the working hypothesis that cytokines are the major mediators of cancer-associated cachexia and anorexia (Fig. 6.4).[26,37,41] These cytokines may be elaborated from the tumor but are more likely to be derived from the patient's response to the tumor. Following an acute trauma, these cytokines act beneficially to

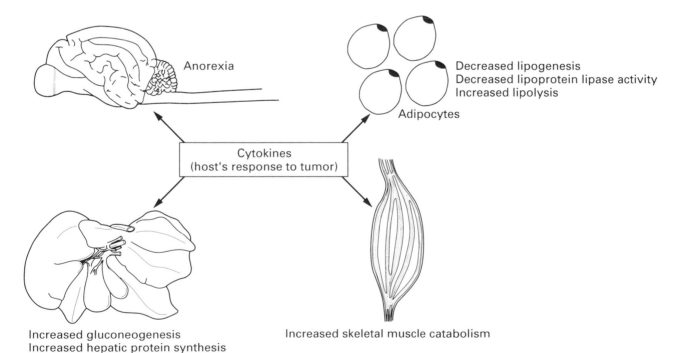

Anorexia

Decreased lipogenesis
Decreased lipoprotein lipase activity
Increased lipolysis

Adipocytes

Cytokines
(host's response to tumor)

Increased gluconeogenesis
Increased hepatic protein synthesis

Increased skeletal muscle catabolism

FIG. 6.4 The proposed role of cytokines such as TNF and IL-1 as humoral mediators of tumor cachexia and anorexia.

mobilize peripheral muscle proteins, fat and glucose, but chronic, sustained release of these cytokines results in wasting through excessive fat and protein mobilization and anorexia. Thus, in addition to the energy-depleting effects of abnormal carbohydrate metabolism induced by the tumor, a humoral cachectic syndrome is present to some degree in most patients with cancer.

Approach to the Cancer Patient with Weight Loss

The first step is to identify through physical examination and history any therapy-related toxicities, gastrointestinal obstruction/infiltration or tumor-associated pain that may be causing anorexia and weight loss. For example, radiotherapy for nasal tumors frequently results in painful mucositis and oral ulcers. Temporarily halting therapy for several days to permit healing may be necessary. Cleansing the mouth and use of broad-spectrum antibiotics may hasten resumption of eating. Chemotherapy-induced anorexia and nausea are usually temporary phenomena. Learned aversion to foods has not been documented in veterinary patients. Decreasing the dose may be an option for subsequent treatments. Unfortunately, decreasing the dose may also compromise the likelihood of successful management of the cancer. Thus, the clinician is faced with balancing the need to continue chemotherapy with the risks of continued anorexia and weight loss with their attendent complications, decreased immune function and delayed healing. Obviously, this is a decision that must be made in consultation with the owner. Use of various drugs to stimulate the appetite and combat nausea may permit continued chemotherapy. These drugs include cyproheptadine, nandrolone decanoate, metoclopramide and prochlorperazine (Fig. 6.5). Anorexia associated with tumor pain is more difficult to identify in veterinary patients and relief through surgery or analgesics may increase food intake.

If the tumor is causing gastrointestinal obstruction, then specific therapy directed at relieving the obstruction is indicated. People with esophageal neoplasia exhibit the most marked weight loss of all cancer patients due to the inability to intake nutrients in addition to the humoral cancer-associated cachexia. If specific therapy for the tumor is delayed or cannot be performed, supportive nutritional management involving bypassing the tumor is indicated. A number of advances have been made in recent years in delivery of nutrients to the veterinary patient. These include nasogastric tubes, percutaneous endoscopically placed gastrostomy tubes, jejunostomy tubes and parenteral nutrition. Specific diets for nutritional support of the veterinary cancer patient have not been formulated. The energy requirement for animals with cancer is thought to be as high as twice the basal energy requirement of a normal animal (see Chapter 3). In general, 40–50% of the nonprotein calories should be fat. Diets relying on simple carbohydrates for the majority of calories should not be used.

If there are no therapy-related causes for weight loss and direct physical involvement of the tumor does not account for anorexia and weight loss, then tumor-associated anorexia and cachexia must be considered.

| DRUGS USED TO CONTROL NAUSEA AND STIMULATE APPETITE ||
Drug	Dose
Metoclopramide	0.2–0.4 mg/kg tid PO
Prochlorperazine	0.2–0.4 mg/kg bid PO
Cyproheptadine	8 mg/m² sid PO
Nandrolone decanoate	3 mg/kg weekly IM
Diazepam	0.5–1.0 mg IV just prior to feeding
Oxazepam	5–10 mg bid PO

FIG. 6.5 Drugs used to control nausea and stimulate the appetite.

Unfortunately, there are no accepted diagnostic tests for this syndrome. In the future, evaluation of the patient with regard to circulating TNF or IL-1 and measurements of insulin or lactate, or lipid profiles may aid in identifying this syndrome. Most cancer patients, however, will experience some degree of humoral cachexia/anorexia, making it difficult to determine if that is the major cause of weight loss. Management of this syndrome is most directly done by treating the tumor. In the only studies to date in domestic animals, treatment of dogs with doxorubicin for lymphoma did not correct lipid or carbohydrate metabolic abnormalities.[13,48]

As discussed above, anorexia may be induced by humoral mechanisms by the tumor. Cats in particular represent a considerable challenge to maintain food intake. More work is necessary not only on the prevalence and mechanism of humoral anorexia in veterinary patients, but also on the intervention and correction of this disorder. There is some evidence that serotonin may be a mediator of this effect in the CNS.[29] At The Ohio State University Veterinary Teaching Hospital, we have used nandrolone decanoate and cyproheptadine with some success in stimulating the appetite of cancer patients (Fig. 6.5). A controlled trial of cyproheptadine in human cancer patients, however, failed to demonstrate any improvement in appetite or weight gain.[23] Diazepam and oxazepam have also been used to stimulate appetite, particularly in cats; however, the short-lived appetite stimulation and the sedative properties limit the effectiveness of these compounds. It is important to monitor and document increased food intake so that if a treatment measure is ineffective, an alternative drug can be considered.

Hydrazine sulfate is a gluconeogenic enzyme inhibitor and has been used in a preliminary double blind trial of humans with cachexia and cancer.[7] There was improved glucose tolerance but no change in survival. A larger randomized, double blind study also failed to demonstrate improved survival.[28] Megestrol acetate, commonly used in hormonal management of breast cancer in women, has been noted to cause weight gain and increased appetite in women with metastatic breast cancer and in patients with other forms of advanced cancer.[6,34,58,59,60] The mechanism of anticachectic activity of megestrol acetate is unknown. Neither of these agents has been evaluated in dogs and cats with cancer and cachexia. Glucocorticoids are known to cause polyphagia in animals, but their effects on carbohydrate, fat and protein metabolism would tend to exacerbate the already existing abnormalities and are not recommended at this time without further studies.

Nutritional support by itself does not correct the underlying metabolic abnormalities and although the weight loss may be corrected, this does not necessarily translate into increases in survival. In fact, in a randomized trial of humans with metastatic sarcoma, those receiving total parenteral nutrition (TPN) had poorer survival compared to people with conventional nutrition.[52] The benefits of aggressive nutritional management remain controversial, as some investigators report improved surgical healing and immune function that they believe result in decreased morbidity, while others believe aggressive nutritional support may enhance tumor growth.[44,61]

Attempts have been made to feed the patient selectively in preference to the tumor. Because the tumor does not directly utilize fats, dogs with lymphoma were randomized to high carbohydrate diets or high fat diets.[49] Following single-agent chemotherapy with doxorubicin, it was found that the dogs receiving the high fat diet had higher lactate and insulin concentrations after remission. Those dogs receiving the high carbohydrate diet experienced lower remission rates than the other dogs but had longer remission durations than the dogs receiving the high fat diet. This study demonstrates the complexity of feeding the cancer patient and indicates the necessity of some intervention to correct the metabolic abnormalities in order to maximize the usefulness of nutritional therapy.

Exercise is one mechanism that causes preferential uptake of nutrients by muscle rather than tumor.[24,45] Insulin therapy also reverses tumor anorexia, improves patient body composition and inhibits protein wasting.[42,43] Both of these potential therapies for cancer cachexia seem applicable if one considers the metabolic abnormality to be similar to noninsulin dependent diabetes mellitus. Demonstration that exercise or insulin therapy results in decreased morbidity or increased survival has yet to be proven clinically. Use of anticachectin antibodies or regulation of the TNF receptor site has also been considered for future investigations in cachectic patients.

In summary, supportive nutritional therapy remains vital in the management of the dog or cat with cancer, but it appears that we will not be able to feed our way out of these metabolic abnormalities. Instead, intervention in the form of drugs, hormones or exercise appears to be necessary to regulate the metabolism of the patient with cancer. In addition, specific therapy directed against the tumor will be needed to aid in correcting the metabolic abnormalities.

Prevention of Cancer Through Diet

Dietary carcinogens are believed to play a greater role in causing cancer than was previously thought. Most of the specific carcinogens have not yet been characterized. Aflatoxins, generated by molds in grain, are one example of dietary carcinogens strongly linked to liver cancer in countries where food quality control is poor. Another example of the role diet plays in carcinogenesis can be found in first-generation Americans of Japanese descent. Gastric cancer is prevalent in Japan due to the consumption of pickled food products; however, the incidence of colorectal cancer is low. Adoption of a Western lifestyle by immigrant Japanese results in a higher fat diet and a higher incidence of colorectal cancer similar to Caucasian Americans. Gastric cancer, however, decreases to a prevalence rate similar to that seen in Western society, presumably also due to dietary changes.

Dietary carcinogens are difficult to identify and even more difficult to avoid completely;

an alternative approach is to increase the cancer-preventing compounds in the diet. These compounds include β-carotene, vitamin A, vitamin C, vitamin E and fiber. Many of these compounds act to scavenge oxygen radicals or inhibit formation of carcinogens in the intestinal tract. β-Carotene, vitamin A and vitamin E have been associated with a protective effect in preventing lung cancer in that serum concentrations of vitamins A and E were lower in persons with lung cancer than the control population.[3,20,31,38,54] Vitamin C is not only an oxygen radical scavenger but is also thought to inhibit N-nitrosoamine compound formation.[30,40] A decreased risk of esophageal and gastric cancer is believed to be associated with increased vitamin C consumption.[16,39] Increased dietary fiber consumption lowers the risk of colon cancer by enhancing the excretion of carcinogens and decreasing the formation of mutagens. All of these findings have led the National Cancer Institute to adopt a "Five Times A Day" program to encourage consumption of five servings of fruit and vegetables per day to lower the risk of cancer by increasing the amount of cancer-preventing compounds and fiber in the diet.

Many of the tumor types in humans that are targeted for prevention by dietary manipulation are uncommon in dogs and cats (e.g. lung, esophagus, stomach, colon). The extent to which dietary carcinogens add to the cancer incidence in animals is unknown and therefore the extent to which a cancer prevention diet would decrease the incidence of cancer is also unknown. This is certain to become an area of intense research in the future because dietary manipulation of disease conditions in the dog and cat have been successful in the past.

References

1. Beutler, B., Greenwald, D., Hulmes, J. D., Chang, M., Pan, Y. C. E., Mathison, J. Ulevitch, R. and Cerami, A. (1985) Identity of tumour necrosis factor and the macrophage-secreted factor cachectin. *Nature*, **316**, 552–554.

2. Bishops, J. S. and Marks, P. A. (1959) Studies on carbohydrate metabolism in patients with neoplastic disease: II. Response to insulin administration. *Journal of Clinical Investigation,* **38,** 668–672.

3. Bjelke, E. (1975) Dietary vitamin A and human lung cancer. *International Journal of Cancer,* **15,** 561–565.

4. Borai, B. and DeWys, W. (1980) Assay for presence of anorexic substance in urine of cancer patients. *Proceedings of the American Association of Cancer Research,* **21,** 378.

5. Brennan, M. F. (1981) Total parenteral nutrition in the cancer patient. *New England Journal of Medicine,* **305,** 375–382.

6. Bruera, E., Macmillan, K., Kuehn, N., Hanson, J. and MacDonald, R. N. (1990) A controlled trial of megestrol acetate on appetite, caloric intake, nutritional status and other symptoms in patients with advanced cancer. *Cancer,* **66,** 1279–1282.

7. Chlebowski, R. T., Bulcavage, L., Grosvenor, M., Tsunokai, R., Block, J. B., Heber, D., Scrooc, M., Chlebowski, J. S., Chi, J., Oktay, E., Akman, S. and Ali, I. (1987) Hydrazine sulfate in cancer patients with weight loss: a placebo-controlled clinical experience. *Cancer,* **59,** 406–410.

8. Copeland, E. M., Daly, J. M. and Dudrick, S. J. (1977) Nutrition as an adjunct to cancer treatment in the adult. *Cancer Research,* **37,** 2451–2456.

9. Daly, J. M., Reynolds, J., Thom, A., Kinsley, L., Dietrick–Gallagher, M., Shou, J. and Ruggieri, B. (1988) Immune and metabolic effects of arginine in the surgical patient. *Annals of Surgery* **208,** 512–523.

10. DeWys, W. D., Begg, C., Band, P. and Tormey, D. (1981) The impact of malnutrition on treatment results in breast cancer. *Cancer Treatment Reports,* **65** (Suppl.), 87–91.

11. DeWys, W. D., Begg, D., Lavin P. T., Band, P., Bennett, J., Bertino, J., Cohen, M., Douglass, H., Engstrom, P., Ezdinli, E., Horton, J., Johnson, G., Moertel, C., Oken, M., Perlia, C., Rosenbaum, C., Silverstein, M., Skeel, R., Sponzo, R. and Tormey, D. (1980) Prognostic effect of weight loss prior to chemotherapy in cancer patients. *American Journal of Medicine,* **69,** 491–497.

12. DeWys, W. D. (1974) Abnormalities of taste as a remote effect of a neoplasm. *Annals of the New York Academy of Science,* **320,** 427–434.

13. Ford, R. B., Babineau, C. and Ogilvie, G. K. (1989) Serum lipid profiles in dogs with lymphosarcoma. *Proceedings of the 9th Annual Veterinary Cancer Society,* Raleigh, North Carolina.

14. Gold, J. (1968) Proposed treatment of cancer by inhibition of gluconeogenesis. *Oncology,* **22,** 185–207.

15. Gutman, A., Thilo, E. and Biren, S. (1969) Enzymes of gluconeogenesis in tumour-bearing rats. *Israeli Journal of Medical Science,* **5,** 998–1001.

16. Haenszel, W. and Correa, P. (1975) Developments in the epidemiology of stomach cancer over the past decade. *Cancer Research,* **35,** 3452–3459.

17. Hammer, A. S., Couto, G. C., Ayl, R. D., Shank, K. A. and DeMorais, E. (1990) Actinomycin D— a clinical study. *Proceedings of the 10th Annual Conference of the Veterinary Cancer Society,* Auburn, AL.

18. Hammond, K. D. and Balinsky, D. (1978) Activities of key gluconeogenic enzymes and glycogen synthase in rat and human livers and hepatoma cell cultures. *Cancer Research,* **38,** 1317–1322.

19. Henkin, R. I., Mattes–Kulig, D. and Lynch, R. A. (1983) Taste and smell acuity in patients with cancer. *Federation Proceedings,* **42,** 550.

20. Hirayama, T. (1979) Diet and cancer. *Nutrition and Cancer,* **1,** 67–81.

21. Holroyde, C. P., Skutches, C. L., Boden, G. and Reichard, G. A. (1984) Glucose metabolism in cachectic patients with colorectal cancer. *Cancer Research,* **44,** 5910–5913.

22. Jasani, B., Donaldson, L. K., Ratcliff, E. D. and Sokhi, G. S. (1978) Mechanism of impaired glucose tolerance in patients with neoplasia. *British Journal of Cancer,* **38,** 287–292.

23. Kardinal, C. G., Loprinzi, C. L. and Schaid, D. S. (1990) A controlled trial of cyproheptadine in cancer patients with anorexia and/or cachexia. (Abstr. 1258). *Proceedings of the American Society of Clinical Oncology,* **9,** 325.

24. Karlberg, H. I., James, J. H. and Fischer, J. E. (1983) Branched chain amino acid enriched diets and exercise prevent muscle wasting and tumour cachexia. *Surgery Forum,* **34,** 437–439.

25. Kawakami, M., Pekala, P. H., Lane, M. D. and Cerami, A. (1982) Lipoprotein lipase suppression in 3T3–LI cells by an endotoxin-induced mediator from exudate cells. *Proceedings of the National Academy of Science, U.S.A.,* **79,** 912–916.

26. Kern, K. A. and Norton, J. A. (1988) Cancer cachexia. *Journal of Parenteral and Enteral Nutrition,* **12,** 286–298.

27. Knoll, J. (1979) Satietin: a highly potent anorexogenic substance in human serum. *Physiology of Behaviour,* **23,** 497–502.

28. Kosty, M., Fleishman, S., Herndon, J., Coughlin, K., Duggin, D., Morris, J., Mortimer, J. and Green, M. R. (1992) Cisplatin, vinblastine and hydrazine sulfate in advanced non-small cell lung cancer (NSCLC): a randomised, placebo-controlled, double blind phase III study. *Proceedings of the American Society of Clinical Oncology,* **11,** 294.

29. Krause, R., Humphrey, C., Von Meyerfeldt, M., James, H. and Fischer, J. (1981) A central mechanism

for anorexia in cancer: a hypothesis. *Cancer Treatment Reports,* **65** (Suppl. 5), 15–23.

30. Krytopoulos, S. A. (1987) Ascorbic acid and the formation of *N*-nitroso compounds: possible role of ascorbic acid in cancer prevention. *American Journal of Clinical Nutrition,* **45**, 1344–1350.

31. Kvale, G., Bjelke, E. and Gart, J. J. (1983) Dietary habits and lung cancer risk. *International Journal of Cancer,* **31**, 397–405.

32. Landel, A. M., Hammond, W. G. and Mequid, M. M. (1985) Aspects of amino acid and protein metabolism in cancer-bearing states. *Cancer,* **55**, 230–237.

33. Lawson, D. H., Nixon, D. W., Kutner, M. H., Heymsfield, S. B., Rudman, D., Moffitt, S., Ansley, J. and Chawla, R. (1981) Enteral versus parenteral nutritional support in cancer patients. *Cancer Treatment Reports,* **65**, 101–106.

34. Loprinzi, C. L., Ellison, N. M., Schaid, D. J., Athman, L. M., Dose, A. M., Mailliard, J. A., Johnson, G., Ebberts, L. and Geeraerts, L. (1990) Controlled trial of megestrol acetate for the treatment of cancer anorexia and cachexia. *Journal of the National Cancer Institute,* **82**, 1127–1132.

35. Lundholm, K., Bennegard, K., Eden, E., Svaninger, G., Emery, P. W. and Rennie, M. J. (1982) Efflux of 3-methylhistidine from the leg in cancer patients who experience weight loss. *Cancer Research,* **42**, 4807–4811.

36. Lundholm, K., Holm, G. and Schersten, T. (1978) Insulin resistance in patients with cancer. *Cancer Research,* **38**, 4665–4670.

37. McNamara, M. J., Alexander, H. R. and Norton, J. A. (1992) Cytokines and their role in the pathophysiology of cancer cachexia. *Journal of Parenteral and Enteral Nutrition,* **16**, 50S–55S.

38. Menkes, M. S., Comstock, G. W., Vuilleumier, J. P., Helsing, K. J., Rider, A. A. and Brookmeyer, R. (1986) Serum beta-carotene, vitamins A and E, selenium and the risk of lung cancer. *New England Journal of Medicine,* **315**, 1250–1254.

39. Mettlin, C., Graham, S., Priore, R., Marshall, J. and Swanson, M. (1981) Diet and cancer of the esophagus. *Nutrition and Cancer,* **2**, 143–147.

40. Mirvish, S. S., Wallcave, L., Eagen, M. and Shubik, P. (1972) Ascorbate–nitrite reaction: possible means of blocking the formation of carcinogenic *N*-nitroso compounds. *Science,* **177**, 65–68.

41. Moldawer, L. L., Rogy, M. A. and Lowry, S. F. (1992) The role of cytokines in cancer cachexia. *Journal of Parenteral and Enteral Nutrition,* **16**, 43S–49S.

42. Moley, J. F., Morrison, S. D. and Norton, J. A. (1985) Insulin reversal of cancer cachexia in rats. *Cancer Research,* **45**, 4925–4931.

43. Moley, J. F., Peacock, J. E., Morrison, S. D. and Morton, J. A. (1985) Insulin reversal in cancer induced protein loss. *Surgery Forum,* **36**, 416–419.

44. Muller, J. M., Dienst, D., Brenner, U. and Pichlmaier, H. (1982) Preoperative parenteral feeding in patients with gastrointestinal carcinoma. *Lancet,* **1**, 68–72.

45. Norton, J. A., Lowry, S. F. and Brennan, M. F. (1979) Effect of work-induced hypertrophy on skeletal muscle of tumour- and nontumour-bearing rats. *Journal of Applied Physiology,* **46**, 654–657.

46. Norton, J. A., Maher, M., Wesley, R., White, D. and Brennan, M. F. (1984) Glucose intolerance in sarcoma patients. *Cancer,* **54**, 3022–3027.

47. Ogilvie, G. K., Vail, D. M., Wheeler, S. L. and Czarnecki, G. L. (1988) Alterations in fat and protein metabolism in dogs with cancer. *Proceedings of the Veterinary Cancer Society,* Estes Park, CO.

48. Ogilvie, G. K., Vail, D. M., Wheeler, S. L., Fettman, M. J., Salman, M. D., Johnston, S. D. and Hegstad, R. L. (1992) Effects of chemotherapy and remission on carbohydrate metabolism in dogs with lymphoma. *Cancer,* **69**, 233–238.

49. Ogilvie, G. K. (1992) Personal communication.

50. Rouzer, C. A. and Cerami, A. (1980) Hypertriglyceridaemia associated with *Trypanosoma brucei* infection in rabbits: role of defective triglyceride removal. *Molecular Biochemistry of Parasitology,* **2**, 31–38.

51. Schein, P. S., Kisner, D. D., Haller, D., Blecher, M. and Hamosh, M. (1979) Cachexia of malignancy. Potential role of insulin in nutritional management. *Cancer,* **43**, 2070–2076.

52. Shamberger, R. C., Brennan, M. E., Goodgame, J. T., Lowry, S. F., Maher, M. M., Wesley, R. A. and Pizzo, P. A. (1984) A prospective randomised study of adjuvant parenteral nutrition in the treatment of sarcomas: results of metabolic and survival studies. *Surgery,* **96**, 1–13.

53. Shaw, J. H. F. and Wolfe, R. R. (1987) Fatty acid and glycerol kinetics in septic patients and in patients with gastrointestinal cancer. The response to glucose infusion and parenteral feeding. *Annals of Surgery,* **205**, 368–376.

54. Shekelle, R. B., Liu, S., Raynor, W. J. Jr, Lepper, M., Maliza, C., Rossof, A., Paul, O., Shryock, A. M. and Stamler, J. (1981) Dietary vitamin A and risk of cancer in the Western Electric Study. *Lancet,* **2**, 1185–1190.

55. Shike, M. and Brennan, M. F. (1989) Supportive care of the cancer patient. In *Cancer: Principles and Practice of Oncology.* Eds. V. T. DeVita, S. Hellman, and S. A. Rosenberg. 3rd edn. pp. 2029–2044. Philadelphia: J. B. Lippincott.

56. Shils, M. E. and Coiso, D. (1979) Report to the

medical board on nutritional assessment of hospitalized adult patients in Memorial Hospital.

57. Spiegel, R. J., Schaefer, E. J., Magrath, I. T. and Edwards, B. K. (1982) Plasma lipid alterations in leukemia and lymphoma. *American Journal of Medicine,* **72**, 775–782.

58. Tchekmedyian, N. S., Halpert, C., Ashley, J. and Heber, D. (1992) Nutrition in advanced cancer: anorexia as an outcome variable and target of therapy. *Journal of Parenteral and Enteral Nutrition,* **16**, 88S–92S.

59. Tchekmedyian, N. S., Hickman, M., Siau, J., Greco, A., Keller, J., Browder, H. and Aisner, J. (1992) Megestrol acetate in cancer anorexia and weight loss. *Cancer,* **69**, 1268–1274.

60. Tchekmedyian, N. W., Tait, N., Moody, M. and Aisner, J. (1987) High-dose megestrol acetate. A possible treatment for cachexia. *Journal of the American Medical Association,* **257**, 1195–1198.

61. Torosian, M. H. (1992) Stimulation of tumour growth by nutritional support. *Journal of Parenteral and Enteral Nutrition,* **16**, 72S–75S.

62. Torti, F. M., Dieckmann, B., Beutler, B., Cerami, A. and Ringold, G. M. (1985) A macrophage factor inhibits adipocyte gene expression: an *in vitro* model of cachexia. *Science,* **229**, 867–869.

63. Vail, D. M., Ogilvie, G. K., Fettman, M. J. and Wheeler, S. L. (1990) Exacerbation of hyperlactatemia by infusion of lactated Ringer's solution in dogs with lymphoma. *Journal of Veterinary Internal Medicine,* **4**, 228–232.

64. Vail, D. M., Ogilvie, G. K., Wheeler, S. L., Fettman, M. J., Johnston, S. D. and Megstad, R. L. (1989) Alterations in carbohydrate metabolism in canine lymphoma. *Journal of Veterinary Internal Medicine,* **3**, 8–11.

65. Vlassara, H., Spiegel, R. J., Doval, D. S. and Cerami, A. (1986) Reduced plasma lipoprotein lipase activity in patients with malignancy-associated weight loss. *Hormone Metabolism Research,* **18**, 698–703.

66. Warren, R. S., Jeevanandam, M. and Brennan, M. F. (1987) Comparison of hepatic protein synthesis *in vivo* versus *in vitro* in the tumour-bearing rat. *Journal of Surgery Research,* **42**, 43–50.

67. Warren, R. S., Jeevanandam, M. and Brennan, M. F. (1985) Protein synthesis in the tumour-influenced hepatocyte. *Surgery,* **98**, 275–281.

68. Waterhouse, C., Jeanpetre, N. and Keilson, J. (1979) Gluconeogenesis from alanine in patients with progressive malignant disease. *Cancer Research,* **39**, 1968–1972.

CHAPTER 7

Dietary Fiber: Perspectives in Clinical Management

JOHN E. BAUER and IAN E. MASKELL

Introduction

As the importance of veterinary clinical nutrition develops, the need for clarity of understanding in primary nutritional areas will continue to grow. The topic of dietary fiber has burgeoned in recent years and although extensive research has raised more questions than answers, the subject has been substantially demystified. Furthermore, the intense interest of human nutritionists has generated sophisticated methodology geared specifically toward the needs of simple-stomached monogastrics. The subsequent evolution of clear definitions and practical recommendations now enables the veterinary clinician to assimilate dietary fiber into an informed strategy of dietary management.

In this chapter the subject is approached in three sections, a broad perspective on dietary fiber, an evaluation of its physiological effects and an assessment of its clinical applications.

Perspective on Dietary Fiber

The term dietary fiber is simply a convenient shorthand expression that covers a wide variety of entities. Indeed, it may be better to consider fiber as a concept rather than a single substance. It has been advocated recently that the term dietary fiber be dropped altogether from scientific discourse[11] and be replaced instead with specific reference to the material in question. For example, precise terms like wheat bran, isphagula husk, citrus, pectin and guar gum/psyllium are much more descriptive. This unambiguous approach is probably more applicable in veterinary than human nutrition because most canine and feline diets contain little dietary fiber and those materials that do add roughage are less diverse than in the human diet. This approach encourages recommendations for specific dietary sources rather than vague references to high fiber foods.

Definition of Fiber

Fiber is a simple term used throughout history to describe the most complex and least definable component of vegetative foods (i.e. of plant origin), encompassing a diverse group of plant polysaccharides, mucilages and phenolic compounds. Fiber can be defined in a number of ways:

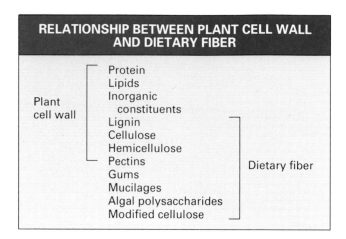

FIG. 7.1 The relationship between the plant cell wall and dietary fiber.

Botanically: the plant cell walls of parenchymous, lignified and cutinized tissues of vegetables, fruits and cereals together with the outer protective layer of seeds.[85]

Physiologically: the plant polysaccharides and lignin resistant to hydrolysis by digestive enzymes.[94]

Chemically: plant nonstarch polysaccharides (NSP) and lignin.[16]

The relationship between plant cell wall and dietary fiber as shown in Fig. 7.1.[6]

Indigestible Nature of Dietary Fiber

Comparison of starch and cellulose illustrates the fundamental difference between digestible and indigestible polysaccharides. Both are long chain polymers comprized entirely of glucose, the *only* difference in their chemical structure being the type of chemical link between each glucose unit; this is illustrated in Fig. 7.2. Mammals can produce the appropriate enzyme (amylase) to hydrolyze the α-1,4 linkages of starch but do not produce the enzyme(s) cellulase(s) required to break the β-1,4 linkages of cellulose. Hence, the two materials succumb to different fates in the small intestine and consequently exert very different physiological effects. All plant derived dietary fibers share the feature that the linkages that bind their structural units are resistant to degradation by mammalian digestive enzymes.

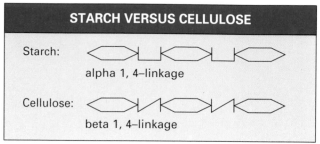

FIG. 7.2 Starch versus cellulose, similar but very different.

The main structural components of dietary fiber are six individual sugar moieties (rhamnose, arabinose, xylose, mannose, galactose, glucose), uronic acids and lignin. The permutations for the formation of matrices from these subunits are infinite: it is this feature that imbues dietary fiber with its truly diverse properties. Certain structural combinations are known to confer specific properties; for example, solubility is very much a function of highly branched macromolecular polysaccharides, whereas soluble-fiber viscosity is influenced by the location o side chains.

The above definitions view fiber from different angles, each satisfying the specific needs of certain individuals. The clinician, for example, may be primarily interested in physiological effects. The biochemist, however, requires an accurate definition based on chemical analyses. To comprehend the mechanisms that underlie the diverse effects of dietary fiber, an appreciation of both definitions is necessary. Furthermore, a botanical awareness allows a perspective on the diversity of dietary fiber sources.

Analysis of Dietary Fiber

Fiber can largely be defined by its method of analysis. Traditionally *gravimetric* methods were used in the analysis of animal feedstuffs, the fiber content being defined as the insoluble residue remaining after chemical solubilization of nonfiber constituents. As

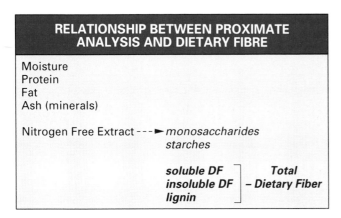

FIG. 7.3 Relationship between proximate analysis and dietary fiber.

scientific interest in "unavailable carbohydrates" has increased, more sophisticated analyses have been developed enabling accurate characterization of specific fractions of dietary fiber.

Proximate Analysis

All foods can be analyzed by *proximate* analysis to reveal macronutrient composition; that is, moisture, protein, fat, ash (minerals) and nitrogen free extract (NFE; i.e. "carbohydrate"). The regular procedure involves quantification of the moisture, protein, fat

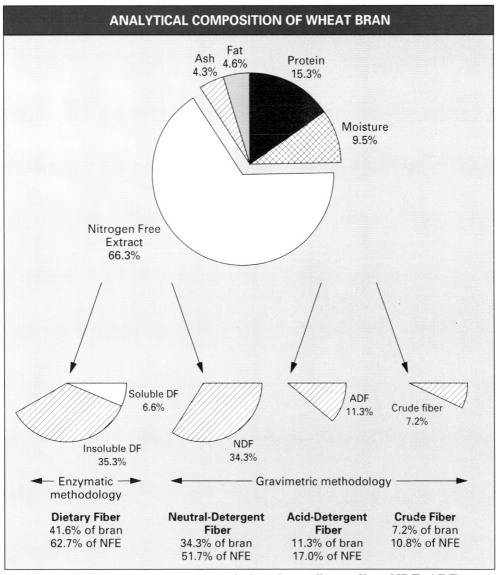

FIG. 7.4 Analytical composition of wheat bran: dietary fiber, NDF, ADF and crude fiber in NFE. The NFE fraction not accounted for by dietary fiber comprises mainly "available carbohydrate" and lignin. Data are from the analytical laboratory at Pedigree Petfoods Ltd, Melton Mowbray.

and ash fractions and then estimation of NFE "by difference." This technique accounts for all nutrients by relying on a "catch all" principle. Minerals are assumed to constitute the ash fraction and vitamins are presumed to be split between fat and moisture. NFE inevitably contains all the fractions not already identified, but it is normally assumed to comprise mainly carbohydrate and some lignin and tannins. The carbohydrate fraction contains "available" polysaccharides (e.g. starch), unavailable polysaccharides (e.g. cellulose) and some simple sugars (e.g. fructose). Figure 7.3 illustrates how dietary fiber fits into the proximate analysis. NFE is the residue unaccounted for by the other analyses and is normally considered as the "carbohydrate" fraction. Figure 7.4 shows the analytical composition of wheat bran and illustrates how the fiber fractions are characterized by the different analytical techniques. It is clear that NFE comprises the bulk of the material and that considerable discrepancies can be expected between the different methods of fiber analysis.

Gravimetric Analysis

Crude fiber (CF) is the residue remaining after *extraction* of a food material with dilute acid and alkali. Although CF is still widely quoted, it has been reported to underestimate unavailable carbohydrate (in human foods) by up to 40% or more in materials with high levels of soluble fiber.[6] CF is the figure most widely quoted on petfood labels, primarily to meet legislative requirements. It is the source of much confusion, has little clinical relevance and is largely obsolete in the field of human nutrition.

Acid-detergent fiber and neutral detergent fiber are modifications of the CF technique that yield more accurate estimations of the unavailable carbohydrate fraction.

Acid-detergent fiber (ADF) is determined using a detergent solution that removes nitrogenous material and, by omitting the alkali step, reduces loss of carbohydrate; losses of hemicellulose, however are still considerable.[6]

Neutral-detergent fiber (NDF) uses buffered sodium dodecyl sulfate and ethylenediamine tetraacetate (EDTA) which retains hemicelluloses, but *not pectins* which are extracted in EDTA.[97]

These methods do not measure botanically definable fractions, but rather the components insoluble in the reagents used and therefore, nutritionally, it is difficult to define what they actually mean. ADF and NDF are, however, used together in the analysis of animal feeds enabling hemicellulose estimation by difference (NDF minus ADF). Cellulose and lignin can be determined by acid hydrolysis of the ADF residue.[6]

Chemical Analysis

Rather than digesting the available fraction and weighing the residue, chemical techniques quantify the individual sugars that comprise the polysaccharides of dietary fiber. These methods also partition the dietary fiber into soluble and insoluble fractions, a reflection of their properties in an aqueous media.

Two chemical methods are commonly used today, AOAC (Association of Official Analytical Chemists)[74] and Englyst.[24] The Englyst method, for example, was primarily developed for the analysis of human food (essentially lignin-free) and allows accurate, separate determination by gas liquid chromatography (GLC) or colorimetric analysis of the constituent individual sugars, of NSPs, non-cellulose polysaccharides (NCP), cellulose and resistant starch (i.e. starch avoiding small intestinal degradation, which may enter the large intestine and hence can be considered as a component of dietary fiber).

Dietary Sources of Fiber

Different vegetative foods comprise very different levels of total dietary fiber and characteristic proportions of the soluble and insoluble fractions. A key observation is that most foods contain more insoluble than soluble

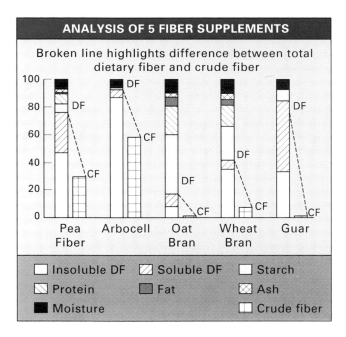

FIG. 7.5 Analysis of 5 fiber supplements, demonstrating the poor correlation between total dietary fiber and crude fiber. Data are from the analytical laboratory at Pedigree Petfoods Ltd, Melton Mowbray.

fiber; even those often quoted as soluble fibers (e.g. guar gum) contain substantial amounts of insoluble fiber. Fruits, oats, beans and lentils tend to contain more soluble fiber than cereals and some vegetables, although there are exceptions. Extensive analysis of the dietary fiber composition of 114 cereals and cereal products and 178 fruits, vegetables and nuts has been published by Englyst *et al.*[25,26] Figure 7.5 shows the analytical composition of pea fiber (purified pea hulls), arbocell (purified cellulose), oat bran, wheat bran and guar gum. These materials are of particular relevance in companion animal nutrition, either as common ingredients in commercial petfood or as dietary fiber supplements (see also Fig. 7.6). Of particular note is that CF substantially *underestimates* total dietary fiber in all the materials. Furthermore, the level of dietary fiber in these so-called "fiber supplements" varies from 17%, in oat bran, to over 90% in arbocell. The work of Fahey,[27,28] who formulated canine diets with different fiber supplements, further highlights the unreliability of CF in estimating actual dietary fiber content.

Fiber as a Nutrient

After ingestion dietary fiber is in a dynamic state, responding to the ever changing environment of the gastrointestinal tract. Although it is generally considered that fiber has no nutrient value, there are some exceptions. For example, the normally tightly bound fibrous matrix may unravel during its gastrointestinal journey, releasing nonfiber nutrients from the core of its structure. In addition, certain fibers are known to contribute to dietary minerals and energy.

Energy

The energy content of a food can be estimated from its composition of energy giving nutrients, that is protein, fat and carbohydrate. The principle assumes that carbohydrate (CHO), or rather the NFE fraction, can yield a fixed amount of energy per gram (i.e. 3.5 kcal(14.7 kJ)/g CHO is assumed for dogs). The assumption, however, breaks down if the NFE contains a proportion of dietary fiber. Furthermore it *cannot* be assumed that the fiber in the NFE contributes nothing to dietary energy. Although component sugars in fibrous polysaccharides cannot be digested and absorbed in the same way as available sugars, it is possible for them to yield available energy in the form of short chain fatty acids (SCFA), if fermented by bacteria. SCFA can be absorbed into the blood and provide energy to peripheral tissues; the process is about 15% less efficient than the production of ATP from glucose metabolism in mammalian tissue.[11] The contribution to digestible energy (DE) depends on the fermentability of the fiber source and type. Thus the amount of energy yielded by dietary fiber is a function of its origin and form of processing prior to ingestion. A highly lignified insoluble fiber, totally resistant to bacterial fermentation (e.g. peanut husks) will indeed yield no energy for the host animal. Conversely, a highly soluble fiber (e.g. pectin) entirely fermented in the colon can potentially yield nearly as much energy as starch digested in the small intestine. Figure 7.7 illustrates

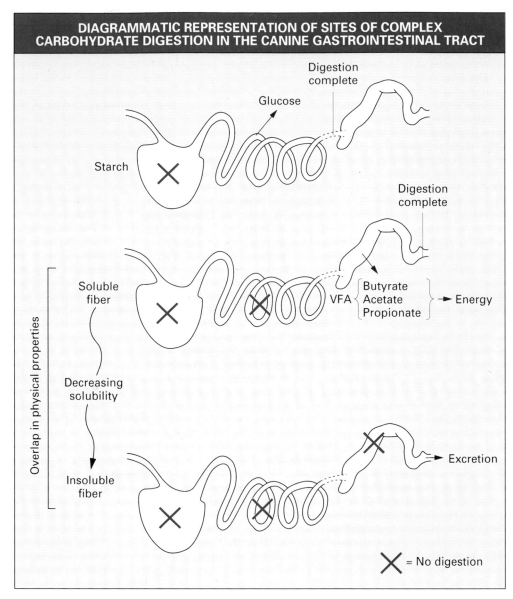

DIAGRAMMATIC REPRESENTATION OF SITES OF COMPLEX CARBOHYDRATE DIGESTION IN THE CANINE GASTROINTESTINAL TRACT

FIG. 7.6 Diagrammatic representation of sites of complex carbohydrate digestion in the canine gastrointestinal tract. VFA = volatile fatty acids.

the sites of complex carbohydrate degradation in the canine digestive tract.

Poorly fermentable fibers such as wheat bran and cellulose have been quoted as contributing about 1 and 0 kcal/g of DE, respectively.[75] The actual amount is probably dependent on the level of other fermentable dietary components. More soluble fibers, like guar gum and gum arabic have been estimated to contribute between 2.4 and 2.9 kcal/g DE.[70] Acetate and propionate are absorbed into the circulation and yield energy to peripheral tissues; however, most butyrate does

not complete its passage across the colonic mucosa but is used as a major energy source by the cells of this tissue.[80] A high butyrate supply is, therefore, thought to be important for colonocyte welfare and may be important in the management of inflammatory bowel conditions.

The effect of dietary fiber on energy intake is contentious and unresolved. More than 25 years ago it was suggested that refined carbohydrates were responsible for much of the obesity seen in highly developed societies.[15] This hypothesis was later modified, placing additional emphasis on a protective role of food fiber for colonic health.[34,93] This fiber–obesity hypothesis, however, is often misunderstood.[35] Merely increasing the intake of dietary fiber in the absence of concurrent calorie reduction will *not* reduce the risk of obesity. The essence of the theory is that foods from which dietary fiber has been removed have similar characteristics to those prone to inducing obesity.

Conversely, it remains to be established whether increasing dietary fiber will aid a weight reduction program. Only preliminary evidence supporting this latter concept has been presented.[32,35] The potentially satiating benefits of increasing dietary fiber, however, appear to be due more to its physical characteristics such as coarseness, hardness and/or textural qualities rather than its chemical properties. The end result may be longer chewing of smaller mouthsful and more rapid satiety.[35] A modest decrease in body weight has been observed in some human studies with wholemeal bread and cellulose[57] and with guar gum.[49] Comparison of these studies indicates that the soluble fiber was more effective at reducing hunger ratings and influencing carbohydrate and lipid metabolism but it appears that additional efforts must be made to reduce caloric intake in order to sustain any effective weight reduction program. A study with nonobese humans found that a moderate intake of dietary fiber (approximately 15 g/400 kcal), in the form of high fiber foods like wholemeal bread and pasta (rather than supplements), led to some suggestion of increased satiety in the first 3 hr following a meal.[13] It was also proposed that the primary impact was on energy intake 5–6 hr after consumption. This information tends to support the theory that dietary fiber has a physiological rather than a mechanical effect on satiety. Moderate levels of dietary fiber are therefore likely to be more effective than high levels that tend to reduce palatability. This is undesirable when intake is already restricted during a program of reduced energy consumption. Additional scientifically controlled studies are needed to understand

better the effects of dietary fiber in this regard.

Effects of Processing

The manufacturing processes to which dietary fiber is subjected will influence its properties. Whether in a canned or dry food or added as a supplement, most dietary fiber consumed will have been processed to some extent prior to ingestion. Soluble fibers hydrated prior to consumption (as would occur during the canning process) may be less physiologically active when compared to consumption of gums in a dehydrated form, which go on to form gels in the gut.

Purification or refining, by definition, isolates fiber from the rest of the food matrix, thereby reducing its potential to interact with other nutrients. Severe processing can physically destroy the macromolecular structure of dietary fiber, having a profound effect on its physiological properties. The gel-forming properties of a soluble fiber may be diminished, but the digestibility of an insoluble fiber may be increased because microbes can become more intimately associated with the smaller particles. Awareness of such modifications and their effect on the final product is thus necessary during formulation of products to meet particular needs.

Physiological Effects of Dietary Fiber

Chemical analyses, by whatever method, give only limited information about the physiological impact of a given fiber as it travels along the gastrointestinal tract (GIT). Despite attempts to categorize fiber into soluble and insoluble fractions, specific physiological effects cannot be ascribed simply to each of these chemical entities.

Moreover, although chemists can make a clear distinction between polysaccharides that are water soluble and those that are not, the GIT is far more discerning. The relative impact of these two fractions depends on a range of factors, primarily the dietary source and form of processing prior to ingestion.

Gastric Emptying, Transit Time and Digestibility

Dietary fiber affects gastric emptying, intestinal transit time, digestion and subsequent absorption. Insoluble fiber (cereal bran and cellulose) has variable effects on gastric emptying times (ranging from normal to more rapid) that appear to depend upon source, dose and the study methodology.[21,39,76] Soluble fiber (guar gum, carboxymethylcellulose) reportedly slows gastric emptying time due to its gel-forming capacity.[43,76] Whether soluble fibers behave in this fashion when mixed with digesta in the GIT is unclear. It seems reasonable to assume, however, that the extent to which a given material hydrates in the GIT and its subsequent viscosity will affect any net physiological effects. For example, conflicting reports on the efficacy of guar gum in reducing postprandial hyperglycemia may be partially explained by different rates at which specific guar preparations become hydrated.[61] Once suspended, some soluble fibers (e.g. psyllium and isphagula) are more viscous than others and are more effective in reducing small intestinal absorption than those exhibiting less viscosity (e.g. pectin). It is important not to oversimplify this concept as viscosity is affected by a range of environmental factors including concentration, temperature, ionic concentration, pH and particle size.[60] The physical state of soluble fiber alters as it travels along the GIT with much of the structure becoming degraded through fermentation in the colon.

Physiological effects of dietary fiber are also a function of the blend, type and physical state of the fiber. Transit time may be further altered via specific mechanical or biochemical interactions. Hence, fiber that decreases the rate of absorption or increases rate of transit of digesta through the small intestine may reduce nutrient availability. Conversely, if transit rate were decreased, availability may be unaffected. In the latter instance nutrients would be absorbed over a longer period due to a longer residence in the gut.

As an example, the addition of insoluble fiber to canine diets may affect nutrient digestibility by decreasing retention times in the GIT, hence affecting nutrient absorption. Most of the evidence in support of this phenomenon comes from studies in rats and humans.[11] In a veterinary setting where petfoods are analyzed and specified by either CF methods[6] or total dietary fiber (TDF) methods,[74] it is difficult to separate insoluble and soluble fiber. The use of the more precise terms discussed earlier may help to avoid confusion in the future.

It is in the above context that the recent studies of Fahey and coworkers merit comment. Two reports from this group described the effects of selected dietary fiber types using isonitrogenous diets.[27,28] One study[28] used diets with varying levels of different fiber supplements (Fig. 7.7) to produce a constant 12.5% dietary fiber in dry matter (approximately 12 g/400 kcal). The high lignin content of peanut hulls and tomato pumice is noteworthy because lignin elevates crude fiber measurement and tends to impair palatability.

Only small differences were observed in wet fecal weights and no differences were noted in frequency of defecation or mean retention time between the diets. They concluded that the diets were all utilized similarly, regardless of the diversity of fiber types.

In another study[27] beet pulp (BP) was included in diets over the range 0–12.5% (5.4–13.7% TDF in DM on analysis), and mean retention times decreased linearly with increased BP. DM and organic matter digestibilities decreased by about 6% through addition of BP, but did not decline significantly over the inclusion range. DE and metabolizable energy (ME) decreased linearly with BP inclusion when expressed as a percentage of gross energy (GE). Wet feces weight increased with BP, but frequency of defecation was only elevated significantly when 12.5% BP was fed. It was concluded that 7.5% BP (12% TDF in DM) is well tolerated by adult dogs without severe reductions in nutrient digestibility and energy utilization. Even at a level of 12.5% dietary BP (13.7% TDF in DM) only feces volume and frequency of defecation were increased; residence time in

FIBER SUPPLEMENT ANALYSIS			
	TDF (g/100g DM)	Lignin (g/100g DM)	Inclusion* (g/100g 'as is')
Peanut hulls	86	27	6.7
Tomato pumice	67	32	8.7
Beet pulp	77	3.3	7.5
Wheat bran	45	4.3	12.8
Treated wheat straw	93	7.6	6.2
* Inclusion of fiber supplement in diet to achieve 12.5% TDF in dry matter			

FIG. 7.7 Total dietary fiber (TDF) and lignin in five fiber supplements used by Fahey *et al.*[28] for formulation of canine diets. The final column shows how much of the supplement in its natural form needs to be added to the diet in order to give TDF content of 12.5% DM.

the gut decreased. This work supports the conjecture that insoluble fibers exert little physiological effect in the gut and that dogs can tolerate fairly high dietary levels. Further studies with soluble fibers would be of interest and would probably reveal greater effects at lower dietary inclusions. The studies also revealed that the digestibility of TDF was at least 30% in all the diets. The SCFA absorbed as a result of this process would be equivalent to about 6% of daily ME intake.

Fiber and Mineral Balance

Fiber–mineral interactions may also play a role in the physiological effects of dietary fiber. Zinc, calcium, iron and phosphorus are commonly bound within the polysaccharide matrix or adsorbed to the surface of dietary fiber.[90] Minerals may also be nutritionally unavailable as long as the fiber remains undegraded. In this case fiber will artificially increase the apparent mineral content of the diet. There is a converse situation where mineral binding occurs readily at neutral pH but is inhibited in an acidic environment.[43] Increased intake of negatively charged (i.e. ionized) polysaccharides may thus have implications on the availability of essential minerals as they can participate in electrostatic binding of minerals carrying a positive charge. Many of the ionic subunits of polysaccharides retain

their charge within the macromolecular structure. Specific electrostatic properties can also depend on the pH; for example, a polysaccharide may be unionized at gastric pH but ionized at the more alkaline pH of the small intestine. These new interactions may be quite different from those in either the food product or in the more proximal portions of the GIT thereby exerting a different physiological effect than that anticipated.

In spite of numerous studies of the effects of fiber-rich foods and isolated fibers on mineral balance in animals and man, the data are contradictory and depend on the duration of individual experiments and presence of other dietary components.[48] Generally, moderate amounts of dietary fiber do not appear to exert a significant effect on adult mineral balance. The potential for imbalances exists, however, in animals on marginal diets, and in young or very old animals. In these cases special attention should be paid to zinc, calcium and iron status.

Colonic Effects, Fermentation and SCFAs

SCFAs, principally acetate, propionate and butyrate, are the main end products of bacterial fermentation. They have three major effects: (1) they are absorbed from the colon and contribute to the energy balance of the host, (2) they acidify the colonic environment and (3) by virtue of their osmotic action they draw water into the stools and thereby increase stool bulking. The physiological implications of SCFA are the subject of much ongoing research aimed at unravelling the diversity of their likely effects.

Fermentation of dietary fiber in the colon results in alteration of colonic flora, changes in bile acid metabolites and production of varying amounts of SCFA, methane, carbon dioxide, hydrogen and water. Alteration of the pH and ammonia concentrations can result.[17] SCFA, butyrate in particular, contribute significantly to colonocyte nutrition[47,80,84] and propionate is metabolized by hepatocytes[86] although the significance of this is not known precisely. Future studies are required to elucidate the metabolic and physiological consequences of the SCFA.

The SCFA butyrate, valerate and octanoate are increased in blood and cerebrospinal fluid (CSF) of patients with hepatic encephalopathy[9] and may participate as synergists in coma production. Therefore fermentable fiber may be contraindicated in liver disease. In spite of this possibility, fiber in the form of lactulose, which is fermentable, has become an important part of the treatment of hepatic encephalopathy. The rationale for its use is that, on fermentation, it may alter pH thereby minimizing ammonia uptake, regulating bacteria numbers and acting as an osmotic laxative. Presumably, the fermentation of lactulose does not contribute significantly to SCFA pools in blood or CSF to any appreciable extent. Alternatively, SCFA may be utilized by colonocytes under this condition prior to portal uptake. Furthermore, studies using vegetable protein diets have also been effective in the management of human liver disease patients.[44,95]

GI Morphology and Function: Implications for Colon Cancer

Numerous studies have suggested a relationship between dietary fiber and the morphological development of the small intestine that appears to be related to altered DNA, RNA and protein content of the mucosa and to specific enzyme activities associated with the mucosal cell and brush border.[96] Such functional alterations, however, do not always occur and species differences exist.[79]

In adult rats pectin supplements have been shown to increase intestinal length and weight.[12] Diets containing alfalfa fiber have been found to enhance growth of the rat small intestine.[101] These effects were more notable when mixed animal and vegetable fiber sources were used rather than semipurified diets containing oats or cellulose. Insoluble fibers (i.e. alfalfa, cellulose and bran:pectin 3:1) caused a slight shortening of small intestinal length of rats. Pectin and psyllium (soluble) increased length while guar (also soluble) reduced it.[96]

Ultrastructural studies have shown that dietary fibers such as bran, cellulose, pectin and alfalfa all can produce morphological alterations.[37] Fiber-induced alterations probably involve proliferation, migration, differentiation and eventual exfoliation of epithelial cells. The significance of these effects is that they may relate to mechanisms of colon carcinogenesis by hyperplasia and increased cell proliferation.[36] Conflicting results (e.g. no effect, protection, enhancement) on the role of dietary fiber in colon carcinogenesis may relate to the lack of standard protocols employed.[46] Studies differ in the species and strain of animals, basal diet, carcinogen type, amount and properties of fibers used, and so forth.

Dietary fiber may play a role in reducing the concentration of fecal mutagens or their formation via altered bacterial metabolism, by diminishing their contact time through rapid colonic transit or by decreasing colonic pH.[30,47,81] Dietary fibers are protective, dietary fats are promotive.[46] Fat is thought to promote colon cancer by stimulating the bile acid pool. Enteric bacteria may then convert these bile acids to secondary bile acids that have been shown to be potent tumor promoters in animal studies.[100] The relationship between dietary fats and fibers is, however, more complex than initially believed due to the fiber–fat interactions. Work in rats using colonic cell proliferation as a risk factor for colon carcinogenesis has found that the fiber effect also depends on dietary fat type and that it is site-specific.[47] Indeed insoluble fiber (6% cellulose vs. 6% pectin) fed in conjunction with beef tallow promoted proximal colon health as evidenced by colonic cell proliferation, while distal colon health was better with fish oil irrespective of fiber types.[50] The combination of soluble fiber (i.e. pectin) with corn oil was least desirable.[50] Functional changes involving increased arachidonic acid and prostaglandin production have been observed in colon tumor cells.[58,62] Prostaglandin inhibition also seems to reduce tumor incidence experimentally.[64,65] Recent work with fish oils has found that prostacyclin and prostaglandin production are both decreased independent of fiber, but that beef tallow with pectin and corn oil with cellulose increased these eicosanoids.[50] Finally, although a negative correlation of prostacyclin production and cell proliferation was found in the proximal colon, a positive correlation was seen in the distal colon (the site of most colon tumors). These data indicate that dietary fiber and fat may affect colon carcinogenesis in an interactive and site-specific manner. For the present, however, there is little firm evidence to support the hypothesis that the development of colorectal cancer can be linked to a diet low in dietary fiber. The limited evidence does however suggest that dietary fiber, particularly of vegetable origin, may have a protective effect.[11] Additional studies with dogs are necessary to answer similar questions in veterinary medicine. For the present, however, the relationship between fiber and cancer remains unclear.

Lipid, Bile Acid and Endocrine Effects

Studies of the effects of fiber on lipid and bile acid metabolism suggest that soluble fiber (oat bran, pectin, legumes and several gums) can help decrease serum cholesterol concentrations in man and laboratory animals.[5,7,14] Wheat bran, cellulose and soy hulls, however, do not have a hypocholesterolemic effect.[83] Hypocholesterolemia may be due to bile acid binding and decreased enterohepatic recirculation with subsequent increases in the demand for hepatic cholesterol, a precursor of hepatic bile acid synthesis. Low-density lipoprotein receptors are also up-regulated in this process. Thus, increased lipoprotein cholesterol is taken from the plasma compartment into hepatocytes, resulting in a serum cholesterol lowering effect.[29] The hypocholesterolemic effect of pectin in cholesterol-fed rats is probably due to increased bile acid excretion.[53] Low fat diets also have a similar cholesterol-lowering effect.[7] Decreased fat absorption in the presence of a high fiber diet may also partially explain this effect. While it is possible to induce hypercholesterolemia and lipoprotein alterations in hypothyroid dogs by fat and cholesterol feeding, athero-genic lesions are observed only when serum cholesterol concentrations exceed 750 mg/dl (20 mM).[55] Whether serum cholesterol concentrations in the 400–750 mg/dl range pose any health risk in dogs (or cats) however, has not been determined and awaits further study. Therefore, the use of dietary fiber for cholesterol lowering purposes in these species is arguable. This phenomenon, however, does not necessarily preclude the use of dietary fiber for other hyperlipidemias such as diabetes mellitus.

Plasma levels of triglyceride have been unchanged in most reports of dietary fiber and hyperlipidemia. Some studies have observed a reduction especially in hyperlipidemic or diabetic patients when leguminous seed fiber sources were fed.[4,40,78] Of greater interest is the possibility that soluble fiber may affect postprandial serum triglyceride concentrations. This may occur by altering the absorption site and/or resultant chylomicron composition and subsequent utilization.[2] Consequently, slower delivery of triglyceride fatty acids to tissues would occur. This effect may, therefore, be of benefit in diets for diabetic animals.

Other reasons for recommending soluble fiber containing diets for animals with diabetes mellitus include alterations of the endocrine response.[2,66] Experiments in dogs, rats and humans have shown that viscous polysaccharides can modify glucose absorption,[23] enhance insulin receptor binding[72,98] and improve glycemic response.[87] High fiber diets cause changes in gut hormones that may contribute to this effect by lowering concentrations of gastric inhibitory peptide[56] or glucagon.[59,63]

Fiber and Pancreatic Activity

Fiber appears to impair the activity of the exocrine pancreas. In humans with pancreatic insufficiency, ingestion of a high fiber diet containing beans, wheat bran and broccoli caused significant increases in fecal weight, fecal fat excretion and abdominal flatulence.[20] *In vitro* studies also demonstrated impairment of enzyme activity and the possibility of a net impairment of pancreatic function.[20] Other effects relating to the gel-forming properties of soluble fibers may cause adsorption of pancreatic enzymes that render them less available for digestion.[62] High fiber diets are therefore contraindicated in pancreatic insufficiency.

In cats a strong association exists between pancreatic insular amyloid deposits and diabetes.[71] High fiber diets containing cellulose have been used to effect glycemic control and aid weight loss in obese diabetic cats. Constipation is a possible complication and additional supplementation of the diet with a sugar-free fiber supplement has been recommended[67] to combat this. The value of dietary fiber and starch in the management of canine and feline diabetes is discussed in Chapter 21. It should be noted, however, that cats and dogs cannot be assumed to respond in the same way to such diets, considering the cat's inefficiency in coping with dietary polysaccharides.[45]

Indications for Dietary Fiber in Clinical Medicine

Few scientific investigations have been specifically designed to study effects of dietary fiber in dogs and cats with gastrointestinal disease. Recommendations are therefore based on the clinical experience of a number of authors and pathophysiological projections from experimental work. As mentioned, fiber has a variety of influences on the bowel and the type and nature of fibers employed can modify the net effect to varying extents.

With respect to supplementing a complete and balanced diet with fiber, the following precautions must be borne in mind. One must first assume that the fiber contributes no energy to the ration. If, however, the fiber contributes to daily energy in a significant way through fermentation, at some point nutrient balance might become impaired. Thus if supplements are used in this way, it should be ensured that the basal diet is not marginal especially where ionic materials are concerned. Where mineral status of a fiber-supplemented diet is in doubt, a mineral supplement should be considered, with emphasis on the cations iron, calcium and zinc. Some evidence also suggests that vitamin B_{12} absorption may be impaired in the presence of excess dietary fiber.[42]

Constipation

One benefit of highly insoluble fiber (e.g. coarse wheat bran) diets is the relief of constipation. This is thought to be the result of increased fecal bulk that increases colonic motility by stretching colonic muscles resulting in a more forceful, albeit less frequent, contraction. It has been suggested that polysaccharides that resist breakdown (i.e. insoluble) increase fecal weight, as shown in dogs by Fahey[27] (discussed earlier), while those that are fermented (i.e. soluble) increase transit times. Thus fibers that combine soluble and insoluble properties may be optimal for correcting constipation. One precaution with the use of fiber for this purpose over the long term is that the hydration status of the patient must be monitored. Humans consuming 30 g of a dietary fiber supplement per day are encouraged to drink at least 2 liters of water per day. An initial supplemental dose for dogs would be 2–4 g wheat bran, oat bran or isphagula per 100 kcal ME.

Gastric Disease: Inflammatory or Ulcerative

A so-called bland diet (moderate protein and fat, high digestibility) is recommended for gastric disorders, usually in frequent, small feedings. Gastric acid secretions may be reduced by liquefied diets and/or moderate protein restriction due to decreased retention times in the stomach.[31] In the case of gastric dumping disorders (i.e. unusually rapid gastric transit), it has been speculated that diets high in soluble fiber may be beneficial due to gel formation in the stomach, although no specific dosages or fiber types have been mentioned.[31] If so, careful supplementation of a quality, adult maintenance petfood may be an adjunct to therapy as long as dietary imbalances are minimized.

Small Intestinal Diarrhea

Small intestinal diarrheas are usually managed on a short-term basis with soluble fiber preparations (i.e. kaolin–pectin). The rationale for their use appears to be based on the adsorbent properties of kaolin and the gelling properties of pectin. If malabsorption is part of the problem, however, long-term use may impair nutrient assimilation and mineral bioavailability and is not recommended.

Chronic small bowel diarrheas are best managed with highly digestible diets low in fiber content. The possibility of some potential benefit, however, by virtue of the binding and gelling properties of soluble fiber has been raized[31] but few reports to date have described any specific recommendations. Supplementation of bland diets with psyllium or oatmeal sources may be a useful starting point in an effort to examine the use of gel-forming fibers. Interference with digestion and absorption, however, may present a major contraindication for their use in small bowel diarrhea.

Finally, with regard to lymphangiectasia there is no evidence that an enriched fiber diet will help. It is unlikely that studies reported using this dietary modification showed improvement due to the low fat content of the diet used rather than to any direct effect of fiber.[77]

Large Bowel Diarrhea

Generally, adding a poorly digestible fiber (containing both soluble and insoluble forms) to the diet can be beneficial in the symptomatic treatment of diseases of the large bowel. Fiber helps to normalize transit time and fecal water content. Generally 15–30 g of a fiber supplement (e.g. isphagula or oat bran) per 400 g of diet is recommended. Therapy of large bowel diarrhea using dietary fiber has been evaluated to some extent in dogs,[51,52] but more studies are needed. Fiber can be a significant nutrient in the large bowel by virtue of its partial fermentation to SCFA. Specifically, butyrate is metabolized by the colonic epithelium[81] and a reasonable supply should therefore be beneficial for colonocyte

health. Roediger has speculated that a low intake of complex carbohydrates may result in "butyrate starvation" in the colonic mucosa and that this may be a factor in the etiology or maintenance of inflammatory bowel disease.[76] Furthermore, mixed bacterial communities exhibit "colonization resistance," that is, they can protect their ecosystem from invading organisms and as such may help prevent bacterial overgrowth.

One study in dogs with idiopathic large bowel diarrhea observed improvement of seven of eight animals using a bland diet with fiber supplementation (1–2 tbsp sid).[52] Some clinicians anecdotally report that psyllium fiber supplements (31.5% soluble DF) are effective while others prefer wheat bran[31,99] (6.6% soluble DF). It seems that partially fermentable fibers that retain their structural integrity exhibit the most useful properties in the management of large bowel conditions. Finally, it should be noted that some controversy over the use of fiber in large bowel diarrhea exists because low residue commercial and home-prepared diets have been successfully used as sole therapy.[54,88] Such diets may be especially useful during the acute phase of some large bowel diarrheas. Once the initial phase has been managed, fiber probably has a prophylactic role in helping to create and maintain an environment less susceptible to remission via its effect on normalization of transit times and gastrointestinal health.

Although dietary protein need not be reduced, moderate protein content should suffice unless protein-losing enteropathies are present. This reduces the potential for colonic ammonia generation that might exacerbate gastrointestinal or metabolic sequelae. Serum albumin should be assessed regularly and protein content adjusted accordingly.

Potential Adverse Effects

Gastrointestinal obstruction in association with gel-forming fiber supplements and wheat bran has been reported in humans.[8,41] This problem appears to be related to inadequate water intake especially in the case of soluble

TYPICAL INGREDIENTS CONTRIBUTING TO THE FIBER FRACTION OF PETFOODS

i) Vegetative ingredients

Materials of vegetative origin that are included primarily as a source of carbohydrate or protein, or as a gelling agent but which also contain dietary fiber and therefore contribute to the final level in the diet. Although they provide less DF than supplements, the DF they do provide is more intimately associated with the nutrient matrix and therefore potentially has a greater physiological impact.

ii) Fiber supplements

Materials added with the specific intention of increasing the dietary fiber increment of the diet. They tend to be by-products of other food manufacturing processes, and as such are purified or fiber enriched. In general they contain relatively high levels of insoluble DF. Fiber supplements are less intimately associated with the food matrix, and as such may have less physiological impact.

Presented below are some common examples of materials from both categories, showing the amount of DF and the percentage of that DF that is insoluble.

Vegetative ingredients	%DF	%insol.	Fiber supplements	%DF	%insol.
Rice	0.4–1	99	Powdered cellulose	90–95	95
Barley	10–14	70	Peanut hulls	70–77	70
Wheat	9–10	80	Pea fiber	70–75	65
Maize meal	3–5	84			
Oat meal	6–7	41			
Wheat bran	35–38	90			
Vegetable polysaccharides (gels)	75–85	39			

FIG. 7.8 Ingredients contributing to the dietary fiber fraction of petfoods.

fiber supplements. Also the presence of phytates in high fiber foods may reduce mineral bioavailability or shift their absorption site.[38,69] This phenomenon may be important in geriatric nutrition or if marginal mineral status is suspected and, although uncommon, should be kept in mind.[10,46] Colonic volvulus and "allergic" reactions have also been noted in humans[82,92] but the extent to which similar problems may occur in dogs is unknown.

Conclusions

Dietary fiber is integral to clinical nutrition. The subject has previously been nonspecific in the veterinary field but is set to play an important role in dietary management. Human nutritionists have had a long-running interest in the subject, by virtue of the naturally diverse human diet and the interest in diseases associated with "refined Western diets." By concentrating on the physical properties of specific fiber types, such as solubility, viscosity or particle size, and with some experience of practical application, different dietary fiber supplements should become a useful adjunct to the management of a range of veterinary clinical conditions.

At present evidence exists for the effects of some dietary fiber types on colonic fermentation and on glucose absorption and metabolism, hence the clear value of fiber in the management of large bowel diseases, constipation and diabetes (see Chapter 21). Further specific recommendations for the use of dietary fiber in the prevention or therapy of chronic bowel syndromes, gastrointestinal tumors and other disorders awaits investigation. It should be noted that all dietary fibers exert their effects as an integral part of a mixed diet, in which they are associated with other dietary constituents. It is important to remember that dietary fiber supplements will not behave in the same way as fiber-rich diets, but that each may have virtues under different circumstances.

References

1. Allbrink, M. J. and Ullrich, I. H. (1984) In *Dietary Fiber, Basics and Clinical Aspects*. Eds. G. V. Vahouny and D. Kritchevsky. p. 324. New York: Plenum.
2. Anderson, J. W. (1980) Dietary fiber and diabetes. In *Medical Aspects of Dietary Fiber*. Eds. G. A. Spiller and R. M. Kay. pp. 193–221. New York: Plenum.
3. Anderson, J. W. (1986) Fibre and health: an overview. *American Journal of Gastroenterology*, **81**, 892–897.
4. Anderson, J. W. and Ward, K. (1979) High carbohydrate, high fibre diets for insulin-treated men with diabetes mellitus. *American Journal of Clinical Nutrition*, **32**, 2312–2321.
5. Anderson, J. W., Chen, W. L. and Sieling, B. (1980) Hypolipemic effect of high carbohydrate, high fibre diets. *Metabolism*, **29**, 551–558.
6. Asp, N.-G. and Johansson, C. G. (1984) Dietary fibre analysis. *Nutrition Abstracts and Reviews*, **54**, 735.
7. Bauer, J. E. (1990) In *Proceedings of the Eighth Annual Veterinary Forum, ACVIM*. pp. 354–356. Madison, WI: Omnipress.
8. Berman, J. I. and Schultz, M. J. (1980) Bulk laxative ileus. *Journal of the American Geriatric Society*, **28**, 224–226.
9. Bernardini, P. and Fischer, J. E. (1982) Amino acid imbalance and hepatic encephalopathy. *Annual Review of Nutrition*, Vol. 2. pp. 419–454. Palo Alto, CA: Annual Reviews.
10. Bright–See, E. and McKeown–Eyssen, G. (1984) Estimation of per capita crude and dietary fibre supply in 38 countries. *American Journal of Clinical Nutrition*, **39**, 821–829.
11. British Nutrition Foundation (1990) *Complex Carbohydrates in Foods: The Report of the BNF Task Force*. London: Chapman and Hall for BNF.
12. Brown, R. C., Kelleher, J. and Loslwsky, M. A. (1979) The effect of pectin on the structure and function of the rat small intestine. *British Journal of Nutrition*, **43**, 357–365.
13. Burley, V. J. and Blundell, J. E. (1990) Time course of the effects of dietary fibre on energy intake and satiety. In *Dietary Fibre: Chemical and Biological Aspects*. Eds. D. A. T. Southgate, K. Waldron,

I. T. Johnson and G. R. Fenwick. pp. 91–102. Royal Society of Chemistry, Special Publication No. 83.
14. Chen, N. L., Anderson, J. W. and Jennings, D. (1982) Propionate may mediate the hypocholesterolemic effects of certain plant fibers in cholesterol-fed rats (41791). *Proceedings of the Society of Experimental Biology and Medicine*, **175**, 215–218.
15. Cleave, T. L. (1956) The neglect of natural principles in current medical practice. *Journal of Royal Navy Medical Service*, **42**, 55.
16. Cummings, J. H. (1981) Dietary fibre. *British Medical Bulletin*, **37**, 65–70.
17. Cummings, J. H. and Branch, W. J. (1984) In *Dietary Fiber, Basic and Clinical Aspects*. Eds. C. V. Vahouny and D. Kritchevsky. pp. 131–149. New York: Plenum.
18. Dimski, D. S. (1991) Dietary fibre in gastrointestinal diseases. In *Proceedings of the Ninth Annual Veterinary Forum, ACVIM*. pp. 633–635. Madison, WI: Omnipress.
19. Dimski, D. S. and Buffington, C. A. (1991) Dietary fiber in small animal therapeutics. *Journal of the American Veterinary Medical Association*, **199**, 1142–1146.
20. Dutta, S. K. and Hlasko, J. (1985) Dietary fibre in pancreatic disease: effect of high fibre diet on fat malabsorption in pancreatic insufficiency and *in vitro* study of the interaction of dietary fibre with pancreatic enzymes. *American Journal of Clinical Nutrition*, **41**, 517–525.
21. Eastwood, M. A. and Brydon, W. A. (1985) Physiologic effects of dietary fibre on the alimentary tract. In *Dietary Fibre, Fibre-Depleted Foods and Disease*. Eds. H. Trowell and K. Heaton. pp. 105–132. London: Academic Press.
22. Eastwood, M. A., Brydon, W. G. and Tadesse, K. (1980) Effect of fiber on colon function. In *Medical Aspects of Dietary Fiber*. Eds. G. A. Spiller and R. M. Kay. pp. 1–26. New York: Plenum.
23. Edwards, C. A. and Read, N. W. (1989) Fibre and small intestine function. In *Dietary Fibre Perspectives 2–Reviews and Bibliography*. Ed. A. Leeds. London: John Libbey.
24. Englyst, H. N. and Cummings, J. H. (1984) Simplified method for the measurement of total non-starch polysaccharides by gas–liquid chromatography of constituent sugars as alditol acetates. *Analyst*, **109**, 938–942.
25. Englyst, H. N., Bingham, S. A., Runswick, Collinson, E. and Cummings, J. H. (1988) Dietary fibre (non-starch polysaccharides) in fruit, vegetables and nuts. *Journal of Human Nutrition and Dietetics*, **1**, 247–286.
26. Englyst, H. N., Bingham, S. A., Runswick, Collinson, E. and Cummings, J. H. (1989) Dietary

fibre (non-starch polysaccharides) in cereal products. *Journal of Human Nutrition and Dietetics*, **2**, 253–271.

27. Fahey, G. C., Jr., Merchen, N. R., Corbin, J. E., Hamilton, A. K., Serbe, K. A., Lewis, S. M. and Hirakawa, D. A. (1990) Dietary fibre for dogs: I. Effects of graded levels of dietary beet pulp on nutrient intake, digestibility, metabolisable energy and digesta mean retention time. *Journal of Animal Science*, **68**, 4229.

28. Fahey, G. C., Jr, Merchen, N. R., Corbin, J. E., Hamilton, A. K., Serbe, K. A. and Hirakawa, D. A. (1990) Dietary fibre for dogs: II. Iso-total dietary fibre (TDF) additions of divergent fibre sources to dog diets and their effects on nutrient intake, digestibility, metabolisable energy and digesta mean retention time. *Journal of Animal Science*, **68**, 4229–4235.

29. Goldstein, J. L. and Brown, M. S. (1982) LDL receptor defect in familial hypercholesterolemia. *Medical Clinics of North America*, **66**, 335.

30. Greenwald, P., Lanza, E. and Eddy, G. A. (1978) Dietary fibre in the reduction of colon cancer risk. *Journal of the American Dieticians Association*, **87**, 1178–1188.

31. Guilford, W. G. (1992) Nutritional management of gastrointestinal tract diseases. In *Proceedings of the Tenth Annual Veterinary Forum, ACVIM*. pp. 66–69. Madison, WI: Omnipress.

32. Hand, M. S. (1988) Effects of low fat/high-fiber in the dietary management of obesity. In *Proceedings of the Sixth Annual Veterinary Forum, ACVIM*. pp. 702–703. Madison, WI: Omnipress.

33. Harland, B. F. and Morris, E. R. (1985) *In Dietary Fibre Perspectives*. Ed. A. R. Leeds. London: John Libby.

34. Heaton, K. W. (1973) Food fibre as an obstacle to energy intake. *Lancet*, **2**, 1418.

35. Heaton, K. W. (1980) Food intake regulation and fiber. In *Medical Aspects of Dietary Fiber*. Eds. G. A. Spiller, and R. M. Kay. pp. 223–238. New York: Plenum.

36. Jacobs, L. R. (1983) Effects of dietary fibre on mucosal growth and cell proliferation in the small intestine of the rat: a comparison of oat bran, pectin and guar with total fibre. *American Journal of Clinical Nutrition*, **37**, 954–960.

37. Jacobs, L. R. (1986) Gastrointestinal epithelial cell proliferation. In *Dietary Fiber: Basic and Clinical Aspects*. Eds. G. V. Vahouny and D. Kritchevsky. pp. 211–228. New York: Plenum.

38. James, W. P. T. (1984) Dietary fiber and mineral absorption. In *Medical Aspects of Dietary Fiber*. Ed. G. A. Spiller and R. M. Kay. pp. 239–259. New York: Plenum.

39. Jenkins, D. J. A., Wolever, T. M. S., Leeds, A. R., Gassuel, M. A., Haisman, D. V., Dilawari, G., Metz, G. L. and Alberti, K. G. M. (1978) Dietary fibres, fibre analogs and glucose tolerance, importance of viscosity. *British Medical Journal*, **1**, 1392.

40. Jenkins, D. J. A., Wong, G. S., Patten, R., Bird, J., Hall, M., Buckley, G. C., McGuire, V., Reichart, R. and Little, J. A. (1983) Leguminous seeds in the dietary management of hyperlipidaemia. *American Journal of Clinical Nutrition*, **38**, 567–573.

41. Kang, J. Y. and Doe, W. F. (1979) Unprocessed bran causing intestinal obstruction. *British Medical Journal*, **1**, 1249.

42. Kasper, H. (1986) Effects of dietary fiber on vitamin metabolism. In *Handbook of Dietary Fiber in Human Nutrition*. Ed. G. A. Spiller. pp. 201–208. Baton Rouge, LA: CRC Press.

43. Kelsay, J. L. (1986) Update on fiber and mineral availability. In *Dietary Fiber, Basic and Clinical Aspects*. Eds. G. V. Vahouney and D. Kritchevsky. pp. 361–372. New York: Plenum.

44. Keshauassian, A., Meck, J., Sutton, C., Emery, V. M., Hughes, E. A. and Hodgsen, H. J. F (1984) Dietary protein supplementation from vegetable protein sources in the management of chronic portal systemic encephalopathy. *American Journal of Gastroenterology*, **79**, 945–955.

45. Kienzle, E. (1988) Investigations on intestinal and intermediary metabolism of carbohydrates (starch of different origin and treatment, mono- and disaccharide) in the domestic cat (*Felis catus*). Hannover, Tierarztliche Hochschule, Habilschrift, 1989.

46. Klurfeld, D. M. (1987) The role of dietary fiber in gastrointestinal disease. *Journal of the American Dieticians Association*, **87**, 1172.

47. Kripke, S. A., Fox, A. D., Berman, J. M., Settle, R. G. and Rombeau, J. L. (1987) Stimulation of mucosal growth with intracolonic butyrate infusion. *Surgery Forum*, **38**, 47–49.

48. Kritchevsky, D. (1988) Dietary fiber. *Annual Review of Nutrition*, **8**, 301–328.

49. Krotkiewski, M. (1984) Effect of guar gum on body weight, hunger ratings and metabolism in obese subjects. *British Journal of Nutrition*, **52**, 97–105.

50. Lee, D. Y. K. (1992) Dietary modulation of biomarkers of colon carcinogenesis: interactive effect of different types of fiber and fat. PhD Dissertation, Texas A&M University.

51. Leib, M. S. (1988) Dietary management of inflammatory bowel disease. In *Proceedings of the Sixth Annual Veterinary Forum, ACVIM*. pp. 711–712. Madison, WI: Omnipress.

52. Leib, M. S. (1990) Fiber-responsive large bowel diarrhea. In *Proceedings of the Eighth Annual Veterinary Forum, ACVIM*. pp. 817–819. Madison, WI: Omnipress.

53. Levielle, G. A. and Sauberlich, H. E. (1966) Mechanisms of the cholesterol depressing effect of pectin in the cholesterol-fed rat. *Journal of Nutrition,* **88**, 209–214.

54. Lewis, L. D., Morris, M. L. Jr. and Hand, M. S. (1987) Gastrointestinal, pancreatic and hepatic diseases. In *Small Animal Clinical Nutrition.* 3rd edn. pp. 1–65. Topeka, KS: Mark Morris.

55. Mahley, R. W., Weisgraber, K. H. and Innerarity, T. (1974) Canine lipoproteins and atherosclerosis. II. Characterization of the plasma lipoproteins associated with atherogenic and nonatherogenic hyperlipidemia. *Circulation Research,* **35**, 722–733.

56. Marks, V. and Turner, D. S. (1977) The gastro-intestinal hormones with particular reference to their role in the regulation of insulin secretion. *American Journal of Clinical Nutrition,* **20**, 462–474.

57. Mickelsen, O., Makdani, D. D., Cotton, R. H., Titcomb S. T., Colmey, J. C. and Gatty, R. (1979) Effects of a high fibre bread diet on weight loss in college-age males. *American Journal of Clinical Nutrition,* **32**, 1703–1709.

58. Minoura, T., Takata, T., Sakaguchi, M., Takata, H., Yamamura, M., Hioki, K. and Yamammoto, M. (1988) Effect of dietary eicosapentaenoic acid on azoxymethane-induced colon carcinogenesis in rats. *Cancer Research,* **48**, 4790–4794.

59. Miranda, P. M. and Horwitz, D. L. (1978) High fibre diets in the treatment of diabetes mellitus. *Annals of Internal Medicine,* **88**, 482–486.

60. Morris, E. R. (1986) Molecular origin of hydro-colloid functionality. In *Gums and Stabilisers for the Food Industry,* Vol. 3. Eds. G. O. Philips, D. J. Wedlock and P. A. Williams. pp. 3–16. London: Elsevier.

61. Morris, E. R. (1990) Physical properties of dietary fibre in relation to biological function. In *Dietary Fibre: Chemical and Biological Aspects,* Eds. D. A. T. Southgate, K. Waldron, I. T. Johnson and G. R. Fenwick. pp. 91–102. Royal Society of Chemistry, Special Publication No. 83.

62. Moser, E. (1989) Fibre types and their physiologic effects. In *Proceedings of the Seventh Annual Veterinary Forum, ACVIM.* pp. 342–345. Madison, WI: Omnipress.

63. Munoz, J. M., Sanstead, H. H., Jacob, R. A., Logan, G. M., Reck, S. J., Klevay, L. M., Dinitzis, F. R., Inglett, G. F. and Shuey, W. C. (1979) Effects of some cereal brans and TVP on plasma lipids. *American Journal of Clinical Nutrition,* **32**, 580–592.

64. Narisawa, T., Sato, M., Tani, M., Kudi, T., Takahashi, T. and Goto, A. (1981) Inhibition of development of methyl nitrosurea-induced rat colon tumors by indomethacin treatment. *Cancer Research,* **41**, 1954–1957.

65. Narisawa, T., Takahashi, M., Niwa, M., Fukaura and Wakizaka, A. (1987) Involvement of prosta-glandin E in bile acid-caused promotion of colon carcinogenesis and antipromotion by cyclooxygenase inhibitor indomethacin. *Japanese Journal of Cancer Research,* **78**, 791.

66. Nelson, R. W. (1988) Dietary therapy for canine diabetes mellitus. In *Proceedings of the Sixth Annual Veterinary Forum, ACVIM.* pp. 54–56. Madison, WI: Omnipress.

67. Nelson, R. W. and Feldman, E. C. (1992) Noninsulin-dependent diabetes mellitus in the cat. In *Proceedings of the Tenth Annual Veterinary Forum, ACVIM.* pp. 351–353. Madison, WI: Omnipress.

68. Nicholson, M. L., Neoptolemes, J. P., Clayton, H. A., Talbot, I. C. and Bell, P. R. F. (1991) Increased cell membrane arachidonic acid in experi-mental colorectal tumors. *Gut,* **32**, 413–418.

69. Nicklin, S. and Miller, K. (1984) Effect of orally administered food grade carrageenans on antibody mediated and cell mediated immunity in the inbred rat. *Food Chemistry and Toxicology,* **22**, 615–621.

70. Nyman, M., Asp, N.-G., Cummings, J. H. and Wiggins, H. (1986) Fermentation of dietary fibre in the intestinal tract: comparison between man and rat. *British Journal of Nutrition,* **55**, 487–496.

71. O'Brien, T. D., Hayden, D. W., Johnson, K. H. and Fletcher, T. F. (1986) Immunohistochemical morphometry of pancreatic endocrine cells in diabetic, normoglycaemic glucose intolerant and normal cats. *Journal of Comparative Pathology,* **96**, 357–369.

72. Orskov, H. (1982) Acetate—inhibitor of growth hormone hypersecretion in diabetic and non dia-betic uraemic subjects. *Acta Endocrinology,* **99**, 551–558.

73. Pilch, S. M. (1987) In *Federal Drug Administration, Center for Food Safety and Applied Nutrition,* Washington, DC, pp. 1–230.

74. Prosky, L., Asp, N.-G., Furda, L., Devries, J. W., Schweizer, T. F. and Harland, B. F. (1985) Determination of total dietary fibre in foods and food products: collaborative study. *Journal of the Association of Official Analytical Chemists,* **68**, 677.

75. Prosky, L., Asp, N.-G., Furda, L., De Vries, J. W., Schweizer, T. F. and Harland, B. F. (1984) The determination of total dietary fibre in foods, food products and total diets: interlaboratory study. *Journal of the Association of Official Analytical Chemists,* **67**, 1044–1052.

76. Read, N. W. (1984) In *Dietary Fiber Basics and Clinical Aspects.* Eds. G. V. Vahouny and D. Kritchevsky. pp. 81–100. New York: Plenum.

77. Remillard, R. L. (1989) Dietary management of intestinal lymphangiectasia. In *Proceedings of the*

Seventh Annual Veterinary Forum, ACVIM. pp. 357–358. Madison, WI: Omnipress.

78. Rivellses, A., Riccardi, G., Giaco, A., Pacioni, D., Gebivesem, S., Mattioli, P. L. and Mancini, M. (1980) Effect of dietary fibre on glucose control and serum lipoproteins in diabetic patients. *Lancet, 2,* 447–450.

79. Rowland, I. R., Mallet, A. K. and Wise, A. (1985) The effect of diet on the mammalian gut flora and its metabolic activities. *Critical Reviews in Toxicology,* **16**, 31–103.

80. Roediger, W. E. W. (1982) The effect of bacterial metabolism on the nutrition and function of the colon mucosa: a symbiosis between man and bacteria. In *Colon and Nutrition.* Eds. H. Goebbel and H. Kaspar. pp. 11–26. Lancaster, U.K.: M. T. P. Press.

81. Roediger, W. E. W. (1982) Utilization of nutrients by isolated epithelial cells of the rat colon. *Gastroenterology,* **83**, 424–429.

82. Rosenberg, S., Landay, R., Klotz, S. D. and Fireman, P. (1982) Serum IgE antibodies to psyllium in individuals allergic to psyllium and English plantain. *Annals of Allergy,* **48**, 294–298.

83. Schneeman, B. O. and Lefevre, M. (1986) Effect of fibre on plasma lipoprotein composition. In *Basic and Clinical Aspects.* Eds. G. V. Vahouny and D. Kritchevsky. pp. 309–321. New York: Plenum.

84. Settle, R. G. (1988) Short chain fatty acids and their potential role in nutritional support. *Journal of Parenteral and Enteral Nutrition,* **12**, 104S.

85. Selvendran, R. R., Stevens, B. J. H., O'Neil M. A. and DuPont M. S. (1982) Special Report No. 8. Chemistry of plant cell wall and dietary fiber. Biennial Report AFRC Food Research Institute, Norwich. pp. 14–20.

86. Scheppach, W., Bartram, P., Richter, A. Liepold, H. and Kasper, H. (1990) Enhancement of colonic crypt proliferation in man by short-chain fatty acids. In *Dietary Fibre: Chemical and Biological Aspects.* Eds. D. A. T. Southgate, K. Waldron, I. T. Johnson and G. R. Fenwick. pp. 233–237. Royal Society of Chemistry, Special Publication No. 83.

87. Schwartz, S. E., Levine, R. A., Weinstock, R. S., Petokas, S., Mills, C. A. and Thomas, F. D. (1988) Sustained pectin ingestion: effect on gastric emptying and glucose tolerance in non-insulin dependent diabetic patients. *American Journal of Clinical Nutrition,* **48**, 1413–1417.

88. Simpson, J. W., Maskell, I. E. and Markwell, P. J. Clinical application of a hypoallergenic diet in the management of canine colitis. *Journal of Small Animal Practice* (in press).

89. Sinkeldam, E. F., Kuper, C. F., Boslan, M. C., Hollanders, V. M. H. and Vedder, D. M. (1990) Interactive effects of dietary wheat bran and lard on N-methyl-N′-nitro-N-nitrosoguanidine induced colon carginogenesis in rats. *Cancer Research* **50**, 1092.

90. Southgate, D. A. T. (1987) Minerals, trace elements and potential hazards. *American Journal of Clinical Nutrition,* **45**, 1256–1266.

91. Story, J. A. (1984) In *Dietary Fiber, Basic and Clinical Aspects.* Eds. G. V. Vahouny and D. Kritchevsky. pp. 139–142. New York: Plenum.

92. Strobel, S., Ferguson, A. and Anderson, D. M. (1982) Immunogenicity of foods and food additives —*in vivo* testing of gums arabic, karayn and tragacanth. *Toxicology Letters,* **14**, 247–252.

93. Trowell, H. (1975) Diabetes mellitus and obesity. In *Refined Carbohydrate Foods and Disease. Some Implications of Dietary Fibre.* Ed. D. P. Burkitt and H. C. Trowel. pp. 227–249. London: Academic.

94. Trowell, H., Southgate D. A. T., Wolever, T. M. S., Leeds, A. R., Gassull, M. A. and Jenkins, D. J. A. (1976) Dietary fibre redefined. *Lancet,* 967–970.

95. Uribe, M., Marquez, M. A., Ramos, G. G., Ramos–Uribe, M. H., Vargras, F., Villalobos, A. and Ramos, C. (1982) Treatment of chronic systemic encephalopathy with vegetable and animal proteins. *Digestive Disease Science,* **27**, 1109–1116.

96. Vahouny, G. V. and Cassidy, M. M. (1986) Dietary fiber and intestinal adaptation. In *Dietary Fiber: Basic and Clinical Aspects.* Eds. G. V. Vahouny and D. Kritchevsy. pp. 181–209. New York: Plenum.

97. Van Soest, P. J. and Wine, R. H. (1967) Use of detergents in the analysis of fibrous feeds. IV. Determination of cell wall constituents. *Journal of the Association of Official Analytical Chemists,* **50**, 50–55.

98. Ward, G. M., Simpson, R. W., Simpson, H. C. R, Naylor, B. A., Mann, J. I. and Turner, R. C. (1982). Insulin receptor binding increased by high carbohydrate low fat diet in noninsulin dependent diabetics. *European Journal of Clinical Investigations,* **12**, 93–96.

99. Willard, M. D. (1988). Dietary therapy in large intestinal diseases. In *Proceedings of the Sixth Annual Veterinary Forum, ACVIM.* pp. 713–714. Madison, WI: Omnipress.

100. Wynder, E. L. and Reddt, B. S. (1983) Dietary fat and fibre and colon cancer. *Seminars in Oncology,* **10**, 264–272.

101. Younoszai, M. K., Adedoyin, M. and Ranshaw, J. (1978) Dietary components and gastrointestinal growth in rats. *Journal of Nutrition,* **108**, 341–350.

CHAPTER 8

Interactions of Clinical Nutrition with Genetics

W. JEAN DODDS and SUSAN DONOGHUE

Introduction

Nutrition interacts with genetics in several ways. First, genetic differences between individuals lead to quantitative variations in dietary requirements for energy and nutrients. Also, genetic defects may result in inborn errors of metabolism, blocking pathways involving nutrients or their metabolites. Diet may affect the health of individuals genetically susceptible to certain disorders. Each of these interactions is discussed below.

Clinical presentation of genetic interactions with nutrition range from severe illness to apparent health. Many of the inborn errors of metabolism are fatal. Other disorders in dogs and cats, perhaps less dramatic in their presentation and certainly less well documented, point toward genetic involvement and show clinical improvement with nutritional management. Even the most common decision in nutritional management, the daily amount of food offered to a dog or cat, is partly determined by genetic predisposition. The daily energy requirement varies widely among healthy dogs and cats and those with energy needs beyond the statistical norms are considered to be normal, although predisposed to becoming too thin or too fat.

Genetic disorders with nutritional involvement appear to be more common in purebred families of inbred or closely linebred dogs and cats. Most documented problems involve purebreds, but close linebreeding also occurs among non-purebreds, such as with competition or performance dogs, by intentional crossbreeding and selection and genetic problems with nutritional involvement have also been documented in these animals (see below). Inbreeding does not necessarily produce the problem, but permits the expression of deleterious recessive genes or increases the frequency of existing mutations within these animal families.[44]

Genetic Variation Influences Requirements for Energy and Nutrients

Minimal nutrient requirements are usually presented as averages in tabular form, such as in National Research Council (NRC) publications of nutrient requirements for dogs and cats.[41,42] Averages may create the impression that requirements are the same for all individuals within a species. In fact, "there is a continuum of individuals in a population with genetically determined variations in their nutrient requirements that extend over a wide range."[14]

Minimal and Maximal Nutrient Requirements

Minimal requirements for most vitamins and trace elements have been sharply defined by means of distinctive clinical signs experimentally determined. The lowest daily dose needed to prevent clinical signs, however, may be well below that which promotes buoyant health. For example, 10 mg of ascorbic acid may prevent scurvy in humans, but 100 mg might be needed to facilitate iron absorption and perhaps 1000 mg or more might maximize immune competence.

Maximal limits or toxic thresholds have been established for most of the trace elements and fat-soluble vitamins, again by means of distinctive clinical signs documented in relatively short-term experiments. Insidious subclinical intoxications may occur below these thresholds. For example, a chronic mild excess of iron may cause poor food intake and weight loss in dogs without more specific clinical signs such as gastrointestinal irritation and anemia.[31] Guidelines for nutritional requirements[41,42] represent minima and maxima that have not taken into consideration the needs of those individuals known or believed to exhibit marked genetic variation.

Energy

For healthy dogs and cats, there is wide individual variation in energy needs: within a group of seven animals of the same breed, age, sex and size, one in seven will need 20% fewer calories and another in the same group needs 20% more calories to maintain body weight.[28] The differences are due to genetic variation and are met by adjusting energy intake appropriately. Clients with large numbers of animals are usually quick to learn the differences in their pets' energy needs. Occasionally, a client must be reminded that dogs are best fed "by eye" and not by a calculation.

Genetic differences in energy metabolism and metabolic rate are likely to have a role in determining body condition and affect the incidence of obesity. Energy needs have been shown to vary with the breed of dog.[29] In a hospital survey, the prevalence of overweight or underweight dogs varied with breed.[34] These preliminary data suggest a need for a more detailed study of genetic variation in sick dogs and cats.

Nutrients

The concept of individual variation in nutrient requirements is well known for humans. A report by the U.S. Department of Agriculture in 1939 recommended that nutrient intakes should be increased by 50% because of individual variation: "The allowance margin of 50 percent above the average minimum . . . is an estimate intended to cover individual variations of minimal nutritional need among apparently normal people . . ."[21] Today, the coefficient of variation approximates 30–45% of the average requirement for several nutrients that have been studied in humans.[21] Coefficients of variation are lacking for most nutrients however, and so for these the coefficient of variation in energy requirement of ±15% is used. The mean plus twice the coefficient of variation yields a *recommended daily allowance* (RDA) that covers 98% of the population, excluding only the top 2%. The RDA is 130% of the mean minimal requirement for humans.

Comparable RDAs have not been proposed for dogs and cats. If the same principles were followed, pet RDAs would be about 140% of the corresponding mean minimal requirements, because the coefficient of variation in energy requirement is about 20% for these species, compared to 15% for humans. Moreover, the selective breeding programs of purebred dogs and cats can promote and reveal individual differences in nutrient requirements.

A key factor in the 30–50% average recommendations for humans is the recognition of increased needs in *apparently normal people* (our emphasis).[21] Individuals with increased nutritional needs may appear normal if their dietary requirements are met, may present with mild or vague signs if part of their needs are met or may show overt signs of disease if their diet is deficient. Although the technology exists to diagnose these patients using specific functional tests, the methodologies are costly and rarely operative in clinical laboratories. Another deterrent is the absence of control data for domestic animals. Practitioners and owners noting improvements in dogs and cats with, say, a vitamin–mineral supplement or a diet change may be demonstrating the "placebo effect." Alternatively, they may have witnessed responses of one of the "apparently normal" animals with increased nutritional needs. Vitamin C is considered here as an example.

Vitamin C Insufficiency

Ascorbic acid clearly is not required for growth and reproduction in most cats and dogs.[41,42] A case may be argued, however, that dietary ascorbic acid might be beneficial under certain circumstances, such as stress, or in certain individuals.

Pathophysiology

Ascorbic acid is a hexose derivative, an antioxidant and a cofactor of several hydroxylases. Ascorbate affects the metabolism of amino acids, other vitamins and drugs. It is required for the synthesis of collagen, steroids and carnitine.

Vitamin C is synthesized in the liver of most mammals and the kidneys of most nonmammalian animals. Those species that require a dietary source of vitamin C, including humans, certain nonhuman primates, guinea pigs, fruit-eating bats, red-vented bulbul birds, other Passeriformes, Coho salmon, rainbow trout and carp, lack the enzyme L-gulano-γ-lactone oxidase.

The average rate of hepatic synthesis of ascorbic acid in dogs and cats is only half that in most mammals.[7] If these rates follow a normal distribution, then one dog in a thousand may have a limiting reserve ability to synthesize ascorbic acid, especially in times of greater need, such as lactation or stress.

The major deficiency syndrome is termed *scurvy*. It is characterized by muscle weakness, fatigue, anorexia, myalgia and immunosuppression in early stages. Signs in later stages include tachypnea, dyspnea, anemia, weakened collagen, bleeding gums, tooth loss, capillary fragility, ecchymoses and hematomas.

Although relatively small doses, 10–60 mg/day in humans, reverse scorbutic signs, larger doses appear to be necessary to overcome marginal deficiencies.[26] Also, larger doses facilitate iron absorption and perhaps, optimal immune function.

Canine milk contains four times the concentration of ascorbic acid in blood.[40] The RDA of ascorbic acid in women is increased from 60 mg/day to 80 mg/day during pregnancy and 100 mg/day during lactation. Perhaps lactating bitches would benefit from supplementation with dietary ascorbic acid.

Stress tends to deplete ascorbic acid in the blood and adrenal cortex. Low plasma ascorbate may occur in dogs with liver disease[52] and in response to pain[53] or stress.[12,33]

Vitamin C in Sled Dogs

A decline in plasma ascorbate that occurred in sled dogs during the racing season was largely prevented by dietary supplementation of ascorbic acid, 1 mg/kcal.[12,33] Supplementary vitamin C also may help racing dogs by facilitating oxidation of fatty acids by mitochondria in working muscles.

Ascorbic acid enhances the activity of hydroxylases involved in the synthesis of carnitine. Carnitine levels in muscle fall before the appearance of typical signs of scurvy. One of the first symptoms of scurvy is muscular weakness.

Mitochondrial oxidation of fatty acids is increased by training and by feeding a high fat diet. Thus racing sled dogs trained on a high fat diet may have unusually high requirements for carnitine, and hence ascorbic acid.

Carnitine is also used to store acetyl groups

as acetylcarnitine. This storage reduces the concentration of acetylhydroxide that inhibits pyruvate dehydrogenase, thereby directing pyruvate to lactate. Accumulation of lactic acid is associated with fatigue and anxiety.

Scurvy in Dogs

Sled dogs in the Antarctic have shown signs of scurvy when fed meat frozen for long periods. These signs abated with the feeding of fresh meat. Several case reports of conditions similar to skeletal scurvy have suggested favorable responses to treatment with vitamin C.[3,18-20,38,57] Even though these clinical experiences lacked untreated controls we believe they deserve more credence than was assigned by the NRC.[42]

An experimental report claimed to exclude a possible role of ascorbic acid in canine hypertrophic osteodystrophy, but was difficult to interpret because of the experimental design.[53] Recently, ascorbic acid (average dose 63 mg/kg body weight) was shown to have little effect in controlling pruritus in dogs with allergic skin disease.[39] When ascorbic acid was combined with a commercial supplement of ω-3 and ω-6 essential fatty acids, only 4.3% responded synergistically to the supplements.

Clinical experiments indicating that large doses of ascorbic acid will help to prevent or ameliorate canine distemper and kennel cough have lacked untreated controls.[42] In controlled studies, with few dogs, vitamin C supplementation failed to offer significant protection against challenges with canine herpes virus, canine hepatitis virus, or a combination of canine adenovirus [type II], parainfluenza virus [SV5], canine mycoplasma and *Bordetella bronchiseptica*.[42] Protection at that time (1972) was evaluated in terms of clinical signs, mortality and gross or microscopic lesions. New data that also evaluate immunoglobulin responses and cellular immune function tests would be more convincing. Numerous studies have shown that ascorbic acid influences both cellular and humoral immunity in various species.

Many drugs and toxins are inactivated by mixed function oxygenase systems in hepatic mitochondria. These hydroxylases require reducing agents and are influenced by the availability of ascorbic acid.[24] Hepatic oxygenase activity drops by over 50% during scurvy. For these reasons, vitamin C supplementation is considered for humans that are intoxicated or receiving medication. This consideration should also be applicable to dogs and cats, especially those that are inbred or closely linebred and are engaged in breeding, working or showing, subject to other stresses or fed high fat diets.

Further Examples

Recommendations for efficacious and safe intakes of nutrients vary and depend in part on the signs and lesions studied. Very high doses of vitamin A, for example, cause acute illness in dogs, evidenced by vomiting, severe depression, coma and death; lower intakes cause insidious intoxication evidenced by vague signs of inappetence and bone pain.[42]

Interactions with environmental, dietary and genetic factors complicate nutrient requirements further. Vitamin E provides an example, increasing phagocytosis, antibody titers and T-cell-mediated immune responses.[53] Vitamin E deficiency has also been associated with decreased lymphocytic blastogenesis and decreased antibody titers after immunization of dogs with viral vaccines.[47,54] Vitamin E status, however, is highly dependent on dietary fat, selenium, vitamin A and other antioxidants.

Inborn Errors of Metabolism

All genetic disorders involve metabolism in some way, but the term *inborn errors of metabolism* was used originally to describe blocks in the flow of metabolism. Today it generally applies to blocks due to the absence of an enzyme that manifests as a genetic defect. Signs of disease may arise from accumulations of intermediate metabolites, deficiencies of end-products or the formation of

EXAMPLES OF GENETIC DISORDERS WITH NUTRITIONAL COMPONENTS IN DOGS AND CATS		
Condition	**Breeds**	**Nutritional components**
Diabetes mellitus	Dachshund, Miniature Poodle	Carbohydrate, fat
Goiter	Fox Terrier	Iodine
Hyperlipoproteinemia	Beagle, Miniature Schnauzer	Fat, fiber
Hypercholesterolemia	Sled dogs	Fat
Dilated cardiomyopathy	Boxers	Carnitine
Glycogen storage disease	Toy breeds	Carbohydrate
Enteropathy	Irish Setters	Protein (gluten)
Fanconi syndrome	Basenjis	Carbohydrate
Exertional myopathy	English Springer Spaniel, Old English Sheepdog	Lactate
Hyperammonemia	Golden Retriever, Beagle	Protein
Portosystemic venous shunts	Many breeds	Protein
Hepatitis	Bedlington Terrier, West Highland White Terrier, Dobermann Pinscher	Copper
Pancreatic insufficiency	German Shepherd	Protein, fat, carbohydrate
Cystinuria	Many breeds	Amino acids
Tyrosinemia	German Shepherd	Amino acids
Uric acid stones	Dalmatian	Purines
Calcium oxalate stones	Miniature Schnauzer	Calcium, alkalinogenic ingredients
Methylmalanic aciduria	Giant Schnauzers, cats	Vitamin B_{12}
Multifactor coagulopathy	Devon Rex cats	Vitamin K
Hip dysplasia	Large breed dogs	Energy
OCD	Large breed dogs	Calcium

Fig. 8.1 Examples of genetic disorders with nutritional components in dogs and cats.

new products through alternative pathways. Metabolic pathways involve nutrients and fuels or their metabolites and therefore nutrition has a role in the diagnosis (through biochemical tests of blood or urine for specific metabolites) or, occasionally, in the management (through special diets) of inborn errors of metabolism (Fig. 8.1).

Over 200 congenital or genetic disorders have been reported for dogs[25,43]; some of these disorders may be classified as inborn errors of metabolism. Many dogs and cats with inherited metabolic diseases die as neonates, often without necropsies or diagnoses. Owners may note just a failure to thrive, signalled by poor growth. Death rates for purebred puppies and kittens range from 15 to 40% and one report cites the death rate of purebred Beagles as increasing linearly with the degree of inbreeding.[44]

Nutritional management has a role in managing certain inherited disorders. For example, alkalinogenic ingredients are used in diets for Miniature Schnauzers with calcium oxalate urolithiasis and the vitamin A derivative, etretinate, is used in Cocker Spaniels with idiopathic seborrhoea.[46] Other genetic disorders, such as diabetes mellitus and copper-storage disease, are managed with drugs and diet (Figs 8.2 and 8.3). A few inherited disorders, such as vitamin B_{12}-responsive methylmalonic aciduria, are treated with specific nutrient replacement.

Copper Intoxication

There is wide variation between domestic species regarding tolerance to dietary copper. For example, rats, goats and ponies are

INGREDIENTS USED IN HOME-MADE DIETS FOR NUTRITIONAL MODIFICATION

| Item | Content in foods | | |
	Low	Medium	High
Purine	Cottage cheese, rice	Meat, fish, poultry	Liver, heart, kidney, tongue, brain, anchovies, sardines, gravy, broth, bouillon, meat extracts
Oxalate	Milk, eggs, cheese, beef, lamb, pork, poultry, fish, shellfish, rice, noodles, bread, bacon, vegetable oils, butter	Sardines, carrots, corn, tomato	Tofu, peanut butter, celery, parsley, greens, sweet potatoes, spinach, citrus peel, grapes, grits, soy crackers, wheat germ, peanuts, almonds, tomato soup, vegetable soup, marmalade
Calcium	Meat flesh, poultry flesh, fish fillet, shellfish flesh, rice, noodles	Legumes	Milk, cream, cheese, yogurt, cottage cheese, ice cream, sherbet, custard pudding
Copper	Beef, cheese, cottage cheese, eggs, oils, fats, rice, fresh tomatoes	Turkey, chicken, peanut butter, olives, potatoes, breads, peas, canned soups	Lamb, pork, pheasant, quail, duck, goose, salmon, liver, heart, kidney, shellfish, meat gelatin, soybean meal, tofu, nuts and seeds, avocado, wheat germ, bran breads, bran cereals, granola, mushrooms, broccoli, raisins, brewers yeast, mineral water

FIG. 8.2 Ingredients used in home-made diets for nutritional modification.[45] Examples of protein, fats and carbohydrate restriction are provided in the Appendix.

relatively resistant to copper poisoning, sheep and rabbits are very susceptible and swine and cattle are moderately susceptible.[5] Bedlington Terriers are susceptible to copper accumulation in hepatic lysosomes with subsequent hepatitis and cirrhosis.[55] A hereditary defect in these dogs affects copper excretion.

To treat the disorder, copper chelating

INGREDIENTS USED TO MODIFY URINE pH

| Urine pH Effect | | |
Acid	Neutral	Alkaline
Meat, fish, fowl, shellfish, egg, cheese, peanuts, corn, rice, noodles, bread, lentils	Butter, margarine, fats, oils, sugar, honey, tapioca	Milk, milk products, almonds, coconut, beets, greens, kale, spinach, molasses

FIG. 8.3 Ingredients used in home-made diets can be modified to alter urine pH.[45]

drugs, such as D-penicillamine, are given to bind copper and remove it from the body via urinary excretion. Use of the drug may, however, be associated with anorexia, nausea and vomiting.

Nutritional therapy of these cases aids their management. For example, home-made diets can be designed to fit the metabolic needs of specific patients. Hepatitis may affect metabolism of any or all of the fuel sources and home-made diets can be designed to be very low in fat, protein or carbohydrate as needed by individual patients. Affected Bedlingtons should receive oral zinc supplementation for two reasons. First, copper-chelating drugs, such as D-penicillamine, bind zinc in addition to copper and may deplete tissue zinc. Secondly, zinc binds copper and removes it from the body and does not produce the side effects of D-penicillamine (anorexia, nausea and vomiting) when given at physiological doses. Indeed, zinc is used as a replacement for penicillamine for treatment of children with Wilson's Disease (inherited copper accumulation).[23]

We provided nutritional support for an 11-year-old male Bedlington Terrier that had been given D-penicillamine for 10 years (Fig. 8.4). The dog presented with signs of hepatitis and zinc deficiency (shrunken testicles, dermatitis and anorexia) and initially showed signs of protein intolerance. He was placed first on a home-made low copper, low protein diet. As the hepatitis resolved, protein was increased to about 30% of calories and low copper levels were maintained. The dog was also supplemented with 15 mg zinc sulfate daily and penicillamine was discontinued. Figure 8.4 shows the Bedlington at presentation and following 6 months on the home-made diet and zinc therapy. The dog maintained good condition and liver function for 3 years. More recently, a protocol has been developed for use of high levels of zinc acetate (200 mg daily in divided doses initially, then 50–100 mg daily) in the management of this disorder.[4]

Cobalamin Deficiency

Vitamin B_{12} deficiency occurs in dogs and cats following loss of the ileum through surgery or illness and as an inherited metabolic defect.

Pathophysiology

Vitamin B_{12} is cyanocobalamin, a porphyrin-like substance containing cobalt. The term is sometimes extended to other cobalamins. In the veterinary literature, the generic term *cobalamin* is sometimes used specifically for vitamin B_{12}, however cobalamin also refers to methylcobalamin.

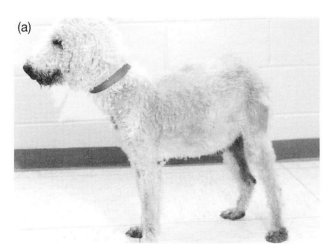

FIG. 8.4 An 11-year-old male Bedlington Terrier with copper intoxication. He was treated for 10 years with D-penicillamine and diagnosed as having zinc deficiency (a). Six months after institution of a home-made low copper diet with daily supplemental zinc sulfate signs resolved (b).

Vitamin B_{12} is synthesized only by micro-organisms, including those in the large intestine. The vitamin B_{12} present in animal tissue, such as the liver, has been synthesized by microorganisms. Deficiency in humans is termed pernicious anemia. The early investigators used dogs, and, in 1934, their research led to a Nobel Prize.

Vitamin B_{12} deficiency occurs in humans with gastric or ileal disease. A glycoprotein, termed *intrinsic factor*, binds to vitamin B_{12} in the stomach and upper small intestine, then attaches to specific receptors in ileal enterocytes. Intrinsic factor is secreted by the stomach in humans and by the stomach and pancreas in dogs.[2,37] Vitamin B_{12} is absorbed solely in the ileum, thus disorders affecting ileal function, or indeed surgical removal of the ileum, lead to depletion in dogs.

Vitamin B_{12} deficiency also occurs as an inherited metabolic defect, as reported in Giant Schnauzers.[15] In dogs, vitamin B_{12} is malabsorbed due to an ileal receptor defect. Cobalamin deficiency has been reported in a cat[56] but more work is needed to determine a genetic component.

Diagnosis and Treatment

Signs of vitamin B_{12} deficiency include anorexia, failure to thrive and weight loss. Hematological findings include nonregenerative anemia, neutropenia, anisocytosis, poikilocytosis, hypersegmented neutrophils and giant platelets.[15]

Diagnosis of vitamin B_{12} deficiency is made by urinary excretion of methylmalonic acid and low serum vitamin B_{12} concentrations. Normal serum levels of vitamin B_{12} are above 200 pg/ml; upper limits differ by laboratory and range from 400 to 700 pg/ml.

Supplemental vitamin B_{12} is often administered IM or IV, as the bioavailability after oral ingestion may be as little as 16% in humans with doses above 10 µg.[1] Dogs with inherited cobalamin deficiency (or with loss of the ileum) are maintained on periodic parenteral administration of 1 mg vitamin B_{12}.

Adverse reactions to administration by IV, IM or enteral routes appear to be rare, although a few humans are allergic to preservatives in vitamin B_{12} preparations. We know of two fatal anaphylactic reactions in horses given vitamin B_{12} preparations intravenously.

Nutrition and Genetically Susceptible Animals

Certain breeds of dogs have been shown to have increased susceptibility to specific diseases. For example, Beagles have been noted to be susceptible to distemper, demodectic mange and thyroiditis, Boxers to neoplasia, Dobermann Pinschers and Rottweilers to parvovirus enteritis and German Shepherds to lupus erythematosus.[16,43] These trends are due to genetic differences that influence susceptibility to disease.

Clearly, not all individuals at risk for a specific disease are affected. One hypothesis to explain why some inbred and closely linebred dogs or cats become ill with specific disorders while others remain normal is the *threshold model*. In this model, genetically susceptible individuals develop disease following the additive effects of inducing agents[10] (Fig. 8.5). Examples of inducing agents include intake of drugs, exposure to toxic or noxious substances, hormonal imbalance and diet (Fig. 8.5). Examples of clinical disorders in inbred and linebred animals affected by inducing agents include such diverse problems as autoimmune disease and musculoskeletal disorders.[9,11] Autoimmune disorders in animals parallel those in humans where the susceptibility is determined genetically by the major histocompatibility complex and reflects the sum of the genetic and environmental factors that induce failure of self-tolerance (autoimmunity).[48] Other examples include vitamin K-dependent multifactor coagulopathy in Devon Rex cats following vaccination,[35] hip dysplasia in pups fed excess calories,[27] osteochondritis dissecans in dogs fed high calcium[50] and hypercholesterolemia in inbred sled dogs fed high fat diets.[32]

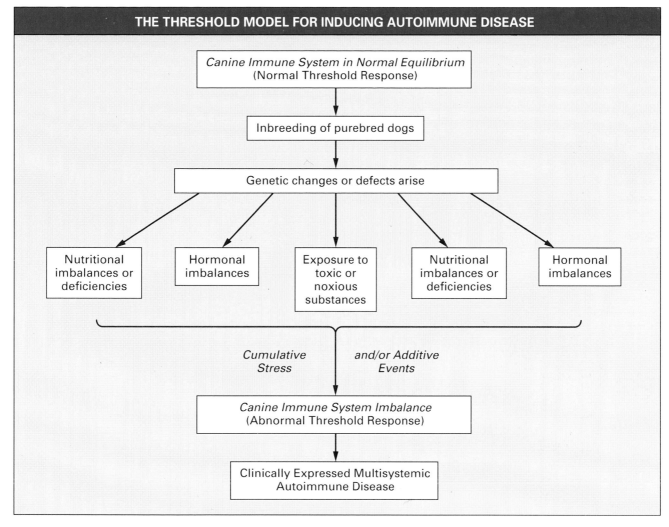

FIG. 8.5 The threshold model for inducing autoimmune disease.

Hip Dysplasia and Energy

Hip dysplasia is an inheritable disease with a phenotypic expression that appears to be affected at least in part by environmental factors, including diet. Nutritional interests focus on "overfeeding," usually defined as excessive intake of energy but sometimes interpreted as excessive intake of specific nutrients, such as calcium or amino acids.

A recent experimental study supports the contention that *ad libitum* feeding of a relatively palatable dog food to puppies genetically predisposed to development of hip dysplasia increases the incidence of hip dysplasia diagnosed by radiography.[27] However, care must be taken to avoid iatrogenic malnutrition by overzealous food restriction.

Prevention of overfeeding is sound nutritional practice; intentional underfeeding that results in stunted growth is unwarranted.

Osteochondritis Dissecans and Calcium

A recent analytic epidemiological study of pet dogs with naturally occurring osteochondrosis dissecans (OCD) found a link with diet.[50] Dogs with OCD were compared to normal dogs matched for breed, sex and age and risk factors were identified. Risk factors included play with other dogs (a likely source of trauma), drinking well water (often a source of minerals and also an indicator of rural lifestyle with increased exercise) and high dietary calcium. Conversely, feeding premium dense-dry dog foods decreased the risk of OCD.

Hypercholesterolemia and Fat

Most dogs develop a transient hypercholesterolemia when changed from low fat (plant-based) to high fat (meat-based) diets. However, one in 10 Alaskan Huskies in a racing sled dog kennel developed persistent, extremely high plasma cholesterol concentrations when fed meat-based diets.[32] All affected dogs traced back to two founding bitches. Inheritance of diet-induced hypercholesterolemia was consistent with an autosomal dominant gene.[32]

Reducing Dietary Stressors

Clinical management of inbred and line-bred dogs believed to be suffering from the effects of cumulative inducing agents often includes dietary adjustments. For example, puppies susceptible to musculoskeletal problems such as hip dysplasia and OCD are fed measured amounts of premium dense-dry dog foods; *ad libitum* feeding is avoided.

Antioxidants

Antioxidants are added to petfoods to protect fats from rancidity. Fat becomes rancid especially when exposed to air, heat and perhaps bacteria. Fat breaks down in two steps. Hydrolysis yields fatty acids that may improve the flavor up to a point (no more than 10%). Oxidation of unsaturated fatty acids yields ketones and free fatty acids that are usually unpalatable and potentially injurious. It takes as little as 0.05% of the fat to react with oxygen to produce rancidity.

All commercial petfoods are preserved by some means. The dryness of dry petfoods provides a hostile environment for most bacteria, fungi and other potential disease-causing organisms. Canned products are sterilized by heat; the metal barrier keeps out pathogens. Frozen dog foods are popular in some countries because low temperatures stop many pathogens. Preservatives include the antioxidants BHT (butylated hydroxytoluene), BHA (butylated hydroxyanisole), ethoxyquin (1,2–dihydro-6–ethoxy-2,2,4–trimethylquinoline), forms of vitamin E (tocopherols) and vitamin C. Petfoods devoid of antioxidants added at the time of processing often contain ingredients (such as animal tallow and fish meal) that are preserved with antioxidants.

Ethoxyquin

Many petfood manufacturers use ethoxyquin because of its excellent antioxidant qualities, high stability and reputed safety.[13,22] However, an ongoing controversy surrounds issues related to its safety when repeatedly fed at permitted amounts in dog foods,[6,11,13] particularly when fed to genetically susceptible breeds of inbred or closely linebred dogs.[11] Toy breeds may be particularly at risk because they ingest proportionately more food and preservative for their size in order to sustain their energy needs.

For human consumption, ethoxyquin is permitted in certain spices to prevent loss of color. Ethoxyquin is permitted in petfoods, fats and oils at levels not exceeding 0.015% in the finished product (e.g. 0.015% as-fed basis).[22] It is readily absorbed, metabolized and excreted in urine and feces,[17] with residual levels in liver, gastrointestinal tract and adipose liver.[30,49]

Ethoxyquin is assigned a toxicity rating of 3 or "moderately toxic," indicating the probable oral lethal human dose is 0.5–5 g/kg,[17] 3- to 33-times the maximum allowed level in petfoods. This toxicity rating is slightly greater than ratings for tetracycline and penicillin, lower than for aspirin and caffeine. Susceptibility of laboratory animals to antioxidant toxicity increases with the nutritional stress of variable dietary constituents. Increased dietary fat, for example, increases susceptibility to toxicity of ethoxyquin fed to chickens and BHT (as well as DDT) fed to rats.[36] The response in chickens to increased dietary fat appeared to be due to the resultant lowered protein. Chickens fed 17 vs. 23% protein showed increased susceptibility to ethoxyquin toxicity. Ethoxyquin levels fed to chickens were, however, almost 17-times the maximum allowable level for petfoods.

In laboratory animals, ethoxyquin increased hepatic vitamin A levels two- to fivefold, and, at levels 3 times that found in petfoods, increased blood vitamin E levels twofold.[8,51] These data suggest that ethoxyquin assumes some *in vivo* antioxidant activities and thus spares natural antioxidants such as vitamin E.

Since the late 1980s, the incidence of chronic disorders in purebred dogs appears to have increased. These disorders include dysfunction of liver, kidney and thyroid, reproductive problems, autoimmune diseases and other immune dysfunction, birth defects in pups, increased stillbirths and neonatal mortalities, neoplasia, allergies and problems with skin and coat condition.[6,10,13] Most concerns have focused on inbred or closely linebred dog families.

Suspicions about the safety of ethoxyquin and any association with these disorders would be difficult to corroborate because the affected animals may have received drugs or other medications to treat their symptoms and other diseases may be present. Furthermore, ethoxyquin has been used in some animal feeds since 1959, some 30 years before the current controversy arose. Nevertheless, the additive or cumulative effects of several environmental insults, as illustrated by the threshold model (Fig. 8.5), could explain the increasing frequency of debilitating illnesses in these dogs. Cumulative effects of metabolites and their interactions may place inbred or closely linebred dogs exposed to other inducing agents at significantly increased risk. The Food and Drug Administration of the U.S.A. Center for Veterinary Medicine states, however, that there is insufficient scientific evidence to show that ethoxyquin is unsafe when used at approved levels or to warrant action against its use in petfoods.[11,13] Future studies incorporating modern toxicological techniques, appropriate medical and epidemiological assessment of cases and consideration of multifactorial interactions in inbred or closely linebred dogs, should help

to clarify the issue. Indeed, for the majority of dogs, health risks from the ingestion of inadequately preserved rancid fats might be more harmful than risks from the potential adverse effects of ethoxyquin.

Low-preservative Diets

Dog owners wishing to omit synthetic additives from their pets' diets can be counselled regarding their options. Commercial dry petfoods usually contain preservatives, dyes and flavoring agents. Certain dry dog foods use natural substances (lecithins, tocopherols and ascorbic acid) as the main antioxidant system, but these petfoods may contain synthetic antioxidants in certain ingredients (tallow, fish meal). Commercial canned petfoods may contain coloring and flavoring agents but often no added antioxidants, although trace amounts will be present in fats.

Home-made diets should be prepared using wholesome ingredients intended for human consumption. Such diets, when balanced and prepared properly, provide alternative nutrition (see Appendix, Home-Made Diets). If preservatives are omitted, clients must be counselled about the relatively short "shelf-life" of home-made diets and the necessity of refrigerating prepared diets.

References

1. Bailey, L. B. (1984) Vitamin B_{12}. In *Clinical Guide to Parenteral Micronutrition*. Ed. T. G. Baumgartner. Melrose Park, IL: Educational Publications.
2. Batt, R. M., Horadagoda, N. U., McLean, L. *et al.* (1989) Identification and characterization of a pancreatic intrinsic factor in the dog. *American Journal of Physiology*, **256**, G517.
3. Bosch, F. B., Court, A. and Vivano, A. C. (1961) A case of Barlow's disease in the dog. *Veterinary Medicine Review*, **2**, 371.
4. Brewer, G. J., Dick, R. D., Schall, W. *et al.* (1992) Use of zinc acetate to treat copper toxicosis in dogs. *Journal of the American Veterinary Medical Association*, **201**, 564–568.
5. Brewer, N. R. (1987) Comparative metabolism of copper. *Journal of the American Veterinary Medical Association*, **190**, 654–658.

6. Cargill, J. C. (1991) A look at the ethoxyquin controversy: it's still the consumer's choice. *Dog World,* **76(2),** 14–15,111–113.

7. Chatterjee, I. B. (1973) Evolution and the bio-synthesis of ascorbic acid. *Science,* **182,** 1271.

8. Combs, G. F., Jr. and Scott, M. L. (1974) Antioxidant effects on selenium and vitamin E function in the chick. *Journal of Nutrition,* **104,** 1297–1303.

9. Dodds, W. J. (1983) Immune-mediated diseases of the blood. *Advances in Veterinary Science and Comparative Medicine,* **27,** 163–196.

10. Dodds, W. J. (1992) Unraveling the autoimmune mystery. *Dog World,* **77(5),** 44–48.

11. Dodds, W. J. (1992) Genetically based immune disorders. Parts 1 and 2. Autoimmune diseases. Part 3. Other autoimmune diseases. Part 4. Immune deficiency diseases. *Veterinary Practice Staff,* **4(1),** 8–10, **4(2),** 1, 26–31, **4(3),** 35–37, **4(5),** 19–21.

12. Donoghue, S., Kronfeld, D. S. and Banta, C. A. (1987) A possible vitamin C requirement in racing sled dogs trained on a high fat diet. In *Nutrition, Malnutrition and Dietetics in the Dog and Cat.* Ed. A. T. B. Edney. pp. 57–59. Waltham Centre Press.

13. Dzanis, D. A. (1991) Safety of ethoxyquin in dog foods. *Journal of Nutrition,* **121,** S163–S164.

14. Elsas, L. J., II and Acosta, P. B. (1988) Nutrition support of inherited metabolic disease. In *Modern Nutrition in Health and Disease.* Eds. M. E. Shiles and V. R. Young. 7th edn. p. 1337. Philadelphia: Lea and Febiger.

15. Fyfe, J. C., Jezyk, P. F., Giger, U. *et al.* (1989) Inherited selective malabsorption of vitamin B12 in Giant Schnauzers. *Journal of the American Animal Hospital Association,* **25,** 533–539.

16. Glickman, L. T., Domanski, L. M., Patronek G. J. *et al.* (1985) Breed-related risk factors for canine parvovirus enteritis. *Journal of the American Veterinary Medical Association,* **187,** 589–594.

17. Gosselin, R. E., Smith, R. P. and Hodge, H. C. (1984) *Clinical Toxicology of Commercial Products.* 5th edn. pp. 2–406. Baltimore: Williams & Wilkins.

18. Gratzl, E. and Pommer, A. (1941) Moeller-barlowsch krankheit beim Hund. *Wiener Tierarztlicher Monatschrift,* **28,** 481.

19. Gregoire, C. (1938) La vitamine C et la maladie de Barlow chez le chien. *Annales de Médecine Vétérinaire,* **83,** 366.

20. Grøndalen, J. (1976) Metaphyseal osteopathy (hypertrophic osteodystrophy) in growing dogs: a clinical study. *Journal of Small Animal Practice,* **17,** 721.

21. Harper, A. E. (1985) Origin of recommended dietary allowances—an historic overview. *American Journal of Clinical Nutrition,* **41,** 140–148.

22. Hilton, J. W. (1989) Antioxidants: function, types and necessity of inclusion in pet foods. *Canadian Veterinary Journal,* **30,** 682.

23. Hoogenraad, T. U., Van Hattum, J. and Ven den Hamer C. J. A. (1987) Management of Wilson's disease with zinc sulphate. *Journal of Neurological Science,* **77,** 137–146.

24. Hornig, D., Glathaat, B. and Mosur, U. (1984) General aspects of ascorbic acid function and metabolism. In *Ascorbic Acid in Domestic Animals.* Eds. I. Wegger, F. J. Tagwerker and J. Moustgaard. Copenhagen: Royal Danish Agricultural Society.

25. Hoskins, J. D. and Taboada, J. (1992) Congenital defects of the dog. *Compendium of Continuing Education for Small Animals,* **14,** 873–887.

26. Jaffe, G. M. (1984) Vitamin C. In *Handbook of Vitamins.* Ed. L. J. Machlin. New York: Marcel Dekker.

27. Kealy, R. D., Olsson, S. E., Monti, K. L. *et al.* (1992) Effects of limited food consumption on the incidence of hip dysplasia in growing dogs. *Journal of the American Veterinary Medical Association,* **201,** 857–863.

28. Kendall, P. T., Blaza, S. E. and Smith P. M. (1983) Comparative digestible energy requirements of adult Beagles and domestic cats for bodyweight maintenance. *Journal of Nutrition,* **113,** 1946.

29. Kienzle, E. and Rainbird, A. (1991) Maintenance energy requirement of dogs: what is the correct value for the calculation of metabolic body weight in dogs? *Journal of Nutrition,* **121,** S39–S40.

30. Kim, H. L. (1991) Accumulation of ethoxyquin in tissue. *Journal of Toxicology and Environmental Health,* **33,** 229–236.

31. Kronfeld, D. S. (1989) *Vitamin and Mineral Supplementation for Dogs and Cats.* Santa Barbara, CA: Veterinary Practice Publishing Company.

32. Kronfeld, D. S., Johnson, K. and Dunlap, H. L. (1979) Inherited predisposition of dogs to diet-induced hypercholesterolemia. *Journal of Nutrition,* **109,** 1715–1719.

33. Kronfeld, D. S. and Donoghue, S. (1988) A role for vitamin C in work-stressed dogs. *Proceedings of the Sixth Annual Veterinary Medical Forum, ACVIM.* p. 537. Madison, WI: Omnipress.

34. Kronfeld, D. S., Donoghue, S. and Glickman, L. T. (1991) Body condition and energy intakes of dogs in a referral teaching hospital. *Journal of Nutrition,* **121,** S157–S158.

35. Maddison, J. E., Watson, D. J., Eade, I. G. *et al.* (1990) Vitamin K-dependent multifactor coagulo-pathy in Devon Rex cats. *Journal of the American Veterinary Medical Association,* **197,** 1495–1497.

36. March, B. E., Biely, J. and Coates V. (1968) The influence of diet on toxicity of the antioxidant 1,2-dihydro-6-ethoxy-2,2,4-trimethylquinoline. *Canadian Journal of Physiology and Pharmacology,* **46**, 145–149.

37. Marcoullis, G., Rothenberg, S. P. and Labombardi, V. J. (1980) Preparation and characterization of proteins in the alimentary tract of the dog which bind cobalamin and intrinsic factor. *Journal of Biological Chemistry,* **255**, 1824.

38. Merrilat, L. A. (1936) Barlow's disease of the dog. *Veterinary Medicine,* **31**, 304.

39. Miller, W. H., Jr., Scott, D. W. and Wellington, J. R. (1992) Investigation on the antipruritic effects of ascorbic acid given alone and in combination with a fatty acid supplement to dogs with allergic skin disease. *Canine Practice,* **17(5)**, 11–13.

40. Naismith, D. J. and Pellett, P. L. (1960) The water-soluble vitamin content of blood serum and milk of the bitch. *Proceedings of the Nutrition Society,* **19**, xi.

41. National Research Council (1986) Nutrient requirements of cats. Washington, DC: National Academy Press.

42. National Research Council (1985) Nutrient requirements of dogs. Washington, DC: National Academy Press.

43. Patterson, D. F. (1980) A catalog of genetic disorders of the dog. In *Current Veterinary Therapy, VII.* Ed. R. W. Kirk. pp. 82–193. Philadelphia: W. B. Saunders.

44. Patterson, D. F., Haskins, M. E., Jezyk, P. F. *et al.* (1988) Research on genetic diseases: reciprocal benefits to animals and man. *Journal of the American Veterinary Medical Association,* **193**, 1131–1144.

45. Pemberton, C. M., Moxness, K. E., German, M. J. *et al.* (1988) *Mayo Clinic Diet Manual.* 6th edn. Toronto: B. C. Dekker.

46. Power, H. T., Ihrke, P. J., Stannard, A. A. *et al.* (1992) Use of etretinate for treatment of primary keratinization disorders (idiopathic seborrhea) in Cocker Spaniels, West Highland White Terriers and Basset Hounds. *Journal of the American Veterinary Medical Association,* **201**, 419–429.

47. Sheffy, B. E. and Schultz, R. D. (1979) Influence of vitamin E and selenium on immune response mechanisms. *Federation Proceedings,* **38**, 2139–2143.

48. Sinha, A. A., Lopez, S. M. and McDevitt, H. O. (1990) Autoimmune diseases: the failure of self-tolerance. *Science,* **248**, 1380–1388.

49. Skaare, J. U. and Nafstad, I. (1990) The distribution of ^{14}C-ethoxyquin in the rat. *Acta Pharmacologica et Toxicologica,* **44**, 303–307.

50. Slater, M. R., Scarlett, J. M., Donoghue, S. *et al.* (1992) Diet and exercise as potential risk factors for osteochondritis dissecans in dogs. *American Journal of Veterinary Research,* **53**, 2119–2124.

51. Spruzs, J. (1971) Stabilizing effect of ethoxyquin and diludin on carotene in grass meal and mixed feeds. In *Regulyatory Prosta i Metabolizma Zhivotnykh.* Akademiya Nauk Latvilskol SSR, Institut Biologii, Riga, pp. 111–124.

52. S. TROMBECK, D. R., Harold, D., Rogers, Q. R. *et al.* (1983) Plasma amino acid, glucagon and insulin concentrations in dogs with nitrosamine-induced hepatic disease. *American Journal of Veterinary Research,* **44**, 2028.

53. Teare, J. A., Krook, L. A., Kallfelz, F. A. *et al.* (1979) Absorbic acid deficiency and hypertrophic dystrophy in the dog: a rebuttal. *Cornell Veterinarian,* **69**, 384.

54. Tengerdy, R. P. (1980) Effect of vitamin E on immune responses. In *Vitamin E. A Comprehensive Treatise.* Ed. L. J. Machlin. pp. 429–444. New York: Marcel Dekker.

55. Twedt, D. C., Sternlieb, I. and Gilbertson, S. R. (1979) Clinical, morphologic and chemical studies on copper toxicosis of Bedlington Terriers. *Journal of the American Veterinary Medical Association,* **175**, 269–275.

56. Vaden, S. L., Wood, P. A., Ledley, F. D. *et al.* (1992) Cobalamin deficiency associated with methylmalonic acidemia in a cat. *Journal of the American Veterinary Medical Association,* **200**, 1101–1103.

57. Watson, A. D. J., Blair, R. C., Farrow, B. R. H. *et al.* (1973) Hypertrophic osteodystrophy in the dog. *Australian Veterinary Journal,* **49**, 433.

CHAPTER 9

Feeding Behaviors

VICTORIA L. VOITH

Introduction

The ingestive behavior system encompasses more than the simple physical acts of eating and drinking. It is a complex and dynamic system that is continuously influenced by numerous internal and external variables. Feeding behaviors include searching, hunting and caching of food as well as post-consummatory behaviors such as grooming and sleeping. Feeding behaviors are integrated in reproductive, parental and social relationships. Maintaining accessibility to food resources can lead to territorial behaviors. "Internal states" such as hunger and appetite influence feeding behaviors and social relationships. Clinicians should be aware that the dietary regimen of an animal influences many behaviors that do not immediately appear to be related to the feeding system. A change in diet may result in the appearance of unexpected behaviors.

Social Systems and Feeding Behaviors

Availability of food is one of the most important factors that shapes the social structures of animals. Cats tend to live solitary lives when food resources are in small units (e.g.

mice, voles, insects) scattered over a large area or space; but when food resources are plentiful and concentrated (e.g. households, provisioned farm yards or garbage dumps) cats are often found in large groups.[19,21,22,42]

Coyotes (*Canis latrans*) provide an example of how the distribution of food affects social groupings. Bekoff and Wells compared the sociality of coyotes in three different locations in the western U.S.A.[2] In one wildlife area, during the summer when the food supply consisted of small rodents scattered over a large area, coyotes were usually solitary. During the winter, however, when dead ungulates (deer, elk, moose) were sometimes available, coyotes were often seen in company. In a nearby elk refuge, where there was a plentiful supply of carrion during the winter, coyotes were seen in the largest groups and tended to stay together during the summer. In the elk refuge, where large, concentrated sources of food were fairly abundant, coyotes manifested territorial behaviors not seen elsewhere. In a third wildlife park, where there was virtually no ungulate carrion at any time of the year, most of the coyotes lived solitary lives throughout the year. The same animal, depending on availability and distribution of the food supply, could alternate between group living or a solitary lifestyle.

Natural Feeding Behaviors

Domestic Cats

Although it is commonly believed that cats predominantly hunt at night, recent studies report that domestic cats are just as likely to hunt in daylight.[37] Domestic cats fed *ad libitum* will eat many small meals over a 24-hr period.[27,32] They also, at least temporarily, prefer novel foods that are not radically different from those to which they are accustomed.[27,32] Hunting is a common activity for the domestic cat. There is no convincing evidence that cats engage in cooperative hunting, even among cats that live in groups.

Fed cats spend less time hunting.[37] Unfortunately, under the assumption that unprovisioned animals are more effective in controlling vermin, some farmers traditionally do not feed their barn cats. It may, however, be more cost-effective to feed barn cats than not. The provisioned cat may be a more effective hunter, more resistant to disease and parasites, and, very importantly, not journey as far from the owner's property or abandon the site altogether. Travelling cats, even in rural areas, are at high risk of being hit by cars. Female cats are as likely as males to roam in search of food.

Several ecological studies indicate that cats are effective in suppressing populations of rodents, lagomorphs and insectivores.[21] Cats do not appear to affect bird populations, except on islands where they have been recently introduced by man and the resident birds had not evolved predator defenses against cats. It has been pointed out that cats and song birds have coexisted for hundreds of generations in Britain without any evidence of cats causing the demise of any avian species.[37] Cats catch few birds in comparison to other prey, but because this predation is more likely to be seen, there is an impression that cats kill many birds. Birds might be warned if a bell is attached to the collar of a cat, but this technique would also reduce the cat's predation on other animals that the owner may wish to be controlled.

Cats frequently bring prey back to households, popularly referred to as "bringing the owner a present" or attributed to maternal motivation, for example, provisioning kittens. This author, however, believes that such prey retrieval is more likely to be an incomplete sequence of food caching. Both male and female cats, regardless of whether they are neutered or if kittens are present, bring prey back to households. Left undisturbed in a house, retrieved dead or weak animals are often consumed within the next few days. The author has found collections of small fur toys (representative of prey items) in holes in the house. One particular fur toy, that looks something like an octopus, is actively played with by most cats, aggressively guarded by some, and has been eaten by a few cats. Cats rarely guard commercial diets but when they possess live prey (and this specific fur toy) they may threaten owners as well as other cats that try to take the item away. The author has also found that cats will deposit these toys and dead mice in empty cat food bowls.

Domestic Dogs

Domestic dogs usually overeat when first presented with a large amount of very palatable food, but when given unrestricted access to most commercial diets, they will eat several smaller meals throughout the day.[17,27] If food is not continuously available, dogs are often socially facilitated to eat.[27,34] These behavior patterns are not unlike those of wolves. When a fresh carcass is available, wolves eat a large amount at one time. A long time may lapse between such meals, therefore this feeding is probably socially facilitated, regardless of how hungry an individual is. Much of the time, however, wolves rely on small, frequent meals throughout the day. They eat small mammals, reptiles, insects, berries and a variety of other vegetable matter.

There is an interesting relationship between appetite, aggression and dominant status in wolves and dogs. The dominant individual in a social group generally has undisputed access to valued resources such as food, shelter and mates, although a hungry subordinate may aggressively guard food it already possesses.[26]

A satiated dominant dog might not defend a steady supply of free choice dry dog food whereas it may be extremely aggressive over a piece of meat or rawhide chew toy, which are highly desirable and relatively rare resources. If water is scarce or restricted, it also may be guarded aggressively.

Occasionally dogs will manifest aggression to people in situations similar to those involving dominance aggression between dogs.[4,23] Dominant dogs may demonstrate aggression to owners when petted, hugged, disturbed when resting, threatened verbally or physically or if the owner tries to take food away. These dogs may not be aggressive over their regular food, but the owners are at risk if they try to take away a stolen bone, piece of garbage or rawhide treat.

Effects of Domestication

Dogs and cats were among the earliest of domesticated animals.[44] In ancient Egypt cats were so prized that they were raized to deity status. Exporting or killing a cat was punishable by death. Cats were highly valued for their abilities to control vermin.

For millennia, the primary economic value of cats has been their predatory capabilities and until recently there has been little or no selective breeding to alter their behavior or morphology. Domestic cats are relatively uniform in size and shape (particularly in comparison to dogs) and are likely to have retained many of the skills exhibited by their early ancestors. Unfortunately, because of these abilities many people believe cats can do well on their own without any human care. Consequently, cats are often neglected or abandoned. Today's environment, which includes an increased concentration of automobiles, dogs, toxic chemicals and feline infectious diseases, is not congenial to the unattended cat.

Unlike cats, dogs have been subjected to controlled breeding and artificial selection by mankind for many centuries. For hundreds if not thousands of years, there has been selection for specific behaviors and the opti-mum morphology to execute those behaviors. Consider the behaviors and shapes of dogs that can track game, run-down gazelle, follow small burrowing animals, attack other animals, intimidate strangers, etc.

Major domestic selection pressures in dogs have been for the initial segments in the hunting sequence and against the final components.[7] For example, in many bird dogs, the freezing component in stalking is selected for, but chasing prey and eating the game is selected against. Herding dogs manifest many of the initial components of a hunting sequence, such as stalking, circling, chasing and even inhibited bites, but a full-blown attack and killing is strongly selected against.[7] Hounds are selected for tracking and chasing prey; whether they catch or eat is often irrelevant. If fact, it would be logical to assume that hunting without being hungry and not eating the prey would be traits selected for.

More recently, morphological extremes independent of function have been selected for by dog fanciers. Misshapen heads, necks and jaws can result in breathing and eating difficulties. Long, pendent, silky ears, highly esthetic to owners, become soiled as the dog eats and cannot be self-cleaned by the dog. Decomposing particles of food can serve as primary irritants, as well as attracting flies. Deep-chested dogs are believed to be predisposed to gastric torsion, especially if vigorous play closely follows a large meal.

The physical examination of dogs, as well as cats, shoud include examination of the mouth as well as a discussion with the owner concerning prevention and recognition of problems that could occur related to feeding and the specific morphology of their pet.

Effect of Food Availability

Most animals, including cats and dogs, fed *ad libitum* regulate the amount that they eat to maintain a set weight.[17,32] Following a period of deprivation, animals will overeat until their previous weight set point has been achieved. If force fed, animals will eat little or not at all until their original weight is attained. Dogs will also adjust their eating to maintain a set protein intake.[33] When the diets of dogs and cats are diluted with water, they will eat a greater volume of food to achieve sufficient energy intake to maintain their homeostatic *status quo*. If their diets are diluted with inert materials such as cellulose or kaolin, however, the animals will not compensate for the lower energy content of the food and will lose weight. It is thought that the reduced intake is due to reduced palatability.[17]

Dogs with access to palatable foods establish higher weight set points, and apparently there are individual and breed differences in this regard.[27] Beagles, for example, gain more weight on palatable diets than do terriers.[17]

Fluctuations in Appetite and Anorexia

A decrease or increase in food consumption may be indicative of a disease process. There are also many nonpathological factors that influence eating behaviors.

Palatability, texture, temperature, novelty and familiarity are all features of food that can be manipulated to encourage and discourage food consumption.[17,27,32] Animals in strange environments prefer familiar foods. If a dog or cat refuses a new food, repeated exposure to the food will often lead to sampling and subsequent consumption. Cats prefer warm food, approximately 41°C.

Estrus is correlated with depressed appetite in dogs.[17] In fact, a reduction of food intake might be the first sign noticed by owners when a bitch starts coming into heat. Exogenously administered estrogen can cause anorexia in rats and endogenous estrogen has been hypothesized as a component in the etiology of anorexia nervosa in women.[13] Ovariectomized dogs eat more, exercise less and as a result gain weight.[18]

Dogs and cats eat more when the ambient temperature is cool and less when it is hot.[17] It is not uncommon for animals to eat little or nothing when the ambient temperature is very high.

Anorexia in response to temporary environmental conditions, such as the absence of the owner or being in an unfamiliar location, usually has no adverse consequences in healthy animals. They will compensate later.

Animals with lesions of the ventral medial hypothalamus (VMH) will overeat and become obese. They do not overeat indefinitely, but until they reach a new, much higher set point of weight. What is particularly intriguing about the VMH-lesioned animals, as well as many obese people, is that compared to normal individuals they are less motivated to "work" for food and only eat excessive amounts of palatable foods.[36] VMH-lesioned, and many overweight, individuals apparently do not overeat because they are hungry but because the proximity and palatability of food stimulates eating.

Conditioned Taste Aversion (CTA)

Animals have been observed to develop aversions to foods associated with bitter tastes and illnesses, and diets deficient in specific nutrients.[12,13,14,31,32,35] The adaptive significance of CTA is obvious.

In the 1960s Garcia, a psychologist with a biological and natural history background, and his colleagues initiated a series of experiments that demonstrated that animals would avoid food or flavored water consumed prior to experiencing an illness. This aversion would occur even if the onset of the illness was hours after eating or drinking, if the illness was induced by radiation exposure or by intraperitoneal injections, or if the agents that induced the illness were administered while the animal was anesthetized. Conditioned taste aversion is most likely to occur in response to the most recent, novel and/or salient (conspicuous) flavor consumed prior to illness and if the illness is prolonged or severe.[12,13,29]

Coyotes and wolves, made ill by lithium chloride after ingesting mutton wrapped in sheep/lamb wool, are subsequently repelled by the taste, odor and even sight of sheep and lambs. Gustavson reported significant reduction in lamb loss due to coyotes following field trials conducted in western Canada.[14]

Although other investigators have not been as successful in replicating CTA in coyotes,[3,6,20] this difference may well be due to differences in methodology and cooperation of livestock ranchers.[8,29,30]

Understanding CTA can be helpful in overcoming anorexia associated with illnesses, medication and radiation therapy. For example, a unique approach to reducing anorexia in children receiving chemotherapy is to have the child eat a strongly flavored sweet prior to treatment.[5] Subsequent to the chemotherapy the child may find that flavor unpalatable but not the nutritional foods eaten previously that day. There is no reason why this technique should not work with cats and dogs receiving chemotherapy. CTA might also be employed to treat many ingestive behavior problems of animals such as pica and hunting behaviors of dogs.

Self-Selection of Diets

Conditioned taste preference, the converse of CTA, has been proposed, but the concept is not universally accepted.[10,11] Reportedly, animals can acquire an increased preference for flavors consumed just before or during the recovery of an illness. Thiamine-deplete rats and cats have been reported to refuse thiamine-deficient diets and develop a preference for thiamine-rich diets.[31,32,35] If thiamine-deficient rats receive thiamine injections coincident with drinking saccharin-flavored water, the rats will increase their intake of saccharin-flavored water above baseline levels, even though the water contains no vitamins or calories.[12,35]

With the exception of salt, it is not believed that a depleted animal can immediately recognize missing nutrients by taste, nor is it believed that animals can immediately identify a nutritionally deficient or balanced diet.[10,35] It is instead theorized that a homeostatic process exists that assesses the internal state of well-being and alters the palatability of foods associated with the onset of those states.[11] Flavors associated with the onset of illnesses become less palatable and flavors associated with recovery become more palatable.

It cannot be assumed that domestic and captive animals will balance their own diets. There are contradictory interpretations of the "cafeteria choice" experiments that report optimal self-selection by laboratory animals.[10] It must also be considered that omnivores such as rats are probably better able to balance their diets than are obligate carnivores such as cats.[32] Veterinary clinicians are well aware of the devastating effects of nutritional secondary hyperparathyroidism that can occur in companion cats and captive primates. Perhaps for cats "in nature" the benefits of ingesting heart, liver and kidney are so valuable and the availability of these items so limited, that the palatability of these foods is protected from malaise and metabolic disturbances. The degree to which an animal is programed to be able to regulate its diet is undoubtedly related to availability of foods in the environment in which it evolved.

Feeding Problems

Coprophagia

Coprophagia is a common phenomenon in canids. It is normal for bitches to ingest feces of puppies for several weeks after they are born. This behavior has obvious hygienic benefits. Dogs of all ages and both sexes are extremely attracted to the manure of herbivores and also consume human and cat feces. Baby diapers are frequently scavenged by pet dogs. Van Heerden has commented that dogs appear to be used to clean up human feces around villages in third-world countries.[38]

Apparently, feces of other species are very palatable to dogs and the phenomenon is so ubiquitous that it must be considered a normal behavior. What benefit, if any, there is of such coprophagia is unknown.

Some dogs eat their own feces or those of other adult dogs (Fig. 9.1). This is considered an abnormal behavior. First, all possible medical disorders that may result in malabsorption should be investigated. The following suggestions, unsupported by any published

FIG. 9.1 A dog eating its own feces or that of other adult dogs. This is considered to be abnormal behavior.

data, have been recommended to treat coprophagia of healthy dogs that are apparently normal in other respects. If the dog is eating its own feces, feed it a highly digestible, predominantly meat diet or put a substance in the food, such as *Adolph's*®, a meat tenderizer (Adolph's Ltd., N. Hollywood, CA) or *FOR-BID*™ (Alpar Laboratories, Inc., La Grange, IL). Practitioners believe that these products work some of the time but it is not known how the products work. Additional bulk in the diet can help in some cases. Applying bitter substances, such as cayenne pepper or quinine, to feces may work to some degree but dogs can quickly learn to differentiate between tainted and untainted feces. Perhaps, conditioned taste aversion would be an effective way to treat this socially unacceptable habit.

Owners can combat coprophagia by immediately cleaning up after their dogs and cats.

Litter trays for cats should be kept in places inaccessible to household dogs. When on a walk the dog could wear a muzzle or head-halter that allows owners to close the dog's mouth easily or pull it away from feces.

Plant Eating

Both dogs and cats eat grass and other plants. Sometimes this material serves as an emetic, perhaps facilitating regurgitation of hair balls or contaminated food. Often, however, the greenery is not regurgitated. Perhaps the plant material is of some nutritional or medicinal value. There is increasing evidence that animals have evolved learning mechanisms by which they can associate flavors that coincide with the onset of recovery or illnesses.[11] To date no one knows why cats and dogs eat plants and grass. Perhaps the substances just taste pleasant.

Plant and grass eating is a normal behavior. The most effective way to keep cats from eating plants is to keep the plants out of reach of the cat. Consistently scaring the cat with a loud noise or bothering it with a spray of water might keep the cat away from the plants, at least when a person is present. A punishment that is activated when the cat approaches or touches the plant, independent of the presence of the owner, would be more effective. "Booby-traps" that might work are motion detectors that emit frightening sounds. Conditioned taste aversion might be effective in preventing ingestion of specific plants. One can also try providing indoor cats with one or more plants that they are allowed to eat.

Pica

Eating undigestible items is not uncommon in young puppies and kittens. It is part of the process of exploring the environment. This is a normal behavior and does not persist. Kittens usually sample kitty litter the first time they find it, prompting frantic calls from owners. Such exploratory behavior necessitates that owners keep valuable and potentially dangerous items out of reach of puppies and kittens.

FIG. 9.2 Wool-eating Siamese. Photograph courtesy of J. Bradshaw, Anthrozoology Institute, University of Southampton, U.K.

Pica, the persistent ingestion of nonnutritive substances, is often attributed to a dietary deficiency, has been associated with zinc toxicosis and can be a manifestation of central nervous system disorders such as hepatic encephalopathy.[16,28]

Lithophagia (ingestion of stones and gravel) by dogs is not rare, although this may be more an accident associated with obsessive retrieval and carrying of stones rather than true pica. Providing dogs with hard balls will reduce rock carrying. Scolding the dog whenever it begins to pick up a rock and instead encouraging it to carry sticks is also helpful. The use of head-halters on walks can prevent a dog from picking up objects. Perhaps the psychotropic drugs used to treat obsessive–compulsive behaviors in people would suppress dogs' compulsions to carry rocks.

Wool and fabric ingestion by cats is truly an obsessive ingestive disorder and primarily occurs in Siamese, Siamese hybrids and breeds derived from the Siamese (Fig. 9.2). Large portions of socks, draperies, bedspreads, sweaters and suits have disappeared down the throats of such cats. Some owners report that rendering fabric bitter and then periodically applying small amounts of the bitter tasting substance on the material will act as a repellent, at least temporarily. The only reliable control is by denying the cat access to the materials it is eating.

Predatory and Hunting Behaviors

In the course of domestication, there has been considerable selection for hunting behaviors. Is it any wonder that most dogs are highly motivated to chase and sometimes attack livestock, domestic fowl and small mammals?

Confinement and keeping pet dogs on leashes is the most effective way to prevent unwanted livestock chasing. Remote-controlled shock collars are often used improperly, resulting in unwarranted infliction of pain and the development of anxiety and fearful behaviors by the dog. The application of Gustavson's conditioned taste aversion work to stop coyote predation on sheep would be worth exploring to treat unwanted hunting behaviors in the domestic dog.[14]

Inappetence

It is important that utility dogs, animals in prolonged transit, show animals and most injured or ill animals take in nutrients. Palatability, familiarity and temperature of the food are variables that can be manipulated to stimulate feeding. Eating can also be facilitated socially by other dogs or people.

Often owners can socially facilitate eating by feeding a dog small amounts by hand. After two or three handsful, if the owner lowers a food-filled hand into the bowl, the dog or puppy will eat from the hand in the bowl. As the dog is eating, the food is then dropped from the hand into the bowl and, as the dog eats from the bowl, the hand is slowly withdrawn. If the owner stays there the dog will usually eat from the bowl. Over a few sessions, hand feeding should gradually stop and the owner begin leaving the dog for short intervals while it is eating; at first leaving the dog for only a few seconds, then gradually extending the time the dog is left alone. It is important that the owner transfer feeding to the bowl within a few days. Feeding by hand can be conditioned, establishing an inconvenient ritual for the owner.

An animal that is recovering from an illness may resist eating its normal diet because that diet coincided with the onset of illness. If the patient is offered a palatable food with a salient novel flavor, the animal may eat it. Thereafter small amounts of the animal's normal diet can be gradually incorporated into the newly flavored food. Once an animal "discovers" its former diet does not result in illness, the flavor of that food ceases to be unpalatable and the animal resumes its normal diet.

Aggression and Competitive Feeding

Palatable food or coveted food-related items (e.g. rawhides, cows' hooves) can precipitate aggression between dogs that are otherwise amicable. Puppies are subordinate to adults, but not necessarily "wise." The author is aware of adult male dogs that have crushed the heads of puppies that approached the males while they were eating or chewing on bones. These males had previously tolerated the same puppies in other situations. These incidents involved puppies that had been introduced into the males' households. Perhaps the adults would not have responded this way if the puppies had been whelped and reared in the males' households, in which case the dogs might have "considered" the approaching puppies their own offspring and inhibited their aggression. This is, however, an unadvisable risk.

If dogs are aggressive to each other over food, owners should feed them separately and spaced apart. Owners should ensure that the dogs are a safe distance from each other when given special treats and monitor the dogs while the treats are consumed. Owners also should be careful about leaving lasting, *coveted* food related objects (such as rawhides or bones) with *some* dogs. If there is a clear delineation in subordinate/dominant roles, the subordinate usually will not compete for the items. Serious fights may, however, ensue if the dogs are close in their dominant status, are vying for a dominant position or if both dogs are highly motivated to possess the items.

When the diets of dogs with aggressive tendencies are reduced or the dogs are on medication that increases their appetite, they may become aggressive over food not previously defended, manifest aggression in areas associated with food (e.g. the kitchen and locations where the dogs are usually fed) and become more aggressive in other circumstances. There are anecdotal reports that dogs on reducing diets are irritable in situations unrelated to food. The irritability subsides, when the dog's food intake is increased. Interestingly, dogs with a dominant aggressive profile are often rated by owners as having excessive appetites.

Obesity

Obesity is common in the pet animal. It has been estimated that 20–30% of companion dogs are overweight (see Chapter 10).[1,17] Obesity is also sometimes a problem in household cats.[17,32] The predisposition to maintain a set weight, easily accessible and highly palatable food, little exercise, breed predispositions, ovariectomy, progestins and the propensity of owners to share food with their companions all contribute to overweight pets.

When the diet of an animal is altered or restricted, the animal may manifest behaviors not exhibited previously. For example, a cat that was accustomed to being fed early in the morning would spray on the bedroom dresser only on mornings when the owners slept late. Excessive vocalization and urine marking have frequently been correlated with suddenly implemented and strict weight reduction programs for cats. When the cats were allowed more food, the behavior problems subsided.[24,43] Sudden weight loss and "starvation diets" also have been associated with hepatic lipidosis in cats.[15]

On restricted diets, dogs may engage in aggravating behaviors, such as pestering the owner, restlessness, garbage raiding, possessiveness over food and irritability.

When a clinician advises owners to reduce the weight of their pets, it would be wise to obtain a complete behavioral profile of the animal. Particularly important information regarding a dog would include its present appetite level, a description of all other animals and people with whom it interacts, its social status among other dogs and people, circumstances in which it may already be exhibiting aggression, where it is fed and accessibility to long-lasting treats.

The clinician should be prepared to provide a daily management plan for dieting pets. Exactly what to recommend would depend on the individual situation. For example, if two dogs already engage in little squabbles, it would be unwise to leave treats with them when unsupervised. If a dog already has a tendency to growl and mutter at family members, it would be very important to keep small children away from these dogs while the dogs are being fed and perhaps also while the family meals are being prepared or eaten. Sometimes specific behavior modification programs may have to be developed to accommodate changes, or potential changes, in behavior. Collaboration with a behavior specialist would be beneficial in some cases.

Treats and Food Rewards

Giving treats and other food to the pet needs to be addressed. One survey in the U.S.A. reported that 86% of dog owners and 68% of cat owners shared their snacks with their pets; 64% of dog owners and 67% of cat owners shared food from the table with the pet. People may be biologically predisposed to share food with humans and non-human animals to whom they are attached.[39,40] Contrary to popular belief, owners who share food with their dogs are not more likely to report that their dogs engage in behavior problems than are owners who do not give treats to their dogs.[41]

Owners are usually determined to share foods with their pets and therefore consideration should be given to accommodation of treat giving in weight-reducing programs. The *type*, *size* and *total amount* of treats should be discussed in detail. Depending upon dietary restrictions, untraditional treats should be considered. It is surprising how many dogs like carrots, celery and even lettuce. Hot air popped popcorn is also a possibility.

Expending calories with exercise would allow some leeway for treats. Cats can learn to walk on leashes. If the owner is incapacitated, dog and cat walkers can be employed. Head-halters permit almost everyone to walk dogs, even dogs that have a tendency to pull.

Food rewards for training and behavior modification programs should also be addressed. If possible, a portion of the animal's regular diet should be used. If another type of food reward is used, it should not compromise the animal's daily intake of appropriate nutrients or result in excessive calorie consumption. A tasty food reinforcer can be very small, about the size of a raisin, and need not be given to the animal every time it responds appropriately. In fact, an intermittently reinforced animal will perform for longer periods of time and more intensely than one that is continuously reinforced. After the animal has acquired the desired behaviors, social reinforcements can usually maintain the behaviors.

References

1. Anderson, R. S. (1984) Nutrition of the dog and cat —an overview. In *Nutrition and Behaviour in Dogs and Cats*. Ed. R. S. Anderson. pp. 3–10. Oxford: Pergamon.
2. Bekoff, M. and Wells, M. C. (1980) The social ecology of coyotes. *Scientific American*, **242**, 130–148.
3. Booth, D. A. (1985) Commentary on "Coyote Control and Taste Aversion": Editor's Report. *Appetite*, **6**, 282–283.
4. Borchelt, P. L. and Voith, V. L. (1986) Dominance aggression in dogs. *Compendium on Continuing Education: Small Animal Practice*, **8**, 36–44.
5. Broberg, D. J. and Bernstein, I. L. (1987) Candy as a scape-goat in the prevention of food aversion in children receiving chemotherapy. *Cancer*, **60**, 2344–2347.
6. Burns, R. J. and Connolly, G. E. (1985) A comment on "Coyote Control and Taste Aversion." *Appetite*, **6**, 276–281.
7. Coppinger, L. and Coppinger, R. (1982) Livestock-guarding dogs that wear sheep's clothing. *Smithsonian*, **13(1)**, 64–73.
8. Ellins, S. R. (1985) Coyote control and taste aversion: a predation problem or a people problem? *Appetite*, **6**, 272–275.
9. Fitzgerald, B. M. (1988) Diet of domestic cats and their impact on prey populations. In *The Domestic Cat: The Biology of Its Behaviour*. Eds. D. C. Turner and P. Bateson. pp. 123–144. Cambridge: Cambridge University Press.
10. Galef, B. G. (1991) A contrarian view of the wisdom of the body as it relates to dietary self-selection. *Psychological Review*, **98**, 218–223.
11. Garcia, J. (1990) Learning without memory. *Journal of Cognitive Neuroscience*, **2**, 287–305.
12. Garcia, J., Hankins, W. G. and Rusiniak, K. W. (1974) Behavioral regulation of the milieu interne in man and rat. *Science*, **185**, 824–831.
13. Gustavson, C. R., Gustavson, J. C., Young, J. L., Pumariega, A. J. and Nicolaus, L. K. (1988) Estrogen induced malaise. In *Neural Control of Reproductive Function: Proceedings of the Fifth Galveston Neuroscience Symposium*. Eds. J. M. Lakoski, J. R. Perez–Polo and D. K. Rassin. pp. 501–523. New York: A. R. Liss.
14. Gustavson, C. R., Jowsey, J. R. and Milligan, D. N. (1982) A 3-year evaluation of taste aversion coyote control in Saskatchewan. *Journal of Range Management*, **35**, 57–59.
15. Hand, M. S., Armstrong, P. J. and Allen, T. A. (1989) Obesity: occurrence, treatment and prevention. *Veterinary Clinics of North America: Small Animal Practice*, **19**, 447–474.
16. Hardy, R. M. (1992) Hepatic encephalopathy. In *Current Veterinary Therapy XI: Small Animal Practice*. Eds. R. W. Kirk and J. D. Bonagura. p. 639. Philadelphia: W. B. Saunders.
17. Houpt, K. A. (1991) Feeding and drinking behavior problems. *Veterinary Clinics of North America*, **21**, 281–298.
18. Houpt, K. A., Coren, B., Hintz, H. F. and Hilderbrant, J. E. (1979) Effect of sex and reproductive status on sucrose preference, food intake and body weight of dogs. *Journal of the American Veterinary Medical Association*, **174**, 1083–1085.
19. Kerby, G. and Macdonald, D. W. (1988) Cat society and the consequences of colony size. In *The Domestic Cat: The Biology of Its Behaviour*. Eds. D. C. Turner and P. Bateson. pp. 67–81. Cambridge: Cambridge University Press.
20. Lehner, P. N. and Horn, S. W. (1985) Research on forms of conditioned avoidance in coyotes. *Appetite*, **6**, 265–267.
21. Liberg, O. (1981) Predation and social behavior in a population of domestic cats: an evolutionary perspective. PhD thesis, University of Lund, Sweden.
22. Liberg, O. and Sandell, M. (1988) Spatial organisation and reproductive tactics in the domestic cat and other felids. In *The Domestic Cat: The Biology of Its Behaviour*. Eds. D. C. Turner and P. Bateson. pp. 83–98. Cambridge: Cambridge University Press.
23. Line, S. and Voith, V. L. (1985). Dominance aggression of dogs towards people: behavior profile and response to treatment. *Applied Animal Behavior Science*, **16**, 77–83.
24. Marder, A. (1992) Feline behavior problems related to diet. *Animal Behavior Consultant Newsletter*, **9**, 1–2.
25. Mead, C. J. (1982) Ringed birds killed by cats. *Mammals Review*, **12**, 183–186.
26. Mech, L. D.(1970) *The Wolf: The Ecology and Behavior of an Endangered Species*. pp. 71–74, New York: The Natural History Press.
27. Mugford, R. A. (1977) External influences on the feeding of carnivores. In *The Chemical Senses and Nutrition*. Eds. M. R. Kare and O. Maller. pp. 25–50. New York: Academic Press.
28. Ogden, L. (1992) Chemical and physical disorders. In *Current Veterinary Therapy XI: Small Animal Practice*. Eds. R. W. Kirk and J. D. Bonagura. p. 198. Philadelphia: W. B. Saunders.
29. Quick, D. L. F., Gustavson, C. R. and Rusiniak, K. W. (1985) Coyote control and taste aversion. *Appetite*, **6**, 253–264.
30. Quick, D. L. F., Gustavson, C. R. and Rusiniak, K. W. (1985) Coyote control and taste aversion: the authors' reply. *Appetite*, **6**, 284–290.

31. Richter, C. P. (1943) Total self regulatory function in animals and human beings. *Harvey Lecture Series,* **38**, 63–103.

32. Robinson, I. H. (1992) A taste for survival. In *Waltham Feline Medicine Symposium.* pp. 44–64. Vernon, CA: Kal Kan Foods.

33. Romsos, D. R. and Ferguson, D. (1983) Regulation of protein intake in adult dogs. *Journal of the American Veterinary Medical Association,* **182**, 41–43.

34. Ross, S. and Ross, J. G. (1949) Social facilitation of feeding behavior in dogs: I. Group and solitary feeding. *The Journal of Genetic Psychology,* **74**, 97–108.

35. Rozin, P. (1977) The significance of learning mechanisms in food selection: some biology, psychology and sociology of science. In *Learning Mechanisms in Food Selection.* Eds. L. M. Barker, M. R. Best and M. Domjan. pp. 557–581. Waco, TX: Baylor University Press.

36. Schachter, S. (1971) Some extraordinary facts about obese humans and rats. *American Psychologist,* **26**, 129–144.

37. Turner, D. C. and Meister, O. (1988) Hunting behaviour of the domestic cat. In *The Domestic Cat: The Biology of Its Behaviour.* Eds. D. C. Turner and P. Bateson. pp. 111–121. Cambridge: Cambridge University Press.

38. Van Heerden, J. (1989) Small animal problems in developing countries. In *Textbook of Veterinary Internal Medicine, Diseases of the Dog and Cat.* Ed. S. J. Ettinger. 3rd edn. pp. 217–226. Philadelphia: W. B. Saunders.

39. Voith, V. L. (1981) Attachment between people and their pets: behaviour problems of pets that arise from the relationship between pets and people. In *Interrelations Between People and Pets.* Ed. B. Fogel. pp. 271–294. Chicago: Charles C. Thomas.

40. Voith, V. L. (1985) Attachment of people to companion animals. *Veterinary Clinics of North America: Small Animal Practice,* **15**, 289–295.

41. Voith, V. L., Wright, J. C. and Danneman, P. J. (1992) Is there a relationship between canine behaviour problems and spoiling activities, anthropomorphism and obedience training? *Applied Animal Behaviour Science,* **34**, 263–272.

42. Wolski, T. R. (1982) Social behavior of the cat. *Veterinary Clinics of North America: Small Animal Practice,* **12**, 693–706.

43. Wright, J. C. (1992) Hunger induced feline inappropriate elimination. *Animal Behavior Consultant Newsletter,* **9**, 2–4.

44. Zeuner, F. E. (1963) *A History of Domesticated Animals.* New York: Harper & Row.

CHAPTER 10

Obesity

PETER J. MARKWELL and RICHARD F. BUTTERWICK

Introduction

The definitions that have been proposed for obesity fall into two broad groupings: one of these is based on a "mathematical" definition of a certain percentage greater than ideal weight, the second on an implication of physiological impairment associated with excessive fat accumulation. The application of a mathematical definition is more appropriate in humans, where extensive data exist on optimal height–weight standards, than in companion animals where data about optimal weights are essentially limited to purebred dogs, with little information on purebred cats and no "standards" being possible in cross-bred animals. Thus the proposed definitions for man that range from 10 to 20% above ideal weight are of very limited application in veterinary practice.[19,20]

Examples of the second type of definition include "excessive accumulation of body fat sufficient to impair body functions" and "pathological condition characterized by an accumulation of fat much in excess of that necessary for optimal body function."[27,54] Implicit in this type of definition is the fact that obesity is a condition detrimental to the health of an individual. This is a valid message for veterinary medicine, as it emphasizes to both the veterinarian and the pet owner the importance of managing the condition. It does not, however, help to assess whether an individual patient is obese or not. The practical assessment of patients is considered in detail later.

Obesity is probably the most common form of malnutrition likely to be observed in companion animal practice. It is important not only in the dog and cat, but also in some cage birds, particularly Amazons and Cockatoos. A number of surveys of the incidence of obesity have been conducted in veterinary practice, mostly with regard to the dog. Data from these studies are summarized in Fig. 10.1. Although these figures imply that approximately one quarter to one third of dogs seen in

INCIDENCE OF OBESITY IN DOGS AND CATS		
	Observed incidence (%)	Ref.
Dogs	28	53
	34	3
	44	74
	*24.3	28
Cats	6–12.5	3
	40	70
	†25	67

* Includes dogs categorized as obese or gross.
† Includes cats categorized as overweight (20%) and obese (5%).

FIG. 10.1 Results of surveys of the incidence of obesity in dogs and cats.

veterinary practice are likely to be obese, it should not be assumed that this is necessarily representative of the population as a whole. The reason for this is that obesity has been linked with a number of serious medical conditions in dogs, and thus it is likely that a biased population is seen in practice, particularly as they are presented primarily for reasons other than obesity. Notwithstanding this, obesity is clearly a highly significant problem in the dog population, and it is interesting to make a comparison with a survey in humans in which 21% of males and 22% of females were judged obese in a study of nearly 1000 subjects.[21]

Among purebred dogs there is evidence that certain breeds are more or less prone to obesity than others; these are summarized in Fig. 10.2.[28] The reasons why certain breeds are apparently prone (or resistant) to obesity are not known, although for one of the breeds (Cavalier King Charles Spaniel), studies have shown poor food selection ability (WCPN, unpublished observations). It is conceivable that limited ability to discriminate between food types may increase the likelihood of an animal overeating and becoming obese. Further clues come from a study comparing feeding behavior of Beagles (shown to be prone to obesity) with that of Irish Setters when faced with foods of varying palatability. The Setters tended to show good regulation of intake despiee wide ranges of energy density and palatability between diets, whereas the Beagles tended to overeat when fed a highly palatable diet, although eating relatively little of the less palatable diets.[56] Similar behavior,

eating large quantities of highly palatable foods, has been reported in obese humans, and thus this apparent overriding of food intake regulation by certain breeds when faced with some types of food may be a contributory factor to their predisposition to obesity. It has also been shown that neutered females were about twice as likely as expected to be obese or gross, and a similar trend was observed in neutered males.[28] These points stress the need for communication with owners about feeding practices following neutering, particularly of breeds known to be prone to obesity.

Fewer data are available regarding the prevalence of obesity in cats. An early study in the U.K. suggested that it appeared to be a much less common condition than in the dog.[3] Recent data, however, suggest that in certain countries obesity may occur in cats with a similar frequency to that observed in dogs.[67,70] Although it is often assumed that cats in general regulate their energy intake in relation to expenditure more accurately than dogs, this view is not supported by these more recent data. It has been suggested that feline obesity may be more of a problem in North America due to differences in lifestyle and feeding patterns of cats.[4] It has been suggested that overweight cats are more likely to be male, neutered and of mixed ancestry.[67]

Clinical Conditions Associated with Obesity

Obesity in humans has been described as one of the most important medical hazards in the U.S.A., shortening the lifespan and particularly increasing the incidence of cardiovascular and gall bladder disease and diabetes.[89] Although parallels between the dog and humans are frequently drawn with regard to the risks of obesity, relatively few clear links have been established in the dog, in part due to the lack of objective assessment of obesity. In cats there are even less data linking obesity with disease, although some important connections have been made.

One of the links between canine obesity

BREED SUSCEPTIBILITY TO OBESITY	
Breeds most likely to become obese	**Breeds least likely to become obese**
Labrador Retriever	German Shepherd
Cairn Terrier	Greyhound
Shetland Sheepdog	Yorkshire Terrier
Bassett Hound	Dobermann
Cavalier King Charles Spaniel	Staffordshire Bull Terrier
Beagle	Lurcher
	Whippet

FIG. 10.2 Breed susceptibility to obesity.

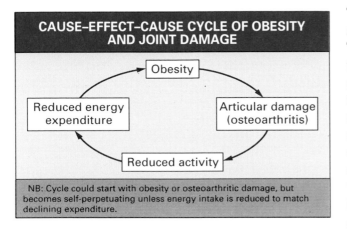

CAUSE–EFFECT–CAUSE CYCLE OF OBESITY AND JOINT DAMAGE

NB: Cycle could start with obesity or osteoarthritic damage, but becomes self-perpetuating unless energy intake is reduced to match declining expenditure.

FIG. 10.3 Cause–effect–cause cycle of obesity and joint damage.[27] The cycle could start with obesity or osteoarthritic damage, but becomes self-perpetuating unless energy intake is reduced to match declining expenditure.

and disease that has been advocated is the association with osteoarthritis. Indeed, Joshua suggested that osteoarthritis was by far the most important pathological effect of over-weight.[38] Edney reported that 24% of the dogs in his series of cases had serious joint complications,[27] and made the point that joint injury, leading to reduced mobility and energy expenditure, could lead to a cause–effect–cause cycle (Fig. 10.3), thus progressively worsening both obesity and joint damage. More recent data fail to confirm such an obvious numerical link, and clearly increased risk of locomotor problems was noted only in dogs that tended to be classified as gross rather than just obese.[28] This survey showed an increased risk of locomotor problems in dogs over 10 years of age, and it is thus possible that the interactions of obesity, age and osteoarthritic problems may have com-plicated interpretation of data in some studies. It does, however, seem reasonable to support the suggestion that the abnormal stress imposed by excess weight on the major weight-bearing joints, especially stifles and elbows, is an important factor in the etiology of canine osteoarthritis.[38]

Respiratory difficulties are associated with obesity in man. The work of breathing is increased if considerable additional weight is carried on the chest wall and excessive adi-pose tissue will complicate body oxygenation.

This results in decreased exercise tolerance and even difficulty with normal breathing.[54] These data may not be particularly applicable to the dog or cat, as gravitational forces act at different angles to the longitudinal axis of the body in quadrupeds and bipeds, thus altering the mechanics of obesity-related dysfunction.[17] Respiratory distress has, how-ever, been noted as a frequent problem in fat dogs and this may be caused by narrowing of the airway in laryngeal and pharyngeal areas by fatty deposits in surrounding tissue.[38] In addition, clinical experience suggests that exercise intolerance may also be an important complication in some individuals.

An increase in circulating blood volume and cardiac output are required to perfuse excess adipose tissue even though it is relatively avascular; the effect of this is to reduce cardiovascular reserves.[17] It has been suggested that this increased workload may be imposed on a heart weakened by fatty infiltration, contributing to congestive heart problems.[49] One clinical report also noted myocardial hypoxia in an obese Beagle, which improved after weight loss.[6] Clinical problems asso-ciated with the cardiovascular system may only become apparent in severely obese animals. Edney and Smith noted an increased risk of circulatory problems only in dogs they categorized as gross, and not in those categorized as obese.[28]

The very close association between obesity and diabetes mellitus in humans has been known for many years,[37] and a similar link has been established in the dog[41] and cat.[64] Insulin resistance has been widely described in human obesity and many studies have investigated its cause. In reviewing these studies Olefsky *et al.* commented that it appears to be a heterogenous phenomenon, arising in some patients from decreased numbers of insulin receptors, and in others from a combination of this with a post-receptor defect.[62] It was considered that a continuum of insulin resistance existed in human obesity, with a relationship between increasing hyperinsulinemia and the severity of insulin resistance. Individuals with mild insulin resistance only show a defect in insulin receptors, whereas obese patients with more severe hyperinsulinemia and insulin resistance also show the postreceptor defect. It has been suggested that the influence of obesity on insulin kinetics could play a role in the transient nature of diabetes in some cats.[59]

Obese dogs may be more at risk from infectious diseases than animals of normal body weight. Studies of experimental infections with canine distemper virus or *Salmonella typhimurium* both resulted in more severe effects in obese than in control dogs.[61,88]

Liver disease may be an important consequence of obesity. Fatty infiltration of the liver has been noted in obese dogs[38,49] that could be associated with decreased function; furthermore, obesity was reported as a major cause of fatty liver in man.[68] Of particular importance is the relationship between liver disease and obesity in cats, as obesity is considered an important predisposing factor to idiopathic hepatic lipidosis, a condition carrying a poor prognosis (see Chapter 15). Typically cats presenting with this condition will have a history of obesity followed by recent weight loss, and will have been partially, or completely, anorectic for a period of time (median: 2–3 weeks). The anorexia may have been caused by some stress, such as moving, or by attempts to change the cat's diet.[15,32] The mechanism by which hepatic lipidosis develops in obese cats has not yet been established, although several hypotheses have been advanced. These include a relative deficiency of arginine (an essential amino acid for the cat), carnitine deficiency or endocrine abnormalities.[32] The importance of this disease necessitates extreme care when weight reduction programs are instituted in cats.

Obesity is also associated with increased surgical risk, both on a physiological and a technical level. Technical difficulties associated with surgery in the fat animal include increased depth of tissue to be penetrated before entering the abdomen, risk of fat necrosis following surgical trauma and an increased potential for slow healing or wound breakdown.[38] If coupled with the increased risk of infection noted above, the potential for postoperative problems is substantially increased in the obese patient. As well as the technical problems associated with surgery in the fat patient, which will lead to prolonged procedures, there are also other factors that influence anesthetic management and potentially increase anesthetic risk. These include altered drug kinetics and the possibility of cardiopulmonary and hepatic disorders.[17] Thus surgery of the obese patient should be approached with considerable caution, and consideration given to weight reduction when possible prior to undertaking procedures. It should be noted, however, that where obesity is a consequence of an underlying condition rather than the cause of it, weight reduction will not necessarily reduce anesthetic risk.[17]

Feline lower urinary tract disease (see Chapter 17) is another condition that has been linked epidemiologically with obesity, with overweight cats suggested as being more at risk of developing lower urinary tract problems.[83,87] These studies did not determine the precise causes of lower urinary tract disorders in affected cats, although in a separate survey of urolithiasis, over 80% of affected animals were considered overweight.[33] Why being overweight should predispose an animal to lower urinary tract disease remains something of an enigma.

A range of other conditions have been considered in association with obesity, but in the extensive survey conducted by Edney and Smith, no clear relationship was established between obesity and skin problems, reproductive problems or neoplasia.[28] Nevertheless, there are clear associations between obesity and serious clinical consequences for dogs and cats, and thus a definition of obesity as a condition detrimental to the health of an individual applies equally to these species, as to humans. At the very least obesity in companion animals will reduce the quality of life, but in addition is likely to reduce the quantity of that life. Thus, there are sound clinical grounds for embarking on programs of obesity management.

Causes of Obesity

Obesity will arise as a result of positive energy balance, that is, intake exceeding expenditure, resulting in formation of excess adipose tissue. Obesity is no longer seen in humans as simply due to overeating, that is due to gluttony or to more serious disorders of personality, but is recognized as a complex phenomenon. Types of human obesity include genetic obesity, obesity of hypothalamic or other central nervous system origin, of endocrine origin and induced by psychic disturbances and cultural and social pressure.[54] Bray suggested that most human obesity was due to the interaction of several factors including genetic and environmental components.[8] Suggested important enviromental factors included the ready availability of tempting foods and a steady decrease in physical activity. The group in which treatable causes could be identified (e.g. hyperadrenocorticism) represented less than 5% of cases of obesity. The extent to which this information can be applied to the dog and cat is not clear, although again it seems likely that treatable causes of obesity will be present in only a small minority of cases.

Obesity is a common familial disorder, and evidence in humans suggests that a genetic component may be responsible for the development of obesity in certain individuals. In reality, it is difficult to delineate a familial aggregation of obesity due to genetic factors, from shared external factors such as diet and lifestyle, although recent studies on identical twins exposed to differing environments have been able to dissociate genetic from environmental factors.[77,78] It has been suggested that the genetic component that predisposes certain individuals to obesity results in a defect in the regulation of energy intake or energy expenditure.[65] In support of this a number of studies have demonstrated a familial aggregation of metabolic rate.[65,66] Although current data are not unequivocal, it appears that individuals with a low or reduced metabolic rate may be predisposed to obesity. It has also been suggested that a reduced thermic effect of food (TEF) may be a contributory factor in the development of obesity.[65] The quantitative importance of this factor in the development of obesity, however, is unclear, because the apparent difference in TEF between obese and nonobese individuals is not great.[58]

Genetic factors may also be important in the dog. Certain breeds may be more prone to obesity than others (Fig. 10.2).[28,56] Imposed upon the genetic base of the pet are environmental and social factors, and the key in terms of feeding is the pet's owner. Indeed, Houpt and Smith commented that most cases of obesity appeared to be related to owner-induced variables.[36] Clearly, no owner deliberately sets out to create an obese pet, but an understanding of some of the circumstances under which this may occur, helps in the formulation of programs for the management of obesity. These behaviors may give rise either to overfeeding, or to underexercising of a pet, or in some cases to both (Fig. 10.4). Some of these behaviors are particularly important and warrant further discussion.

OWNER BEHAVIORS CONTRIBUTING TO OBESITY		
Behavior	Cause of overfeeding	Cause of underexercising
Failure to adjust feeding to individual needs	✓	
Additional energy from snacks or treats or dietary supplements is ignored	✓	
Encouraging appetite as a sign of good health	✓	
Indulging begging behavior	✓	
Providing food as a palliative to the pet when left alone	✓	
Preventing aggressive behavior	✓	
Providing food instead of exercise	✓	✓
Providing inadequate exercise		✓

FIG. 10.4 Owner behaviors contributing to obesity.

Mason commented that dogs fed on home-prepared diets were more likely to become obese than those fed on commercial canned meat.[53] The reasons for this are likely to lie with the choice of raw materials in home-prepared diets (perhaps using ingredients with high fat contents), their lack of day-to-day consistency and their lack of feeding guides. These features combine to cause difficulties in accurate feeding. Conversely, commercial diets provide consistency in formulation and feeding guides for the owner; however these feeding guides do require some interpretation. Although based on the best scientific knowledge at a given time, they represent an average requirement and must, therefore, be adjusted for an individual animal on the basis of body weight change. Although it is considered that the maintenance requirement of an individual dog may be calculated from:

$$523 \ (125) \times \text{body weight (kg)}^{0.75} = \text{kJ (kcal)}$$
$$\text{metabolizable energy/day,}$$

feeding studies have shown that for individual dogs this may vary by as much as ±30% (WCPN, unpublished data). Failure to recognize the need for individual adjustment may be an important factor in the development of obesity in some cases.

A second important factor is failure to take into account additional energy obtained from snacks, treats or other supplementary foods given by the owner. Snacks and treats help to fulfil an important aspect of the pet–owner relationship, and are an acceptable part of the diet provided that their contribution to the total is limited, and is taken into account as part of the complete energy intake. Some supplements that may be added to the diet, for example vegetable oil, can also add significant quantities of energy, which may not be taken into account during feeding the main meal.

Provision of inadequate exercise may result from inability of the owner to devote adequate time to a pet, or could result from changed circumstances such as illness or infirmity making it impossible for an owner to walk their dog. In cats, confinement to the home could be a contributory cause to decreased exercise, and in both species orthopedic injury could result in enforced confinement.

Most of these owner–pet interactions are of greater relevance to dogs than to cats.

One further owner-related factor that has been associated with obesity in dogs, is obesity in the owner.[53] A number of reasons may explain this, including lack of exercise by the owner (and hence the dog), perhaps consumption of excess high energy foods by the owner with scraps going to the dog, possible failure of recognition of a problem situation developing in the pet or even acceptance of a fat dog as "normal."

An understanding of the factors that may have contributed to the development of obesity in an individual dog or cat, including possible genetic aspects, factors such as neutering, and environmental or social factors will enable weight reduction programs to be tailored toward individual cases, with a consequent increase in the likelihood of success.

Diagnosis of Obesity

Obesity is generally a simple condition to recognize in the dog, and is unlikely to be confused by the veterinarian with other clinical problems that could cause abdominal distension. In the cat, the condition may be more difficult to recognize unless an individual is grossly obese, as excess fat tends to be stored intraabdominally and in the inguinal region. In both species, assessing the degree of obesity and differentiating between uncomplicated obesity and that arising from a primary disease is more difficult.[85]

Assessment of Obesity

As has been noted, the application of optimal height–weight standards used in humans that allow definitions of obesity of 10–20% above ideal weight are not readily applicable to veterinary practice because of limited information about ideal weights of dogs and cats.[19,20] In addition to this, the value of these types of tables may be limited as they do not account for body composition, thus potentially assessing individuals with substantial muscle development as overweight.[82]

An alternative and simple method of obesity assessment used in humans is the measurement of subcutaneous fat by skinfold calipers.[25] This method assumes that the thickness of the subcutaneous fat layer is representative of total body fat. This technique has been evaluated in dogs, and unfortunately has proven useless, because the skinfold appears to lift off the underlying subcutaneous tissue and as a result does not correlate with the degree of obesity.[3]

A variety of other, essentially laboratory based, methods of assessment of body composition have been reported in humans and/or dogs. These include densitometry, a technique that requires underwater weighing of a compliant subject. This technique provides good precision, and is often considered as the "standard" method,[18] although it is clearly not appropriate to a companion animal clinical situation. Other techniques include total body water (isotope dilution),[69] total body potassium; dual photon and dual energy X-ray absorptiometry, neutron activation analysis and electrical conductivity,[34,50] computed tomography[5] and magnetic resonance imaging.[30] These techniques are complex and require expensive equipment and currently have little application to clinical assessment of obesity in dogs or cats.

One technique that may show greater promise is the measurement of subcutaneous fat by ultrasound. This has proven successful in humans,[43] and has also been used for estimating body fat in dogs.[1,86] In the latter study, using A-scan mode ultrasonography, it was found that subcutaneous fat could be estimated reliably, and that if measurements were taken from the midlumbar area they could be used to predict total body fat (Fig. 10.5). Thus this technique may provide a means by which obesity can be quantified, at least in the dog.

Until such techniques are developed fully and applied widely, assessment of obesity in practice has to remain largely subjective, based essentially on observation and palpation. The amount of tissue overlying the

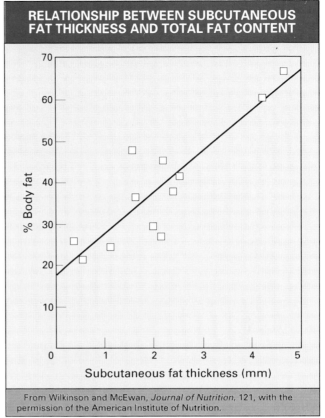

RELATIONSHIP BETWEEN SUBCUTANEOUS FAT THICKNESS AND TOTAL FAT CONTENT

From Wilkinson and McEwan, *Journal of Nutrition*, 121, with the permission of the American Institute of Nutrition.

FIG. 10.5 Relationship between ultrasonic measurement of subcutaneous fat thickness at the lumbar area and total fat content in 12 dogs.[86]

ribcage represents the most practical means of assessing obesity in dogs.[85] Joshua suggested that if a layer of tissue of 0.5 cm or more overlays the ribs, although they were still palpable, then a degree of overweight was probable.[38] Inability to palpate the ribs was indicative of frank obesity. Edney and Smith tried to improve quantification of assessment by listing criteria appropriate to a five-point condition scale.[28] Obese was defined by the ribcage not being visible when the dog moved, the bones of the chest being barely palpable and body weight noticeably more than normal for the type. Even with these criteria, this still remained largely a subjective assessment.

The situation with cats is more difficult, owing to the different fat distribution in this species. Thus assessment of an obese cat is, perhaps, even more subjective. Some guidance may be gained from body weight, where it is generally accepted that a healthy nonobese domestic cat should weigh 3.5–4.5 kg, although some domestic cats with larger frames may be heavier.[13]

Differential Diagnosis

The terms uncomplicated and complicated obesity differentiate between adiposity resulting from overeating/underexercising, and that resulting from other causes.[17] The most important causes of complicated obesity are endocrine disorders and drug-induced polyphagia, although it is possible that hypothalamic lesions may be responsible for very rare cases.[22] Although treatable disorders certainly can result in the development of obesity in humans, it has been estimated that these account for less than 5% of cases.[8] Similar data do not exist for the dog, but it is unlikely that the situation is substantially different. Even if these underlying conditions are relatively uncommon, it is necessary to determine their presence or absence to allow for appropriate therapy.

Drug-induced polyphagia can be diagnosed from the animal's history. Drugs that are appetite stimulants include megestrol acetate and some anticonvulsants, especially primidone.[47]

Endocrine diseases that can result in complicated obesity include hypothyroidism and hyperadrenocorticism, and these are the two main differential diagnoses.[22] History and physical examination may allow differentiation (Fig. 10.6), but if necessary confirmation may be obtained through thyroid and adrenal function testing (Figs. 10.7 and 10.8). Adrenal function can be assessed using an ACTH stimulation test (Fig. 10.7), a dexamethasone suppression test or urine cortisol.

Treatment of Obesity

The first stage in the treatment of obesity is a thorough clinical evaluation with appropriate hematological and biochemical testing to rule out underlying causes of the condition. The presence of coexisting diseases such as heart or osteoarthritic problems that may complicate some aspects of therapy should also be evaluated. Although endocrine and other diseases may be the primary cause of some cases of obesity, the genetic and environmental/social factors discussed above are likely to be responsible for the majority of cases seen in practice. The cornerstone of obesity management is to create a situation of negative energy balance and hence body tissue mobilization in the animal; however, identifying and addressing the underlying cause or causes on an individual basis is clearly a key factor both in obtaining and maintaining successful weight reduction. Without this, weight loss may not occur, or at best is likely to be transitory. It follows that these issues must be considered when deciding on the most appropriate therapeutic regimen for a particular case.

SYMPTOMS AND SIGNS THAT AID DIFFERENTIATION OF OBESITY FROM HYPERADRENOCORTICISM AND HYPOTHYROIDISM*			
	Obesity	Hyperadrenocorticism	Hypothyroidism
Age at occurrence	All ages	Middle-age	<5 years
Overweight	++	±	±
Location of fat	Trunk	Belly	Trunk
Polydipsia/polyuria	−	++	−
Polyphagia	±	++	−
Absence of estrus	−	+	+
Lethargy	±	+	++
Decreased exercise tolerance	±	+±	++
Alopecia	−	+	±
Heat intolerance	−	+	−
Sensation of cold	−	−	+
* From de Bruijne and Lubberink, 1977, *Current Veterinary Therapy* VI, with the permission of Saunders, Philidelphia.			

Fig. 10.6 Symptoms and signs that aid differentiation of obesity from hyperadrenocorticism and hypothyroidism.

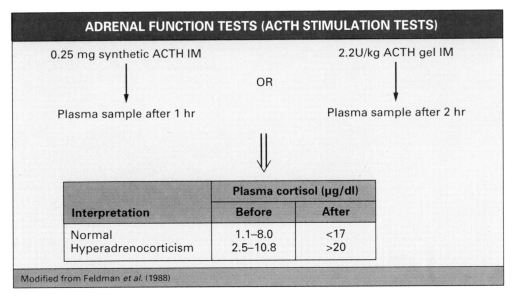

FIG. 10.7 Adrenal function tests (ACTH stimulation tests).

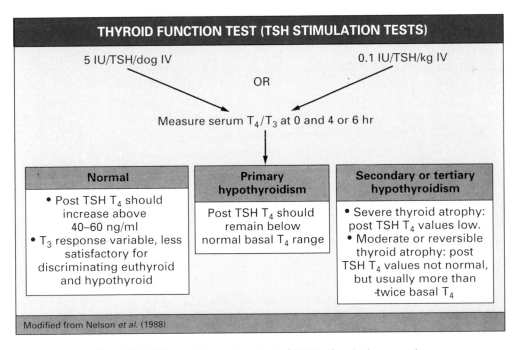

FIG. 10.8 Thyroid function tests (TSH stimulation tests).

The practical options for the management of obesity vary between dogs and cats. In dogs the options are starvation or controlled calorie reduction, although pharmacological management may present a possible future alternative or adjunct to therapy.

In cats the only option currently available to the clinician is a program of controlled calorie reduction, as starvation is contra-indicated because of the risk of hepatic lipidosis. Pharmacological management of obesity has yet to be evaluated in this species.

Starvation of dogs has been advocated as a means of management of obesity.[22,49] The regimen used for starvation is to hospitalize the patient and, following complete examination

WEIGHT LOSS RECORDED DURING STARVATION IN DOGS	
Week	% Weight loss*
1	7
2	12
3	17
4	20
5	23
* From starting weight.	

FIG. 10.9 Weight loss recorded during starvation in dogs.[22,49]

to eliminate endocrine–metabolic disorders, to withdraw all food from the animal. Water is then available *ad libitum* and vitamin and mineral supplements are given daily to meet the animal's requirements. Once the desired weight is reached, commercial food is introduced gradually and the animal returned home, with regular follow-ups to adjust feeding levels.[49] Rapid weight loss has been recorded by this method (Fig. 10.9). Concerns that have been raised about weight reduction by starvation include risk of tissue damage and the possibility that the dog may not start eating again. The humanity of the method has also been questioned.[27] Although the latter point must remain a matter for personal judgement, the former areas have not been reported as problems in studies of starvation. No severe clinical changes were reported in patients with starvation periods as long as 8 weeks, and it was considered that starvation was a simple and reliable method of obesity management, with the only disadvantage being the need for prolonged hospitalization.[22] Thus starvation may be an appropriate method for the management of some cases of canine obesity, and could at least be considered as an option where "emergency" weight reduction is required. It is, however, probably not appropriate for the management of the majority of cases for the following two reasons.

First, implicit in the need for hospitalization is a lack of owner involvement in the process of weight reduction. A program of controlled calorie reduction allows "retraining" of the owner with regard to feeding of their pet, and modification of appropriate

components of the pet–owner interaction; this does not occur if the animal is fasted in a hospital situation and thus the underlying causes associated with the obesity will not have been addressed. Under these circumstances it seems unlikely that the reduction in body weight that has been achieved will be maintained.

The second disadvantage relates to the composition of the weight lost by the animal. Studies in humans involving short-term starvation have shown that significantly greater weight loss occurs via this route, compared with that obtained using calorie-reduced diets. Triglyceride loss was also approximately 50% greater; however, protein losses were 2.8–5 times greater with starvation and water loss was also much higher.[90] Whether similar body composition changes occur in dogs during starvation is not clear; however, the potential for increased loss of lean body mass (presumably resulting from lack of dietary protein as well as the greater energy deficit)[90] would be an undesirable consequence of starvation. In addition, much of the weight loss due to water would be expected to be replaced during refeeding. Thus for these reasons starvation is not considered an appropriate method for the management of the majority of cases of canine obesity.

A program of controlled calorie reduction represents a better alternative for the dog, and is the only viable alternative for the cat. The program advocated involves a number of simple stages (Fig. 10.10). The first two of these have already been considered, and will result in identification of suitable cases for weight reduction. The next stage, client counselling, is probably the key to obtaining and maintaining successful weight reduction. It is important through discussion of the clinical implications of obesity to convince the owner of the value to their pet of weight reduction. Client counselling should be used to try to identify any social/environmental problems that may have contributed to the development of obesity, and may thus be considered as something of an equivalent to the behavioral and psychological therapy advocated as a

STAGES IN CONTROLLED CALORIE REDUCTION

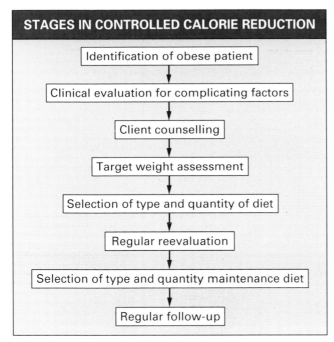

FIG. 10.10 Stages in controlled calorie reduction.

component of obesity management programs in humans.[84]

Assessment of a final target weight for an individual patient may be difficult on initial presentation, given the limited data on ideal weights. In addition, if the patient is grossly obese the reduction in food intake associated with a one-step target to the individuals ideal weight may be very severe. For these reasons an initial target weight of up to 15% less than the individual's current weight is recommended.[13,27,52] This approach provides a target that is attainable within a reasonable time period (an important psychological boost to the owner), and then provides the veterinarian with the opportunity to reassess more accurately a final (or next) target weight for the patient.

The veterinarian is faced with two basic choices in terms of diet to prescribe for the patient: either a modification of the existing diet, or substitution of this diet by one of several commercially available veterinary diets. The former option offers several disadvantages. First, the diet may have contributed to the patient's current problem, perhaps through containing many inappropriate food items, or being home-prepared where consistency in energy content is hard to achieve.

Second, even if the diet is a commercial one, as the nutrient content of diets is balanced to their energy content, feeding at levels significantly below maintenance may result in deficiencies of essential nutrients. These difficulties are largely overcome by the use of commercial low calorie diets designed for weight reduction. The diets need to be consistent in energy content and formulated to minimize the risk of dietary deficiencies. Complete replacement of the animal's existing diet in conjunction with precise feeding instructions is also likely to be easier for the owner than modifications to feeding levels of one (or more) existing foods. Thus use of a veterinary diet is likely to be associated with greater compliance.

The quantity of diet to be fed initially should be calculated as a proportion of the animal's maintenance requirement at its *target weight*. For dogs it is recommended that maintenance is calculated from the equation given previously.

The proportion of this maintenance requirement that is fed should be between 40 and 60%. Weight losses averaging approximately 1% per week (Fig. 10.11) have been achieved by feeding 40% of the maintenance energy requirements.[51] The overall weight loss in this study, which reported data from 20 dogs, averaged $11.7 \pm 5.1\%$ over 12 weeks.

Care should be taken with calorie restriction in obese cats, because of the risks of hepatic lipidosis. Thus the health of cats on weight reduction programs should always be closely monitored. In a recent study weight reductions averaging $13.5 \pm 6.3\%$ were

PERCENTAGE WEIGHT CHANGE IN DOGS DURING CONTROLLED CALORIE REDUCTION	
Weeks	**Percentage change**
0–2	−1.1
2–4	−2.4
4–6	−2.2
6–8	−2.5
8–10	−2.7
10–12	−1.2

FIG. 10.11 Percentage weight change in dogs during controlled calorie reduction.[51]

obtained over 18 weeks in a group of cats fed a commercial low calorie diet.[13] The success of this program was underlined by the absence of signs associated with liver disease in the cats. The food allowance in this study was based on 60% of maintenance at target weight (a 15% reduction from the cats' current weights, with maintenance calculated from 80 kcal ME/kg/day.[57] Intake data from the study indicated that the food consumption during weight loss was somewhat lower than the allowances. In addition, recent data on maintenance energy requirements of very inactive cats (which perhaps many obese ones would tend to be) suggests that they may require considerably less energy than the 80 kcal/kg/day noted above.[26] Taken together, these data suggest that a food allowance based on 60% of an assumed maintenance requirement of 60 kcal ME/kg/day at the target weight, should be appropriate for most cats.[51]

Regular reevaluation of cases is a very important component of the management program. This should take place every 2 weeks, ideally at the same time of day on each occasion. It provides an opportunity for reweighing of the animal and for discussion of progress with the owner. If weight loss is not occurring, then possible reasons for this, particularly in relation to compliance by the owner or the opportunities for theft of food by the animal, should be discussed. Wide individual variation in rates of weight loss must be expected, and thus each case should be evaluated individually before a decision is made on whether to change the food allowance. It is interesting to note that many dogs that showed an apparent levelling off of weight loss in one study, later lost further weight without any reduction in food allowance.[52]

Once weight loss has been achieved (in one or more stages) it is important to provide an appropriate maintenance diet for the animal, and to continue to follow-up cases regularly to adjust food allowances as necessary, particularly over the first few months. Depending on the diet fed for weight reduction it may be appropriate for the animal to continue on an increased allowance of the same food, or it may be more economical for the client to revert to a standard, high quality commercial petfood. The advantages of a commercial food over a home-prepared diet are again consistency and the ability to regulate intake accurately. The amount to be fed is difficult to predict, and will almost certainly have to be adjusted for the individual case. It has been suggested that the amount fed should not exceed 10% more than the final amount fed during treatment,[27] but an alternative is to base the initial feeding level on the maintenance requirements given above. It is worth remembering that particularly wide variability has been noted in dogs, with some individuals requiring 30–35% less than the average. It is thus logical to err initially on the conservative side of the maintenance allowance. Cases should then be followed at 2–4 week intervals until body weight has stabilized, with visits subsequently declining in frequency. Avoidance of the recurrence of obesity is as desirable a goal as its successful management.

Appetite in the companion animal may be manifested in the form of begging, and this type of behavior is well established in many dogs (Butterwick and Markwell, unpublished observations). Begging is a learned trait and may operate independently of the animal's apparent hunger. Presumably in these animals, social or other factors have overcome normal satiety mechanisms, contributing to the development of begging behavior and subsequently to obesity. It is unlikely that any diet would be satiating from the point of view of controlling this behavioral trait in these individuals. Behavioral modification of the companion animal and a change in habits of the owner are, therefore, essential to the success of long-term maintenance of weight loss.

In human and small animal medicine one approach to overcome hunger during weight reduction programs has been to include dietary fiber. A recent review indicated a variable response of different dietary fibers on satiety in humans.[46] Although a number of studies in humans have shown a short-term effect of fiber on intake and on the perception of hunger,[7,48] there is no evidence to suggest that these effects are maintained, or whether compensation in energy intake occurs in the long term. There is some evidence to suggest that the type of fiber ingested may be important with regards to satiety. There are few data to support the feeding of high levels of insoluble fiber for the treatment of obesity in the companion animal. Observations in the dog suggest that inclusion of raw materials containing high levels of insoluble fiber to a commercial low calorie diet has no apparent beneficial effect on satiety (Butterwick and Markwell, unpublished observations). In contrast a number of studies have suggested that soluble fibers may be more effective than insoluble fibers in inducing satiety.[23,75] Long-term studies need to be conducted in order to evaluate the potential benefits of soluble fiber in the obese companion animal. In addition a high intake of soluble fiber is associated with gastrointestinal signs (bloating and flatulence) and loose stools that may limit its application in the treatment of obesity.[42,75]

In cats, dilution of the energy content of the diet with dietary fiber has been shown to reduce daily energy intake.[35,39] This effect, however, is probably due to the negative effect of fiber on palatability because later studies have shown that cats are able to adjust their dietary intake if presented with diets that have a variable energy density.[14,40] Studies with pigs have shown that when offered high fiber diets that physically limit energy intake, these animals display unmistakable evidence of hunger.[63] It is clear that such a simplistic approach to satisfying hunger during weight reduction is unlikely to achieve its goal due to the complexity of control mechanisms governing food intake and because, by definition, energy intake will be deficient.[76,91]

The suggestion that provision of a bland or tasteless food to companion animals on weight reduction programs will be beneficial because it will limit intake is clearly unhelpful.[49] It could be argued that any mechanism by which palatability is reduced could be used as a method in designing a low calorie diet. It is important that any diet used as part of a weight reduction program is palatable, because

long-term acceptance will influence compliance and ultimately the success of the program. This is particularly important in cats who are recognized as "finicky" eaters,[14] because poor acceptance of an unpalatable diet can lead to withdrawal from weight reduction studies, or worse still, prolonged anorexia and the risk of hepatic lipidosis.[12] Clearly a balance between the provision of a palatable and acceptable diet and that which ensures adequate and safe weight loss must be achieved.

In addition to restriction of energy intake, a second potential route of stimulating weight loss is through manipulation of the TEF. TEF can be manipulated in response to diet and there is some evidence that obese individuals have a reduced thermogenic response to certain macronutrients.[79] TEF, however, represents approximately only 10% of total daily energy expenditure.[65] Consequently, dietary manipulation will have only a relatively minor effect on total energy expenditure or weight loss.

In addition to dietary management and psychological/behavioral therapy, exercise has also been recommended in the management of obesity in humans. The exercise program should promote increased energy expenditure, and promote fat loss and the maintenance of lean body mass.[84]

The influence that exercise has on the energy requirement of dogs has been estimated using a factorial approach.[11] It was calculated that energy requirements were increased by approximately 6.7, 11.3 and 14.8% above resting energy expenditure in dogs of 10, 25 and 40 kg body weight, respectively, with the typical activity levels of pet dogs. This energy expenditure is roughly equivalent to the dog covering a total of 5 km/day. Thus, although exercise can clearly influence energy expenditure and assist in weight reduction, its effect should not be overestimated. The additional energy requirements for exercise in the cat are difficult to quantify, but are unlikely to exceed those of an equivalent-sized dog.

The type of exercise program that has been advocated for humans involves sessions of at least 30 min 3–5 times weekly.[84] This type of program could be applied relatively easily to the dog in the form of walking, although it is difficult to apply to cats. In this species the only real options for trying to increase exercise are through allowing access to the outside if possible and through encouraging play. Clearly the extent of any exercise program for an obese animal will be determined by the physical capability and health of the pet (and owner), and an important criterion for any exercise program is that it should be safe for an individual. The benefits of exercise as a component of obesity management programs have not been quantified in dogs and cats; however, it seems logical to advocate their use as part of a weight reduction program, provided that the animal is physically capable of undertaking the chosen regimen.

Pharmacological Management

Limited data exist on the use of drugs in the management of canine obesity, and with the exception of recent studies on dehydroepiandrosterone (DHEA), results are not particularly encouraging.

The use of anorectic drugs in humans has been described as controversial, because of their potential for abuse.[8] In one study an anorectic, fenfluramine, was found to be ineffective in reducing weight in a group of spayed bitches.[10]

Use of thyroid hormone is also controversial. At low doses it replaces endogenous thyroid hormone; at higher doses its effects include increased lipolysis and protein breakdown. This protein catabolism may account for a considerable proportion of the weight loss during treatment and is an undesirable consequence.[8] In addition, it has been suggested that thyroid hormone administration at pharmacological doses to euthyroid dogs may predispose them to future hormone imbalances.[2]

Growth hormone has been evaluated in obese individuals due to its lipolytic properties. Studies to date have reported no increase in fat loss in obese, energy restricted individuals treated with growth hormone, although treat-

ment did result in a reduction in the loss of lean body mass.[16,71,72]

Perhaps the most promising pharmacological route currently appears to lie in the use of DHEA, a 17-ketosteroid of adrenal and gonadal origin. Administration of exogenous DHEA for a 3-month period resulted in moderate weight losses of 3% per month in 68% of a group of 19 spontaneously obese dogs. Of particular interest was that this occurred without a reduction in food intake.[44] Further longer-term studies may demonstrate whether these initially promising results for DHEA can be applied more widely.

Pharmacological intervention may be a method increasingly used in the treatment of individuals that are considered to be obese due to a low metabolic rate (see above). It has been suggested that weight reduction results in a decrease in resting metabolic rate (RMR)[55,81] and this may predispose subjects to rapid weight regain following realimentation leading to the so-called "yo–yo" phenomenon.[24,73] Although evidence supporting a long-term reduction of RMR following weight loss is not conclusive,[55,80] a short-term reduction in RMR during weight loss may slow progress. Concurrent dietary and pharmacological treatment is a potentially synergistic approach that may be employed in the future.

Other pharmaceutical approaches have focused on β-adrenergic agonists and α_2-adrenergic antagonists. The suitability of these agents for weight reduction has been evaluated in humans and the dog.[8,31] The most promising to date appears to be the α_2-antagonist yohimbine.[45] The majority of these drugs, however, with the exception of yohimbine,[31] have adverse side effects that precludes their routine use in the management of obesity.

Although pharmaceutical management of obesity is an emerging field that may provide useful and safe treatments in the future, the long-term benefits of current treatments are unproven. In addition, the cost and ethical issues surrounding pharmaceutical management render this approach, at least at present, unsuitable for most subjects.

Prognosis

Theoretically the prognosis for uncomplicated obesity should be good, as the owner can regulate the animal's food intake, both to ensure weight loss and subsequent maintenance at a normal body weight. This means that prognosis is very dependent on the degree of owner compliance in a particular case, and the extent to which this can be obtained will be determined by both the veterinarian–client interaction and by the underlying cause of the obesity in the pet. Some clients may be unwilling to modify their behavioral interactions with their pet with regard to feeding and exercise, and in these cases the likelihood of success will be very low.

The initial consultation with a client is a key factor in obtaining compliance, as has already been discussed. This provides an opportunity to review the problem with the client, stressing the benefits of weight reduction to the quality of life of the pet. Regular follow-up consultation during the program of weight loss will allow for reinforcement of the message and use of a graphical record of weight loss will help to show the progress that is being made. Another important factor is provision of a palatable diet. If an animal refuses to eat the food provided, or eats only reluctantly, the owner is likely to offer other food in place of the prescribed diet to the detriment of progress. It is also very important to try to understand the reason why obesity has occurred in an individual pet, and to try to address that issue with the client. If this does not occur, weight loss may be achieved, but success is likely at best to be short lived. Follow-up of cases subsequent to weight loss should also be considered as an integral part of the program of obesity management. This will enable adjustments to be made to food intake and will help to avoid return of the problem.

To be successful in achieving and maintaining weight reduction it is necessary to ensure that a permanent change takes place in the feeding/exercise relationship of the pet and owner, and only through this route will a cycle of weight gain followed by weight loss be avoided.

References

1. Anderson, D. B. and Corbin, J. E. (1982) Estimating body fat in mature beagle bitches. *Laboratory Animal Science*, **32**, 367–370.
2. Anderson, G. L. and Lewis, L. D. (1980) Obesity. In *Current Veterinary Therapy VII*. Ed. R. W. Kirk. pp. 1034–1039. Philadelphia: W. B. Saunders.
3. Anderson, R. S. (1973) Obesity in the dog and cat. In *Veterinary Annual*. 14th edn. pp. 182–186. Bristol: J. Wright.
4. Armstrong, R. S. and Hand, M. S. (1989). Nutritional disorders in the cat, disease and management. In *The Cat—Diseases and Clinical Management*. Ed. R. G. Sherding. pp. 141–161. New York: Churchill Livingstone.
5. Ashwell, M., Cole, T. J. and Dixon, A. K. (1985) Obesity: new insight into the anthropometric classification of fat distribution shown by computed tomography. *British Medical Journal*, **290**, 1692–1694.
6. Baba, E. and Arakawa, A. (1984) Myocardial hypoxia in an obese beagle. *Veterinary Medicine/ Small Animal Clinician*, **79**, 790–791.
7. Bolton, R. P., Heaton, K. W. and Burroughs, L. F. (1981) The role of dietary fibre in satiety, glucose, and insulin: studies with fruit and fruit juice. *American Journal of Clinical Nutrition*, **34**, 211–217.
8. Bray, G. A. (1972) Clinical management of the obese adult. *Postgraduate Medicine*, **51**, 125–130.
9. Bray, G. A. (1992). Drug treatment of obesity. *American Journal of Clinical Nutrition*, **55** (Suppl.), -538S–544S.
10. Bromson, L. and Parker, C. H. L. (1975) Effect of fenfluramine on overweight spayed bitches. *Veterinary Record*, **96**, 202–203.
11. Burger, I. H. and Johnson J. V. (1991) Dogs large and small: the allometry of energy requirements within a single species. *Journal of Nutrition*, **121**, S18–S21.
12. Burrows, C. F., Chaipella, A. M. and Jezyk, P. (1981). Idiopathic hepatic lipidosis: the syndrome and speculations of its pathogenesis. *Florida Veterinary Journal*, **10**, 18–20.
13. Butterwick, R. F., Wills, J. M., Sloth, C. and Markwell, P. J. A study of obese cats on a calorie-controlled weight reduction programme. *Veterinary Record* (in press).
14. Castonguay, T. W. (1981). Dietary dilution and intake in the cat. *Physiology Behaviour*, **27**, 547–549.
15. Center, S. A. (1986) Feline liver disorders and their management. *Compendium on Continuing Education*, **8**, 889–903.
16. Clemmons, D. R., Snyder, D. K., Williams, R. and

Underwood, L. E. (1987) Growth hormone administration conserves lean body mass during dietary restriction in obese subjects. *Journal of Clinical Endocrinology and Metabolism,* **64**, 878–883.

17. Clutton, R. E. (1988) The medical implications of canine obesity and their relevance to anaesthesia. *British Veterinary Journal,* **144**, 21–28.

18. Coward, W. A., Parkinson, S. A. and Murgatroyd, P. R. (1988) Body composition measurement for nutrition research. *Nutrition Research Reviews,* **1**, 115–124.

19. Craddock, D. (1969) *Obesity and its Management.* Edinburgh: Livingstone.

20. Craig, L. S. (1969) Anthropometric determinants of obesity. In *Obesity.* Ed. N. L. Wilson. pp. 13–23. Philadelphia: Davis.

21. Dawes, M. G. (1984). Obesity in a Somerset town: prevalence and relationship to morbidity. *Journal of the Royal College of General Practitioners,* **34**, 328–330.

22. de Bruijne, J. J. and Lubberink, A. A. M. E. (1977) Obesity. In *Current Veterinary Therapy.* Ed. R. W. Kirk. pp. 1068–1070. Philadelphia: W. B. Saunders.

23. Delargy, H. J., Burley, V. J., Blundell, J. E., O'Sullivan, K. R. and Fletcher, R. V. (1993) The effects of fibre in the breakfast upon short-term appetite control: a comparison of soluble and insoluble fibre. In *Proceedings of the Nutrition Society,* Summer Meeting, July 1993. PC 52.

24. Dulloo, A. G. and Girardier, L. (1990). Adaptive changes in energy expenditure during refeeding following low-calorie intake: evidence for a specific metabolic component favoring fat storage. *American Journal of Clinical Nutrition,* **52**, 415–420.

25. Durnin, J. V. G. A. and Womersly, J. (1974) Body fat assessed from total body density and its estimation from skin fold thickness: measurement of 481 men and women aged from 16–72 years. *British Journal of Nutrition,* **32**, 77–97.

26. Earle, K. E. and Smith, P. M. (1991) Digestible energy requirements of adult cats at maintenance. *Journal of Nutrition,* **121**, S45–S46.

27. Edney, A. T. B. (1974) Management of obesity in the dog. *Veterinary Medicine/Small Animal Clinician,* **49**, 46–49.

28. Edney, A. T. B. and Smith, P. M. (1986) Study of obesity in dogs visiting veterinary practices in the United Kingdom. *Veterinary Record,* **118**, 391–396.

29. Feldman, E. C., Schrader, L. A. and Twedt, D. C. (1988) Diseases of the adrenal gland. In *Handbook of Small Animal Practice.* Ed. R. V. Morgan. pp. 537–549. New York: Churchill Livingstone.

30. Fowler, P. A., Fuller, M. F., Glasbey, C. A., Foster, M. A., Cameron, G. G., McNeill, G. and Maughan, R. J. (1991) Total and subcutaneous adipose tissue in women: the measurement of distribution and accurate prediction of quantity by using magnetic resonance imaging. *American Journal of Clinical Nutrition,* **54**, 18–25.

31. Galitzky, J., Vermorel, M., Lafontan, M., Montastuc, P. and Berlan, M. (1991). Thermogenic and lipolytic effect of yohimbine in the dog. *British Journal of Pharmacology,* **104**, 514–518.

32. Hardy, R. M. (1989) Diseases of the liver and their treatment. In *Textbook of Veterinary Internal Medicine.* Ed. S. J. Ettinger. pp. 1479–1527. Philadelphia: W. B. Saunders.

33. Hesse, A. and Sanders, G. (1985) A survey of urolithiasis in cats. *Journal of Small Animal Practice,* **26**, 465–476.

34. Heymsfield, S. B. and Waki, M. (1991) Body composition in humans: advances in the development of multicompartment chemical models. *Nutrition Reviews,* **49**, 97–108.

35. Hirsch, E., Debose, C. and Jacobs, H. L. (1978). Dietary control of food intake in the cat. *Physiology and Behaviour,* **20**, 287–295.

36. Houpt, K. A. and Smith, S. L. (1981) Taste preferences and their relation to obesity in dogs and cats. *Canadian Veterinary Journal,* **22**, 77–81.

37. Hundley, J. M. (1956) Diabetes–overweight: US problems. *Journal of the American Dietetic Association,* **32**, 417–422.

38. Joshua, J. O. (1970) The obese dog and some clinical repercussions. *Journal of Small Animal Practice,* **11**, 601–606.

39. Kanarek, R. B. (1975). Availability and caloric density of the diet as determinants of meal patterns in cats. *Physiology and Behaviour,* **20**, 287–295.

40. Kane, E., Rogers, Q. R., Morris, J. G. and Leung, P. M. B. (1981). Feeding behaviour of the cat fed laboratory and commercial diets. *Nutrition Research,* **1**, 499–507.

41. Krook, L., Larsson, S. and Rooney, J. R. (1960) The interrelationship of diabetes mellitus, obesity and pyometra in the dog. *American Journal of Veterinary Research,* **21**, 120–124.

42. Krotkiewski, M. (1984) Effect of guar gum on bodyweight, hunger ratings and metabolism in obese subjects. *British Journal of Nutrition,* **52**, 97–105.

43. Kuczmarski, R. J., Fanelli, M. T. and Koch, G. G. (1987) Ultrasonic assessment of body composition in obese adults: overcoming the limitations of the skinfold caliper. *American Journal of Clinical Nutrition,* **45**, 717–724.

44. Kurzman, I. D., MacEwen, E. G. and Haffa, A. L. M. (1990) Reduction in body weight and cholesterol in spontaneously obese dogs by dehydroepiandrosterone. *International Journal of Obesity,* **14**, 95–104.

45. Lafontan, M., Berlan, M., Galitzky J., and Montastruc, J.-L. (1992). Alpha-2 adrenoceptors in lipolysis: α_2 antagonists and lipid-mobilizing strategies. *American Journal of Clinical Nutrition,* **55** (Suppl.), 219S–227S.
46. Leeds, A. R. (1990). Mechanisms of action of dietary fibre in weight loss. In *Progress in Obesity Research.* Eds. Y. Oomura, S. Tarui, S. Inoue and T. Shimazu. pp. 519–522. London: John Libbey and Company.
47. Legendre, A. M. (1983) Anorexia and polyphagia. In *Textbook of Veterinary Internal Medicine.* Ed. S. J. Ettinger. 2nd edn. pp. 139–141. Philadelphia: W. B. Saunders.
48. Levine, A. S., Tallman, J. R., Grace, M. K., Parker, S. A., Billington, C. J. and Levitt, M. D. (1989). Effect of breakfast cereals on short-term food intake. *American Journal of Clinical Nutrition,* **50,** 1303–1307.
49. Lewis, L. D. (1978) Obesity in the dog. *Journal of the American Animal Hospital Association,* **14,** 402–409.
50. Lukaski, H. C. (1987) Methods for the assessment of human body composition: traditional and new. *American Journal of Clinical Nutrition,* **46,** 537–556.
51. Markwell, P. J., Butterwick, R. F., Wills, J. M. and Raiha, M. Clinical studies in the management of obesity in dogs and cats. *International Journal of Obesity* (in press).
52. Markwell, P. J., van Erk, W., Parkin, G. D., Sloth, C. J. and Shantz–Christienson, T. (1990) Obesity in the dog. *Journal of Small Animal Practice,* **31,** 533–537.
53. Mason, E. (1970) Obesity in pet dogs. *Veterinary Record,* **86,** 612–616.
54. Mayer, J. (1973) Obesity. In *Modern Nutrition in Health and Disease.* Eds. R. S. Goodhart and M. E. Shils. pp. 625–646. Philadelphia: Lea and Febiger.
55. Melby, C. L., Schimdt, W. D. and Corrigan, D. (1990). Resting metabolic rate in weight-cycling collegiate wrestlers compared with physically active, non-cycling control subjects. *American Journal of Clinical Nutrition,* **52,** 409–414.
56. Messent, P. R. (1980) Breed of dog and dietary management background as factors affecting obesity. In *Over and Under Nutrition.* Ed. A. T. B. Edney. pp. 9–16. Melton Mowbray: Pedigree Petfoods.
57. National Research Council (1986) *Nutrient Requirements of Cats.* Washington, DC: National Academy Press.
58. Nelson, K. M., Weinsier, R. L., James, L. D., Darnell, B., Hunter, G. and Long, C. L. (1992). Effect of weight reduction on resting energy expenditure, substrate utilization, and the thermic effect of food in moderately obese women. *American Journal of Clinical Nutrition,* **55,** 924–933.
59. Nelson, R. W. (1989) Disorders of the endocrine pancreas. In *Veterinary Internal Medicine* Ed. S. J. Ettinger. pp. 1676–1720. Philadelphia: W. B. Saunders.
60. Nelson, R. W., Meric, S. M., Hawkins, E. C. and Turrel, J. M. (1988) Diseases of the thyroid glands. In *Handbook of Small Animal Practice.* Ed. R. V. Morgan. pp. 507–520. New York: Churchill Livingstone.
61. Newberne, P. M. (1966) Overnutrition on resistance of dogs to distemper virus. *Federation Proceedings,* **25,** 1701–1710.
62. Olefsky, J. M., Kolterman, O. G. and Scarlett, J. A. (1982) Insulin action and resistance in obesity and noninsulin-dependent type II diabetes mellitus. *American Journal of Physiology,* **243,** E15–E30.
63. Owen, J. B. (1990). Weight control and appetite: nature over nurture. *Animal Breeding Abstracts,* **58(7),** 583–591.
64. Panciera, D. L., Thomas, C. B., Eicker, S. W. and Atkins, C. E. (1990) Epizootiologic patterns of diabetes mellitus in cats: 333 cases (1980–1986). *Journal of the American Veterinary Medical Association,* **197,** 1504–1508.
65. Ravussin, E. and Bogardus, C. (1992). A brief overview of human energy metabolism and its relationship to essential obesity. *American Journal of Clinical Nutrition,* **55,** 242S–245S.
66. Ravussin, E., Zurlo, F., Ferraro, R. and Bogardus, C. (1990). Energy expenditure in man: determinants and risk factors for body weight gain. In *Progress in Obesity Research.* Eds. Y. Oomura, S. Tarui, S. Inoue and T. Shimazu. pp. 175–182. London: John Libbey and Company.
67. Scarlett, J. M., Donoghue, S., Saidla, J. and Wills, J. Overweight cats: prevalence and risk factors. *International Journal of Obesity* (in press).
68. Scheuer, P. J. (1973) *Liver Biopsy Interpretation,* p. 128. London: Balliére Tindall.
69. Sheng, H.-P. and Huggins, R. A. (1979) A review of body composition studies with emphasis on total body water and fat. *American Journal of Clinical Nutrition,* **32,** 630–647.
70. Sloth, C. (1992). Practical management of obesity in dogs and cats. *Journal of Small Animal Practice,* **33,** 178–182.
71. Snyder, D. K., Clemmons, D. R. and Underwood, L. E. (1988). Treatment of obese, diet-restricted subjects with growth hormone for 11 weeks: effects on anabolism, lipolysis and body composition. *Journal of Clinical Endocrinology and Metabolism,* **67,** 54–61.
72. Snyder, D. K., Underwood, L. E. and Clemmons, D. R. (1990). Anabolic effects of growth hormone

148   *Peter J. Markwell and Richard F. Butterwick*

in obese diet-restricted subjects are dose dependent. *American Journal of Clinical Nutrition,* **52,**
431–437.

73. Steen, S. N., Opplinger, R. A. and Brownell, K. D.
(1988) Metabolic effects of repeated weight loss
and regain in adolescent wrestlers. *Journal of the
American Medical Association,* **260,** 47–50.

74. Steininger, E. (1981) Die Adipositas und ihre Dia-
tetische Behandlung. *Wiener Tierarztlicher Monatschrift,* **68,** 122–130.

75. Stevens, J., Levitsky, D. A., VanSoest, P. J.,
Robertson, J. B., Kalkwarf, H. J. and Roe, D. A.
(1987) Effect of psyllium gum and wheat bran on
spontaneous energy intake. *American Journal of
Clinical Nutrition,* **46,** 812–817.

76. Stricker, E. M. and Verbalis, J. G. (1990). Control
of appetite and satiety: insights from biologic and
behavioral studies. *Nutrition Reviews,* **48(2),** 49–56.

77. Stunkard, A. J., Harris, J. R., Pedersen, N. L. and
McClearn, G. E. (1990). The body-mass index of
twins who have been reared apart. *New England
Journal of Medicine,* **322,** 1483–1487.

78. Stunkard, A. J., Sorensen, T. I. A., Hanis, C.,
Teasdale, T. W., Chakraborty, R., Schull, W. J.
and Schulsinger, F. (1986). An adoption study of
human obesity. *New England Journal of Medicine,*
314, 193–198.

79. Swaminathan, R., King, R. F. G. J., Holmfield, J.,
Siwek, R. A., Baker, M. and Wales, J. K. (1985).
Thermic effect of feeding carbohydrate, fat, protein
and mixed meal in lean and obese subjects.
American Journal of Clinical Nutrition, **42,** 177–181.

80. van Dale, D. and Saris, W. H. M. (1989) Repetitive
weight loss and weight reduction, resting metabolic
rate, and lipolytic activity before and after exercise
and/or diet treatment. *American Journal of Clinical
Nutrition,* **49,** 409–416.

81. Van Gaal, L. F., Vansant, G. A. and De Leeuw,
I. H. (1992). Factors determining energy expenditure during very-low-calorie diets. *American Journal
of Clinical Nutrition,* **56** (Suppl.), 224S–229S.

82. Volz, P. A. and Ostrove, S. M. (1984) Evaluation
of a portable ultrasonoscope in assessing the body
composition of college-aged women. *Medicine and
Science in Sports and Exercise,* **16,** 97–102.

83. Walker, A. D., Weaver, A. D., Anderson, R. S.,
Crighton, G. W., Fennell, C., Gaskell, C. J. and
Wilkinson, G. T. (1977) An epidemiological survey
of the feline urological syndrome. *Journal of Small
Animal Practice,* **18,** 283–301.

84. Weinsier, R. L., Wadden, T. A., Ritenbaugh, C.,
Harrison, G. G., Johnson, F. S. and Wilmore,
J. H. (1984) Recommended therapeutic guidelines
for professional weight control programmes.
American Journal of Clinical Nutrition, **40,** 865–872.

85. Wilkinson, M. J. and Mooney, C. T. (1991) *Obesity
in the Dog: a Monograph.* Melton Mowbray:
Waltham.

86. Wilkinson, M. J. A. and McEwan, N. A. (1991)
Use of ultrasound in the measurement of subcuta-
neous fat and prediction of total body fat in dogs.
Journal of Nutrition, **121,** S47–S50.

87. Willeberg, P. and Priester, W. A. (1976) Feline
urological syndrome: associations with some time,
space and individual patient factors. *American
Journal of Veterinary Research,* **37,** 975–978.

88. Williams, G. D. and Newberne, P. M. (1971)
Decreased resistance to salmonella infection in
obese dogs. *Federation Proceedings,* **30,** 572.

89. Wilson, N. L., Farber, S. M., Kimbrough, L. D.
and Wilson, R. H. L. (1969) The development and
perpetuation of obesity: an overview. In *Obesity.*
Ed. N. L. Wilson. pp. 3–12. Philadelphia: Davis.

90. Yang, M.-U. and van Itallie, T. B. (1976) Composition of weight lost during short-term weight reduction. *Journal of Clinical Investigation,* **58,** 722–730.

91. York, D. A. (1990). Metabolic regulation of food
intake. *Nutrition Reviews,* **48(2),** 64–70.

CHAPTER 11

Natural Food Hazards

IVAN H. BURGER

Introduction

Food hazards may be defined as the intrinsic properties of ingested material that, in a particular set of circumstances, can cause adverse health effects. There are many substances that represent a potential hazard to dogs and cats when eaten and it is impossible to include details of all of them here. This chapter, therefore, is restricted to naturally occurring food hazards, particularly those that are likely to give rise to symptoms of acute poisoning. Chemical toxicities are not discussed here; the reader is referred to the general reading list for further information.

Apart from the accidental consumption of chemicals (such as herbicides or pesticides), other ingested materials may have hazards as shown in Fig. 11.1.

The hazards associated with inappropriate feeding methods, which may result in nutrient deficiency or toxicity syndromes, are described in Chapter 1, whereas idiosyncratic reactions of individual animals to normal foods (food allergy and intolerance) are covered in depth in Chapter 12.

The true incidence of poisoning in dogs and cats is difficult to assess and varies widely between countries, as does the relative importance of individual poisons. It is thought that poisoning accounts for between 1 and 2% of all cases seen in small animal veterinary practice in the U.S.A.[28] In the U.K., it is likely that poisoning is responsible for no more than 1% of the total losses in all animal species,[23] whereas in Israel, up to 27% of all deaths in dogs may be due to poisoning.[14] However, most cases of poisoning in dogs and cats are a result of chemical toxicities. Being carnivores, they are at lower risk of plant poisoning than herbivores, but they are at greater risk of secondary toxicities due to the ingestion of poisoned prey, such as birds, rodents and other mammals.

Dogs, in particular, are renowned for their scavenging habits, greed and lack of discrimination, all qualities which combine to increase the likelihood of ingesting toxic material. The harmful effects of such materials are offset to some degree, however, by the ability of dogs to vomit readily and by the relative efficiency of normal detoxification mechanisms in this species.[24]

Cats are less likely to ingest poisons that have a noticeable taste or smell because their close inspection of potential food will often deter them from eating it. Although it is often stated that cats chew their food more thoroughly than dogs (and so are more likely to detect unusual-tasting material), this

HAZARDS ASSOCIATED WITH THE INGESTION OF VARIOUS MATERIALS

Consumed material	Hazard
Food	Specific nutrient imbalances
	Food components with inherent toxicity
	Food additives
	Contaminants of food • mycotoxins
	• bacteria and bacterial toxins
	• toxic residues in animal tissue
	Food allergy or intolerance
	Mechanical injury
	Parasitic infection
Plants	Ornamental plants
	Fungi e.g. toadstools
	Toxic algae
Animals	Venomous toads
	Poisoned prey
Minerals	Heavy metals

FIG. 11.1 Ingested materials with associated hazards.

is by no means a universal characteristic; many cats will eat their food very rapidly, with minimal mastication. The cat's strong hunting instincts make it particularly prone to secondary intoxications through ingestion of poisoned prey, but the most significant behavioral pattern that increases the incidence of ingestion of poisons is self-grooming of the coat. Cats have an inquisitive nature and although this often stops short of eating poisonous substances, their close examination of the material will often result in contamination of the fur. The poison may therefore be ingested in the course of subsequent grooming. Aside from its behavioral characteristics, the cat has an increased susceptibility to some materials compared to the dog, owing to differences in some of its metabolic pathways.

Overview

General Clinical Signs

The spectrum of clinical signs associated with natural food hazards is very broad. Nutrient imbalances and dietary sensitivity reactions may give rise to specific syndromes and these are described elsewhere. Where the ingestion of material results in mechanical injury, the animal will exhibit signs that are related to the nature of the injurious substance and to the site of injury. There are, however, certain clinical signs that may be associated with food-related toxicities and these are summarized in Fig. 11.2.

When presented with an animal that is showing signs of acute toxicity, it is important to differentiate these from signs of acute trauma, acute infectious disease (particularly gastrointestinal infections) and acute manifestations of long-term metabolic or other disease (such as nephritis, hepatitis, epilepsy, diabetes, autoimmune disease or neoplasia).

Diagnosis

Unless a definite history of contact with a poison is available, a general diagnosis of poisoning is usually made with difficulty and diagnosis of a specific toxicosis is often impossible. Sudden onset of clinical illness involving neurological disturbances, vomiting and diarrhea, or respiratory distress may all lead to a suspicion of poisoning. All of these, however, may be the acute manifestation of a long-standing illness. On the other hand, many cases of poisoning will go unrecognized, appearing only as a general malaise and poisoning is unlikely to be suspected in animals with a history of chronic illness. Diagnosis of poisoning must, therefore, be

CLINICAL SIGNS COMMONLY ASSOCIATED WITH POISONING	
Clinical signs	**Poison**
Abdominal pain	Castor seed, *Amanita muscaria*, *Amanita phalloides*, lead
Anemia	Onion, propylene glycol, aflatoxins, lead
Ataxia	Staphylococcal enterotoxins, aflatoxins, castor seed, laburnum, cannabis, *Amanita pantherina*, magic mushrooms, *Clitocybe*, venomous toads, mercury
Cardiac arrhythmias	Cyanogenic plants, oleander, venomous toads
Coma	Laburnum, *Amanita muscaria*
Convulsions	Roquefortine, staphylococcal enterotoxins, oleander, cyanogenic plants, *Brunfelsia*, venomous toads, some blue–green algae, lead
Cyanosis	Cyanogenic plants, *Oscillatoria*
Depression/weakness	Aflatoxins, castor seeds, calcium oxalate-containing plants, cyanogenic plants, oleander, staphylococcal enterotoxins, *Salmonella*, tobacco, cannabis, *Amanita phalloides*, *Clitocybe*, venomous toads, blue–green algae, lead, mercury
Diarrhea	Cocoa, aflatoxins, ochratoxins, staphylococcal enterotoxins, *Salmonella*, *Campylobacter*, calcium oxalate-containing plants, oleander, castor seed, laburnum, *Amanita muscaria*, *Amanita phalloides*, lead
Dyspnea	Aflatoxins, calcium oxalate-containing plants, cyanogenic plants, tobacco, blue–green algae (some strains)
Excitement	Benzoic acid, cocoa, *Amanita pantherina*, magic mushrooms, *Oscillatoria*, lead
Jaundice	Aflatoxins
Liver disease	Aflatoxins, indospicine, blue–green algae
Muscle tremors	Roquefortine, laburnum, cannabis, *Oscillatoria*, lead
Paralysis	Botulinum toxin, venomous toads, some blue–green algae, lead
Salivation	Calcium oxalate-containing plants, tobacco, *Amanita muscaria*, magic mushrooms, *Clitocybe*, *Oscillatoria*, venomous toads
Vomiting	Cocoa, ochratoxins, staphylococcal enterotoxins, *Salmonella*, oleander, laburnum, *Amanita muscaria*, *Amanita phalloides*, *Clitocybe*, venomous toads, blue–green algae, lead

FIG. 11.2 Clinical signs commonly associated with poisoning.

made only after all of the possible etiologies of the condition have been considered.

It is important, however, to collect and record as much evidence as possible for future reference so that recognition and treatment can be continually improved. There are four main areas that can provide this evidence:

(1) *Clinical signs* are useful for the record but are not definitive on their own and may be caused by other illness totally unrelated to poisoning.

(2) *Circumstantial evidence* is probably the most useful and is best obtained by asking about recent changes to the environment of the pet or about its behavior.

(3) *Analytical evidence* can provide indisputable proof of the poison but it is practicable only if the general type of poison is known or strongly suspected. Although analytical techniques have advanced dramatically over recent years, there is no single, easy test that can screen a wide range of different substances; only analysis within certain groups is possible.

Also, advice must be taken from the analyst on the types of samples required for testing, storage conditions and so on.

(4) *Pathology* can give much useful information but it must be appreciated that many poisons do not give characteristic postmortem changes. Furthermore, it is obviously to the animal's advantage if this particular evidence *cannot* be obtained!

Treatment

In cases where a definite history of poisoning exists, treatment can be tailored to suit the individual poison and, if available, an antidote administered. In many cases, this will not be possible, either because the specific poison is not known, or there is no specific antidote. In these cases, symptomatic treatment and supportive therapy are all that can be given.

Treatment for any poisoning follows three main courses of action.

First of all, further absorption of the poison should be prevented and this involves removing suspected food and washing off skin contamination. The latter is best achieved by first clipping or shaving the fur then washing with running water where the poison is water-soluble, and paper towels (or rags) for oily material followed by treatment with clean cooking oil again wiped off with paper. Soap or detergent can assist absorption through the skin and are probably best avoided until other measures have been carried out. If complete removal is impossible, or the animal's condition precludes such manipulations, then further licking must be prevented either by bandaging the area or by use of an Elizabethan collar until complete removal can be achieved. Following cleaning of the coat, the animal should be thoroughly dried to help prevent hypothermia.

Suspect material should be retained and this includes anything vomited. The use of emetics is controversial but should not be attempted if the animal is comatose or convulsing, or has ingested petrol, paraffin or corrosive liquids. In dogs, apomorphine may be used (0.1 mg/kg SC) as an emetic. A salt or mustard solution (about 2 tsp of either salt or mustard in a cup of warm water) or a crystal of washing soda (placed on the back of the tongue) will have the same effect. Emetics

are not particularly useful for cats as different agents can vary widely in their effects and may merely hinder the provision of proper treatment, for example by delaying transfer of the patient to the surgery.

Gastric lavage is a useful procedure in both dogs and cats if the poison has been ingested relatively recently (within 2 hr), vomiting has not occurred and if the animal is a suitable candidate for anesthesia. The procedure is described in Fig. 11.3. If clinical signs of poisoning have developed, then the risk of anesthesia must be balanced against the need for effective gastric lavage. If it is decided that anesthesia is undesirable, then activated charcoal and water may be administered by mouth. Great care must be taken in administering this charcoal mixture in animals with central nervous system (CNS) depression, because accidental inhalation of the material can easily occur.

The second line of treatment is nonspecific or supportive therapy that reinforces the body's own defense and aids elimination of the poison. This treatment includes replacement of fluids lost by vomiting, controlling convulsions and keeping the animal warm and undisturbed. Of primary importance in most instances is to ensure that respiratory function is adequate. If possible, the animal's respiratory rate should be monitored using a suitable

PROCEDURE FOR GASTRIC LAVAGE

- The animal should be anesthetized and intubated using a cuffed endotracheal tube (which extends at least as far forward as the incisor teeth) to prevent inhalation of gastric contents
- A soft, large bore rubber tube should be used for stomach tubing. In the cat, the tube should be about 8 mm external diameter and about 30 cm in length; in the dog, the tube should be of the same diameter as the endotracheal tube
- Mark on the tube the distance from the animal's nose to the xiphoid cartilage, then pass the tube and gradually introduce about 5–10 ml/kg body weight of tap water into the stomach
- Gently massage the stomach through the body wall and lower the animal's head and the end of the tube to allow the gastric contents to escape
- If very little material is removed in this way, then the contents should be gently aspirated with a 50 ml syringe while massaging the stomach
- The procedure should be repeated until clear water is aspirated from the stomach tube
- The stomach should then be filled with a mixture of activated charcoal and water (20 ml in 100 ml of water), drained 10 min later and refilled with the same mixture to attempt to absorb any remaining poison

Fig. 11.3 Procedure for gastric lavage.

apnea alarm to provide adequate warning of respiratory failure.

Respiratory depression should be treated by the use of respiratory stimulants such as doxapram at 5 mg/kg body weight. Repeated dosage every 20–30 min may be necessary. If a suitable infusion pump is available, then doxapram can be given by continuous infusion. If cyanosis of the visible mucous membranes is detected, oxygen should be administered and ventilation assisted by manual compression of the thorax. Supplementation of inspired air with oxygen is also often beneficial in cases with less severe respiratory depression. A simple oxygen tent for cats and smaller dogs can be produced by placing a wire cat basket in a large clear plastic bag. Oxygen is introduced at one side and an outlet made at the other end. If prolonged severe respiratory depression occurs, or if there is blockage of the pharynx with mucus or vomit, then the animal should be intubated, or exceptionally, a tracheostomy performed.

As an emergency measure respiration can then be maintained using a Rees-modified T-piece with periodic ventilation using the reservoir bag. When using a T-piece in this way, it is important to increase the gas flow rate to three or four times that normally used, to try to minimize any rebreathing. Prolonged artificial ventilation requires the use of a mechanical ventilator and careful monitoring of the patient.

A second major complication in treating poisoning is disturbance of body temperature. Careful monitoring is important in a sedated or unconscious patient. This is best achieved by using an electronic thermometer and rectal probe to provide a continuous updated readout. Hyperthermia can be controlled by the use of ice packs, but this should be undertaken carefully as it is easy to cool the animal excessively. Hypothermia should be treated by providing supplemental heating and insulating the animal. Insulation alone is relatively ineffective in reversing the fall in body temperature. Care should be taken in rewarming an animal that shows signs of shock, because the rapid cutaneous capillary dilatation that results can cause a precipitous fall in blood pressure, which may be fatal. In these cases, procedures to maintain fluid balance and support circulatory function must first be instituted.

If the animal has a history of fluid or blood loss, then appropriate replacement therapy will be necessary. Even in cases where no noticeable loss has occurred, adequate fluid intake should be ensured by the intravenous administration of lactated Ringer's solution (40 ml/kg body weight sid), at rates not exceeding 1 ml/kg body weight per minute. If the ingested poison is excreted by the kidneys, then diuretics (frusemide, 2–4 mg/kg body weight IM) should be administered together with intravenous fluids. It is essential to monitor urine production, ideally by catheterizing the bladder, to ensure that renal function is adequate to cope with the additional fluid load.

If the animal appears to be in pain, then an analgesic should be administered (Fig. 11.4). Pethidine should be used with caution if respiratory depression is present.

Convulsions, hyperactivity or excitement can be controlled by the administration of diazepam or midazolam. Intravenous injection is the preferred method of administration, but both agents are rapidly absorbed following IP injection (in about 10–15 min). Repeated dosage may be required every 3–4 hr, depending on effect. If the animal survives the acute phase of poisoning, supportive therapy must be continued until normal body functions are restored. Recumbent animals should be turned every 4 hr to help prevent hypostatic congestion of the lungs, and any pressure points (e.g. elbows and hocks) massaged to prevent the development of pressure sores.

If severe diarrhea has developed, this can be controlled by a mixture of activated charcoal (to continue to absorb the toxin), kaolin and water. If damage to the gastrointestinal tract is suspected, then antibiotics should be administered to help prevent systemic infection with gut organisms. Fluid therapy should be continued. If irritant or caustic material has been ingested, the damage to the oral membranes may cause prolonged inappetence and in

DOSAGE OF DRUGS USED IN THE TREATMENT OF POISONING

Purpose	Drug	Dosage
Adsorption of poison	Activated charcoal	20 ml in 100 ml of water
Emesis	Apomorphine	0.1 mg/kg SC
Respiratory stimulant	Doxapram	5 mg/kg IV
Pain relief	Buprenorphine	Cat: 0.005–0.01 mg/kg SC or IM (tid)
		Dog: 0.01 mg/kg SC, IM or IV (repeated every 4 hr)
	Pethidine	Dog: 2 mg/kg SC or IM repeated every 2–4 hr
		Cat: 2.5 mg/kg SC or IM repeated every 4 hr
	Meperidine	Dog: 10 mg/kg SC
		Cat: 2 mg/kg SC
Fluid therapy	Lactated Ringer's solution	40 ml/kg sid
Sedation	Diazepam	2 mg/kg IV or IP
	Midazolam	2 mg/kg IV or IP
Diuresis	Frusemide	2–4 mg/kg IM
Treatment of cardiac arrhythmias	Propranolol	Dog: 0.5 mg/kg (up to 8 kg/mg reported for toad venom)
	Diphenylhydantoin	Dog: 15 mg/kg
Treatment of cyanide poisoning	Sodium nitrite (1% solution) followed by sodium thiosulfate (25% solution)	Dog: 25 mg/kg IV
		1.25 g/kg IV
		Combined therapy may be repeated as necessary, with half initial dose
Control of salivation (antidote to anticholinesterase - or parasympathomimetic - type drugs)	Atropine	0.05 mg/kg SC or slow IV (up to 0.4 mg/kg reported in the dog)
Anticholinesterase (antidote to atropine-like drugs)	Physostigmine	Cat: 0.25–0.5 mg SC (in total)
		Dog: 0.1–0.6 mg/kg
Treatment of lead poisoning	CaEDTA	75–100 mg/kg/day in 2–4 bid–qid by slow IV or by SC (diluted to 1–4% solution) injection for 5 days
	D-Penicillamine	12.5 mg/kg qid PO
Treatment of mercury poisoning	Dimercaprol	3 mg/kg IM qid for the first 1–2 days, reducing to qid – tid the next day and then bid for a further 10 days (maximum) or until recovery is complete
	N-Acetyl-D-1-penicillamine	3–4 mg/kg qid PO

C — cat; D — dog.

FIG. 11.4 Dosage of drugs used in treatment of poisoning.

these cases, use of a pharyngostomy tube to supply the animal's nutritional and fluid requirements should be considered.[7] Even when the animal appears to have recovered from the harmful effects of the poison, a regular, thorough clinical examination should be undertaken over the following few weeks to enable early detection of any chronic toxic effects.

In addition to these measures, a specific antidote can be given, if one is available and if the identity of the poison is known.

Specific Hazards

Food Components

There are very few foods that are inherently poisonous to dogs and cats; most cases of food poisoning in these species arise as a result of accidental contamination of wholesome food with a poisonous substance. Nevertheless, there are a few reports of poisoning in which food itself has been implicated.

The hazards associated with nutrient

imbalances due to inappropriate feeding practices are described in Chapter 1. The best-known case of an inherent food toxicity is that of hypervitaminosis A, which is most common in cats, although it may also occur in dogs. It is normally produced by feeding excessive quantities of liver, or oversupplementation of the diet with cod (or other fish) liver oil, both of which are particularly rich in vitamin A. Oversupplementation with cod liver oil may also result in vitamin D toxicity. Growing dogs and cats are particularly susceptible to nutritional errors, especially those that may result in developmental abnormalities of bone, such as nutritional secondary hyperparathyroidism (Chapter 22). This is caused by a Ca:P imbalance and is often associated with an "all-meat" diet, which contains very little calcium and has a very adverse Ca:P ratio (approximately 1:32).

Nutrient imbalances may also occur as a result of nutrient interactions in the diet. For example, diets high in polyunsaturated fatty acids (such as oily fish) or rancid, oxidized fat increase the animal's requirement for vitamin E; high levels of phytate (in fiber rich diets) will reduce the availability of zinc, copper and iron; the protein avidin, in raw egg white, reduces the availability of biotin, although this effect is reversed by heat. Adequate cooking is also required to destroy potentially harmful components of certain foods, such as the enzyme, thiaminase, in raw fish.

Onion Poisoning

Onion poisoning has been described in many species, including farm animals and dogs, but has only relatively recently been reported in cats.[30] A hemolytic anemia and an increase in Heinz bodies in erythrocytes was reported in cats that had consumed onion soup. Hemoglobinuria was also observed when large amounts of soup were given. Nevertheless, cats normally have a small percentage of erythrocytes with Heinz bodies and unless the proportion of affected cells is very high, this is unlikely to be a useful diagnostic tool. This is in contrast to the situation in dogs. Onion poisoning is not considered to be a major clinical problem in cats unless ingestion is continuous.

In dogs, onion ingestion equivalent to 30 g of raw onion once daily for 3 days produced hemolytic anemia with hemoglobinemia and hemoglobinuria.[38] It is thought that onions contain oxidizing agents, or stimulate the production of such agents, which cause damage to erythrocytes, resulting in the production of Heinz bodies and methemoglobinemia. Although recovery is possible once intake has ceased, fatalities have occurred, often where there is a marked methemoglobinemia. A phenolic compound has been isolated and incriminated as a major causative agent of the disease[36]; previous reports had suggested *n*-propyldisulfide as another agent, although this has not been confirmed. It appears that some dogs with a hereditary high level of anti-oxidizing agents in erythrocytes are more resistant to onion poisoning.

Cocoa Poisoning

Poisoning from cocoa and related products has often been reported in dogs. The toxic principle is the methylxanthine derivative theobromine and symptoms of poisoning are vomiting, diarrhea, sudden collapse and death. Other presenting signs include hyperactivity, tachycardia, ataxia and depression.[21] Sudden death following the consumption of 300 g of chocolate has been reported in Dachshunds, and a Springer Spaniel also died some 12 hr after eating 250 g of household cocoa.[24] In another incident a Springer Spaniel died 15 hr after the ingestion of 2 lbs (about 900 g) of chocolate.[13] In relation to body weight, the intakes of chocolate were similar for the two breeds, in the region of 40 g/kg. The same authors reported that the serum concentration of theobromine in a mixed-breed dog (given a limited amount of chocolate) declined at an extremely slow rate, much slower than in man.[13] This may indicate a long half-life and, hence, an inefficient metabolism of this substance. The oral LD_{50} for theobromine in cats is quoted as 200 mg/kg body weight. As chocolate typically contains about 0.2% theobromine, this value is equivalent to a chocolate intake of about 100 g/kg body weight.

There is no specific antidote to theobromine. Treatment is symptomatic and, in the dog, induction of vomiting is probably beneficial to remove any remaining chocolate, thereby assisting the decline of the serum theobromine level.

Food Additives

Benzoic Acid

The best known incidence of food additive toxicity in the cat is to benzoic acid, which is used as a human food preservative but is not permitted in any animal feeds within the European Community (EC). Bedford and Clarke found that the highest intake that could be fed safely was 200 mg/kg body weight.[6] An intake of 300 mg/kg body weight caused mild hyperesthesia from which the cat recovered and higher doses caused aggression, hyperesthesia and death. The susceptibility of cats to benzoic acid arises from their inability to detoxify this substance sufficiently quickly, a deficiency also responsible for their sensitivity to aspirin, paracetamol and other phenol compounds. Poisoning has been reported in cats that received lactated Ringer's solution containing benzyl alcohol as a preservative.[12] On entering the body, benzyl alcohol is rapidly oxidized to benzoic acid and is, hence, toxic to the cat. At the time of writing, none of the commercial preparations of lactated Ringer's solution available in the U.K. contain benzyl alcohol.

Propylene Glycol

Propylene glycol (PG) has been used as a humectant in semimoist dog and cat foods, where it exerts a preservative action by binding water, thereby making it unavailable for food spoilage organisms. In 1986, the U.S.A. National Research Council stated that high levels of PG caused Heinz body formation in cats and might result in hemolytic anemia.[37] Some recent research now indicates that the main action of PG is on erythrocyte survival time, whereas hemolytic anemia was not observed.[10,20] These effects were seen with dietary concentrations of PG similar to those used in semimoist products (typically 5–15%). Whether this effect is actually detrimental to the health of the cat is open to question but it was recently reported that the erythrocytes of cats fed a commercial diet containing 8.3% PG were more susceptible to oxidant stress. It must remain a possibility, therefore that PG can decrease the cat's ability to respond to challenges posed by certain diseases or chemicals.[51] As a result of the uncertainties surrounding its safety-in-use, PG was delisted as a permitted additive for cat foods in the EC in 1991.

Food Contaminants

Food that constitutes the normal diet of dogs and cats may be contaminated in various ways. Microbial proliferation on food (usually as a result of poor methods of storage or preparation) can result in disease due to the pathogenic effects of the agent itself or of its toxins. An additional food hazard is provided by a number of parasites that complete their life-cycle when the dog or cat eats infected tissue from the intermediate host of the parasite. Examples of such parasites include *Toxoplasma gondii*, *Echinococcus granulosus*, *Taenia* spp., *Isospora* spp. and *Sarcocystis* spp. Food may also be contaminated with nonbiological compounds. Toxic residues may be present in the tissues of animals that were themselves intoxicated prior to death, or contamination may occur accidentally (or maliciously) during storage or preparation.

Mycotoxins

The adverse effects of fungal toxins (mycotoxins) have been well documented in many species of animals, particularly farm livestock. These toxins include substances such as ergot, ochratoxin and aflatoxin. The adverse effects can be one of three general forms: acute, chronic or secondary. Acute poisoning usually affects the liver and can result in hepatitis, jaundice and related disturbances. The chronic and secondary effects are not usually manifested as overt disease, but in farm animals result in general loss of performance, such as reduction of feed and reproductive efficiency and an increased susceptibility to infection.[39]

Aflatoxin is probably the most toxic of this group of compounds with an oral LD_{50} of 0.55 mg/kg reported for the cat, which makes it one of the most susceptible species, along with the dog, pig, rabbit and duck. In dogs, 5 μg/kg body weight per day fed for 10 weeks (5 days per week) did not cause morphological changes, but the authors estimated, using regression analysis data from higher intakes, that even this low intake would induce adverse effects in about 3 years.[1] For a dose of 1 μg/kg body weight per day the induction time was estimated to be beyond the dog's normal lifespan. There does not seem to be any specific antidote to aflatoxin, but the area is the subject of current research. Activated charcoal, anabolic steroids and oxytetracycline are among the substances reported to be beneficial in aflatoxin poisoning.

An outbreak of naturally occurring cases of aflatoxicosis in dogs in South Africa was associated with a contaminated batch of commercial dog food.[3] Some dogs died suddenly without prior clinical signs and others died following a short clinical course. Clinical signs were related to liver pathology and included icterus, hemorrhagic diathesis and peripheral edema (ascites, hydrothorax, pulmonary edema, hydropericardium, anasarca). Cases could be classified as acute, subacute or chronic. Fatty degeneration of the liver was present in all three stages and was the prominent feature of the acute cases. Subacute cases were characterized by extensive bile duct proliferation and chronic cases had extensive liver fibrosis. Fatal aflatoxicosis has also been reported in dogs that received a ration supplemented with contaminated cornmeal.[33]

Over-ripe (moldy) blue cheeses are also a potential source of poisoning for both humans and pet animals. A fatal case of roquefortine poisoning in a dog has been reported, in which recumbency, shivering, tetanic spasms, opisthotonos and convulsions occurred within 2½ hr of consumption of *very* over-ripe blue cheese.[42] Roquefortine is a neurotoxic mycotoxin which is produced by *Penicillium roquefortii* (and by other *Penicillium* sp.). This organism is used in the production of Roquefort and other similar cheeses.

Bread has a high moisture content and is therefore a good substrate for the growth of molds. Ochratoxins, aflatoxins and patulin are recognized as natural contaminants of moldy bread and an outbreak of lethal gastroenteritis in rabbits, poultry and dogs was reported in Italy, due to the feeding of moldy bread.[49] The bread was contaminated extensively by *Aspergillus ochraceus* and high levels of ochratoxins were detected in the sample.

Bacterial Contamination

Dogs commonly scavenge putrefying material and this may result in gastrointestinal and systemic disease, often due to the presence of staphylococcal enterotoxins. In the U.S.A., this condition is often referred to as "garbage disease" and may be associated with convulsions, depression and weakness, diarrhea, incoordination and vomiting.[24] Staphylococcal food poisoning has also been shown to affect cats, in fact kittens have been used as the test animal for investigations into this toxin.

Salmonella infections may occur in both animals and man as a result of eating contaminated foodstuffs. Food may become contaminated by prior infection of the animal from which the food (meat, milk, eggs) is derived, through contamination of the carcass after slaughter (usually with fecal material) or where food is prepared in conditions of poor hygiene. The organism multiplies rapidly in moist foods, but is destroyed by adequate heat treatment, desiccation and most disinfectants. Infected dogs and cats may become asymptomatic carriers, but some will show signs of acute gastroenteritis and in severe cases, septicemia. Fluid therapy is essential in all cases to prevent dehydration, and in cases of septicemia, it may be necessary to administer antibiotics, especially ampicillin or trimethoprim/sulfonamide combinations. The use of antibiotics in enteric salmonellosis in animals is controversial and should be avoided where possible to avoid the risk of the development of drug resistance. Although humans rarely acquire *Salmonella* infections from their pets, the zoonotic potential should be noted.

Campylobacter jejuni is recognized as a major cause of food poisoning in humans and may also be responsible for a number of food-borne infections in the dog and cat. In these species, infections cause diarrhea (with tenesmus and the production of clear mucus, which is sometimes blood-stained), dullness and inappetence. Contaminated water and fresh foods such as raw liver, chicken, or tripe and unpasteurized milk are sources of infection, although in pet animals, the disease is most commonly spread by direct and indirect contact with infected feces. The organism is resistant to chilling and freezing, but is readily inactivated by heat and desiccation. It is excreted in large numbers in the feces and this represents an important potential zoonotic risk. Treatment of affected animals is symptomatic and may involve fluid therapy and antibiotics, especially erythromycin, neomycin, tylosin and metronidazole.

Botulism is another form of food poisoning, although its occurrence in dogs and cats is very rare. The causal agent, *Clostridium botulinum*, is a soil organism that multiplies under anaerobic conditions (such as may be found in decaying carcasses) and produces an extremely potent neurotoxin. This causes stiffness and progressive muscle paralysis in affected animals. Dogs and, to a lesser extent, cats are much more resistant to botulism than are other animals, including humans[41] and often respond well to treatment with antitoxins and general supportive therapy.

Toxic Residues in Animal Tissues

Secondary poisoning in dogs and cats may occur following the consumption of animal tissues that contain toxic residues. The most common example of this is where the dog or cat consumes prey (birds, insects and small mammals) that were poisoned. Dogs that are fed the flesh of horses (or other species)

euthanized by means of a barbiturate or chloral hydrate overdose, may suffer profound CNS depression and occasionally death, due to the presence of drug residues in the meat. Estrogenic substances in the ears of slaughtered cattle and in turkey meat may represent a food hazard for cats and dogs.[24] Secondary poisoning with organic mercurial compounds may occur following the consumption of contaminated fish, particularly tuna.

Dogs have developed fatal hepatotoxicity after consuming horseflesh contaminated with indospicine.[19] This is a toxic alkaloid found in the legume *Indigofera innaei* (a native of the desert areas of the Northern Territory in Australia, in particular) and causes a neurological condition, Birdsville disease, in horses. Restrictions have now been imposed on the use of horsemeat from areas where Birdsville disease is endemic.

Food Sensitivity

There are many reports of allergic reactions to food; however, the overall scale of the problem is difficult to ascertain and the true incidence is generally thought to be fairly low. The usual sites affected in food sensitivity reactions in the dog and cat are the skin and/or the alimentary tract and incriminated substances include fish, beef and cow's milk. Use of an elimination diet for diagnosis is the recommended course of action and therapy is usually successful if the offending allergen(s) can be excluded from the diet. The subject is covered in depth in Chapter 12.

Mechanical Injuries

Ingested material may have adverse effects due to physical injury to the animal. Sharp objects, such as bone splinters, may cause irritation (or even perforation) of the alimentary tract and large amounts of relatively indigestible material (including bones) may result in obstructions. Certain plants may also produce injury if eaten. Those with thorns (such as roses or blackberries) can cause lacerations of the oral mucosa. The awns and seed heads of grasses and other crops may lodge in the pharynx, causing irritation, or migrate deeper into the tissues where abscessation may occur. Dogs that chew and play with twigs are also liable to injury, often as a result of penetrating wounds.

Poisonous Plants

Although dogs and cats are natural carnivores and do not normally consume large amounts of plant material, cases of plant poisoning do occasionally occur. Cats are less likely to eat plants than dogs and poisoning in adults of either species is rare. Young animals are, however, very inquisitive and puppies, in particular, will often "mouth" objects they encounter, making them vulnerable to accidental ingestion of toxic plants. This is also true, to a lesser extent, of kittens and of older animals and some individuals who suffer from boredom or behavioral idiosyncrasies that prompt them to eat plant material.

It should be noted that many cats (particularly those housed mainly or solely indoors) habitually chew the leaves of house plants, some of which contain irritant or poisonous substances. Owners should be advized to provide an alternative source of foliage, such as grass, to minimize this problem. An additional hazard for cats is that they often scratch unfamiliar plants and if these are toxic, poisoning may occur when the cat licks its claws.[9]

Ornamental Plants

Although plant poisoning may occur in both dogs and cats, there seem to be relatively few references to plant poisoning in the cat. However, laurel, ivy, laburnum and the pine needles of Christmas trees have been reported in poisoning incidents in cats. It is most unlikely that there is a specific antidote to the majority of plant poisons and treatment involves removal of as much material as possible from the gastrointestinal tract, the use of an adsorbent such as activated charcoal and general supportive therapy.[18,22] The signs associated with plants most likely to cause

toxicity in the dog and cat are summarized in Fig. 11.5.

It has been suggested[18] that the Araceae family contains the houseplants most commonly recognized as toxic. This includes species such as *Philodendron* spp. and *Dieffenbachia* spp. (dumb cane). These plants have explosive ejector cells that inject calcium oxalate crystals and soluble oxalates into the buccal mucosa when bitten or chewed. This causes pain, stinging and irritation of the lips, mouth, tongue and throat, often with edema of the mucous membranes, salivation and dysphagia. Occasionally, airway obstruction and respiratory distress may accompany the edema and there may be vomiting and diarrhea. Treatment, which involves washing the mouth out with a solution of sodium bicarbonate and the administration of painkillers and corticosteroids, usually results in complete recovery within a few days, although parenteral nutritional support may be necessary in the interim if the animal cannot eat or drink. Other calcium oxalate-containing plants include *Anthurium, Arum* (cuckoo pint, lords and ladies), *Caladium, Monstera* (Swiss cheese plant), *Schleffera* (umbrella tree) and *Begonia*. Rhubarb *(Rheum)* leaves contain soluble oxalates that have more serious systemic effects and may induce hypocalcemia, but animals are unlikely to eat sufficient quantities to be seriously affected.

Oleander *(Nerium oleander)* contains digitalis-type glycosides and is highly toxic to most species of animal. Although consumption is rare due to its bitter taste, the dried plant is more palatable and poisoning may occur following the consumption of discarded clippings or where the plant has been killed in freezing weather. Poisoning has been reported in puppies that suffered convulsions and later died.[43] It has also caused continued vomiting, diarrhea and cardiac arrhythmias in the dog.[24] Treatment with atropine and propranolol may be beneficial in some cases.

The seeds of the castor oil plant *(Ricinus communis)* contain the toxic protein ricin and are poisonous to all species. Ingestion of the seed or seed residue may produce violent gastroenteritis (with abdominal pain, vomiting

TOXICITY OF ORNAMENTAL PLANTS

Plant	Toxic effects	Therapy
Calcium-oxalate-containing plants e.g. *Anthurium; Arum* (Cuckoo pint, lords and ladies); *Caladium; Dieffenbachia; Monstera* (Swiss cheese plant); *Philodendron; Schleffera* (Umbrella tree); *Begonia*	Stinging and intense irritation of the lips, mouth, tongue and throat Edema of muscosal surfaces, profuse salivation and dysphagia. Rarely, edema may cause airway obstruction and respiratory compromise. Occasionally, vomiting and diarrhea. Systemic absorption unlikely	Washing the mouth out with a solution of sodium bicarbonate and the administration of painkillers and corticosteroids
Soluble oxalate-containing plants e.g. *Rheum* (rhubard leaves); *Amaranthus; Oxalis* (wood sorrel); *Calendrinia; Portulaca; Rumex* (docks)	Large quantities of plants would have to be consumed for toxic effects to occur. Nausea, rapid respiration and stupor then vomiting, bloody diarrhea, coma and tetany due to hypocalcemia. Renal lesions due to the formation of Ca oxalate crystals in the kidney (oliguria, oxaliuria, albuminuria, hematuria)	Administration of parenteral Ca salts, and hydration to avoid deposition of insoluble Ca salts in urine
Digitalis purpurea (foxglove)	Cardiotoxic effects due to the presence of digitalis glycosides. Depression, anorexia, vomiting, bradycardia, heart block ventricular tachycardia	Demulcents to counteract gastrointestinal irritation. Ventricular arrhythmias may be treated with potassium chloride and propranolol
Nerium oleander	Effects as for *Digitalis*	Treatment with atropine and propranolol may be effective
Ricinus communis (castor oil plant, especially the seeds)	Violent gastroenteritis (with abdominal pain, vomiting and hemorrhagic diarrhea), CNS disturbances, collapse and death	Ideally, the antiserum for ricin should be administered, but this is unlikely to be available. Otherwise, sedatives and general treatment for shock should be beneficial
Solanine-containing plants e.g. Green, sprouted potatoes; wild (*Solanum dulcamara*) and garden (*Solanum nigrum*) nightshade; ornamental pepper; Japanese cherry (*Solanum pseudocapsicum*)	Depression, vomiting, diarrhea, salivation	Removal of remaining plant material (purgatives, etc.) plus supportive therapy. The use of stimulants may be tried, administration of physostigmine may be beneficial
Allium spp. including, the onion and the garlic	Hemolytic anemia if eaten in sufficient quantity	Supportive therapy
Laburnum (*Cytisus laburnum* or *Laburnum anagyroides*)	Vomiting, diarrhea and muscle tremors	Supportive therapy. Sedation may be required
Cyanogenic plants e.g. *Nandina; Acacia; Aquilegia; Euphorbia; Hydrangea; Lotus; Nerium oleander; Passiflora; Prunus; Trifolium*	Excitement, salivation, convulsions, pulmonary edema, pale and cyanotic mucous membranes, bradycardia, vomiting and sudden death	Symptomatic supportive therapy in mild cases. More severe cases may require the traditional therapy for cyanide poisoning (Fig. 11.4)
Brunfelsia species	Convulsions and other neurological disturbances, constipation	Treatment with diazepam may be successful in some cases
Cycads e.g. seeds of *Zamia floridiana*	Persistent vomiting and icterus. Generalized hemorrhagic syndrome associated with hepatic necrosis	Supportive therapy (may require euthanasia)
Rhododendron and azalea	Presence of grayanotoxin causes salivation, vomiting, weakness, ataxia, seizures, coma, bradycardia	Nonspecific supportive therapy. The animal should be kept warm and stimulants may be administered
Delphinium (larkspur)	Gastrointestinal upset, CNS disturbances, cardiotoxic effects	Nonspecific therapy. Sedation or anesthesia may be required
Tobacco plant (*Nicotiana tabacum*)	Depression, salivation, intestinal hyperperistalsis and dyspnea	Symptomatic. Recovery usually spontaneous
Cannabis sativa (marijuana)	Ataxia, muscle tremors and weakness	Symptomatic. Recovery usually spontaneous

FIG. 11.5 Toxicity of ornamental plants.

and hemorrhagic diarrhea), CNS disturbances, collapse and death.

All parts of the laburnum *(Cytisus laburnum* or *Laburnum anagyroides)* plant are poisonous to all species and fatalities have been reported in dogs.[32] Vomiting, diarrhea and muscle tremors were observed in affected animals with severe gastroenteritis evident at postmortem.

Some plants are considered hazardous because of their cyanide content; poisoning may be rare, but is potentially very serious when it occurs. Members of the family *Rosaceae* are particularly implicated.[35] *Nandina domestica Thumb* (heavenly bamboo), a cyanogenic plant, has caused convulsions, pulmonary edema, pale and cyanotic mucous membranes, bradycardia, vomiting and frequent defecation in a puppy.[8] This case was successfully treated with symptomatic supportive therapy only, but more severe cases may require the traditional therapy for cyanide poisoning. In the dog, this involves the intravenous injection of a 1% solution of sodium nitrite (25 mg/kg) followed by 25% sodium thiosulfate at 1.25 g/kg. The treatment may be repeated as necessary, using half the initial dose.

Convulsions and other neurological disturbances were noted in a dog that ate *Brunfelsia pauciflora* ("yesterday, today and tomorrow").[2] Constipation was also a feature. Other *Brunfelsia* species are similarly toxic, although some species have been used as medicinal compounds in many Amazonian Indian tribes. Treatment with diazepam may be successful in some cases.

Ingestion of the tobacco plant *(Nicotiana tabacum)* in the form of cigarettes (a packet of 20) led to depression, salivation, intestinal hyperperistalsis and dyspnea in a puppy, which later recovered.[24]

Cannabis sativa (marijuana) intoxication may follow ingestion of the plant or inhalation of the smoke.[46] Affected animals appear ataxic and "glassy eyed," with muscle tremors and weakness, but usually recover spontaneously over the next 24 hr. If the possibility of cannabis intoxication is not admitted by the owner, severe neurological disease may be suspected and could result in destruction of the animal. Diagnosis is possible using the appropriate laboratory methods to detect cannabinoids in the urine.

Case,[9] Leroux[31] and Hoskins[22] all provide comprehensive lists of houseplants that can be implicated in poisoning incidents in pets.

Mushroom Poisoning

Mushroom (*Amanita* spp.) poisoning has been reported in cats and can be of two main types. With *A. muscaria* (fly agaric), the toxic principal is muscarine, that produces excessive cholinergic activity manifested as profuse salivation, vomiting and diarrhea. Treatment is with atropine. In the second type of poisoning, for example with *A. pantherina* (false blusher), the toxic agents resemble atropine in their mode of action and the use of atropine is therefore contraindicated.[45] Signs include dilated pupils, dryness of the mouth, incoordination, apparent blindness and CNS excitement that may lead to convulsions, paralysis and death. In this case, gastric lavage, activated charcoal and general supportive measures are recommended together with carefully controlled doses of physostigmine (0.25–0.5 mg total SC in cats). The latter is a cholinesterase inhibitor that overcomes the action of atropine, in effect, the muscarine treatment in reverse.

A. muscaria poisoning also occurs in dogs, producing dullness, salivation, anorexia, aggression, abdominal pain, diarrhea, dyspnea, constriction of the pupils and posterior paralysis, which may be followed by coma.[24] Atropine (0.05 mg/kg) may be administered subcutaneously as therapy.

Amanita phalloides (death cap) has resulted in fatal toxicity in dogs after consumption, or simply playing with the mushrooms.[26] Clinical signs of depression, anorexia, vomiting, tenesmus and bloody diarrhea were reported, which may easily be confused with acute parvovirus enteritis. Gastric ulceration and hepatic and renal cell necrosis were found at postmortem. Therapy is supportive and symptomatic, involving fluid therapy, gastric lavage and the use of adsorbents (activated charcoal). Thiocitic acid (α lipoic acid) has been administered as an antidote for amanitin toxicity in man. If ingestion of *Amanita phalloides* is suspected, vomiting should be induced immediately.

Intoxication following the consumption of the hallucinogenic "magic mushrooms" *Psilocybe semilanceate* and *Panaeolus foenisecii* has been reported in a dog.[29] The dog displayed overt aggression initially, with ataxia, nystagmus, hypersalivation and vocalization. Body temperature was initially very high (>42.2°C), but later became subnormal (<34.4°C). Recovery was spontaneous and completed by the third day following consumption of the mushrooms.

Poisoning due to the fungus *Clitocybe* spp. has been reported in a dog.[44] Within 45 min of chewing and playing with the fungus, the dog was reported to be ataxic, lethargic, vomiting, the hind legs were in spasm and copious amounts of clear frothy mucus were passed from both the mouth and anus. Later, diarrhea was evident, but the animal's condition improved during the ensuing couple of days. The fungus has a strong odor of anise and so may be attractive to dogs.

Algal Poisoning

Poisoning may occur in dogs and other species following the ingestion of water that is contaminated with certain strains of blue–green algae. Hepatotoxins and neurotoxins produced by these algae can rapidly prove lethal if ingested in sufficient quantities. Dr.inking of contaminated water may result in poisoning, but in dogs, it is most likely to occur when the animal licks its coat after swimming.

In still fresh waters, the algae can multiply sufficiently during the summer months to produce a blue–green scum on the surface of the water. Not all blue–green algal blooms are toxic and even those that are toxigenic must be ingested in sufficient amounts to produce symptoms of poisoning. Traditionally, the most commonly incriminated algal species were *Microcystis aeruginosa* (also known as *Anacystis cyanea* or *Microcystis toxica*) in fresh water and *Nodularia spumigena* in brackish water, both of which produce hepatotoxicosis, with massive hemorrhage into damaged liver. Recently, however *Anabaena flosaquae* has been associated with toxicoses produced primarily by the postsynaptic neuromuscular blocking agents, anatoxin-a(s) and anatoxin-a.[5] These give rise to vomiting, muscle tremors, weakness, convulsions and paralysis. In some cases, death may follow within minutes of the onset of the clinical signs.

Algal blooms usually occur after a period of hot weather, when the water temperature is between 15 and 30°C and often where there is a high concentration of nitrogen (in the form of nitrates or ammonia) and phosphorus in the water. Warm and windy conditions will encourage water evaporation and thus concentrate the algae and may help to propel the algal scum to the shore, where it is more likely to be eaten. The incidence of algal blooms on surface waters may have increased recently owing to higher nutrient levels in still waters. This has been associated with the widespread use of fertilizers.

An outbreak of acute illness and deaths has been reported in dogs after swimming in Rutland Water, a reservoir in the U.K.[11,27] Affected dogs were acutely ill within a short period of being in contact with contaminated water. A detailed case report was given for one of these dogs, which had severe hepatic, gastric and renal lesions, resulting in azotemia, anuria, anemia and hemoptysis.[27] Post-mortem revealed widespread petechial hemorrhages, extensive coagulative hepatic necrosis, renal tubular and glomerular necrosis, focal mucosal hemorrhages of the stomach and small intestine, and necrotizing arteritis.

Signs associated with hepatotoxicity should be treated symptomatically, although in some cases, euthanasia may be required. However, there is some evidence to suggest that the effect of the postsynaptic neuromuscular blocking agents, anatoxin-a is reversible with time, so there may be some chance of recovery if affected animals are given prolonged artificial respiration, together with gastric lavage and activated charcoal.[4]

The bottom-dwelling (as opposed to plank-tonic) cyanobacteria *Oscillatoria* sp. has been reported as the cause of death in four dogs in the Highland region of Scotland, due to the presence of anatoxin-a.[17] Variable clinical signs of convulsions, cyanosis, rigors, limb twitching and hypersalivation were apparent within 1 hr of contact with water in the contaminated loch; death occurred between 10 and 30 min after the onset of clinical signs. These cyanobacteria may be present from summer until late in the year.

Poisonous Animals

Venomous Toads

The common toad in the U.K., *Bufo vulgaris*, secretes various noxious substances from the glands in the head and may represent a hazard to both cats and dogs. Dogs and cats that have bitten or mouthed toads, show symptoms of poisoning such as excessive salivation and general distress. However, the animal usually recovers fairly rapidly and, in general, no specific treatment is necessary, other than rinsing the mouth with copious amounts of running water. Dogs may occasionally ingest the toad, in which case clinical signs are more severe and may include prostration, cardiac arrythmia, collapse and possibly death. In these cases, atropine may be administered (1.0–2.0 mg) to control salivation and the cardiac effects may be treated with propranolol (0.5 mg/kg).

More toxic species of toad are found in other countries. Poisoning from the parotid glands of *Bufo marinus* in Queensland, Australia is most commonly seen in dogs of the Terrier breeds, but cats are rarely affected.[34] The toxin is rapidly absorbed through the buccal and gastric mucosae, so clinical signs develop within a short time of contact with the toad. Initially, there is profuse salivation, with hyperemia of the buccal mucosa and head shaking. This usually progresses to ataxia and is often accompanied by vomiting, polypnea and diarrhea, which is frequently bloody.

Convulsions may follow in severe cases and these generally precede death if therapy is not immediately instituted. Swabbing or hosing the animal's mouth (with the head down) helps to dilute the toxin. Intravenous injections of diazepam and atropine are useful in mild cases, but induction of general anesthesia using pentobarbital may be required if the animal is convulsing. Ventricular fibrillation may be a complication and this may be treated with intravenous propranolol at 2 mg/kg body weight up to 15 kg live weight. The corneal surfaces of the eyes may be rinsed with isotonic saline and protected with ophthalmic ointment. Mild cases usually respond well to early treatment, but for those animals that are convulsing, the prognosis is guarded to poor.

The same species of toad is also responsible for cases of intoxication and death in dogs in Central America.[48] The cardiac effects of the venom were found to be greatly reduced by the antiarrhythmic drugs, diphenylhydantoin (at 15 mg/kg) and propranolol (at 8 mg/kg). Atropine administered at a dose of 0.4 mg/kg had a moderate therapeutic effect.

Minerals

Intoxications by heavy metals are unlikely to occur as a result of feeding prepared pet-foods because levels of these substances are strictly controlled by legislation.

Lead

Although lead poisoning is relatively common in dogs (particularly puppies), it appears to be comparatively rare in cats. This may be due to the more fastidious eating habits of cats, but some cases may go undetected because of the lack of obvious specific signs. The most common source of lead is paint (particularly old paint that contains a high proportion of lead), but linoleum, putty and metallic lead objects can also be hazardous.

Lead inhibits hemoglobin synthesis and thus causes anemia, but it is the presence of lead in the liver, kidneys and the CNS that causes the signs of lead poisoning. The signs associated with acute toxicity can be divided into two main categories: gastrointestinal and nervous. Some reports describe only pronounced neurological signs in cats including hyperactivity, frantic behavior and with dilated pupils in some affected cats. Two reports, however, both in Persian cats, describe symptoms of lethargy, poor appetite and intermittent vomiting.[25,50] In both cases, the probable source of the lead was old paint being sanded or scraped off preparatory to redecoration. In one cat blood lead concentration was increased to 0.7 parts per million (ppm) (mg/l). Blood lead levels are not always the best indicator of lead toxicity and there is some disagreement as to the normal blood lead levels in the cat, with reported values ranging from 0.05–0.2 ppm.[40] The urinary concentration of delta-amino levulinic acid (DALA) can be a useful indicator of lead poisoning, particularly to monitor the effectiveness of chelation therapy. DALA is a metabolite that accumulates as a result of impairment or blocking of hemoglobin synthesis. Normal values are reported to be less than 38 μM, which can rise to several hundred during lead poisoning. Basophilic stippling of erythrocytes is also a frequent, although not invariable, feature of lead poisoning.

The standard treatment is calcium disodium ethylenediamine tetraacetate (CaEDTA), which chelates the lead and removes it from the body. The recommended total daily dose is 75–100 mg/kg body weight given in two to four divided doses by slow intravenous injection or subcutaneously for 5 days. The CaEDTA should be diluted in a 1–4% solution for subcutaneous injection as it is painful when given by this route. D-Penicillamine at 12.5 mg/kg body weight PO qid was reported to be superior to CaEDTA in dogs given lead in a laboratory investigation.[15] This may be useful if antilead therapy is to be given in the client's home, but it does have some undesirable side effects such as vomiting and anorexia. In severe lead poisoning, supportive therapy with diazepam to control convulsions and fluid therapy to combat dehydration are also indicated. Severe convulsions are usually indicative of cerebral edema, which will require treatment with mannitol and dexamethasone. There are no characteristic postmortem signs.

Mercury

Mercury poisoning has been reported in animals, particularly cats, as a result of the build up of mercury in foodstuffs (although it may also occur following the use of mercury salts as fungicides for seed grains). This may arise when metallic mercury (which is of relatively low toxicity) is converted to more toxic compounds, particularly organic derivatives such as ethyl- or methylmercury, by the action of microorganisms. The compounds can then accumulate in animals (particularly fish) that are ultimately used as food sources. In regions where methylmercury concentrations in fish are very high, poisoning may occur in domestic cats if fish constitutes an appreciable part of their diet.[16]

The effects of methylmercury intoxication may be acute or chronic, depending on the exposure dose. Acute effects, associated with large doses (>1 mg/kg), seem to be confined mainly to the lung vascular structure. Chronic effects are neurological in nature and include impairment of motor coordination (loss of balance and the righting reflex, abnormal gait), muscle weakness (especially of the hindlimbs) and changes in temperament and behavior. It is these chronic signs that are more likely to be encountered in the cat.

The standard treatment for mercury poisoning is dimercaprol (Fig. 11.4) and/or *N*-acetyl-*d*-*L*-penicillamine (3–4 mg/kg body weight qid PO), but related and more effective substitutes such as 2,3–dimercaptosuccinic acid are under investigation. The postmortem signs of mercury poisoning are confined to microscopic changes in the brain and dorsal root ganglia and consist primarily of neuronal degeneration. The highest levels of mercury are found in the liver and kidney with elevated levels also found in the brain and blood.

References

1. Armbrecht, B. H., Geleta, J. N., Shalkop, W. T. and Durbin, C. J. (1971) A subacute exposure of

Beagle dogs to aflatoxin. *Toxicology and Applied Pharmacology*, **18**, 579–585.

2. Banton, M. I., Jowett, P. L. H., Renegar, K. R. and Nicholson, S. S. (1989) *Brunfelsia pauciflora* ("Yesterday, Today and Tomorrow") poisoning in a dog. *Veterinary and Human Toxicology*, **31**, 496–497.

3. Bastianello, S. S., Nesbit, J. W., Williams, M. C. and Lange, L. (1987) Pathological findings in a natural outbreak of aflatoxicosis in dogs. *Onderstepoort Journal of Veterinary Research*, **54**, 635–640.

4. Beasley, V. R., Dahlem, A. M., Cook, W. O., Valentine, W. M., Lovell. R. A., Hooser, S. B., Harada, K.-I., Suzuki, M. and Carmichael. W. W. (1989) Diagnostic and clinically important aspects of cyanobacterial (blue–green algae) toxicoses. *Journal of Veterinary Diagnostic Investigations*, **1**, 359–365.

5. Beasley, V. R. (1990) Algal poisoning. *Veterinary Record*, **127**, 243 (Correspondence).

6. Bedford, P. G. C. and Clarke, E. G. C. (1972) Experimental benzoic acid poisoning in the cat. *Veterinary Record*, **90**, 53–58.

7. Böhning, R. H., DeHoff, W. D., McElhinney, A. and Hofstra, P. C. (1970) Pharyngostomy for maintenance of the anorectic animal. *Journal of the American Veterinary Medical Association*, **156**, 611–615.

8. Bradley, M., Neiman, L. J. and Burrows, G. E. (1988) Seizures in a puppy. *Veterinary and Human Toxicology,* **30**, 121.

9. Case, A. A. (1983) Poisoning and injury by plants. In *Current Veterinary Therapy VIII. Small Animal Practice*. Ed. R. W. Kirk. pp. 145–152. Philadelphia: W. B. Saunders.

10. Christopher, M. M., Perman, V and Eaton, J. W. (1989) Contribution of propylene glycol-induced Heinz body formation to anaemia in cats. *Journal of the American Veterinary Medical Association*, **194**, 1045–1056.

11. Corkhill, N., Smith, R., Seckington, M. and Pontefract, R. (1989) Poisoning in Rutland water. *Veterinary Record*, **125**, 356 (Correspondence).

12. Cullison, R. F., Menard, P. D. and Buck, W. B. (1983) Toxicosis in cats from use of benzyl alcohol in lactated Ringer's solution. *Journal of the American Veterinary Medical Association*, **182**, 61.

13. Glauberg, A. and Blumenthal, H. P. (1983) Chocolate poisoning in the dog. *Journal of the American Animal Hospital Association*, **19**, 246–248.

14. Golomb, P., Yacobson, B., Perl, S., Nobel, T. A. and Orgad, U. (1988) Rates of mortality in dogs from various diseases in the years 1981–85 as compared to 1970–75. *Israel Journal of Veterinary Medicine*, **44**, 144–145.

15. Green, R. A., Selby, L. A. and Zumwaldt R. W. (1978) Experimental lead intoxication in dogs: a comparison of blood lead and urinary delta-amino levulinic acid following intoxication and chelation therapy. *Canadian Journal of Comparative Medicine*, **42**, 205–213.

16. Gruber, T. A., Costigan, P., Wilkinson, G. T. and Seawright, A. A. (1978) Chronic methyl-mercurialism in the cat. *Australian Veterinary Journal*, **54**, 155–160.

17. Gunn, G. J., Rafferty, A. G., Rafferty, G. C., Cockburn, N., Edwards, C., Beattie, K. A. and Codd, G. A. (1991) Additional algal toxicosis hazard. *Veterinary Record*, **129**, 391 (Correspondence).

18. Hanna, G. (1986) Plant poisoning in canines and felines. *Veterinary and Human Toxicology*, **28**, 38–40.

19. Hegarty, M. P., Kelly, W. R., McEwan, D., Williams, O. J. and Cameron, R. (1998) Hepatotoxicity to dogs of horse meat contaminated with indospicine. *Australian Veterinary Journal*, **65**, 337–340.

20. Hickman, M. A., Rogers, Q. R. and Morris, J. G. (1990) Effect of diet on Heinz body formation in kittens. *American Journal of Veterinary Research*, **50**, 475–478.

21. Hooser, S. B. (1985) Methylxanthine (theobromine, caffeine, theophylline) toxicoses in animals—a review of 1984 cases at the NAPCC. *Veterinary and Human Toxicology*, **28**, 313. (Abstr.)

22. Hoskins, J. D. (1989) Plants and their clinical importance. *Veterinary Technician*, **10**, 93–102.

23. Humphreys, D. J. (1988) Chapter 1: introduction. In *Veterinary Toxicology*. 3rd edn. pp. 1–14. London: Ballière Tindall.

24. Humphreys, D. J. and Clarke, M. L. (1991) Poisoning. In *Canine Medicine and Therapeutics*, 3rd edn. Eds. E. A. Chandler, D. J. Thompson, J. B. Sutton and C. J. Price. pp. 723–738. Oxford: Blackwell Scientific Publications.

25. Jacobs, G. (1981) Lead poisoning in a cat. *Journal of the American Veterinary Medical Association*, **179**, 1396–1397.

26. Kallet, A., Sousa, C. and Spangler, W. (1989) Mushroom *(Amanita phalloides)* toxicity in dogs. *California Veterinarian*, **42**, 9–11, 22, 47.

27. Kelly, D. F. and Pontefract, R. (1990) Hepatorenal toxicity in a dog following immersion in Rutland water. *Veterinary Record*, **127**, 453–454.

28. Kirk, R. W., Bistner, S. I. and Ford, R. B. (1990) Poisonings. In *Handbook of Veterinary Procedures and Emergency Treatment*. 5th edn. pp. 163–194. Philadelphia: W. B. Saunders.

29. Kirwan, A. P. (1990) "Magic mushroom" poisoning in a dog. *Veterinary Record*, **126**, 149 (Correspondence).

30. Kobayashi, K. (1981) Onion poisoning in the cat. *Feline Practice,* **11(1),** 22–27.
31. Leroux, V. (1986) Poisoning of pets by house plants. *Le Point Vetérinaire,* **18,** 45–55 (French).
32. Leyland, A. (1981) Laburnum *(Cytisus laburnum)* poisoning in two dogs. *Veterinary Record,* **109,** 287.
33. Liggett, A. D., Colvin, B. M., Beaver, R. W. and Wilson, D. M. (1986) Canine aflatoxicosis: a continuing problem. *Veterinary and Human Toxicology,* **28,** 428–430.
34. Macdonald, B. (1990) Terrier toad toxicity. *Australian Veterinary Practitioner,* **20,** 118.
35. Mgbeme, F. N. and Burrows, G. E. (1989) Cyanogenic ornamental plants of Oklahoma. *Toxicon,* **27,** 84–85.
36. Miyata, D. (1990) Isolation of a new phenolic compound from the onion *(Allium cepha L. onion)* and its oxidative effect on erythrocytes. *Japanese Journal of Veterinary Research,* **38,** 65.
37. National Research Council (1986) *Nutrient Requirements of Cats.* p. 35. Washington, DC: National Academy Press.
38. Ogawa, E., Shinoki, T., Akahori, F. and Masaoka, T. (1986) Effect of onion ingestion on anti-oxidizing agents in dog erythrocytes. *Japanese Journal of Veterinary Science,* **48,** 685–691.
39. Pier, A. C., Richard, J. L. and Cysewski, S. J. (1980) Implications of mycotoxins in animal disease. *Journal of the American Veterinary Medical Association,* **176,** 719–724.
40. Prescott, C. W. (1983) Clinical findings in dogs and cats with lead poisoning. *Australian Veterinary Journal,* **60,** 270–271.
41. Prévot, A. R., Brygoo, E. R. and Sillioc, R. (1953) Botulism and its five toxins. *Annales de i Institut Pasteur,* **85,** 559–561 (French).
42. Puls, R. and Ladyman, E. (1988) Roquefortine toxicity in a dog. *Canadian Veterinary Journal,* **29,** 569.
43. Reagor, J. (1985) Increased oleander poisoning after extensive freezes in South/Southeast Texas. *Southwestern Veterinarian,* **36,** 95.
44. Reid, D. A. (1985) Canine poisoning by *Clitocybe* species. *Bulletin of the British Mycological Society,* **19,** 117–118.
45. Ridgway, R. L. (1978) Mushroom *(Amanita pantherina)* poisoning. *Journal of the American Veterinary Medical Association,* **172,** 681–682.
46. Schwartz, R. E. (1989) Comments on cannabis intoxication in pets. *Veterinary and Human Toxicology,* **31,** 262 (Correspondence).
47. Sutton, R. H. (1981) Cocoa poisoning in a dog. *Veterinary Record,* **109,** 563–565.
48. Villalobos–Salazar, J., Velez, M. and Meneses, A. (1989) Effect of antiarrythmic drug and atropine therapy on toad *(Bufus marinus)* poisoning in dogs. *Toxicon,* **27,** 84–85.
49. Visconti, A. and Bottalico, A. (1983) High levels of ochratoxins A and B in moldy bread responsible for mycotoxicosis in farm animals. *Journal of Agricultural and Food Chemistry,* **31,** 1122–1123.
50. Watson, A. D. J. (1981) Lead poisoning in a cat. *Journal of Small Animal Practice,* **22,** 85–89.
51. Weiss, D. J., McClay, C. B., Christopher, M. M., Murphy, M. and Perman, V. (1990) Effects of propylene glycol-containing diets on acetaminophen-induced methemoglobinaemia in cats. *Journal of the American Veterinary Medical Association,* **196,** 1816–1819.

General Reading

Burger, I. H. and Flecknell, P. A. (1994) Poisoning. In *Feline Medicine and Therapeutics.* 2nd edn. Eds. E. A. Chandler, C. J. Gaskell and A. D. R. Hilbery. Oxford: Blackwell Scientific Publications (In press).

Humphreys, D. J. (1988) *Veterinary Toxicology.* 3rd edn. London: Balliére Tindall.

Spoerke D. J. and Smolinske, S. C. (1990) *Toxicity of Houseplants.* Boca Raton, FL: CRC Press.

CHAPTER 12

Dietary Sensitivity

JOSEPHINE M. WILLS and RICHARD E. W. HALLIWELL

Introduction

Dietary sensitivity is widely recognized in human medicine and has been implicated as the cause of a number of clinical conditions, including diarrhea and vomiting, abdominal pain, eczema, asthma, migraine headaches and even hyperactivity in children. Although fewer clinical studies have been carried out in domestic animals, there is, nevertheless, an increasing awareness within the veterinary profession that dogs and cats may also experience adverse reactions to food. Most researchers in this field would agree that such reactions are still comparatively rare and that the spectrum of clinical signs associated with food sensitivity is narrower in these species than that in humans.

Adverse reactions to food may be classified in a number of ways, although in both the veterinary and medical literature the terminology used is often confusing. The most logical approach would seem to be that proposed by Anderson and Sogu[2] depicted in Fig. 12.1. An adverse reaction to food is generally defined as any clinically abnormal response attributed to the ingestion of a dietary component, irrespective of the etiology of the reaction, and may be divided into two groups: food hypersensitivity (which is generally synonymous with allergy) and food intolerance. Definitions applied to the various adverse reactions to food are summarized in Fig. 12.2.

Food intolerance denotes an abnormal response to a food that can result from an inability to digest the food adequately, or from pharmacological, metabolic or toxic reactions. Pharmacological reactions result from a drug-like effect of natural or added chemicals. Metabolic reactions result from the effect of the food on the animal's metabolism; toxicity is the result of toxic substances contained in the food or as a result of contamination.

True food hypersensitivity is an immunologically mediated phenomenon; it can be antibody-mediated, immune-complex mediated or cell-mediated. To confirm a diagnosis of food hypersensitivity, the antigen should be identified, a relationship between antigen exposure and reaction should be shown and the involvement of a known immunopathogenic pathway demonstrated. If these criteria have not been fulfilled, then strictly speaking, the term food hypersensitivity should not be used; the terms adverse reaction or food sensitivity should be used instead. However, common usage in veterinary medicine has been to use food allergy or hypersensitivity for any abnormal reaction to food, regardless of the

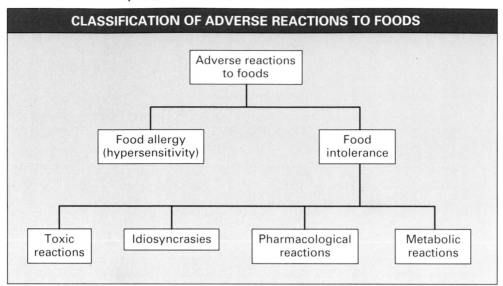

FIG. 12.1 Classification of adverse reactions to foods. (After Anderson and Sogu[2].)

etiology. The inciting, or sensitizing, agent is commonly referred to as an allergen although hypersensitivity may not always be involved.

In some cases, the picture may be confused by evidence of immunological reactivity to an allergen at the same time as biochemical abnormalities are noted. It can then be unclear as to which abnormality triggered the process and the relative contribution of each to the pathogenesis of the disease may be difficult to assess.

Reactions to ingested food components in dogs and cats can affect many body systems and can produce signs involving the skin,[3,8,28,33,39,50,63,67] the gastrointestinal tract,[11,36,40,60,65,66] the respiratory tract[64,65] and the central nervous system.[39] The most

TERMINOLOGY APPLIED TO MECHANISMS OF ADVERSE FOOD REACTIONS	
Term	**Definition**
Adverse reaction or sensitivity to food	Any clinically abnormal response to a dietary component
Food hypersensitivity	An adverse reaction to a food component, which is immune-mediated and not related to any physiological sensitivity
Food intolerance	An abnormal nonimmune physiological response to a food component; of an idiosyncratic, metabolic, pharmacological or toxic nature
Food idiosyncrasy	An abnormal response to a food component in a sensitive individual, resembling hypersensitivity but not involving immune mechanisms. Often due to an enzyme defect
Pharmacological reaction to food	An abnormal response to a pharmacological compound in food, or to a component that causes release of a pharmacological substance from the body
Food poisoning	Adverse reaction to a specific toxin, e.g. botulism
Metabolic reaction to food	Adverse reaction as the result of an effect on the recipient's metabolism
Enterometabolism	The absorption of metabolites produced as a result of bacterial fermentation in the colon

FIG. 12.2 Definitions of terminology applied to potential mechanisms of adverse reactions to food.

common presenting signs, however, are associated with the skin and gastrointestinal tract.

Incidence

Most authors agree that food hypersensitivity and intolerance is rare in dogs and cats.[39,47] The incidence of the condition is difficult to establish and differing figures have been cited in the literature. It has been estimated that food sensitivity accounts for 1% of all canine and feline dermatoses in general practice[63] and is the second most common allergic skin disorder in the cat.[47] It has been stated that approximately 11% of feline miliary dermatitis cases may be due to food sensitivity.[58] The proportion will vary according to the geographical location and in particular the contribution of flea allergy. Food sensitivity may account for 10% of all nonseasonal dermatitis[4] and 10% of all canine allergic skin diseases (including parasitic allergy[56]). Nevertheless, food hypersensitivity appears to be much less common than atopy (allergic inhalant dermatitis), the ratio of incidence being approximately 1:10, and probably occurs with about the same frequency as allergic contact dermatitis.[22] In a recent report of food sensitivity in cats, only 14 cases were seen in a 5 year period.[68]

A possible explanation for this apparent variation in incidence may be that skin disease is often multifactorial. Two or more contributory factors may combine so that the clinical threshold is exceeded, resulting in the production of disease. In a study of food allergy in cats, three of the 14 cats had concurrent flea-bite hypersensitivity, flea-collar hypersensitivity or atopy that all contributed to the pruritus and skin lesions.[68] Removal of one of the factors may be sufficient to drop below the clinical threshold, thereby causing an improvement in clinical signs. Having rendered the animal asymptomatic, the clinician would assume that a correct diagnosis had been made and may not consider further investigation in order to identify additional etiological factors. The variation in published incidence may therefore be a reflection of the order in which the possible etiologies are investigated by the clinician during diagnostic work-up.

Reports of the incidence of dietary sensitivity as a cause of gastrointestinal disease are scant. The true incidence is difficult to ascertain because in many cases, the owner will make the correct association between diet and disease and will remove the offending food item without seeking veterinary advice. Gastrointestinal signs may occur in both dogs and cats, either alone or in combination with dermatological changes. Such a combination has been reported to occur in 10–15% of canine cases of food hypersensitivity.[72] A number of chronic conditions of the gastrointestinal tract in both dogs and cats will respond to dietary manipulation and there is increasing evidence to suggest that food hypersensitivity may have an etiological role in at least some of these.

There appears to be no breed or sex predilection in dogs and cats with food sensitivity. In one study,[49] however, it was reported that Labrador Retrievers, Collies, Cocker Spaniels, Springer Spaniels and Miniature Schnauzers were all represented with at least twice the expected frequency as compared to the base clinic population; the statistical significance of this was not quoted.

In general, it is agreed that there is no age predilection; and the age of onset in both dogs and cats can be from 6 months to 14 years.[39,48,63,67] August,[3] however, states that the clinical signs frequently appear before 9 months old and Rosser[50] reported that 33% of his series of 51 dogs were less than 1 year old at the onset of signs. In a recent study of food sensitivity in 25 dogs in the U.K., 13 (52%) of the dogs were under 12 months old at the onset of clinical signs of diet-sensitive dermatitis.[28] These findings suggest that dietary sensitivity should be considered to be a strong possibility in young animals with cutaneous signs suggestive of an allergic dermatosis.

Food sensitivity is typically perennial but can be intermittent if the patient is on a varied diet, or even seasonal if an animal is sensitive to a seasonally available foodstuff.

Clinical Features

With few exceptions, it is impossible to differentiate between food intolerance and food hypersensitivity on the basis of the presenting clinical signs. In most instances, the clinical signs are compatible with allergic disease and in some cases, food intolerance may occur in combination with a true allergic reaction. Although involvement of other systems is occasionally reported, the bulk of the well documented cases of food sensitivity in dogs and cats involve the skin or the gastrointestinal system. In a small proportion of cases, both are involved and presentation of a patient with both dermatological and gastrointestinal signs is strongly suggestive of dietary sensitivity.

Dermatological Signs

Unfortunately the clinical signs of food sensitivity are highly variable and the condition lacks such characteristic signs as are seen in other dermatoses, such as atopic disease, allergic contact dermatitis and flea allergy, which it can mimic.

The major presenting sign of food sensitivity in both dogs and cats is a pruritic skin disorder. The response to the pruritus results in a gradation of signs ranging from saliva staining of the hair to severe self-trauma and a variety of secondary skin lesions.[36,39,47]

In the dog, pruritus is seen in almost all cases and may be regional or generalized in distribution. A primary papular eruption is seen in approximately 40% of cases with a secondary staphylococcal folliculitis in some 20–30%.[22,24] The latter is extremely difficult to control and although responsive to antibiotics, affected animals relapse within a few days or weeks of cessation of therapy. Some 50% of cases exhibit marked erythema and there are variable secondary lesions of self-trauma and resulting alopecia.[24] Scales and crusts resulting from the secondary superficial pyoderma are commonly seen.[3,39,63] Chronic changes of seborrhea, hyperpigmentation and lichenification are seen in long-standing cases. Urticaria may be seen with or without

other signs in a small number of cases, but when it does occur it is highly suggestive of food hypersensitivity.[21,24] It is often cited that food allergy is relatively resistant to corticosteroid therapy, but this is not a consistent feature. Indeed, in a recent study[50] involving 51 cases with dermatological signs, 39% responded well to anti-allergic doses, 44% exhibited a partial response and in only 19% was there a poor response.

Unilateral or bilateral otitis externa may be a feature and may occur in the absence of other signs of skin disease.[8,28,67] In the series reported by Rosser,[50] the ear region was affected in 80% of 51 cases and 24% presented with otitis externa only. In an earlier study, however, only one of 30 cases had otitis externa.[67] This marked difference can only be reconciled by the selection of case material referred to the dermatology services in the respective institutions. Characteristically, an allergic otitis externa, whether resulting from atopy or food allergy, initially presents as an erythema that involves the pinna and the vertical canal. The horizontal canal is only affected in chronic, long-standing cases, when secondary infection ensues.[21,24] In Rosser's study,[50] the feet were affected in 61% of cases and the axilla in 53%. In 20% of cases, the pruritus was generalized. Occasionally dogs will be nonpruritic and exhibit only seborrhea.[63,67] Some cases may present with recurrent superficial pyoderma only.[28,36,46] Dermatological

SKIN LESIONS COMMONLY ASSOCIATED WITH DIETARY SENSITIVITY IN DOGS	
Primary lesions	Pruritus (regional or generalized) Erythema (regional or generalized) Papular or pustular eruptions Otitis externa Occasionally • seborrhea • urticaria
Secondary lesions	Excoriation Alopecia Secondary pyoderma Seborrhea Hyperpigmentation Lichenification

FIG. 12.3 Skin lesions commonly associated with dietary sensitivity in dogs.

SKIN LESIONS COMMONLY ASSOCIATED WITH DIETARY SENSITIVITY IN CATS

Facial and neck pruritus
Localized or generalized scales or crusts
Miliary dermatitis
Symmetrical or localized areas of alopecia
Eosinophilic granuloma complex (plaques or ulcers)
Otitis externa

FIG. 12.4 Skin lesions commonly associated with dietary sensitivity in cats.

signs commonly observed in cases of canine dietary sensitivity are summarized in Fig. 12.3.

In cats, there is no classic set of skin lesions that is pathognomonic for food sensitivity. However, a pruritus characterized by excessive licking and scratching is evident in over 90% of cases.[70,71] In fact, in two recent studies of food sensitivity, all the affected cats were pruritic.[8,68] Food sensitivity in cats can manifest as one of several cutaneous patterns, which are listed in Fig. 12.4.

Facial pruritus is the most common presenting sign, with crusting areas around the neck and excoriation of the face.[4,57,60,68,70] When miliary dermatitis is present, this also has a predilection for the ears, head and neck. Otitis externa may also be a clinical feature and dietary allergy has also been implicated as a cause of the feline eosinophilic granuloma complex,[8,58,68] particularly eosinophilic plaques.

Gastrointestinal Signs

There is no doubt that occasional and even persistent bouts of diarrhea and/or vomiting can be associated with dietary sensitivity. The true incidence of such conditions is unknown, as many owners make the association without veterinary advice and initiate the necessary dietary changes. Furthermore, the percentage of such cases that represent intolerance as opposed to hypersensitivity is also unknown.

Gastrointestinal signs have been reported in 10–15% of dogs with skin lesions associated with food sensitivity.[72] Some cases of inflammatory bowel disease in dogs respond well to a restricted (elimination) diet[69] and a number of chronic conditions of the gastrointestinal tract

are reported in which dietary hypersensitivity may play a role.

- Canine lymphocytic–plasmacytic enteritis is characterized clinically by soft to liquid stools and increased frequency in defecation. The condition most commonly affects the small bowel, although the stomach and the colon may also be involved. Biopsies reveal a diffuse infiltrate of lymphocytes and plasma cells in the affected mucosa or concentrated in the villi or lamina propria. The condition may be more common in German Shepherds.[30] The histopathological features and the success that is sometimes achieved by dietary manipulation suggest that food hypersensitivity could be an etiological factor.[14]

- Canine eosinophilic gastroenteritis is characterized by intermittent diarrhea, vomiting, anorexia and weight loss.[14] A peripheral eosinophilia is often seen and biopsies of the intestine reveal an infiltrate composed of eosinophils, lymphocytes and plasma cells, features that are quite consistent with a Type I hypersensitivity. Although there is an association with migrating ascarids in some cases,[29] dietary manipulation is effective in controlling the condition in many instances.

- Colitis is commonly diagnosed in dogs and can have many causes. The condition is characterized by intermittent or persistent mucoid or hemorrhagic diarrhea. A recent report described 13 cases of "idiopathic" colitis, all of which became asymptomatic when fed a diet of rice and cottage cheese.[41] Thus dietary hypersensitivity represents a distinct possibility in at least some colitis cases.

Gastrointestinal signs have also been reported in cats with food sensitivity either concurrently with skin lesions or as a separate entity. It can manifest as vomiting,[11] profuse watery diarrhea[60,66] and inflammatory bowel disease including lymphocytic–plasmacytic colitis.[40,61] In the six cases of lymphocytic–plasmacytic colitis reviewed by Nelson *et al.*,[40] diffuse intestinal thickening could be felt in two cats. Fresh blood may be seen in diarrheic feces and there may be an increased frequency of defecation.

Controversy still exists as to the etiology of hemorrhagic gastroenteritis in dogs and it seems likely that this should not be considered a single entity, but rather a manifestation of a number of diseases. It has been noted that recurrent bouts can in some cases be prevented by dietary manipulations, which suggests a possible involvement of dietary sensitivity in some instances.

Gastrointestinal upsets as a result of food intolerance can occur where there is a deficiency of enzymes necessary to digest a dietary component. This can occur when a novel food is introduced into the diet and it is recommended that any dietary changes are made gradually to allow the digestive tract to adapt by increasing specific enzyme production. Lactose intolerance is a common finding in both cats and dogs. After weaning, the intestinal levels of the enzyme lactase are considerably reduced in many individuals and their ability to digest lactose in milk is therefore limited. This results in an osmotic diarrhea in susceptible animals whenever excessive quantities of cow's milk are fed.

Gluten enteropathy of Irish Setters is a malabsorption syndrome, characterized by weight loss and poor condition (despite a ravenous appetite) and is usually, but not always, accompanied by chronic diarrhea. Although the pathogenesis of the condition is still unclear, this is an example of food intolerance in which immune reactions may or may not also have a role (see Chapter 13).

Other Signs

Respiratory signs, such as feline asthma, and neurological signs are rare. Some abnormal behavioral signs, such as hyperesthesia and hyperactivity, may be attributable to a food sensitivity in cats and dogs, but definitive clinical data are lacking.[71,72] Epileptiform seizures have also been reported in three dogs with dietary hypersensitivity.[39] A number of animal behaviorists consider that diets high in protein may be associated with aggression

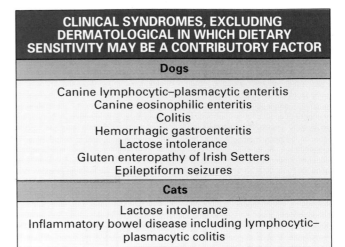

CLINICAL SYNDROMES, EXCLUDING DERMATOLOGICAL IN WHICH DIETARY SENSITIVITY MAY BE A CONTRIBUTORY FACTOR
Dogs
Canine lymphocytic–plasmacytic enteritis Canine eosinophilic enteritis Colitis Hemorrhagic gastroenteritis Lactose intolerance Gluten enteropathy of Irish Setters Epileptiform seizures
Cats
Lactose intolerance Inflammatory bowel disease including lymphocytic–plasmacytic colitis

FIG. 12.5 Clinical syndromes in which dietary sensitivity may be a contributory factor.

in certain breeds of dog but this is unlikely to be a true intolerance.[72]

A number of clinical syndromes have thus been identified in which dietary sensitivity may have an etiological role. These are summarized in Fig. 12.5.

Differential Diagnosis

Dietary sensitivity manifests clinically as a wide range of dermatological or gastrointestinal signs, or a combination of both. However, the most common presenting sign in both dogs and cats is pruritus that usually results in some degree of self-inflicted trauma and a variety of associated secondary lesions. The resulting clinical picture is thus highly variable and may mimic a number of other conditions.

In both dogs and cats, the major differential diagnoses for cases that present with skin disorders are all listed in Fig. 12.6.

Pathophysiology

The pathogenesis of an adverse reaction to a food component involves the interaction of the component with a biological amplification system that leads to inflammation and the development of clinical signs. The amplification

**MAJOR DIFFERENTIAL DIAGNOSES
IN DOGS AND CATS**

- Atopy (allergic inhalant dermatitis)
- Allergic or irritant contact dermatitis
- Flea-bite hypersensitivity
- Ectoparasitic disease – fleas
  – lice
  – *Cheyletiella* sp.
  – *Otodectes* sp.
  – *Trombicula autumnalis*
- Primary bacterial infection
- Dermatophyte infection
- Idiopathic seborrhea
- Autoimmune diseases
- Certain neoplastic conditions
- Drug eruptions

OTHER DIFFERENTIALS IN DOGS

- Sarcoptic mange
- Demodectic mange
- Hookworm dermatitis

OTHER DIFFERENTIALS IN CATS

- Miliary dermatitis related to other causes
- Eosinophilic granuloma complex related to other
  causes
- Notoedric mange
- Flea-collar sensitivity
- Demodectic mange

FIG. 12.6 Major differential diagnoses in dogs and cats.

system may include immunological mechanisms, the complement pathway, eicosanoid synthesis, chemotaxis of phagocytes and production of inflammatory mediators such as kinins, leukotrienes and histamine. On occasion, there may be direct toxicity from the ingested component, without amplification, such as reactions to pharmacological substances. Furthermore, it has been proposed that the products of colonic bacterial metabolism may produce an adverse reaction in humans.[32]

Many cases of food intolerance will therefore produce clinical signs that are indistinguishable from those resulting from hypersensitivity reactions. In addition, histamine-releasing factors or agents from different sources can act in a summative manner. As a result, it can be extremely difficult to implicate correctly a specific cause or source in many instances. The theoretical pathways of interaction are indicated in Fig. 12.7.

Food Hypersensitivity

Throughout life, animals are exposed daily to a great variety of potential dietary allergens and yet most remain refractory to sensitization. After a period varying from a few weeks to several years, however, a small number of animals may develop an immune response against one particular foodstuff that activates one or more immunopathogenic pathways. After this time any subsequent ingestion of this foodstuff will result in the development of clinical signs.

Normal Immune Response to Food

During the course of the animal's lifetime, a vast array of potential immunogens, contained within the digestive products of foods, are presented in great quantities to the gastrointestinal tract. The intestinal mucosa forms a barrier that limits the absorption of macromolecules, but this mechanism is imperfect. A number of lines of evidence indicate that a significant proportion of antigens are absorbed through both normal and abnormal gut and it is perhaps surprising that allergic responses are not more common. Indeed, antibodies to food allergens, usually of the IgG class, are often demonstrable in normal individuals, but these do not ordinarily result in ill effects.[21]

Upon initial presentation of the antigen to the gut mucosa, there is generally an immune response, predominantly involving IgA. This reduces the amount of antigenic material that is absorbed across the mucosal barrier. There is also a "back-up" mechanism by which antigenic material that is absorbed is cleared, again involving IgA. Immune complexes of antigen and IgA antibody are transported from the circulation, across hepatocytes, into the bile and thus find their way back into the intestine. This local IgA response may be followed by a transitory systemic immune response, but in the normal situation, immunological tolerance follows. Thus, there is an apparent paradox of a vigorous local immune response followed by a systemic tolerance.[22]

Absorption of macromolecules can be

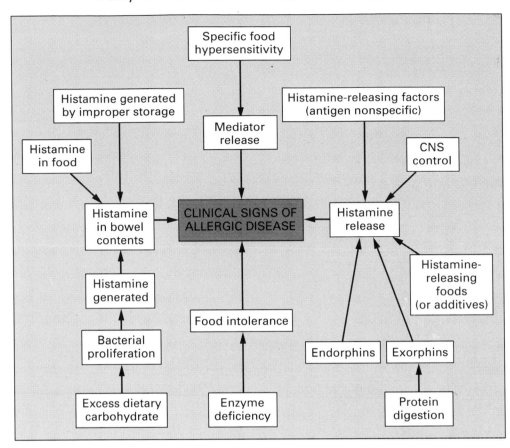

FIG. 12.7 Potential pathways of interaction between immunological and nonimmunological factors in causing symptoms of food allergy. (From Halliwell[22])

altered in either direction by local immunity. Decreased uptake has been demonstrated in experimental animals following oral or parenteral immunization[62] and increased absorption occurs in IgA-deficient humans.[21,24] Absorption is also enhanced by vasodilation in the gut mucosa, such as that resulting from a local allergic reaction. In this case, the patient becomes caught in an immunological vicious circle because local hypersensitivity reactions favor the access of allergen that in turn heightens the antibody response. In cases of gastrointestinal disease that result from food hypersensitivity, systemic absorption may not be required because the clinical signs could result from a local immune response.

Abnormal Immune Response to Food

The factors that lead to the development of a hypersensitivity to ingested antigens are speculative. It is known that allergic responses are mounted more frequently in humans to certain types of allergen. Those most frequently implicated in humans share some common physicochemical characteristics. Most are heat- and acid-stable glycoproteins with molecular weights of 18,000–36,000 Da.[21,22]

The classic experiments of Praustnitz and Kustner[44] established the involvement of reaginic antibody, later designated IgE, in the pathogenesis of food allergy in humans. The condition has also been associated on more circumstantial grounds with short-term sensitizing IgG antibody[42] and experimentally it has been shown that Type III reactions involving IgG antibody and immune complexes can be damaging to the intestinal tract.[16] Finally, although Type IV reactions can occur in the gastrointestinal tract and are readily demonstrable in parasitic immunity, their involvement in allergic reactions to food is speculative.

Thus there is the potential for involvement of a number of different pathways, but extensive immunological studies of humans with defined food allergy lead to the conclusion that IgE is implicated in most instances and the reactions involved include both the classic, immediate Type I reactions and the late-phase IgE-mediated reactions. Although there have been no studies on the immuno-pathogenesis in domestic animals, it can be presumed that the same principles apply.

The immunological mechanisms of food hypersensitivity in dogs and cats probably involve both immediate (Types I and III) and delayed (Type IV) hypersensitivities.[25] Immediate reactions to foods occur within minutes or hours and delayed responses occur within several hours or days. However, it is not at all clear which mechanism, if any, predominates in the pathogenesis of food hypersensitivity and it may be that more than one mechanism operates in some cases of an allergic reaction. IgE has been identified in dogs[20] and although there have been no reports in scientific journals of the immuno-chemical characterization of feline IgE, there is evidence to support the belief that cats do produce an IgE-like (reaginic) antibody[6,43] and indeed *in vitro* assays for the identification of allergen-specific IgE in cat sera are offered by laboratories in the U.S.A.

It is tempting to speculate that the atopic state might predispose to the development of IgE antibodies to ingested allergens in the same way as it apparently does to inhaled allergens and to injected allergens such as the flea antigen.[27] There is no evidence, to date, however, that indicates a higher level of food allergy in atopic dogs as compared to normal animals.

The factors that determine the extent of absorption of allergens by the intestine are not fully understood, although local vasodilation is clearly facilitatory. Likewise, it is unclear how the allergen induces the cutaneous inflammatory response. It is possible that it interacts with mast cell bound antibody in the skin. An alternative possibility is that immune complexes of antigen and antibody are deposited in the skin, where they induce a local inflammatory response. Finally, it is possible that the skin is affected by inflammatory mediators generated by immune interaction in other parts of the body.

The inflammatory mediators involved in food allergy in humans and animals remain speculative. Interleukins, platelet activating factor and the products of mast cells and basophils, both those that are preformed and also the membrane-derived products of the arachidonic acid cascade, are clearly implicated, but no data exist on the relative importance of any of them.

Histamine-Releasing Factors

Interesting data have recently emerged concerning the role of histamine-releasing factors in food allergy in humans.[37,55] These cytokines are derived from mononuclear cells and are generated in a range of allergic diseases. They have an affinity for IgE and cause spontaneous release of histamine, in an antigen-nonspecific manner, from peripheral blood leucocytes of atopic human patients with food allergy. They continue to be induced for some time after the allergen has been removed, although the level of spontaneous release declines over a period of about 6 months. In one patient, however, this level was still significantly greater than controls some 4 months after the patient was placed on a diet free from the allergen. If the same phenomenon was to be demonstrated in dogs and cats, it could account for the relatively slow clinical improvement seen in some cases of food allergy in these species following exclusion of the implicated dietary allergen.

Antigenic Sources

Most basic food ingredients have the potential to induce an allergic response, although the majority of reactions are caused by proteins. Dietary items reported to have caused food sensitivity in the dog and cat are shown in Fig. 12.8. More than 50% of the reported cases in cats are sensitive to cow's milk, beef or fish[63,68] and a similar percentage of the reported cases of dietary intolerance in

DIETARY ITEMS REPORTED TO HAVE CAUSED FOOD SENSITIVITY IN THE DOG AND CAT	
Dog	Cat
Cow's milk	Cow's milk
Beef	Beef
Mutton	Mutton
Pork	Pork
Chicken	Chicken
Rabbit	Rabbit
Horse meat	Horse meat
Fish (variety)	Fish (variety)
Eggs	Eggs
Oatmeal	Canned foods
Wheat	Dry foods
Corn	Cod liver oil
Soy	Mice
Rice flour	Benzoic acid
Potatoes	
Kidney beans	
Canned foods	
Dog biscuits	
Dog foods	
Artificial food additives	

FIG. 12.8 Dietary items reported to have caused food sensitivity in the dog and cat.

dogs were sensitive to milk, beef or cereal, either alone or in combination.[8,28,33,68] It is noteworthy that the vast majority of animals with a reaction to a dietary component react to major components of the diet.

Food Intolerance

Food intolerance is a nonimmunological, abnormal physiological response to a food item and may involve toxic, pharmacological or metabolic reactions or dietary idiosyncrasies, in which the animal is unable to digest or otherwise process a dietary component. With a few exceptions, the underlying pathogenesis of food intolerance in animals has not been demonstrated.

Lactose Intolerance

Food intolerance can occur if an individual lacks a certain enzyme necessary for a physiological or digestive process, for example disaccharides such as lactose have to be hydrolyzed to monosaccharides before they can be absorbed from the intestinal lumen. Puppies and kittens have adequate levels of intestinal lactase to enable them to digest lactose in the mother's milk. In many individuals, however, the brush border disaccharidase activity falls after weaning to a fraction of the levels found in the young animal, and this results in a limited ability to hydrolyze lactose in milk. When excessive amounts of milk are fed, undigested lactose remains within the intestinal lumen, allowing proliferation of lactose-fermenting bacteria and the subsequent development of an osmotic diarrhea.

Certain milk proteins, such as casein, are also allergens and an overlapping spectrum exists with diseases resulting from hypersensitivity to such proteins. In general, smaller quantities of cow's milk or other dairy products are required to induce gastrointestinal signs when the condition results from dietary hypersensitivity rather than lactose intolerance. Furthermore, in the case of allergic reactions, concomitant dermatological signs may also be apparent.[23]

A secondary lactose intolerance and indeed intolerance of other carbohydrates such as sucrose, may occur following inflammation of the intestines from any cause. It results from loss of the brush border with its associated disaccharidase activity. Careful dietary management is thus an important part of therapy of acute or chronic diarrhea, irrespective of the etiology. This phenomenon may, in part, account for the fact that recovery of normal intestinal function in dietary hypersensitivities is often delayed following correct implication of the allergen and its removal from the diet.

A final consideration is that a variety of congenital diseases associated with carbohydrate intolerance are recognized in humans, some of which may cause infant mortality if not correctly diagnosed and dietary corrections made. Neonatal death is not infrequent in puppies and kittens and it is quite possible that a spectrum of disorders similar to those found in man would be uncovered if all such cases were subjected to in-depth investigation.

Gluten Enteropathy of Irish Setters

This condition is a malabsorption syndrome that is similar in many respects to coeliac disease in humans (see Chapter 13). Villus atrophy of the jejunum together with an increased intraepithelial lymphocyte density and a loss of brush border alkaline phosphatase activity may be demonstrated and are indicative of an impaired mucosal barrier. The condition responds to removal of wheat (and hence gluten) from the diet. It is not yet clear whether the condition has an immunological basis or if impairment of the mucosal barrier is attributable primarily to other causes.

There appears to be an underlying permeability defect in susceptible Irish Setters, even when fed a gluten-free diet.[18] Gluten peptides could thus gain access across the intestinal mucosa and thereby initiate a pathological process. Recent studies suggest that the clinical response may be modified by delaying exposure to gluten until the puppy is 1 year of age. Although a response to gluten could be demonstrated histologically in susceptible individuals not previously exposed, this had no adverse clinical effects.[19] The results of immunological studies of affected animals are awaited with interest.

Dermatitis herpetiformis is an intensely pruritic papular skin disease that, in humans, sometimes occurs in association with celiac disease. There is one report in the literature of a dog with clinical signs and histological features compatible with dermatitis herpetiformis.[26] The tentative diagnosis received support from a dramatic response to the sulfone-derivative dapsone, which is used in the treatment of the condition in humans, but is otherwise effective only in a few canine skin diseases.

Vasoactive Amines in the Diet

Another area of food intolerance recognized in humans is reaction to pharmacological substances in food. Either the food itself contains the chemical, in particular the vasoactive amines, such as tyramine or histamine, or the food causes the release of inflammatory mediators such as histamine. Tyramine is believed to be a common cause of migraine. Fish can stimulate the release of histamine from mast cells and may cause gastrointestinal upset in some animals, particularly cats. The clinical signs that are observed are compatible with allergic disease and may include urticaria and a range of gastrointestinal disorders. Individuals that are already suffering from clinical or subclinical allergic disease are more likely to suffer adversely. The importance of these reactions to pharmacological substances in food in dogs and cats is unknown and would be very difficult to ascertain, particularly if the body's response to the substance was, for example, a migraine.

Histamine-Containing Foods

There are no published data on the histamine content of commercial pet foods. Figure 12.9 lists the histamine content of some foods commonly eaten by humans, albeit less commonly by most pets. Improper storage of foods can lead to far higher levels due to the conversion of histidine to histamine. High levels of histamine are often found in poorly stored scombroid fish, such as mackerel, and this phenomenon is responsible for so-called scombroid fish poisoning. Excessive carbohydrate in foods may also increase the bacterial flora, leading to excessive fermentation and production of histamine, with the resultant symptoms of food "allergy."[12]

Histamine-Releasing Foods

Certain foods are known to be able to cause histamine release from mast cells in a non-

FOODS RICH IN HISTAMINE	
Food	Content (μg/g)
Fermented cheeses	up to 1330
Dry pork and beef sausage	225
Pig's liver	25
Tinned tuna	20
Meats	10
Spinach	37.5

FIG. 12.9 Foods rich in histamine.

COMMON HISTAMINE-RELEASING FOODS
Egg white
Shellfish
Chocolate
Fish
Alcohol
Strawberries
Tomatoes

FIG. 12.10 Common histamine-releasing foods.

immune fashion. These foods may produce clinical signs in normal individuals, but they are more likely to exacerbate signs in allergic individuals. The most common offenders are listed in Fig. 12.10.

Endorphins and Exorphins

Opiates are well known as histamine-releasing agents, so it is not surprising that endorphins can likewise cause histamine release both *in vitro* and *in vivo* and this action is blocked by the opiate antagonist naloxone.[9,59] It is indeed possible that those dogs who become pruritic when exhibiting signs of euphoria may in fact be pruritic from histamine release rather than merely exhibiting a behavioral quirk.[23]

Exorphins are proteinase-resistant peptides that are generated during the digestion of several dietary proteins, including those found in wheat and milk. These have endorphin-like activity and are receiving some attention as to a possible pathogenic role. Experiments on a gluten hydrolysate have demonstrated an increase in intestinal transit time that was blocked by naloxone.[38] Whether this mechanism is operative in the gluten-sensitive enteropathies is still speculative.

Food Additives

In human medicine, it is known that certain food additives will produce untoward reactions in a number of patients. Clinical signs attributed to these include hyperactivity, urticaria, rhinitis, irritable bowel syndrome and behavior disturbances (including hyperactivity). Those additives that have been implicated are mono-sodium glutamate, the coloring agents tartrazine and erythrosin and preservatives such as sodium nitrite and butylated hydroxytoluene. The mechanism of action remains speculative; true allergic reactions may be involved in some, but clearly not all, instances.

There are no definitive reports of adverse reactions in animals, although many clinicians have observed occasional cats and dogs who develop pruritic skin disease when fed any commercial diet while being asymptomatic on all home-prepared diets.[23]

Food Toxicity

Food toxicity may result from the presence of toxic substances within the food or as a result of contamination; the mechanism of action varies with each toxin.

In some cases, a particular nutrient may be present in excessive quantities in the diet and this results in toxicity. Toxic levels of vitamins A and D are well recognized examples of this phenomenon and may produce disease in both dogs and cats. Cats regularly fed large quantities of liver or fish liver oil supplements are particularly susceptible to hypervitaminosis A.

Improper storage may result in contamination with microorganisms (bacteria or fungi) or their toxic metabolites. In dogs, poisoning may result from the scavenging of putrefied material. Contamination with nonbiological compounds may also occur. Furthermore, inadequate preparation of food can cause toxic reactions by failing to destroy potential toxins or enzymes that could result in disease, such as thiaminase in raw fish.

Specific foods known to cause poisoning if fed to excess include onions and chocolate. Onion poisoning has been described in both cats and dogs. In one report,[34] cats that consumed onion soup once or several times over a 3-day period developed an increased number of Heinz bodies in their erythrocytes and eventually developed hemolytic anemia and hemoglobinuria. Chocolate and cocoa, which contain theobromine, have caused toxicity in dogs resulting in vomiting, diarrhea, sudden collapse and death. Dachshunds died suddenly following the consumption of 300 g of chocolate and a Springer Spaniel also died some 12 hr after eating 250 g of household cocoa.[15] Although there are no reports of cocoa poisoning in cats, this could theoretically occur, with as little as 40–50 g of cocoa providing a potentially lethal dose.[7] For further details see Chapter 11.

The best known food preservative toxicity in cats is caused by benzoic acid. Cats are unable to detoxify benzoic acid quickly and ingestion of high levels of this compound (300 mg/kg body weight daily) can cause aggression, hyperesthesia and death.[71] Propylene glycol, another preservative used in some semimoist cat foods, has been shown to decrease erythrocyte survival time.[10] For this reason, European petfood manufacturers recently discontinued its use in cat food. Cat food products and treats in the U.S. are no longer allowed to contain propylene glycol.

The Brain–Allergy Axis

The brain can exert a profound effect upon the immune response. It is well known that immune responsiveness may be markedly impaired in times of stress, but more recently it has been demonstrated that the brain can also affect histamine release in a nonimmunological fashion. Experimentally, guinea pigs may show histamine release as the result of a conditioned response. Following sensitization with bovine serum albumin, clinical signs of allergic disease due to histamine release were observed when they were subsequently challenged with the allergen. They were then conditioned with an odor at the same time as antigenic challenge and this resulted in the release of histamine when subsequently exposed to the odor in the absence of antigen.[53] This finding may have far-reaching implications in the understanding of apparent allergic phenomena, in both humans and animals, that are currently hard to explain.

Diagnosis

The diagnosis of food sensitivity is not easy and requires a methodical approach based on an accurate clinical history, full clinical examination and dietary investigation in the form of elimination diets and test meals.[39] A number of additional diagnostic criteria have been used, but with varying degrees of success. The histopathological changes in the skin are usually nondiagnostic, although examination of intestinal biopsies may prove more rewarding. There are no consistent laboratory findings.

Clinical Signs

The clinical signs of food sensitivity in the dog and cat are quite variable and unpredictable depending on the individual response. In the cat, it probably helps to group the differential diagnoses into the four major presenting signs: facial pruritus, scales and crust, miliary dermatitis and self-inflicted hair loss (Fig. 12.11).[13]

Onset of pruritus in a young dog (>1 year old) is suggestive of dietary hypersensitivity because atopy typically takes one or more years of sensitization before clinical signs occur. After elimination of parasitic diseases and obvious causes such as bacterial folliculitis, it should be the first consideration. Nevertheless, some cases may require years of exposure to the allergen to develop a clinical hypersensitivity and the diagnosis should not be discounted where the animal's diet has not recently changed. In about 60% of cases, the animal has been eating the allergen for over 2 years.[64]

Food sensitivity can occur at any time of the year and is not seasonally related. If the clinical signs present initially during the spring or summer, potential differential diagnoses would include atopy, flea-bite sensitivity or infection with *Trombicula autumnalis*.

Observation of urticaria in the dog likewise points to the possibility of food hypersensitivity, as does the coincident development of gastrointestinal signs. When the latter occur on their own, again they are rarely diagnostic of dietary hypersensitivity, but this should be considered as a possibility in any case of persistent diarrhea and/or vomiting.

Where the presenting signs include otitis

DIFFERENTIAL DIAGNOSIS OF DIETARY SENSITIVITY IN THE CAT				
	Facial pruritus	Scales or crust	Miliary dermatitis	Self-inflicted hair loss
Dermatophytosis	*	*.	*	*
Ectoparasites	*			
Flea allergy	*	*	*	*
Cheyletiellosis		*	*	
Trombiculidiasis		*	*	
Demodicosis	*	*		
Pediculosis		*	*	
Notoedric mange	*	*		*
Tick-bite reaction		*		
Otodectic mange	*		*	
Immunology				
Food sensitivity	*	*	*	*
Eosinophilic granuloma	*	*		
Contact dermatitis	*	*		
Drug sensitivity	*	*	*	
Atopy	*	*	*	*
Intestinal parasite hypersensitivity			*	
Viral				
Poxvirus	*	*		
Herpesvirus		*		
Calicivirus		*		
Pseudorabies		*		
Other				
Neurodermatitis		*		*
Cat bite		*		
Pyoderma	*	*	*	
Feline acne	*			
Idiopathic			*	
Biotin deficiency			*	
Fatty acid deficiency			*	

*After Foil.[13]

FIG. 12.11 Differential diagnosis of dietary sensitivity in the cat.

externa, this has the typical appearance of an allergic otitis, involving only the pinna and vertical canal in the initial stages. There may also be a history of recurrent otitis externa.

Contagious disease may be ruled out if the condition affects only a single dog or cat within a group, with no evidence of spread of skin disease or diarrhea to in-contact animals. Scabies can, however, sometimes affect one dog in the household and it may be months before in-contact dogs are affected. Furthermore, there may have been a transient improvement in clinical signs with anti-inflammatory drugs with relapse following the cessation of therapy. A summary of key diagnostic points that may be suggestive of dietary sensitivity is given in Fig. 12.12.

Laboratory Findings

There is no consistent laboratory finding for cats with food sensitivity. One investigator found peripheral eosinophilia ($1.16–5.8 \times 10^7$ eosinophils/l) in two of four cats from a group of 14 with miliary dermatitis attributable to food hypersensitivity.[58] In dogs, there are no consistent changes in the eosinophil count that might implicate an allergic reaction.[24]

Histopathology is nondiagnostic; biopsies of affected skin show changes that are merely indicative of an allergic etiology. Varying degrees of subacute to chronic nonsuppurative perivascular dermatitis with elevated numbers of neutrophils, plasma cells, lymphocytes and mast cells may be observed[71] and a tissue

KEY DIAGNOSTIC POINTS FOR DIETARY SENSITIVITY
• Pruritic skin disorder — not related to parasitic infection — in young animals less than 1 year of age • Nonseasonal occurrence • Generalized urticarial reaction in dogs • Predilection of skin lesions for head and neck regions in cats — but this is not pathognomonic • Concurrent gastrointestinal and dermatological signs • Variable response to corticosteroid therapy — will relapse following cessation of therapy • Recurrent or chronic nature of clinical signs, such as superficial pyoderma, otitis externa, diarrhea or vomiting • Single member of a group affected (contagious disease unlikely) • Response to elimination diet and subsequent dietary challenge (diagnostic)

FIG. 12.12 Key diagnostic points: signs suggestive of dietary sensitivity.

eosinophilia is seen in some cases.[22] Intestinal biopsy is indicated in any chronic disease affecting the gut and will assist in assigning a pathological diagnosis to the case. However, a pathological diagnosis does not equate to an etiological diagnosis and a given pathological entity can result from a number of different causes. For example, eosinophilic gastroenteritis can result from either an endoparasite infection or a food allergy. Nevertheless, the finding of histopathology consistent with any of the chronic gastrointestinal diseases previously mentioned should certainly necessitate the inclusion of dietary hypersensitivity in the differential diagnosis.

Diagnostic Tests

There is no single test available to confirm or refute the presence of food sensitivity. Logically, if food allergy is suspected, the most definitive way to confirm the diagnosis would be to demonstrate the presence of hypersensitivity by means of intradermal skin tests or by *in vitro* measurement of IgE antibody by either the radioallergosorbent test (RAST) or enzyme-linked immunosorbent assay (ELISA).

Intradermal testing with food extracts has been used in dogs as an aid to the diagnosis of pruritic skin disease, but it has been demonstrated to be unreliable as a diagnostic test.[35] False positive reactions are commonplace and, at best, one can state that a negative test implies that dietary hypersensitivity is unlikely to be the cause of the problem.[24]

Both a RAST and an ELISA are commercially available in some countries to evaluate canine serum for food allergens, by identifying antigen-specific IgE. Although there is doubt about their validity as diagnostic aids because false positives are commonly encountered,[1,33] a negative test to the allergen is probably an indication that the allergen is *not* responsible for a hypersensitivity reaction. More recently, similar systems have been introduced for cats, but these have not yet been thoroughly evaluated.[72]

The reason for the limited diagnostic value of skin tests and *in vitro* tests dependent upon IgE antibody measurement is obscure. Three possibilities deserve consideration:

• the conditions are not always immune-mediated • the conditions are immune-mediated, but IgE is not always involved • the allergenic extracts employed in diagnosis differ from those responsible for the disease

In dogs, encouraging results have been reported with the basophil degranulation test, which has been used in the diagnosis of canine atopic disease and would appear to have wide applications.[45] However, the technique demands specialized laboratory facilities and because fresh whole blood is required, this limits its application to institutions or practices close to such facilities.

Recent studies using a differential sugar absorption test suggest that this could prove useful in the differentiation of food hypersensitivity from intestinal permeability abnormalities due to food intolerance or underlying intestinal disease.[54]

Elimination Diets and Dietary Challenges

Undoubtedly, the most useful and reliable aid to diagnosis of dietary sensitivity is the procedure of feeding a restricted, or elimination, diet followed by dietary challenge with a test meal. Elimination diets should be individualized whenever possible on the basis of dietary history.

A detailed study of the animal's diet will allow the identification of foods that have not been fed before and that could be used to formulate a nutritionally balanced elimination diet that will be "hypoallergenic" for the individual. If the animal has not previously been exposed to a particular ingredient, it is unlikely to have mounted an immune response to it, unless antigenically related foods have been fed. If it is not possible to formulate a suitable elimination diet, then a restricted diet may be used that contains only one or two potential allergens, preferably ones that the animal has not eaten in the preceding month.

Elimination diets that have been employed for dogs include lamb, chicken, rabbit, horsemeat and fish as sources of protein and these are typically fed with rice or potatoes as a source of carbohydrates. Chicken and rice diets have been used with success in dogs.[28] In certain cases, the use of vegetable protein may have some merit.[22] The majority of these home-prepared elimination diets are not nutritionally complete unless balanced with suitable vitamins and minerals[52] and it is recommended that immature animals should not be fed such a diet for periods in excess of 3 weeks unless appropriate supplements are given.[72]

Successful elimination diets for cats include boiled, baked or microwaved lamb, chicken, rabbit or venison, with rice. Chicken has been used successfully in elimination diets for cats with food sensitivity.[4,31,47,57] Bearing in mind the cat's special dietary requirements, vitamin and mineral as well as taurine, supplementation may be needed if the diet is not complete and balanced. A diet of 100 g chicken and 100 g rice will provide sufficient calories for a moderately active 4-kg cat for 1 day. If the chicken is microwaved or boiled, 100 g cooked chicken will still provide sufficient taurine for an adult cat in maintenance. If the meat is boiled, the gravy should also be fed, as taurine can leach into the cooking water.

Drinking water should always be available, but distilled or bottled water is preferable to tap water in this diagnostic phase, as fungal contaminants and chemicals in tap water have been cited as a cause of food sensitivity in the dog. No other food should be allowed and other potentially allergenic sources, such as rawhide chews and even toys, should be withheld. Fatty acid supplements containing fish oils should not be fed at this time, as fish is a commonly reported allergen in the cat.[68] In those parts of the world where *Dirofilaria immitis* is endemic in dogs, consideration should be given to using an alternative preventive medication for the duration of the trial. If possible, cats should be confined indoors as this will help to ensure that those with multiple feeding stations and those that hunt are unable to consume alternative foodstuffs.

Feeding of the elimination diet must be of sufficient duration in order to provide a valid trial. It has previously been claimed that a period of 3 weeks was adequate.[39,64,67] However, Rosser (1990)[50] reported a prospective study of 51 dogs and found that 25.5% responded within this period, 33.3% required 4–6 weeks, 23.5% required 6–7 weeks and 17.6% required 8–10 weeks to improve on the elimination diet. Similarly, Rosser (1993) reported a prospective study of 13 cats with food allergy, where times before maximum improvement of clinical signs while being fed an elimination diet were 1–3 weeks (4 cats), 4–6 weeks (7 cats), 7 weeks (1 cat) and 9 weeks (1 cat).[51] The current recommendation is that a restricted diet be fed for 60 days, before a diagnosis of food sensitivity may be ruled out. It is tempting to speculate that the delayed response in some cases could be the result of histamine-releasing factors previously referred to, whose production falls only slowly after withdrawal of the allergen.

It is not recommended that a commercially available restriction diet be used in this diagnostic period because it has been shown that some dogs with food sensitivity that improve with a home-prepared diet of lamb and rice, develop exacerbation of clinical signs when fed the equivalent canned prescription diet.[33,50,67] This occurrence has been recorded in one cat[68] and indeed cats are occasionally encountered that are asymptomatic on any home-prepared diets but exacerbate when any commercial diet is fed. The reason why this occasionally occurs is not fully understood. It may be that the animal is only sensitive to the allergen in the form in which it is presented in prepared petfood, which could be attributable to the manufacturing process.

Occasionally the pruritus may be so severe that it may be necessary to try to break the "itch–scratch–lick" cycle with a short course of glucocorticoids. Cats are more resistant than dogs to the side-effects of glucocorticoids and they can tolerate higher doses. Oral prednisolone can be given to cats at 2.2 mg/kg eod, twice the appropriate dose for the dog. Anti-histamine therapy is of benefit in some pruritic cats and the most commonly recommended is oral chlorpheniramine at 2–4 mg per cat bid. While the pruritus is being controlled with drugs, the animal can be started on the elimination diet protocol. If the allergen has been eliminated, it should be possible to reduce the drug dose after 2–4 weeks without return of the pruritus. The diet should be fed for at least 2 weeks after the medication has stopped, or for a total of 60 days.

In most cases, the major clinical sign being evaluated while feeding the elimination diet is the pruritus. If food sensitivity is the cause of the pruritus, then the degree of the pruritus should decrease markedly at some point during the course of feeding the prescribed diet. Chronic skin changes may take longer to resolve completely. Diarrheic cases may improve in 3–5 days,[72] but where there is chronic GI tract disease some 4–6 weeks may be necessary before substantial and diagnostic improvement is noted. Cottage cheese and rice are often selected as the trial diet.[22]

If there is no improvement while on the elimination diet, this could be because

- the animal is pruritic for a reason other than dietary sensitivity
- the animal has a concurrent hypersensitivity (e.g. flea-bite hypersensitivity or atopy)
- the animal is reacting to the protein source in the elimination diet
- there is poor owner (or family) compliance

Improvement in clinical signs while on the elimination diet is suggestive of food sensitivity. The diagnosis should be confirmed by again feeding the animal the diet it was on at the time it originally presented with clinical signs. Exacerbation of clinical signs should occur within 7–14 days of feeding, with most cases taking 1–3 days for the reappearance of clinical signs.[50] This procedure establishes two things. First, it demonstrates that there is a dietary etiology and the improvement was not coincidental. Second, it allows identification of the duration between challenge and clinical relapse. This information may be useful when producing a calendar detailing challenge items, order of challenge and duration of challenge. Unfortunately, many owners are unwilling to challenge with the original diet.

Having established the diagnosis, it is prudent to carry the investigation further by reintroducing single dietary components sequentially. This procedure is described below and summarized in Fig. 12.13. In this way, it may be possible to define specifically the allergens involved and will facilitate management of the condition, particularly if only a single allergen needs to be eliminated. For those owners that do not want to take the case further, it is worthwhile extending the range of prepared petfoods that the animal can tolerate. A different product may be fed every 14 days and the clinical response assessed.[28,50]

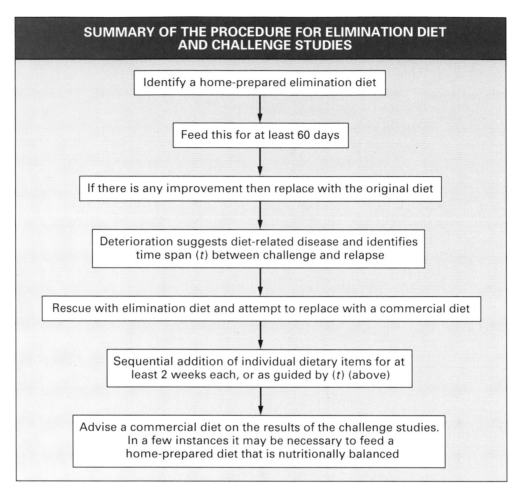

SUMMARY OF THE PROCEDURE FOR ELIMINATION DIET AND CHALLENGE STUDIES

Identify a home-prepared elimination diet

Feed this for at least 60 days

If there is any improvement then replace with the original diet

Deterioration suggests diet-related disease and identifies time span (*t*) between challenge and relapse

Rescue with elimination diet and attempt to replace with a commercial diet

Sequential addition of individual dietary items for at least 2 weeks each, or as guided by (*t*) (above)

Advise a commercial diet on the results of the challenge studies. In a few instances it may be necessary to feed a home-prepared diet that is nutritionally balanced

FIG. 12.13 Summary of the procedure for elimination diet and challenge studies.

Prognosis

The prognosis for food sensitivity in the dog and cat is good if the specific allergen can be avoided. Sometimes it is not possible to identify the offending item, in which case the nutritionally balanced elimination diet can be maintained, or a range of grocery brands tried. Most food-sensitive dogs and cats can be managed on a commercial diet. Rarely, an animal may develop multiple sensitivities in a sequential manner that makes formulation of a suitable balanced diet very difficult. In this situation, continued corticosteroid or other antiinflammatory or immunosuppressive therapy is occasionally necessary.

Therapeutics and Management

Once the diagnosis of food sensitivity has been established unequivocally, the management is usually relatively simple.

Dietary Manipulation

The prime goal in management is to institute a diet that is balanced and on which the patient is asymptomatic. In some instances, the allergen is not present in the animal's main diet, but may be fed as table scraps, treats or other additions to a commercial diet; these are easily avoided and a variety of balanced commercial diets may still be fed. Most cases can be managed with a commercial diet,

Internatic
Companic
S150–S15
19. Hall, E.
introducti
developm
Irish Sett(
Internatio
Companic
S152–S15:
20. Halliwell,
in canine s
of Immun
21. Halliwell,
diet. *Proc*
18, pp. 3–
22. Halliwell,
hypersensi
Practice, **3**
23. Halliwell,
food intole
24. Halliwell, |
in the dog.
25. Halliwell, l
allergy. In
E. W. Hal
W. B. Sau
26. Halliwell, l
H. *et al.*
pruritic der
and derma
the Americ:
697–701.
27. Halliwell, l
G. (1987) *A*
allergy derr
and Immur
28. Harvey, R
intolerance
of Small A:
29. Heyden, D
Eosinophili
dogs and it:
Journal of tl
tion, **162**, 3
30. Heyden, D
(1982) Lymp
Shepherd d
Hospital As
31. Holzworth,
Cat. Medici
Philadelphia
32. Hunter, J.
metabolic d
33. Jeffers, J. (

which may be either a grocery brand or dispensed by a veterinarian.[63]

If the client is cooperative, the presumptive food allergen can be investigated. First, the diet is gradually changed from a home-prepared elimination diet to a commercial diet of selected protein. This not only provides a nutritionally balanced and complete diet, but it is more convenient for the owner and improves owner compliance. This diet acts as the basal diet for test meals, summarized in Fig. 12.13. The clinical response to the diet should be assessed. Individual foodstuffs are subsequently introduced into the diet and the response noted carefully.

For each dietary component, challenge with the test diet should continue for at least 7 days.[3,39] Rosser,[50] however, found that a small proportion (7.7%) of his canine cases required 7–14 days before a positive response to the test meal was noted. Harvey[28] also found that the response to dietary challenge in some dogs could be delayed for up to 14 days and noted that the time required to produce a response to the test diet varied significantly according to the provoking dietary component. Dogs reacted to cereal in a mean time of 8.3 days, whereas reaction to dairy products occurred within a mean of 4.2 days.

If the animal is not pruritic after 6–14 days, then the next item can be introduced. The most commonly reported allergenic foods should be tried first. In the dog, these are cow's milk, cheese, beef, cereal and egg[28] and in the cat, cow's milk, beef and fish.[70,71] If the offending allergen is reintroduced, then there will be a clinical relapse in 1 hr to 14 days, with most cases taking 1–3 days for clinical signs to reappear. When the offending allergen is removed, clinical signs will subsequently improve.

For the small number of cases that cannot tolerate any commercial diet, indefinite maintenance on a home-prepared diet on which they are asymptomatic is clearly required. These must necessarily be formulated to suit each individual case, but care must be taken to ensure that they are nutritionally balanced and complete.

Dietary sensitivity should be regarded as a dynamic and fluctuating condition. It is quite possible for the patient to develop additional sensitivities necessitating a further change of diet at a later date. Similarly, it is possible for the sensitivity to diminish or disappear, as is known to occur in both atopy and flea allergy.[22]

Antiinflammatory Agents

In the diagnostic phase, it may be necessary to employ antiinflammatory agents either to break the initial itch–scratch–lick cycle, or on humanitarian grounds where response to the elimination diet is delayed. Having established a positive diagnosis of dietary sensitivity and following the introduction of a suitable diet, there are few indications for the continued use of these drugs. Their use, however, may be indicated where there is poor owner compliance in either diagnosis or therapy, or in the rare cases where the animal develops multiple sensitivities to one protein after another, making formulation of a suitable, balanced diet exceedingly difficult.

Before resorting to maintenance corticosteroid therapy, the clinician must first ascertain that all other possible causes of the disorder have been eliminated.

As previously indicated, corticosteroids are variably effective. Antihistamines may be of value in dogs when urticaria is a presenting sign, but they are of no value in the vast majority of cases. Likewise, the nonsteroidal antiinflammatory drugs have not proved efficacious. The mast cell stabilizers currently undergoing clinical trials for canine atopic disease would on theoretical grounds hold some promise in cases of dietary allergy. This avenue is certainly worthy of investigation.[24]

In cases of chronic gastrointestinal disease in dogs, such as colitis, lymphocytic–plasmacytic enteritis and eosinophilic gastroenteritis, patients are frequently placed on relatively high doses of corticosteroids with or without concomitant azathioprine. Although some cases will prove to be idiopathic and thus require such therapy, the practice of resorting to the use of such drugs without a diligent search for evidence of dietary allergy cannot be con

effective

positive

the poss

mode of

that the

dine and

tion. It

antipros

The d

chapter

1. Ackerr

but ma

1142–1

2. Anders

Advers

Health

3. August

cutanec

Compe

4. Baker,

Practic

5. Berg, I

intolera

Scandir

6. Bevier,

the intr

passive

test pos

tology,

Halliwe

7. Burger,

In *Feli*

Chandl

335. Ox

Eds. C. von Tscharner and R. E. W. Halliwell. pp. 404–406. London: Ballière Tindall.

49. Rosser, E. J. (1990) Data presented at the 1990 meeting of the American College of Veterinary Dermatology. San Francisco.

50. Rosser, E. J. (1990) Food allergy in the dog: a retrospective study of 51 dogs. *Proceedings of the Annual Meeting of the American College of Veterinary Dermatology.* San Francisco. p. 47.

51. Rosser, E. J. (1993) Food allergy in the cat: a prospective study of 13 cats. In *Advances in Veterinary Dermatology, Vol. 2.* Ed. P. J. Ihrke, I. S. Mason and S. D. White. pp. 33–39. Oxford: Pergamon.

52. Roudebush, P. and Cowell, C. S. (1992) Results of a hypoallergenic diet survey of veterinarians in North America with a nutritional evaluation of home-made diet prescriptions. *Veterinary Dermatology,* **3(1)**, 3–28.

53. Russell, M., Dark, K. A., Cummins, R. W. *et al.* (1984) Learned histamine release. *Science,* **225**, 733.

54. Rutgers, H. C., Hall, E. J., Sorensen, S., Proud, J. and Batt, R. (1992) Differential sugar absorption for the assessment of diet-sensitive intestinal disease in the dog. *Proceedings BSAVA Congress, 1992.*

55. Sampson, H. A., Broadbent, K. R. and Bernhisel–Broadbent, J. (1989) Spontaneous release of histamine from basophils and histamine-releasing factor in patients with atopic dermatitis and food hypersensitivity. *New England Journal of Medicine,* **321**, 228.

56. Scott, D. W. (1978) Immunologic skin disorders in the dog and cat. *Veterinary Clinics of North America,* **8**, 641–664.

57. Scott, D. W. (1980) Feline dermatology, 1900–1978: a monograph. *Journal of the American Animal Hospital Association,* **16**, 380–381.

58. Scott, D. W. (1987) Feline dermatology 1983–1985: The secret sits. *Journal of the American Animal Hospital Association,* **23**, 255–274.

59. Shanahan, F., Lee, T. G. D., Binenstock, J. and Befus, A. D. (1984) The influence of endorphins on peritoneal and mucosal secretion. *Journal of Allergy and Clinical Immunology,* **74**, 499–504.

60. Stogdale, L., Bomzon, L. and Bland van den Berg, P. (1982) Food allergy in cats. *Journal of the American Animal Hospital Association,* **18**, 188–194.

61. Tams, T. R. (1988) Inflammatory bowel disease: an important cause of vomiting and diarrhoea in cats. *Proceedings of the 12th Annual Kal Kan Symposium.* pp. 19–22.

62. Walker, W. A. and Isselbacher, K. J. (1974) Uptake and transport of macromolecules in the intestine. Possible role in clinical disorders. *Gastroenterology,* **67**, 531–550.

63. Walton, G. S. (1967) Skin responses in the dog and cat to ingested allergens. *Veterinary Record,* **81**, 709–713.

64. Walton, G. S. (1977) Allergic responses to ingested allergens. In *Current Therapy VI.* Ed. R. W. Kirk. pp. 576–579. Philadelphia: W. B. Saunders.

65. Walton, G. S. (1979) Responses to ingested allergens in the dog and cat. *Proceedings of the Kal Kan Symposium, September 1979.* pp. 30–32.

66. Walton, G. S., Parish, W. E. and Coombs, R. A. A. (1968) Spontaneous allergic dermatitis and enteritis in a cat. *Veterinary Record,* **83**, 35–41.

67. White, S. D. (1986) Food hypersensitivity in 30 dogs. *Journal of the American Veterinary Medical Association,* **188**, 695–698.

68. White, S. D. and Sequoia, D. (1989) Food hypersensitivity in cats: 14 cases (1982–1987). *Journal of the American Veterinary Medical Association,* **194**, 692–695.

69. Willard, M. D. (1992) Inflammatory bowel disease: perspectives on therapy. *Journal of the American Animal Hospital Association,* **28**, 27–32.

70. Wills, J. M. (1991) Dietary hypersensitivity in cats. *In Practice,* **13**, 87–91.

71. Wills, J. M. (1992) Diagnosing and managing food sensitivity in cats. *Veterinary Medicine,* **87**, 884–892.

72. Wills, J. M. and Harvey, R. (1994) Diagnosis and management of food allergy and intolerance in dogs and cats. *Australian Veterinary Journal* (In press).

PART 2

Clinical Nutrition in Practice

CHAPTER 13

Canine Gastrointestinal Tract Disease

ROGER M. BATT and COLIN F. BURROWS

Introduction

Gastrointestinal disease is a common reason for consultation in small animal practice and can present a considerable challenge to the diagnostic and therapeutic skills of the clinician. One problem is that although signs such as vomiting or diarrhea may draw attention to the gastrointestinal tract, these signs are relatively nonspecific and may also occur with disease of other organ systems. A further problem is that these signs may not be apparent in all cases of gastrointestinal disease, some of which may therefore be overlooked despite weight loss or other evidence of impaired gastrointestinal function. This underlines the need for a logical approach to diagnosis, including not only a detailed history and thorough clinical examination but also the use of relevant diagnostic procedures. Diagnosis is important to initiate rational therapy and this in turn is advancing with emerging knowledge of underlying pathophysiological mechanisms. This chapter summarizes current understanding of some of the most important gastrointestinal diseases in dogs and has focused particular attention on chronic diseases that can prove the most challenging to diagnose and treat.

Diseases of the Stomach

The stomach acts as an adjustable reservoir for ingested food that allows mixing of gastric contents with gastric secretions and the initiation of protein and fat digestion. Gastric diseases of the dog are listed in Fig. 13.1.

Chronic Gastritis

Pathophysiology and Clinical Consequences

The stomach normally protects itself from digestion by means of a complex interaction of physical and chemical factors that are known collectively as the gastric mucosal barrier.[88] The lipid rich hydrophobic apical cell membrane, tight cell junctions, surface mucus and gastric mucosal blood flow comprise the anatomical structures; bicarbonate secretion from the epithelial cells and prostaglandin mediated control of blood flow, mucus production and bicarbonate secretion comprise the chemical events. Working in concert, these various protective mechanisms allow the stomach to withstand the hostile environment of concentrated hydrochloric

DISEASES OF THE CANINE STOMACH: DIAGNOSTIC FEATURES AND TREATMENT			
Condition	**Vomiting**	**Diagnostic features**	**Treatment**
Acute gastritis	Acute vomiting	History and self-limiting features	24–48 hr, fluid and electrolyte therapy
Chronic gastritis			
Lymphocytic–plasmacytic	Intermittent chronic vomiting	Endoscopy, mucosal biopsy	Cimetidine, prednisone, diet change to a different protein source?
Eosinophilic	Intermittent chronic vomiting	Endoscopy mucosal biopsy, peripheral eosinophilia (±), contrast radiography	Diet change, cimetidine, prednisone
Chronic hypertrophic	Intermittent chronic vomiting	Contrast radiography, endoscopy, biopsy	Surgical correction (resection, pyloroplasty) cimetidine
Atrophic	Intermittent chronic vomiting, weight loss, diarrhea	Endoscopy, mucosal biopsy	Diet change, cimetidine, prednisone
Reflux	Intermittent vomiting on an empty stomach. Bile-stained vomitus	Endoscopy, mucosal biopsy	Cimetidine, metoclopramide. increase frequency of feeding
Gastric ulceration	Hematemesis, inappetence or anorexia, occasional vomiting, anemia, weight loss	Contrast radiography, endoscopy, biopsy	Antacids, cimetidine, surgery if bleeding is severe
Gastric outlet obstruction (pyloric stenosis)	Chronic postprandial vomiting of undigested food	Radiography, delayed gastric emptying	Surgical correction
Gastric stasis	Chronic postprandial vomiting of undigested or partly digested food. Gastric distension	Radiography, large flaccid stomach, elimination of other disorders	Metoclopramide, low-fat liquid or blended diet
Gastric tumors	Chronic vomiting usually progressing to hematemesis. melena, anemia, weight loss, inappetence anorexia, anemia, loss of condition	Radiography, endoscopy, biopsy, exploratory surgery	Surgical correction
Gastric dilatation volvulus	Acute onset, non-productive vomiting, gastric distension	Clinical signs, physical findings, radiography	Decompression, shock, surgical correction, increase frequency of feeding
Gastric parasites	Chronic vomiting, vomitus may contain parasite. rural dogs, colony cats	Fecal examination, (Baermann) endoscopy	Anthelmintics, cimetidine
Foreign body	Chronic vomiting, decreased appetite in some patients	Radiography, endoscopy	Endoscopic or surgical removal

FIG. 13.1 Diseases of the canine stomach.

acid and pepsin and prevent autodigestion. The back-diffusion of hydrogen ions from the gastric lumen into the mucosa that is central to the pathogenesis of gastric mucosal damage is impaired by the barrier. Barrier disruption by both endogenous and exogenous agents can, however, result in increased permeability to acid, direct damage to the mucosa and destruction of the subepithelium.

Acute gastritis is a common disease entity in small animals. There are a variety of causes including infectious and physical agents, chemicals and drugs and underlying systemic conditions including uremia, hepatic disease, shock, sepsis and stress. Most cases, however, do not merit detailed investigation because they respond to simple symptomatic therapy and the withholding of food.

Chronic gastritis also has many causes, including repeated exposure to the same agents as those incriminated in acute gastritis and can be divided into different types based on histopathological appearance (Fig. 13.1).[139] Chronic lymphocytic–plasmacytic gastritis is the most common form and involves variable inflammatory cell infiltrate with lymphocytes, plasma cells and less often, neutrophils. There is concomitant mucosal and submucosal fibrosis accompanied by superficial mucosal erosion, edema and hemorrhage.

Eosinophilic gastritis may represent disease confined to the stomach, but can be a manifestation of eosinophilic gastroenteritis in which there is segmental or diffuse infiltration with eosinophils in the mucosa and submucosa at various sites along the gastrointestinal tract.

Chronic atrophic gastritis also has an inflammatory component and is associated with a reduction in the size and depth of gastric glands because chief and parietal cells are reduced in number and may be replaced by mucus-secreting cells. This condition may represent progression of chronic superficial gastritis and can be induced experimentally in dogs by repeated immunization with gastric juice.[103] Achlorhydria, due to loss of parietal cells in atrophic gastritis, may be associated with small intestinal bacterial overgrowth in the dog,[132] a condition considered in detail in the section on chronic small intestinal disease.

Chronic reflux of bile into the stomach can also cause gastritis and the syndrome of reflux gastritis, also known as the bilious vomiting syndrome.[139] Bile reflux is a normal physiological event but when gastric motility is impaired, prolonged contact of the bile with the mucosa results in destruction of surface mucus and disruption of the lipid rich apical surface of gastric epithelial cells. This disrupts the mucosal barrier and results in gastritis.

Vomiting is the main clinical manifestation of chronic gastritis, but this sign is variable in frequency, intensity and duration and may not occur in all affected animals. In many patients bile-tinged fluid rather than partially digested food may be vomited because the emetic reflex begins with jejunal retroperistalsis. In patients with marked gastric mucosal ulceration or erosion, the vomitus may contain fresh or digested blood ("coffee grains"). In advanced chronic gastritis, other signs may include poor appetite, weight loss, depression and cranial abdominal pain. Gastritis also impairs normal patterns of gastric motility with a subsequent impairment of gastric trituration and emptying. Thus, it is not at all uncommon for the vomitus to contain food many hours after a meal.

Diagnosis of Chronic Gastritis

A variety of laboratory investigations need to be performed in order to exclude potential systemic causes of chronic vomiting, especially renal or hepatic disease. Other potential findings that provide important background information include anemia due to chronic blood loss, electrolyte imbalance due to loss of electrolytes in vomitus and panhypoproteinemia due to protein loss from the stomach or from other regions of the gastrointestinal tract in diffuse disease. Contrast radiography and gross appearance at gastroscopy may provide additional useful information, but the lack of obvious abnormalities does not exclude chronic gastritis that requires histological examination of gastric biopsies for definitive diagnosis.[130]

The differential diagnosis of chronic gastritis includes many of the other causes of chronic vomiting listed in Figs. 13.1 and 13.2. Most important however, are uremic gastritis, gastric foreign body, gastric outlet obstruction and hypoadrenocorticism. In cats with chronic vomiting the clinician should also consider hyperthyroidism and in endemic areas, heartworm infection, which commonly causes chronic vomiting in this species (see Chapter 14).

Treatment of Chronic Gastritis

Removal of the causative agent should be the main aim in the management of animals

MAJOR NONGASTRIC CAUSES OF VOMITING	
Abdominal disorders	Enteritis
	Pancreatitis
	Peritonitis
	Colitis
	Pyometra
	Paralytic ileus
	Intestinal obstruction/foreign body
	Prostatitis
	Constipation
	Hepatic disorders
	Acute urinary obstruction
	Pancreatic and intestinal tumors
Systemic or metabolic disorders	Ketoacidosis
	Uremia
	Hepatic encephalopathy
	Adrenal insufficiency
	Drugs (erythromycin, tetracycline, acetaminophen, chloramphenicol, digoxin, adriamycin) or toxins
Neurological disorders	CNS tumors
	Encephalitis
	Central nervous system trauma
	Canine distemper
	Motion sickness
	Autonomic or visceral epilepsy
	Vestibular disease

FIG. 13.2 Major nongastric causes of vomiting.

with chronic gastritis. In most patients however, the cause will not be determined and treatment must be supportive rather than curative. If the biopsy results show infiltration with eosinophils, plasma cells or lymphocytes it is possible that the disease may reflect a sensitivity to dietary protein. In these patients an elimination diet should be fed that excludes the suspected offending constituent. This may prove difficult to identify, particularly in animals fed commercial diets because these contain a variety of foodstuffs and food additives that could potentially affect the gastrointestinal mucosa. Management could therefore involve feeding a home-made or commercially available selected protein or exclusion diet, containing for example chicken, lamb, rabbit or cottage cheese as the main protein source and rice rather than cereal as a source of carbohydrate, as suggested for certain intestinal diseases (Fig. 13.3). Generally, canned meat-based or semimoist diets are preferable to dry high fiber diets that may have adverse effects by causing distention and mechanical irritation. There is some clinical evidence that eosinophilic gastritis may improve with dietary management alone.[132] Dietary change is unnecessary in reflux gastritis; signs can often be ameliorated by increasing the frequency of feeding, particularly if the animal is fed last thing at night and first thing in the morning.

In addition to dietary modification, inhibition of gastric acid secretion with H_2-receptor antagonists is critical in the management of chronic superficial gastritis and may also give a beneficial clinical response in mild mucosal atrophy. Cimetidine (5–10 mg/kg tid) or ranitidine (2 mg/kg bid) should be given for 7–10 days or until signs subside. Oral prednisone (or prednisolone) therapy (0.5–1.0 mg/kg bid, depending on the severity of histological changes) for 2–4 weeks followed by a tapering dose and alternate day therapy may help to control chronic lymphocytic–plasmacytic or eosinophilic gastritis. A secondary motility disorder should be suspected if chronic vomiting persists following feeding. In this case metoclopramide (0.2–0.4 mg/kg) given 30 min before each of three meals a day may help by

HOME-PREPARED DIETS FOR USE IN INTESTINAL DISEASE (APPROXIMATELY 10 kg DOG)		
Highly digestible low fat exclusion diets (single source protein)		
Lamb 6 oz lean lamb No corn oil	**Venison** 6 oz venison 2 teaspoons corn oil 10 oz boiled white rice	**Rabbit** 6 oz rabbit 1 teaspoon corn oil
Add to:	1 teaspoon dicalcium phosphate 1 teaspoon "lite" salt Adult multi-vitamins	
Supplies 675–700 kcal (2822–2926 kJ)	(20–34% protein, 46–48% CHO, 19–22% fat)	
Highly digestible moderate and high fat diets		
Chicken (moderate fat) 6 oz chicken 8 oz boiled white rice 2 teaspoons corn oil		**Beef** (high fat) 6 oz hamburger (lean) 5 oz boiled white rice No corn oil
Add to:	1 teaspoon dicalcium phosphate 1 teaspoon "lite" salt Adult multivitamins	
Supplies 680 kcal (2842 kJ)	(33% protein, 30% fat (chicken)) (33% protein, 43% fat (beef))	
Ingredients for each diet should be well mixed and cooked in a microwave oven or casserole before serving.		

FIG. 13.3 Home-prepared diets for intestinal disease.

stimulating gastric motility and promoting gastric emptying. In countries where it is available, cisapride (1.0 mg/kg tid–qid) may also be helpful. If ulceration is present, then cytoprotective agents such as sucralfate (0.25–1.0 g PO tid) may be effective. Sucralfate, however, should not be given concurrently with other medication and should be administered at least 1 hr before H_2-receptor antagonists because gastric acid is required to break down the tablets. In severe cases and for more rapid action, sucralfate tablets may be ground and mixed with about 30 ml of tap water and dosed as a slurry.

The prognosis for chronic mild lymphocytic–plasmacytic and eosinophilic gastritis is generally good, although prolonged treatment may be required to maintain clinical remission. The prognosis is guarded for animals with atrophic gastritis. There is no place for the routine use of antiemetics in the treatment of animals with vomiting caused by chronic gastric disease.

Gastric Motility Disorders

Although primary motility disorders are poorly documented in the dog and cat because of the lack of appropriate diagnostic techniques, they appear to be an unusual, but distinct, clinical entity. Functional motility disorders represent the major entity, but gastric dilatation–volvulus and some cases of gastric outflow obstruction resulting from "pylorospasm" also probably result from deranged gastric motility. These disorders are, however, sufficiently distinct in their clinical presentations to be considered as separate disease entities.

Pathophysiology and Clinical Consequences

Gastric stasis has been observed both in dogs and cats and there are many potential causes. These include: nervous inhibition, for example due to pain, peritonitis or trauma; metabolic disorders such as electrolyte or acid–base inbalances, uremia, hypothyroidism and hepatic encephalopathy; inflammation or ulceration of the stomach; anticholinergic drugs and narcotic analgesics used to treat vomiting and diarrhea; and idiopathic disorders of gastric motility perhaps due to abnormal function of the gastric pacemaker resulting in gastric dysrhythmia.

Clinical signs of gastric motility disorders include chronic belching and vomiting of variable duration and frequency. In some cases the patient may vomit an undigested meal many hours after feeding. Anorexia, poor body condition and weight loss are other common signs. The normal gastric emptying time in the dog fed once daily is 7–8 hr, thus, vomiting food more than 10 hr after eating should focus diagnostic attention on the stomach.

Diagnosis of Gastric Stasis

Diagnosis is typically made following exclusion of other causes of vomiting. Vomiting an undigested meal more than 10 hr after feeding provides strong evidence for a motility disorder if outflow obstruction has been eliminated. The diagnosis may be supported by retention of a barium meal for more than 10 hr,[28] but diagnosis by such simple means is the exception rather than the rule, because liquid barium is handled by the stomach very differently from food. It may sometimes be possible to diagnose the disorder by feeding canned food mixed with 5%, by weight, barium sulfate and feeding about 50% of the animal's daily energy requirement. Radiographs are then taken every hour until the stomach is empty. The normal stomach should empty this volume of food in 5 hr and retention for longer than this suggests a motility problem. Definitive diagnosis of gastric motility disorders requires specialist techniques that are not generally available, including radioisotope studies and measurement of gastric electrical and contractile activity.

Treatment of Gastric Stasis

Underlying causes such as metabolic and inflammatory lesions should first be identified

and treated. A moderately low fat, canned meat based diet, homogenized with water in a blender to a semiliquid consistency is more easily emptied from the stomach and reduces the frequency of vomiting.[140] This fulfils the need to feed a diet low in fiber, fat and osmolarity and liquid in consistency, in an attempt to promote gastric emptying.

Metoclopramide (0.2–0.4 mg/kg, 30 min before each of three meals a day) increases the tone and amplitude of gastric contractions, relaxes the pyloric canal and increases contraction in the proximal small intestine and could be considered the drug of choice for the management of motility disorders in dogs.[29] Erythromycin (1 mg/kg tid) is a promotility drug at low doses, binding with smooth muscle motilin receptors to increase gastric contractility. It should be tried if metoclopramide alone is ineffective. Cisapride (1.0 mg/kg tid) is another effective promotility drug. Contraindications to drug therapy include gastric outflow obstruction and concurrent phenothiazine anticholinergic or narcotic therapy.

Peptic Ulcer

Peptic ulcer is a term used to describe ulcers in either the stomach or duodenum. Ulcers are defined as circumscribed breaks in the surface of the gastrointestinal mucosa that occur in areas bathed by acid–pepsin. They may be acute or chronic, superficial or deep; almost all superficial ulcers are acute. Diagnosis of a gastric ulcer is uncommon in the dog. Ulceration of gastric or duodenal mucosa is, however, a relatively common postmortem finding in dogs (but not cats) suffering from a variety of severe diseases. Development of ulcers in such patients is believed to be a complication of protein–calorie malnutrition with its associated decrease in immunity and gastric epithelial cell turnover and stress and hypoxia that increase acid secretion and decrease mucosal blood flow. Sepsis, especially peritonitis, is also associated with gastric ulceration in the dog although the mechanisms are unclear.[98] Peptic ulcers are also seen in association with liver disease, after administration of nonsteroidal antiinflammatory drugs and in conjunction with mast cell tumors. Ulcers may also occur in the canine Zollinger–Ellison syndrome, a disease associated with an increase in blood gastrin levels due to a gastrin secreting tumor in the gastric antrum or the pancreas. Mucosal tumors may also become ulcerated and without histological confirmation may pose a problem in diagnosis. Ulcers and erosive gastritis also occur in hypoadrenocorticism and uremia.

The role of stress in the etiopathogenesis of naturally occurring peptic ulcers in the dog is unclear. Critically ill animals, particularly those that are septic, have a high incidence of ulcers, erosions and gastric hemorrhage. Aspirin, acetaminophen, flunixin and other nonsteroidal antiinflammatory drugs frequently cause gastritis or peptic ulcers and should be avoided unless absolutely necessary. Corticosteroids also delay epithelial cell division and exacerbate ulceration. These drugs are contraindicated unless essential for treatment of an unrelated disorder.

Regardless of the underlying cause, ulceration occurs when the gastric mucosal barrier is disrupted to expose unprotected mucosal cells to acid and pepsin.

Signs of peptic ulceration are poorly defined. Chronic vomiting is perhaps the most frequent sign, with or without hematemesis. Unlike humans, the dog and cat do not secrete acid continuously and blood in the vomitus does not always appear digested (the classical "coffee-grounds" appearance). Melena and anemia may also be observed if bleeding is severe. Inappetence or anorexia are common. Diagnosis is best made by gastroscopy, but contrast radiographic studies or exploratory laparotomy and biopsy are acceptable substitutes.

The differential diagnosis includes gastrointestinal tumors, hypoadrenocorticism and uremia. Animals that have received long-term nonsteroidal and/or corticosteroid therapy for chronic arthritis may have chronic gastric ulcers. Treatment is predicated on the elimination of any underlying cause and reducing or neutralizing acid secretion. Antacids, H_2-receptor antagonists, antiemetics and other antisecretory agents, such as propantheline hydrochloride, may also be indicated. Frequent small feedings of a blended, canned moderate to low fat diet are also beneficial.

Gastric Tumors

Primary gastric tumors are rare in both the dog and cat.[40] They include adenomatous polyps, adenomas, leiomyomas, adenocarcinomas, leiomyosarcomas, fibrosarcomas and lymphosarcomas. Adenocarcinomas are most common in the dog, comprising 75% of all tumors with most occurring in the distal stomach or pylorus of older male dogs.

Gastric tumors are characterized by chronic vomiting of gradually increasing severity and frequency, inappetence or anorexia and loss of body weight and condition. Anemia, diarrhea and hematemesis may also be present. The time from onset of signs to presentation varies from as little as 2–3 weeks to as long as 12 months. Signs may be more subtle in some animals with, for example, anorexia of sudden onset as the only presenting sign.

Diagnosis is made by contrast radiography (delayed gastric emptying, filling defects, rigid wall, ulcers), gastroscopy and biopsy, or exploratory laparotomy and biopsy. Some tumors ulcerate and resemble peptic ulcers so that examination of a biopsy specimen is essential for diagnosis. Treatment is by wide surgical excision; chemotherapy may be possible in selected patients depending on the tumor type.[40] Total parenteral or enteral nutrition may be necessary in the immediate postoperative period. The prognosis is guarded to hopeless for malignant gastric tumors.

Gastric Outlet Obstruction

Chronic vomiting as a result of occlusion of the pyloric or antral region of the stomach is a relatively common occurrence in the dog. The disorder occurs most frequently as a result of abnormal pyloric function or congenital or acquired pyloric stenosis. Less frequent causes are tumors of the pylorus or adjacent structures and chronic pancreatitis.

Antral mucosal hypertrophy, a condition that is being diagnosed with increasing frequency is an additional and important cause of gastric outlet obstruction.[146] Congenital and acquired pyloric stenosis both have a similar clinical presentation, but the congenital form occurs in young dogs, usually at the time of changing to solid food, while the acquired form occurs in older animals. Congenital pyloric stenosis occurs most frequently in brachycephalic breeds (especially the Boston Terrier). Both congenital and acquired pyloric stenosis result from hypertrophy of the circular muscle of the pylorus and must always be differentiated from "pylorospasm." The latter, poorly understood disorder appears to be a failure of the pylorus to relax appropriately and occurs mainly in miniature and toy breeds. Afflicted animals appear normal in all respects except for intermittent postprandial vomiting. The vomitus is characteristically undigested or partly digested and occasionally mucoid and may have an acid pH; bile is absent. The precise interrelationships between the two types of pyloric stenosis and pylorospasm are unclear.

Signs of gastric outlet obstruction vary with the degree of obstruction. Vomiting is the predominant sign and may occur at any time after a meal. The time for complete emptying of a normal meal from the stomach in the dog is 7–8 hr. Thus, vomiting of all or part of a meal at periods greater than 10 hr after ingestion suggests delayed gastric emptying and the probability of a gastric pancreatic or proximal duodenal lesion.

Diagnosis is by contrast radiography or gastroscopy, with surgical confirmation an absolute necessity in most patients. In some patients food may be present in the stomach at gastroscopy some 18–24 hr after the previous meal, suggesting delayed gastric emptying. In an otherwise normal animal, this supports the diagnosis of gastric outlet obstruction. Mucosal biopsies are normal in animals with pyloric dysfunction. Chronic vomiting may result in hyponatremia, hypokalemia, hypochloremia and metabolic alkalosis. A variety of neurological abnormalities are associated with the electrolyte abnormalities

and metabolic alkalosis that occur in long standing cases. These signs, however, quickly reverse with appropriate fluid and electrolyte therapy.

Animals with pylorospasm respond well to parasympatholytics such as propantheline bromide, and barbiturates may be useful as adjunctive therapy. Response to treatment may serve to differentiate pylorospasm from pyloric hypertrophy. The only effective treatment for hypertrophic disease is pyloroplasty. Occasionally obstruction may be sufficiently severe that nutritional homeostasis cannot be sustained on any kind of diet and enteral or parenteral feeding becomes essential.

Acute Gastric Dilatation–Volvulus

Acute gastric dilatation–volvulus (GDV) is a sudden, dramatic and often fatal gastrointestinal disorder that affects many breeds, but particularly the large, deep-chested breeds such as the German Shepherd, Irish Setter, Great Dane, St. Bernard, Bloodhound, Boxer, Weimaraner, Collie, Wolfhound, Basset Hound and Dobermann Pinscher.[31] It has been estimated that there are as many as 60,000 cases in the U.S.A. each year with an overall mortality of about 20% depending on time from onset to treatment.[31] Dogs of all ages are afflicted, but there is a predilection for older dogs. Gastric dilatation, which is the rapid distention of the stomach with food, fluid and especially gas (from swallowed air, fermentation or a combination of the two) may progress to volvulus. This occurs because the forces exerted on the distended canine stomach cause it to rotate either to the right or left (most often to the left–clockwise) on an axis at right angles to a line between the the esophageal and pyloric sphincters.

The cause is unknown and the subject of much debate. Risk factors are believed to include diet (dry cereal-based dog food), breed, "stress," overeating, overdrinking, exercise, anesthesia, aerophagia, intragastric fermentation and previous gastric "trauma." Postprandial exercise and excitement are additional, frequently reported risk factors. This is not the whole story, however, as is evidenced by the fact that the disease occurs in kennels managed by knowledgeable breeders who take all possible steps to avoid these risk factors.

GDV should be differentiated from simple acute gastric dilatation by survey radiographs made after a period of initial stabilization (unstable patients can die during the time it takes to take radiographs). Early recognition is important because dilatation precedes volvulus and is associated with a more favorable prognosis. In both instances, however, death from hypovolemic and cardiogenic shock may occur within a few hours of the onset of signs.

Rapid gastric distention adversely affects the function of the lower esophageal sphincter and impairs gastric motility and emptying. It has been postulated that distention occludes the gastroesophageal junction, precluding emptying by either eructation or emesis. Distention decreases gastric motility both by reducing contractility and by reflex nervous inhibition. Following dilatation, the gastric mucosa and later gastric smooth muscle undergoes potentially irreversible ischemic necrosis. There is also an accumulation and sequestration of gastric secretions. The distended stomach occludes venous return from the rear limbs and caudal abdomen and precipitates hypovolemic and cardiogenic shock. Lactic acid and other metabolic by-products accumulate in the hypoperfused hind limbs and viscera to cause severe metabolic acidosis, especially after relief of dilatation. Signs of reperfusion injury, endotoxemia, disseminated intravascular coagulation and fatal cardiac arrhythmias may also occur. Splenic torsion with infarction and necrosis is a common sequela.

The "typical" history is that of rapid ingestion of a large meal late at night in conjunction with consumption of large quantities of water followed by exercise. Dr.y (cereal-based) dog food has also been incriminated, perhaps unjustly because most large-breed dogs are fed these diets for reasons of cost and ease of use. Some dogs are apparently overly susceptible and have no pertinent presenting history. Abdominal distension is associated with progressive restlessness, unproductive retching, salivation, dyspnea, gastric tympany and shock.

Diagnosis is made from the clinical signs; confirmatory radiographs are time consuming, usually unnecessary and cause a potentially fatal delay in treatment. Depending on the type and severity of the signs, treatment may consist of gastric intubation and decompression, vigorous fluid therapy, intravenous cimetidine (10 mg/kg qid–tid) and lidocaine (1 mg/kg IV) and gastrostomy or gastrocentesis. After decompression, shock treatment should be maintained. Decompression should be maintained by tube gastrostomy or pharyngostomy. Contemporary surgical experience suggests that the incidence of recurrence may be reduced by gastropexy. Cardiac arrhythmias occur frequently (usually premature ventricular contraction or ventricular tachycardia) after surgical correction and should be treated appropriately. There is little evidence that diet affects the incidence of recurrence. Despite this however, the approach recommended by most authorities is the feeding of a meat-based, canned, highly digestible diet at least three times daily in conjunction with gastropexy as the best approach to the prevention of recurrence. Pyloroplasty does not influence the rate of recurrence.

Diseases of the Pancreas

Acute Pancreatitis

Pathophysiology and Clinical Consequences

The acinar cells of the pancreas contain inactive forms of proteolytic and phospholipolytic enzymes that are capable of causing substantial tissue destruction, and pancreatitis is thought to be a consequence of factors that allow activation of these enzymes within the pancreas. Clinical manifestations may be minimal in mild edematous pancreatitis, but can be substantial in the severe, but less common, necrotizing form where acute vomiting, depression and abdominal pain can quickly progress to death. Resolution of tissue damage may be complete in animals that survive the initial episode of acute pancreatitis, but the continuing inflammation of chronic pancreatitis is more likely to cause irreversible damage that may be asymptomatic until extensive destruction of pancreatic tissue results in impairment of exocrine or endocrine function.

Diagnosis of Acute Pancreatitis

In dogs, there is a clear association between acute pancreatitis and increased serum activities of lipase and amylase.[27,76,90,93,129,133] Serum lipase is considered to be more reliable than serum amylase, but both originate from extrapancreatic sources such as the gastric and intestinal mucosa and increased activities can be due to conditions such as gastroenteritis and hepatic and renal diseases[75,105,129]; dexamethasone administration can also increase serum lipase.[99] Serum trypsin-like immunoreactivity (TLI) is not derived from extrapancreatic sources and a high concentration may provide an early indication of acute pancreatitis in dogs.[21,126] Because lipase, amylase and TLI normally appear to be cleared from the serum by glomerular filtration, increased activities secondary to impairment of renal function can be a diagnostic problem because it may prove particularly difficult to distinguish between acute renal failure and acute pancreatitis.[43]

The trypsin component of total TLI in pancreatitis circulates exclusively complexed with the plasma inhibitor α_1-antiprotease that can readily be separated from trypsinogen by chromatography.[21] The assay of these trypsin–protease inhibitor complexes therefore offers the possibility of a very specific test in the future for the detection of pancreatitis in the dog and potentially in the cat.

Management of Pancreatitis

Maintenance of fluid and electrolyte balance while withholding food represents a key component of the management of acute pancreatitis and additional supportive measures may include administration of antibiotics, analgesics and corticosteroids. The latter should only be given for a short period to animals in shock with severe pancreatitis because long-term administration may impair the ability of the reticuloendothelial system to remove uninhibited α-macroglobulin-bound proteases. Consumption of plasma protease inhibitors and saturation of available α-macroglobulins in severe pancreatitis can be lifethreatening, so that replacement of α-macroglobulins by transfusion of blood or plasma may prove particularly valuable. This action will also support the plasma albumin concentration, hence maintaining blood volume and minimizing pancreatic edema and ischemia.

Resumption of feeding should begin with nutrients that stimulate pancreatic secretion the least, initially glucose or small polymers of glucose and then starch, for example as boiled rice. Protein of high biological value, such as cottage cheese or lean meat, can then be gradually added to the diet. Amino acids and fatty acids cause release of cholecystokinin (CCK) which is a potent stimulant of pancreatic enzyme secretion and consequently home-cooked or commercial diets that are low in fat and do not contain high protein levels may help to prevent recurrence.

Exocrine Pancreatic Insufficiency

Pathophysiology and Nutritional Consequences

Exocrine pancreatic insufficiency (EPI) represents one of many potential abnormalities affecting the gastrointestinal tract that can result in chronic malabsorption by interfering with either the degradative or absorptive phases in the handling of one or more ingested nutrients (Fig. 13.4). Pancreatic acinar atrophy is the most common cause of EPI in dogs,[1,72,157] and typically presents in fully grown adults between 1 and 5 years-of-age. This occurs in many purebreed and also mix-breed dogs, although there is a genetic predisposition in German Shepherd dogs.[150] Pancreatic hypoplasia and chronic pancreatitis are relatively rare causes of EPI in dogs. EPI is relatively uncommon in cats but when it does occur it is most likely to be due to chronic pancreatitis[124] (see Chapter 14).

CONDITIONS RESULTING IN MALABSORPTION		
Condition	**Deficiency**	**Malabsorption**
Exocrine pancreatic insufficiency	α-Amylase	Starch
	Proteases	Protein
	Lipase	Triglycerides
Small intestinal disease (secondary deficiencies)	Disaccharidases	Disaccharides
	Sugar carriers	Monosaccharides
	Brush border peptidases	Oligopeptides
	Peptide and amino acid carriers	Peptides and amino acids
	Reduced surface area	Lipid
Brush border disease (primary deficiencies)	Lactase	Lactose
	Sucrase/α-dextrinase	Sucrose, α-limit dextrins
	Glucose/galactose carrier	Glucose, galactose
	Enteropeptidase	Protein
Bile salt deficiency	Conjugated bile salts	Lipid

FIG. 13.4 Conditions resulting in malabsorption.

Clinical signs are variable, but typically include polyphagia, weight loss and voluminous output of semiformed feces. Some animals may vomit and watery diarrhea may also occur due to passage of malabsorbed dietary constituents along the intestinal tract and secondary alterations in the transmucosal flux of fluid. Coprophagia and pica are reported in many cases, but affected animals typically show no signs of systemic disease.

Deficiencies of α-amylase, proteolytic enzymes and lipase result in relatively severe malabsorption of starch, protein and triglycerides, respectively. Intestinal function may also be impaired due to reduced protein synthesis[156] and bacterial overgrowth in the proximal small intestine.[160] In addition to deficiencies of fat soluble vitamins, reduced concentrations of serum cobalamin (vitamin B$_{12}$) have been reported.[15,64] This may be related to defective secretion of pancreatic intrinsic factor which promotes ileal absorption of cobalamin,[8,9] compounded by factors such as bacterial overgrowth.

Diagnosis of EPI

Because EPI is a well defined disease that can have secondary effects on the small intestine, it is important to assess exocrine pancreatic function either to diagnose EPI or to eliminate this diagnosis before the investigation of small intestinal disease. There have been many proce-

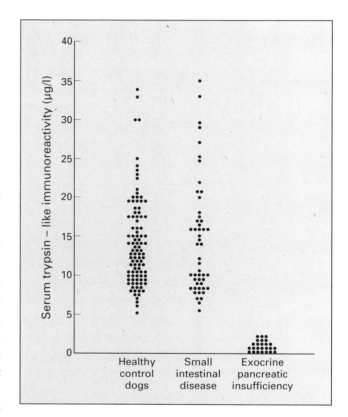

FIG. 13.5 Serum trypsin-like immunoreactivity in 100 clinically normal dogs, 50 dogs with small intestinal disease and 25 dogs with exocrine pancreatic insufficiency. (Reproduced from Ref. 159 with permission.)

dures used over the years, including estimations of fecal proteolytic activity,[10,35,70,109,125,151,159] and the bentiromide test.[10,54,109,128] Assay of canine serum TLI is now used extensively as a sensitive and specific test for the diagnosis of

EPI in dogs,[158] a low concentration clearly identifying affected animals (Fig. 13.5).

Treatment of Exocrine Pancreatic Insufficiency

Medical management: Treatment of EPI typically involves life-long replacement therapy by the addition of pancreatic extract to the diet, immediately before feeding each of three meals a day. Powdered extracts (1 tsp/10 kg body weight with each meal) are preferable to tablets, capsules and enteric-coated preparations.[64] An alternative is to add approximately 100 g of chopped bovine or porcine pancreas stored frozen at $-20°C$ and freshly thawed. Diarrhea, polyphagia and coprophagia should resolve within a few days and weight gain of 0.5–1.0 kg per week can be expected. If this occurs the dose of powdered pancreatic extract should be reduced to that found by the owner to be the least needed to maintain normal body weight and fecal consistency. If response to pancreatic replacement therapy is poor, small intestinal bacterial overgrowth may be suspected and treated with oral antibiotics for at least a month (e.g. with oxytetracycline, tylosin or metronidazole) as detailed later in the section on small intestinal disease. If the patient still fails to improve, H_2-receptor antagonists such as cimetidine (5–10 mg/kg) or ranitidine (2 mg/kg) may be given 20 min before a meal to inhibit acid secretion and minimize degradation of enzymes in the pancreatic extract. If diarrhea and other clinical signs persist, the possibility that the patient needs additional treatment for concurrent small intestinal disease, especially lymphocytic–plasmacytic enteritis, should be entertained. An intestinal mucosal biopsy is the quickest and easiest means of making this diagnosis. Oral multivitamin supplements containing fat-soluble vitamins should also be considered as supportive therapy, but because cobalamin malabsorption is not likely to resolve with pancreatic replacement therapy,[127] cobalamin needs to be given parenterally (e.g. 500 µg/month).

Dietary management: The majority of dogs with EPI can be fed and maintained on a normal diet provided they receive enzyme replacement. Dietary modification can, however, play an important supportive role in some cases and may permit reduction of enzyme supplementation. Fat malabsorption is a major problem in EPI, so that a low fat diet may prove particularly effective and diets containing poorly digestible fat should not be fed. Results of feeding a moderate fat, highly digestible diet have been variable.[100,101,152] Carbohydrate malabsorption may cause osmotic diarrhea and accompanying problems including abdominal discomfort and flatulence. Breath hydrogen testing indicates that absorption of carbohydrate is more complete when rice is fed compared with cereal,[148] so that rice may be preferable as a source of carbohydrate, particularly in cases with persistent watery diarrhea. In all cases of EPI, the diet should comprise a high quality source of protein. In some cases, a restriction or selected protein diet might prove helpful because bombardment of the small intestine by undegraded antigenic macromolecules could potentially result in temporary dietary sensitivity.

Chronic Small Intestinal Disease

Pathophysiology and Nutritional Consequences

Chronic small intestinal disease typically involves interference with the number or function of absorptive epithelial cells (enterocytes) and can result in osmotic diarrhea and loss of body weight due to impaired absorption of solutes. This is due to secondary deficiencies of brush border enzymes and carrier proteins and a reduction in surface area, resulting in interference with the absorption of small carbohydrates, peptides, amino acids and lipids (Fig. 13.4). Interference with absorption of vitamins and minerals can also occur and may contribute to poor condition. When present, accompanying mucosal inflammatory changes may make a further contribution to clinical disease, for example, resulting in secretory and exudative diarrhea. Protein-losing enteropathy, increased utilization of calories and a failure to recycle protein

---

STEPWISE APPROACH TO INTESTINAL DISEASE IN DOGS

History and baseline investigations
- Analysis of feces, blood and urine
- Radiography, ultrasonography if partial obstruction suspected
- Colonoscopy if signs of large bowel disease

Exclude exocrine pancreatic insufficiency
- Assay serum trypsin-like immunoreactivity

Indirect investigation of small intestinal damage
- Serum folate and cobalamin
- Intestinal permeability

Direct examination of small intestine
- Endoscopy or laparotomy
- Histology of biopsies
- Culture duodenal juice

FIG. 13.6 Stepwise approach to small intestinal disease in dogs.

from exfoliated enterocytes may also be a consequence of severe intestinal damage and may potentiate loss of body weight in some animals. Lipid malabsorption in intestinal disease may also be related to deconjugation of bile salts in bacterial overgrowth, defective absorption of conjugated bile salts in ileal disease and pathological dilatation of intestinal lymphatic vessels such as occurs in intestinal lymphangiectasia.

Primary brush border defects (Fig. 13.4) occur in the absence of morphological damage to the mucosa and although some are relatively rare, all have been well described in humans and may occur in small animals. Small intestinal bacterial overgrowth, which is considered in more detail below, reinforces the concept of damage to enterocytes without obvious histological changes in the mucosa and helps explain how intestinal disease may be overlooked by reliance on morphological criteria alone.

The clinical consequences of small intestinal disease in an individual animal depend not only on the severity but also the extent of damage. The small intestine does have a functional reserve and the distal intestine has some capacity to adapt and compensate for proximal damage. Consequently, although

diarrhea may be an important sign, the absence of diarrhea does not exclude the possibility of small intestinal damage and failure to grow normally or loss of body weight may be the only evidence of disease in some patients. Other clinical signs may include vomiting (especially if mucosal damage is severe), polyphagia and coprophagia. Most animals with chronic small intestinal disease typically show no signs of systemic disease, such as lethargy and depression until terminally ill, although some animals may be intermittently anorexic. Patients with severe intestinal disease and protein losing enteropathy can present with edema, ascites and profound cachexia.

Diagnosis of Small Intestinal Disease

A stepwise approach to small intestinal disease in dogs is summarized in Fig. 13.6. Baseline investigations of animals with suspected small intestinal disease include examinations of feces, blood and urine to determine whether intestinal pathogens, or systemic disorders might be present.[34] Findings that may contribute to subsequent investigations of small intestinal disease include eosinophilia, perhaps reflecting parasitism or eosinophilic gastroenteritis, neutrophilia in inflammatory disease, lymphopenia in immunodeficiency or lymphangiectasia and panhypoproteinemia in protein-losing enteropathy. The possibility of partial intestinal obstruction should be considered and can be pursued by radiography or ultrasonography. Some clinicians may use colonoscopy at this stage to examine animals with prominent signs of large bowel disease, such as urgency of defecation, frequency greater than 5 times a day, tenesmus and passage of fresh blood and mucus. The possibility that colonic disease may be part of a diffuse disease also affecting the small bowel, or may be secondary to a primary small bowel disorder should be considered, in which case the following steps are taken.

Once EPI has been eliminated, as described above, indirect tests can provide information not only for the diagnosis but also the management of intestinal disease. To save time and money, many clinicians proceed directly to mucosal biopsy at this stage if small intestinal disease is suspected, but this approach depends on morphological criteria alone and does have limitations which are considered below.

The assay of serum folate and cobalamin (vitamin B_{12}) concentrations represents a helpful initial test that can assist the detection and characterization of small intestinal disease in dogs.[15] Disease of the proximal or distal small intestine can result in reduced serum concentrations of folate or cobalamin, respectively, reflecting the different sites for the normal absorption of these vitamins in dogs.[5,6,15] In addition, proximal small intestinal bacterial overgrowth (SIBO) should be suspected if there is an elevated serum folate and/or reduced serum cobalamin concentration. This is because many enteric bacteria can synthesize folate that is subsequently absorbed in the jejunum and at the same time bind cobalamin making it unavailable for transport.[7,11,16] It is important to be aware that normal serum folate and cobalamin levels do not exclude the possibility of small intestinal disease, because these alterations depend on the nature, extent and duration of mucosal disease

and also on the type and numbers of organisms present in SIBO. It is also important for laboratories to validate and establish their own control ranges for the different commercial assays available that can give quite different results.[14] The validity of such an approach for the detection of small intestinal disease in cats is unclear.

Function tests for the detection of intestinal disease, such as the xylose or glucose absorption test have been largely abandoned due to problems including a lack of sensitivity. Permeability tests provide a new approach to the detection of small intestinal damage and this was demonstrated recently in dogs by determining 24-hr urinary excretion of ^{51}Cr–EDTA following oral administration.[58] The principal is that intestinal damage allows increased passage of this probe across the mucosa into the blood, most likely by a paracellular route due to interference with the integrity of tight junctions between enterocytes, followed by rapid excretion in urine. The use of a sugar such as lactulose or cellobiose as a permeability probe, given with a sugar such as rhamnose or mannitol to provide a simultaneous assessment of surface area,[62,113] overcomes practical limitations of using a radiochemical and should result in these tests becoming available outside academic referral centers.

Histological examination of intestinal biopsies (which can be taken perorally, by use of an endoscopic or a suction biopsy instrument, or at laparotomy) can assist the identification of mucosal damage and early descriptions of intestinal diseases in dogs were largely dependent on this approach.[36,48,51,68,71,80,97,145] Morphological abnormalities can provide a baseline to assess response to treatment, while severity of any changes can contribute to the assessment of prognosis (Fig. 13.7). Indeed, these findings may be the best evidence of intestinal disease that most clinicians have to rely on at the present time. It should, however, be emphasized that there may be minimal or no obvious morphological abnormalities in certain disorders, despite considerable interference with intestinal function. Furthermore, histological descriptions

SUMMARY OF HISTOLOGICAL CHANGES IN CHRONIC SMALL INTESTINAL DISEASES IN DOGS

Condition	Histological changes and incidence of protein-losing enteropathy
Lymphocytic–plasmacytic enteritis	Minimal change to severe villus atrophy and fusion of villi, necrosis and abcessation of crypts. Prominent infiltrate with lymphocytes and plasma cells in lamina propria. Can be protein-losing enteropathy in severe cases
Eosinophilic enteritis	Minimal change to moderate villus atrophy. Prominent infiltrate of intestinal wall with eosinophils, and also lymphocytes and plasma cells. May affect whole or any part of gastrointestinal tract and can result in protein-losing enteropathy
Intestinal lymphangiectasia	Minimal or no villus atrophy. Dilation of lymphatics in mucosa, submucosa and serosa. Severe protein-losing enteropathy
Intestinal lymphosarcoma	Typically a severe villus atrophy. Dense infiltrate of intestinal wall particularly with lymphocytes. May affect any part of gastrointestinal tract. Severe protein-losing enteropathy
Small intestinal bacterial overgrowth	Minimal or no changes in most cases. May be partial villus atrophy and/or lymphocyte and plasma cell infiltrate in lamina propria of proximal small intestine
Gluten-sensitive enteropathy	Patchy partial villus atrophy and increased intraepithelial lymphocytes in proximal small intestine

FIG. 13.7 Summary of histological changes in chronic small intestinal diseases in dogs.

alone provide little information on possible etiology or underlying mechanisms of damage, although this knowledge would clearly assist effective management. These problems may also be relevant to cats where descriptions of enteropathies are also based predominantly on morphological criteria.[92,123,149,155]

The first two conditions described here illustrate an emerging understanding of the etiology of intestinal damage, before considering the diseases still classified solely according to morphological appearance.

Chronic Enteropathies

Bacterial Overgrowth

SIBO in the proximal small intestine is emerging as an important condition in dogs. SIBO in humans can have a variety of causes, but is typically associated with physical or functional interference with normal motility as occurs in the stagnant or blind loop syndrome.[81] Similar conditions may occur in dogs and stasis following partial constriction of the intestine due to tumors, foreign bodies, inflammatory disease and chronic intussusception are further possibilities. SIBO may also be a secondary treatable complication in exocrine pancreatic insufficiency. The most common presentation in dogs is typified by

the naturally occurring enteropathy associated with SIBO that has been identified in German Shepherd dogs.[7,11,13,16] Although the cause of the overgrowth in these dogs is not known, there is no obvious evidence of intestinal stasis, but there is some evidence that defective local immunity may be relevant to the clinical presentation of SIBO in this breed.[4,153]

Typically, SIBO presents in young dogs as chronic intermittent diarrhea and may be accompanied by loss of body weight or failure to gain weight. Clinical signs are variable and some animals for example, may exhibit only intermittent vomiting. Exocrine pancreatic function is normal and xylose absorption may be reduced in only some cases, particularly in those dogs with anaerobic overgrowth. A high serum folate and/or reduced cobalamin concentration may provide indirect evidence for bacterial overgrowth once other conditions such as EPI have been eliminated, but normal results do not exclude SIBO. Microbiological culture of duodenal juice obtained endoscopically or at laparotomy is needed to confirm the diagnosis and should demonstrate $>10^5$ colony forming units per ml. The most frequent isolates typically include enterococci and *Escherichia coli* in dogs with aerobic overgrowth and clostridia in dogs with anaerobic overgrowth.

There is minimal morphological damage in many dogs with SIBO, although biopsies from some animals exhibit partial villus atrophy and/or prominent lymphocyte and plasma cell infiltrates in the lamina propria. Such extreme changes were demonstrated in a dog with aerobic overgrowth in which partial villus atrophy was accompanied not only by increased lymphocytes and plasma cells in the lamina propria but also by considerably increased numbers of intraepithelial lymphocytes, that decreased markedly after treatment with oral oxytetracycline for 4 weeks.[112]

Because villus atrophy is relatively uncommon, a reduction in the number of enterocytes is likely to make a relatively small contribution to nutrient malabsorption in these dogs. Interference with the function of individual enterocytes could, however, have important clinical consequences and definite evidence that bacteria can have a selective effect on the brush border has been provided by investigating the biochemical changes in jejunal biopsy specimens, damage being more severe with anaerobic than aerobic overgrowth.[11,13] Bacterial metabolism of intraluminal constituents may make a further, and in some cases more important, contribution to the clinical signs. For example, bacterial deconjugation of bile salts can interfere with micelle formation resulting in malabsorption of lipid and metabolites such as deconjugated bile salts and hydroxy fatty acids can cause diarrhea by stimulation of colonic secretion.[81]

Gluten-Sensitive Enteropathy

Gluten-sensitive enteropathy in Irish Setters is the best characterized diet-sensitive enteropathy in dogs and provides an insight that should assist with the identification and management of other dietary sensitivities in small animals. This condition has many similarities to celiac disease in humans, but the severity of intestinal damage and clinical signs tends to be less marked. Affected dogs typically present with poor weight gain or weight loss, in most cases accompanied by chronic diarrhea and clinical signs are often first observed between 4 and 7 months-of-age.[12] There may

be some indirect evidence for proximal small intestinal disease, including low serum and red cell folate. In contrast, serum cobalamin concentrations are unaltered and culture of duodenal juice shows no evidence of small intestinal bacterial overgrowth. Histological changes in jejunal biopsies can be variable, but the most consistent abnormalities are partial villus atrophy and increased intraepithelial lymphocytes.

Intestinal lesions are not pathognomonic, consequently morphological criteria alone do not permit the diagnosis of dietary sensitivity. This can only be confirmed by monitoring response to dietary exclusion and subsequent challenge using objective criteria such as intestinal morphology or permeability. Initial studies in Irish Setters showed a relationship between damage and dietary wheat and suggested that there might be an underlying permeability disorder.[12,59–61] Sensitivity to the gluten component of wheat was demonstrated by performing challenge studies on affected animals reared on a normal wheat-containing diet, 6 weeks after introduction of a cereal-free diet at 12 months-of-age.[63] Villus height and intraepithelial lymphocyte counts in jejunal biopsy specimens returned to normal on the cereal-free diet. Intestinal permeability to ^{51}Cr–EDTA also decreased on the cereal-free diet, but remained elevated compared with control dogs providing further support for an underlying permeability disorder.[20] Subsequent gluten challenge reversed these changes and resulted in clinical relapse manifest by diarrhea and weight loss.

These studies illustrate how intestinal permeability can be used to monitor the response to dietary manipulation without the need for sequential biopsies and this approach has now been used to identify dietary sensitivities in other breeds.[113] The main importance of distinguishing between dietary sensitivity and intolerance from a practical point of view is that the cause of gastrointestinal damage has been identified for a dietary sensitivity. In contrast, a beneficial response to a new diet in dietary intolerance may mask underlying disease, such as idiopathic inflammatory bowel disease or SIBO, which may need additional treatment.

Lymphocytic–Plasmacytic Enteritis

Lymphocytic–plasmacytic enteritis is the most common form of inflammatory bowel disease seen in dogs and cats and indeed, is the most commonly diagnosed cause of chronic vomiting and diarrhea in the two species.[41,77,78,107,135,136,161] The disorder is characterized by excessive infiltration of the lamina propria with variable numbers of lymphocytes and plasma cells.[74,110,154] It has been reported as a morphological entity in young adult German Shepherd dogs with relatively mild clinical signs,[69] while a hereditary enteropathy in Basenji dogs may represent the other end of the spectrum and typically presents as a relatively severe protein-losing enteropathy.[22–25,85,97] Nonspecific lymphocytic–plasmacytic intestinal infiltrates may also accompany other chronic enteropathies and have been described in dogs with giardiasis, lymphangiectasia, regional enteritis, lymphosarcoma and canine sprue.[5,42,123]

The cause is unknown, but it has been surmised that lymphocytic–plasmacytic enteritis might represent a common response of the canine intestinal mucosa to more than one provocative agent.[69,107] It is important to remember that this is merely a histological description and therefore may be a single manifestation of diseases with a variety of causes. The infiltrate could represent an immune-mediated inflammatory reaction directed against antigens from the intestinal lumen. Indeed, a clear association between lymphocytic–plasmacytic enteritis and bacterial overgrowth in the proximal small intestine has been demonstrated in a German Shepherd dog,[112] and has been considered above. Furthermore, similar, but less marked, morphological abnormalities have been reported in other German Shepherd dogs with bacterial overgrowth,[11] and a good clinical response to oral tylosin therapy was demonstrated in a series of suspected cases of canine inflammatory bowel disease.[144]

Sensitivity to dietary protein may sometimes also be involved, which is supported by the finding that there can be a clinical response to dietary change in dogs with plasmacytic–lymphocytic colitis.[95,96]

Eosinophilic Enteritis

Eosinophilic enteritis describes a condition in dogs and cats characterized by infiltration of the gastric, intestinal or colonic mucosa or a combination with eosinophils. There is generally minimal or no villus atrophy if the small intestine is involved.[68,71,104,154] Clinically there can be chronic, sometimes bloody, diarrhea and occasionally vomiting, and there may be an associated eosinophilia. This condition has been reported in association with visceral larva migrans,[68] but it seems unlikely that helminthic infections are a common cause of eosinophilic enteritis. A dietary allergen has been suggested to be a more likely possibility,[104,142] but no specific antigen has yet been identified. In cats the condition can also be associated with the hypereosinophilia syndrome.

Intestinal Lymphangiectasia

Lymphangiectasia is an uncommon but important cause of protein-losing enteropathy in dogs.[134] The major morphological abnormality is marked dilatation of intestinal lymphatic vessels in the mucosa, submucosa and serosa and hypoproteinemia is thought to be a consequence of loss of protein-rich fluid from damaged lymphatic vessels. Villus height and enterocyte function may not be affected so that affected animals may not exhibit malabsorption.

Small Intestinal Tumors

Leiomyosarcoma, leiomyoma, lymphosarcoma and adenocarcinoma are the most common tumors of the small intestine in the dog and cat.[55] Signs include vomiting, diarrhea and weight loss and may be related to malabsorption and protein-losing enteropathy secondary to infiltration of the intestinal wall and bacterial overgrowth, or alternatively to obstruction with secondary bacterial overgrowth. Occasionally a tumor will perforate and the patient is presented with signs of an acute abdomen.

The prognosis is guarded with lymphoma because the alimentary type often responds poorly to treatment. Adenocarcinoma also has a very poor prognosis unless diagnosed very early, in which case resection of a localized lesion can produce marked clinical improvement for a period of time; recurrence may be slow, particularly in cats. Leiomyomas and leiomyosarcomas are often slow-growing tumors with an excellent prognosis following complete surgical excision.

Neoplasms have been identified that secrete gastrointestinal regulatory peptides (e.g. vasoactive intestinal peptide, VIP, "vipomas") and cause diarrhea in humans, but with the exception of gastrinomas (see below) they have not been documented in dogs and cats. Clinical signs in these cases usually reflect the secretion of the regulatory peptide rather than the presence of tumor mass itself, although metastasis usually occurs.

Short Bowel Syndrome

This syndrome occurs after extensive surgical resection of the small intestine.[162] The pathophysiology is complex and includes bacterial overgrowth, failure of digestion and absorption, hypersecretion of gastric acid and secondary changes due to malnutrition. The bowel remaining after resection can compensate and increase its absorptive capacity; there can, therefore, be clinical improvement with time. Supportive care will be required in most patients, however, and includes a low fat, highly digestible diet supplemented with medium chain triglyceride oil, vitamin and mineral supplementation (including parenteral replacement in some patients), elemental dietary supplements, pancreatic enzyme supplements, oral antibiotics, frequent small meals and cimetidine to inhibit gastric acid secretion.

Mycotic Enteropathies

These are generally rare, but some may be more commonly encountered in specific geographical regions.[91] *Histoplasma capsulatum* often produces a disseminated infection and in some patients weight loss and diarrhea may be prominent signs. Other findings depend on the organs involved. There may be panhypoproteinemia. Organisms may be seen in various tissues or organs (e.g. blood, respiratory tract) but may also be evident in colonic scrapings or fecal cytological specimens. Treatment with ketoconazole, enconazole, itraconazole and/or amphotericin B may be effective.

Intestinal phycomycosis due to *Pythium* sp. (and others) is fairly common in states bordering the Gulf of Mexico (especially Louisiana). Vomiting and weight loss due to cranial small intestinal involvement are the prominent signs, but more extensive intestinal involvement may occur and result in diarrhea. The intestinal wall may be palpably thickened (abdominal mass apparent in some cases on physical examination). Phycomycosis should be included in the differential diagnosis of dogs with intestinal masses in the Southeastern U.S.A. Diagnosis requires intestinal biopsy and identification of fungal elements; surgical resection of the affected bowel is indicated because successful medical treatment is not reported. The prognosis is poor to hopeless.

Aspergillosis and *Candidiasis* have been reported rarely as gastrointestinal pathogens. Immunosuppression of the host has to be suspected in these patients. Treatment with ketoconazole, itraconazole, nystatin or amphotericin B may be of value.

Bacterial Enteropathies

Bacterial causes of chronic small intestinal disease are not well documented. *Yersinia enterocolitica*, *Campylobacter* sp. and *Salmonella* sp., are generally believed to be causative agents in rare instances of acute diarrhea and have also been cited as causes of chronic diarrhea in some dogs and cats.[131] These cases are of importance because they are potential sources for zoonotic transmission to human beings. *Clostridium difficile* and its cytotoxin have been isolated from dogs with chronic diarrhea but no weight loss; the diarrhea responded to metronidazole therapy, but relapses necessitated repeated therapy. Dogs can also be asymptomatic carriers of *Clostridium difficile*. *Clostridium perfringens* may be a cause of acute hemorrhagic small bowel diarrhea and chronic large bowel diarrhea in dogs.

Treatment of Chronic Enteropathies

Dietary Management

Effective treatment of small intestinal disease depends on the nature of the disorder, although when a specific diagnosis cannot be made treatment may need to be empirical. Diet has an important role to play in the management of small intestinal disease, particularly in dogs. The diverse nature of intestinal diseases illustrated in the previous section serves to emphasize that no single diet is likely to be effective for all small intestinal diseases. For example, interference with the number or functioning of individual enterocytes with subsequent defective absorption of relatively small molecules cannot be compensated by feeding a diet that might be considered highly digestible when fed to a normal animal or an animal with compromized digestive capacity such as occurs in exocrine pancreatic insufficiency. In contrast, dietary sensitivities such as gluten-enteropathy and perhaps Basenji enteropathy, lymphocytic–plasmacytic and eosinophilic enteritis, may need key components of the diet to be excluded. Inflammation may also cause secretory or exudative diarrhea that will not resolve merely by attempting to ensure that dietary components are readily digestible, as for example in some animals with SIBO, lymphocytic–plasmacytic enteritis, canine "sprue" or lymphangiectasia.

Attention should be given to the carbohydrate content of the diet, particularly in conditions that result in reduction in viilus height or functional damage to individual enterocytes because malabsorption of small carbohydrate molecules can lead to osmotic diarrhea. Breath hydrogen testing in dogs has indicated that less carbohydrate remains unabsorbed when rice is fed rather than commercial foods containing cereals,[147,148] suggesting that rice may be a preferable source of carbohydrate in certain dogs with small intestinal disease as long as degradation does not overwhelm absorptive capacity. Certainly, additional monosaccharides or disaccharides, particularly lactose, should be avoided.

Restriction of dietary fat intake may also be valuable in both dogs and cats, particularly when there is malabsorption of fat due to reduced surface area. A low fat diet is also important in SIBO where it can minimize the secretory diarrhea that is a consequence of bacterial metabolism of fatty acids and bile salts. The main problem is that fat normally provides a substantial proportion of daily energy intake, but may be replaced in part by medium chain triglycerides (MCT) that are hydrolyzed more efficiently than long chain triglycerides in fat and reach the circulation through portal rather than lymphatic vessels. A low fat diet and addition of MCT may also help the management of lymphangiectasia.

A good quality protein source is also important because protein malnutrition may have adverse secondary effects for example on the turnover of brush border proteins or on local immunity, perhaps exacerbating intestinal damage. Deficiencies of trace elements and vitamins may also make a major contribution to the clinical outcome of intestinal disease and the potential for appropriate dietary supplementation should not be neglected. A variety of highly digestible or moderate fat diets are commercially available. Specific circumstances, however, still require that clients cook for their pets; suggested recipes are given in Fig. 13.3.

An exclusion diet consisting of a selected protein source should be used as trial therapy in suspected cases of dietary sensitivity and should be fed for a period of at least 3 weeks. Boiled white rice or potato are suitable carbohydrate sources, while lamb or chicken are often used as a protein source, dependent upon the dietary history. Horsemeat, rabbit or venison are acceptable alternatives. Because cereals that contain gluten (wheat, barley, rye, buckwheat and oats) are excluded and replaced with rice as an alternative source of carbohydrate, the fact that this is a gluten-free diet may be the key factor in animals with gluten-sensitive enteropathy. In other cases, the fact that the diet excludes alternative potentially damaging components of a previously fed diet such as soy protein, meat protein and milk products may be important. When it is impossible to find a protein the animal has not been exposed to previously, a protein source that has not been fed recently may be used. Commercial restriction diets may be generally less suitable than home-cooked diets for diagnostic purposes, although they can be good for maintenance.

Medical Management

Specific antibiotic medication for small intestinal disease depends on the condition. Oral broad-spectrum antibiotic therapy with oxytetracycline (10–20 mg/kg body weight tid for 28 days or more) has been used successfully to treat proven cases of SIBO and may be of value in conditions such as Basenji enteropathy and canine sprue. Alternative possibilities are metronidazole (10 mg/kg bid) and tylosin (20 mg/kg tid). Repeated treatment may be necessary.

Oral prednisolone (or prednisone, 0.5–2 mg/kg bid for 2–4 weeks, followed by a reducing dose) may help many animals with chronic enteropathies, for example dogs with lymphocytic–plasmacytic enteritis, Basenji enteropathy, canine sprue, eosinophilic enteritis and lymphangiectasia. In general the dose should be proportional to the severity of the infiltrate. Higher doses (1–2 mg/kg bid) are indicated when the words "moderate" or "severe" appear on the biopsy report. Immunosuppression may be the reason for the beneficial effects, but direct stimulation of the digestive and absorptive functions of enterocytes by glucorticoids may contribute.[17] Betamethasone should probably not be used as an alternative because this longer-acting glucocorticoid may actually cause atrophy.[17] Azathioprine is another important drug (0.1 mg/kg sid) in dogs with severe lymphocytic–plasmacytic enteritis or eosinophilic enteritis, but should be used with caution in cats (0.3 mg/

kg sid). The drug can either be given alone, or in conjunction with oral prednisone. Supplementation with vitamins, particularly oral folic acid (e.g. 5 mg sid for 1–6 months) and parenteral cobalamin (e.g. 5 mg/month for 6 months) may help because deficiencies of these vitamins may contribute to the mucosal lesions.

Inflammatory Disease of the Large Intestine

The Normal Colon

The canine and feline colon are complex and versatile organs, functioning in the maintenance of fluid and electrolyte balance, as a site for nutrient absorption, as a temporary store for excreta and as a reservoir for billions of microorganisms. Electrolyte and water absorption occur primarily in the proximal half and storage in the distal half of the organ. Colitis disrupts these functions in a variety of ways but the almost inevitable sequela is diarrhea.

While colonic fluid and electrolyte transport and patterns of colonic motility are relatively well understood, the complex interrelationships between the colonic mucosa and luminal contents are coming under ever increasing scrutiny. This is because there is an increasing body of knowledge to suggest that short chain fatty acids (SCFA) produced by microbial fermentation of carbohydrate are an important energy source for the colonic mucosa and that their removal causes mucosal atrophy and increased susceptibility to damage.[18] In humans, for example, diversion of colonic flow through an ileostomy can result in colitis that can be resolved by intracolonic infusion of SCFA.[66] Also, rectal infusion of solutions containing propionic and butyric acids has been shown to resolve otherwise intractable ulcerative colitis in humans.[26] In contrast to the major nutritional role played by the colon in herbivores and some omnivores,[18] there is no evidence to suggest that SCFAs produced by colonic microbial fermentation of luminal contents serve as an energy source for the dog and is even less likely in the cat, a true carnivore. Limited intraluminal microbial fermentation of nondigestible fiber and carbohydrate to SCFA does take

CLASSIFICATION OF COLITIS

Etiological

Infectious
Parasitic
  Trichuris vulpis
  Ancylostoma caninum
  Entamoeba histolytica
  Balantidium coli
Bacterial
  Salmonella sp.
  Clostridium sp.
  Campylobacter sp.
Algal
  Prototheca
Fungal
  Histoplasma capsulatum
  Phycomycosis
Traumatic
Uremic
Segmental (secondary to chronic pancreatitis)
Idiopathic acute
Allergic
  Dietary protein
  Bacterial antigens

Histological

Lymphocytic–plasmacytic
Eosinophilic
Histiocytic
Granulomatous

FIG. 13.8 Classification of colitis.

place in the dog, with the extent depending on dietary composition.[2,89]

It is probable that as in other species SCFA influence canine colonic mucosal health.[38] There are no data in support of this hypothesis in the cat.

The Inflamed Colon

Colonic inflammation can result from a variety of causes (Fig. 13.8) and can be either acute or chronic. The cause of acute colitis is usually unknown but the disease is characterized by mucosal infiltration with varying numbers of neutrophils and epithelial disruption and ulceration. Chronic colitis on the other hand, may exhibit some neutrophil infiltration but is more often characterized by the mucosal accumulation of increased numbers of plasma cells and lymphocytes (Fig. 13.9).[111] In both acute and chronic colitis

absorption of water and electrolytes is markedly reduced and motility is disrupted. The result of these changes are small volumes of liquid feces that are passed at a much higher than normal rate.[67,105,115,116]

Prevalence

The prevalence of colitis in the dog is not well documented but appears relatively common based on the number of patients with this disorder referred to a teaching institution.[47] About 30% of dogs with chronic diarrhea referred to the University of Florida Veterinary Medical Teaching Hospital have colitis. In some patients both the small and large intestine are involved in the inflammatory process, with signs of involvement of one of the two organs predominating. The prevalence of colitis in cats appears to be much lower.[41,161]

Mechanisms of Disease

Regardless of the many inciting causes, development of chronic colonic inflammation appears to be due to a defect in mucosal immunoregulation.[163] In human inflammatory bowel disease the data in support of a disruption of normal mucosal immunoregulation are impressive, but there are no equivalent data as far as the authors are aware that apply to the dog or cat. Studies in humans have implicated a number of products of activated immunocytes in the pathogenesis of inflammatory bowel disease, including cytokines, eicosonoids, oxygen radicals, immunoglobulins and proteases.[46,84] It is now apparent that cytokines, protein products of activated immune cells that influence the activity, differentiation and rate of proliferation of other cells, are key mediators in inflammatory bowel disease.[49,117] After initial mucosal damage, submucosal lymphocytes and macrophages are exposed to a variety of luminal antigens of dietary and bacterial origin. The antigens are taken up by macrophages, processed and re-expressed on the cell surface while simultaneously inducing the secretion of a number of cytokines including interleukin-1 which is

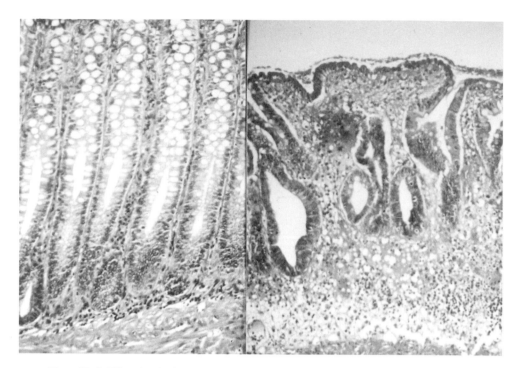

FIG. 13.9 Histological appearance of normal colon (left) compared with lymphocytic–plasmacytic colitis (right) showing a moderate infiltrate of plasma cells and lymphocytes, fusion of crypts and empty goblet cells that are reduced in number.

critical in stimulating T-lymphocytes. A series of multiple complex events amplifies the number of T- and B-lymphocytes with concomitant accumulation of neutrophils and macrophages armed with a variety of destructive enzymes and inflammatory mediators. Net activation or suppression of inflammation is determined by an intricate balance of pro- and antiinflammatory forces, the nature of which remain incompletely understood. The accumulation of inflammatory cytokines, however, probably initiates most of the clinical signs of colitis. Most drugs currently used to treat colitis decrease cytokine production, as well as influencing eicosanoid production and oxygen radical scavenging.[47,117]

Colonic inflammation disrupts the tight junctions between colonic epithelial cells. This reduces the transmucosal electrical potential difference and impairs the ability of the colon to absorb sodium against a strong electrochemical gradient.[44] Some of the cytokines may also stimulate colonic secretion and with the loss of mucosal integrity, protein rich tissue fluid as well as erythrocytes and neutrophils leak into the lumen.[44,86] Tissue architecture is disrupted and mucosal goblet cells are stimulated to secrete large quantities of mucus.

Colitis also disrupts motility. Experimental colitis in dogs inhibits normal segmental contractions and stimulates "giant migrating contractions" (GMC) that sweep down the length of the organ to rapidly eliminate luminal contents and effectively eradicate any storage function.[119] Colitis also abolishes the motor response to a meal in the dog. New ingesta entering the colon also stimulate an excessive number of GMC that are probably related to postprandial discomfort and increased frequency of defecation.[120] The inflamed tissue also becomes more sensitive to stretch. Stretch receptors sensitized to cytokines and distended by normal or even reduced volumes of luminal fluid stimulate peristaltic activity. The inflamed rectum is also more susceptible to stretch with a subsequent augmentation of the urge to defecate.[105]

Classification of Colitis

Colitis can be either acute or chronic with most patients presenting for evaluation of chronic disease. Colitis can be classified on either an etiological or a histological basis (Fig. 13.8), an overlap between the two causing some confusion and disagreement in the literature. Many patients with chronic colitis for example, are said to have "inflammatory bowel disease." In human medicine inflammatory bowel disease refers specifically to either ulcerative colitis or granulomatous ileitis and colitis (Crohn's disease), two enigmatic disorders that may result in some way from defective mucosal immunoregulation. The term inflammatory bowel disease has unfortunately been adopted and expanded in veterinary medicine to encompass any chronic disease of the small or large intestine that is characterized by mucosal inflammatory cell infiltration. Eosinophilic, lymphocytic–plasmacytic, histiocytic and granulomatous colitis are the four best known types of canine inflammatory bowel disease, although other systems of classification have been proposed.[141,143] These histological classifications do little to help our understanding of the disorders. They must suffice however, until more is known about the underlying cause, which could potentially include defective mucosal immunoregulation and sensitivities to luminal antigens.

Clinical Signs

Signs of colitis are classically those of large bowel diarrhea: increased frequency of defecation with small volume liquid feces passed with a sense of urgency, increased fecal mucus and in many patients hematochezia and prolonged tenesmus after defecation.[122,30] Vomiting is reported in about 30% of dogs with colitis[47] and results from vagal stimulation of the emetic center from the inflamed and stretch sensitive colonic wall. The incidence of vomiting in cats with colitis is unknown.

Results of physical examination are usually normal but depend to some extent at least,

---

DIFFERENTIAL DIAGNOSIS OF COLITIS IN THE DOG

Tumors
  Lymphosarcoma
  Adenocarcinoma
  Leiomyosarcoma
  Colonic polyp
Ileocolic intussusception
Cecal inversion
Irritable colon syndrome

FIG. 13.10 Differential diagnosis of colitis in the dog.

on the duration and severity of colonic disease. Loss of body weight and condition are uncommon except in the rare colonic fungal or algal infection and in histiocytic ulcerative colitis of Boxers and related breeds.

The differential diagnosis includes other types of colonic disease (Fig. 13.10), with colonic tumors, cecal inversion and the putative stress-induced "irritable colon" predominating.

Approach to the Patient with Colitis

A logical approach to the patient with signs of large intestinal diarrhea and tenesmus must first include the elimination of infectious agents by appropriate fecal flotation, smear and culture techniques. If the results of these tests are negative, symptomatic dietary or antiinflammatory therapy may be attempted. A thorough endoscopic evaluation of the colon to obtain appropriate colonic mucosal biopsy specimens is, however, preferable to symptomatic therapy. Flexible fiber-optic endoscopy allows examination of the entire organ: however, colonic preparation for flexible endoscopy is time consuming and because most colonic disease is diffuse, the colon can be readily examined and a representative biopsy taken using a rigid colonoscope for a much lower investment of time, money and effort.[30]

Results of the mucosal biopsy will indicate the type and severity of colitis which allows the clinician to give an accurate prognosis and prescribe appropriate treatment. A biopsy report of normal or hyperplastic colonic mucosa in a patient with signs of large intestinal disease suggests the diagnosis of irritable colon, still currently a diagnosis of exclusion. Before diagnosing irritable colon and committing to long-term symptomatic therapy, however, one should be certain that all drug and dietary sensitivities have been excluded. Repeat biopsies are also indicated because there is not always a good correlation between clinical signs, the appearance of the mucosa at endoscopy and the pathologist's interpretation of the biopsy specimen.[110]

Treatment of Colitis

Treatment of colitis depends on the specific diagnosis but usually includes a combination of symptomatic, dietary and antiinflammatory therapies.

Symptomatic Management

Symptomatic treatment includes the use of drugs that decrease fecal water and stimulate colonic segmental contractile activity. The most appropriate of these is loperamide (0.1–0.2 mg/kg bid–qid PO).[79] Unlike other narcotic motility modifiers loperamide does stimulate colonic segmental activity, thereby slowing the speed of colonic transit. It also increases anal sphincter tone and rectal function, decreases colonic secretion and may enhance salt and water absorption.[50,106,118]

Some clinicians recommend the use of antibiotics, such as metronidazole or tylosin, in the treatment of colitis but their routine use remains controversial. Metronidazole for example, has no effect on the progress of human inflammatory bowel disease.[106] It does however, decrease fecal anaerobe concentration while having no effect on the metabolism of salazopyrine.[82,121] Because some cases of large bowel diarrhea in dogs may be associated with clostridial infection or overgrowth,[19,37,138] metronidazole may be warranted at least on a trial basis in some patients with large bowel diarrhea. Whether the drug should be prescribed before antiinflammatory drugs or in conjunction with them is a matter of clinical judgement. Ideally it should be used only if an anaerobic fecal culture in a patient with large bowel diarrhea has revealed a high

concentration of clostridia or a test for fecal clostridial toxin is positive.

Tylosin has been recommended for the treatment of inflammatory bowel disease in dogs[144] but the evidence of any beneficial effect in colitis is unconvincing. Any positive effect is probably due to its effect on small intestinal bacterial overgrowth.

Dietary Management

Dietary therapy of colitis is emerging as one of the most important approaches to treatment and can be divided into three categories: (1) fasting, (2) feeding an exclusion diet and (3) fiber supplementation.

Fasting: Placing the bowel in a state of physiological rest by withholding food for 24–48 hr has long been advocated as part of symptomatic therapy for acute diarrheal disease. This may help not only acute small intestinal disease but also acute colitis. In a controlled study in human patients fasting did not effect the outcome in severe ulcerative colitis.[87]

Fasting for any longer than is required for bowel preparation for colonoscopy is contraindicated in chronic disease, particularly given the need for protein for tissue repair. Enteral feeding with refined formula diets alone or in combination with parenteral nutrition has been of clear benefit in humans with severe inflammatory bowel disease.[137] This approach is likely to be more limited in canine and feline colitis, not only due to cost but also because the degree of protein loss and catabolism is typically less than in humans with inflammatory bowel disease.

Feeding an exclusion diet: Because the apparent role of food sensitivity as a primary cause, or at least a perpetuating factor in the maintenance of canine and feline colitis, is becoming increasingly apparent, feeding the patient a protein to which it has not been previously exposed may be a critical aspect in the treatment of many animals with colitis.

There is now some indirect evidence that some cases of lymphocytic–plasmacytic colitis in the dog may be due to a sensitivity to dietary protein because signs improve if the diet is changed to contain a protein to which the animal has not been previously exposed.[96] Dogs and cats with chronic idiopathic colitis that were fasted for 2 days to allow bowel preparation for colonoscopy and biopsy and that were then fed a diet of cottage cheese and rice[96] or lamb and rice[95] had a complete resolution of signs within 2 weeks. Challenge with the original diet resulted in recurrence of diarrhea in 80% of dogs, strongly suggesting that dietary sensitivity may play a role in the pathogenesis of the disease.[96] This experience, as well as that of others,[83,108] has resulted in a change in the general management of chronic colitis to include not only antiinflammatory drugs but also a diet, where possible, containing proteins to which the animal has not been previously exposed. Home-prepared diets containing cottage cheese and rice or lamb and rice (Fig. 13.3) and various commercial diets containing rice with chicken, mutton or lamb, venison and rabbit that are now available in some countries may prove beneficial. Signs of large bowel diarrhea however, do not always resolve immediately with dietary therapy alone and many clinicians find that client satisfaction is enhanced if antiinflammatory therapy with salazopyrine (sulfasalazine) is initiated concurrently (see below).

Fiber supplementation: Supplementation of the diet with fiber in the form of bran or cellulose has been recommended as symptomatic therapy for large bowel diarrhea.[30] Poorly fermented fiber such as α-cellulose increases fecal bulk, stretches colonic smooth muscle and improves contractility.[33] It also binds fecal water to produce formed feces and thus improves client satisfaction.[156] Supplementation with fiber is seldom sufficient in itself to bring about resolution of signs but is useful as adjunctive therapy. It is possible, however, that this approach may change as our knowledge of colonic function improves. There are many different types of dietary fiber all of which are fermented to a different degree by the colonic microbial flora and probably have different effects on colonic function. It is possible for example, that future dietary treatment of colitis may include fermentable fibers from which colonic bacteria produce SCFAs to improve colonic mucosal nutrition.

Immunosuppressive Therapy

Despite our knowledge of symptomatic and nutritional therapy for colitis, the basis of treatment still involves the judicious use of drugs that modulate mucosal immune function. The most important and widely used of these are salazopyrine (sulfasalazine), prednisone (or prednisolone) and azathioprine. However, drugs such as methotrexate, cyclosporin and some of the newer drugs that specifically influence individual cytokine function are now being used in the treatment of human inflammatory bowel disease.[53] It is possible that such drugs may also be useful in the treatment of canine colitis.[65]

Salazopyrine (sulfasalazine) (25–40 mg/kg tid for 3–4 weeks) is the drug of choice for treatment of chronic lymphocytic–plasmacytic colitis, the most common type of idiopathic colitis. The mechanism of action of the drug, a combination of sulfapyridine and 5-aminosalicylate joined by an azo bond, remains controversial. The combination of the two drugs is merely a delivery mechanism, the result of a serendipitous discovery by Swedish pharmacologists in the 1940s during an experiment to find a drug that could be used to treat rheumatoid arthritis.[73] The azo bond prevents absorption of the drug in the small intestine allowing it to be delivered unchanged to the colon where it is split by colonic bacteria to release sulfapyridine and the active ingredient 5-amino-salicylate.[57] The latter may have a variety of beneficial effects because it has been shown to decrease interleukin-1 production, inhibit 5-lipoxygenase, reduce platelet activating factor, act as a free radical scavenger and inhibit mucosal mast cell histamine release and prostaglandin production.[39,45,52,56,94]

Most patients respond either to sulfasalazine alone or in combination with dietary change within 1–4 weeks of the onset of treatment. The diagnosis, however, must be reexamined if signs fail to improve after 4 weeks of treatment or if they recur when the drug is withdrawn. Long-term therapy is contraindicated because sulfasalazine causes keratoconjunctivitis sicca (KCS) induced by a sensitivity to sulfapyridine.[114] A new drug, a combination of two molecules of 5-amino-salicylate (olsalazine) has recently been introduced into human medicine in an attempt to relieve some of the side-effects associated with sulfasalazine therapy. Veterinary experience with the drug is limited but dogs can apparently also develop KCS in response to a sensitivity to 5-amino-salicylate alone.[3]

Prednisone has been widely used in the treatment of lymphocytic–plasmacytic enteritis in the dog but does not appear to be as effective as sulfasalazine in the treatment of lymphocytic–plasmacytic colitis. Indeed, the indication for corticosteroids in the treatment of canine colitis has not been clarified and remains empirical. Some dogs with colitis reportedly deteriorate when treated with corticosteroids, while others appear to benefit.[128] Because chronic colitis is an inflammatory disease of unknown etiology, corticosteroid use in conjunction with sulfasalazine is occasionally beneficial in patients unresponsive to more conventional treatment.

Eosinophilic colitis, however, seldom responds to dietary change alone and immunosuppressive doses of prednisone (or prednisolone, 2–4 mg/kg sid for 2 weeks and then tapered over 6–10 weeks) is the treatment of choice for this disease. If mucosal changes are severe, prednisone may be combined with azathioprine (1.0 mg/kg sid for 2 weeks). Occasional 7–14 day courses of azathioprine alone can also be used to control signs in patients with either lymphocytic–plasmacytic or eosinophilic colitis that appear resistant to other forms of treatment.

References

1. Anderson, N. V. and Low, D. G. (1965) Juvenile atrophy of the canine pancreas. *Animal Hospital*, **1**, 101–109.
2. Banta, C. A., Clemens, E. J., Krinsky, M. M. and Sheffey, B. E. (1979) Sites of organic acid production and patterns of digesta movement in the gastrointestinal tract of dogs. *Journal of Nutrition*, **109**, 1592–1600.
3. Barnett, K. C. (1986) Keratoconjunctivitis sicca and the treatment of canine colitis (letter to the editor). *Veterinary Record*, **117**, 263.
4. Batt, R. M., Barnes, A., Rutgers, H. C. and Carter, S. D. (1991) Relative IgA deficiency and small intestinal bacterial overgrowth in German Shepherd dogs. *Research in Veterinary Science*, **50**, 106–111.
5. Batt, R. M., Bush, B. M. and Peters, T. J. (1983) Subcellular biochemical studies of a naturally occurring enteropathy in the dog resembling chronic tropical sprue in human beings. *American Journal of Veterinary Research*, **44**, 1492–1496.
6. Batt, R. M., Carter, M. W. and McLean, L. (1984) Morphological and biochemical studies of a naturally occurring enteropathy in the Irish setter dog: a comparison with coeliac disease in man. *Research in Veterinary Science*, **37**, 339–346.
7. Batt, K. M., Carter, M. W. and Peters, T. J. (1984) Biochemical changes in the jejunal mucosa of dogs with a naturally occurring enteropathy associated with bacterial overgrowth. *Gut*, **25**, 816–823.
8. Batt, R. M. and Horadagoda, N. U. (1989) Gastric and pancreatic intrinsic factor mediated absorption of cobalamin in the dog. *American Journal of Physiology*, **257**, G344–G349.
9. Batt, R. M., Horadagoda, N. U., Mclean, L., Morton, D. B. and Simpson, K. W. (1989) Identification and characterization of a pancreatic intrinsic factor in the dog. *American Journal of Physiology*, **256**, G517–G523.
10. Batt, R. M. and Mann, L. C. (1981) Specificity of the BT-PABA test for the diagnosis of exocrine pancreatic insufficiency in the dog. *Veterinary Record*, **108**, 303–307.
11. Batt, R. M. and Mclean, L. (1987) Comparison of the biochemical changes in the jejunal mucosa of dogs with aerobic and anaerobic bacterial overgrowth. *Gastroenterology*, **93**, 986–993.
12. Batt, R. M., Mclean, L. and Carter, M. W. (1987) Sequential morphologic and biochemical studies of naturally-occurring wheat-sensitive enteropathy in Irish setter dogs. *Digestive Disease and Science*, **32**, 184–194.
13. Batt, R. M., Mclean, L. and Riley, J. E. (1988) Response of the jejunal mucosa of dogs with aerobic and anaerobic bacterial overgrowth to antibiotic therapy. *Gut*, **29**, 473–482.
14. Batt, R. M., McLean, L., Rutgers, H. C. and Hall, E. J. (1991) Validation of a radioassay for the determination of serum folate and cobalamin concentrations in dogs. *Journal of Small Animal Practice*, **32**, 221–224.
15. Batt, R. M. and Morgan, J. O. (1982) Role of serum folate and vitamin B_{12} concentrations in the differentiation of small intestinal abnormalities

in the dog. *Research in Veterinary Science*, **32**, 17–22.

16. Batt, R. M., Needham, J. R. and Carter, M. W. (1983) Bacterial overgrowth associated with a naturally occurring enteropathy in the German Shepherd dog. *Research in Veterinary Science*, **35**, 42–46.

17. Batt, R. M. and Scott, J. (1982) Response of the small intestinal mucosa to oral glucocorticoids. *Scandinavian Journal of Gastroenterology*, **17** (Suppl. 74), 75–88.

18. Bergman, E. N. (1990) Energy contribution of volatile fatty acids from the gastrointestinal tract in various species. *Physiological Reviews*, **70**, 567–590.

19. Berry, A. P. and Levitt, P. N. (1986) Chronic diarrhea in dogs associated with *Clostridium difficile* infection. *Veterinary Record*, **118**, 102–103.

20. Bjarnason, I., Peters, T. J. and Veall, N. (1983) A persistent defect in intestinal permeability in coeliac disease demonstrated by a ^{51}Cr-labelled EDTA absorption test. *Lancet*, **i**, 323–325.

21. Borgstrom, A. and Ohlsson, K. (1980) Immunoreactive trypsins in sera from dogs before and after induction of experimental pancreatitis. *Hoppe–Seyler's Zeitschrift für Physiologische Chemie*, **361**, 625–631.

22. Breitschwerdt, E. B., Barta, O., Waltman, C., Hubbert, N. L., Pourciau, S. S. and Liu, W. (1983) Serum proteins in healthy Basenjis and Basenjis with chronic diarrhea. *American Journal of Veterinary Research*, **44**, 326–328.

23. Breitschwerdt, E. B., Halliwell, W. H., Foley, C. W., Stark, D. R. and Corwin, L. A. (1980) A hereditary diarrhetic syndrome in the Basenji characterized by malabsorption, protein losing enteropathy and hypergammaglobulinaemia. *Journal of the American Animal Hospital Association*, **16**, 551–560.

24. Breitschwerdt, E. B., Ochoa, R., Barta, M., Barta, O., McLure, J. and Waltman, C. (1984) Clinical and laboratory characterization of Basenjis with immunoproliferative small intestinal disease. *American Journal of Veterinary Research*, **45**, 267–273.

25. Breitschwerdt, E. B., Waltman, C., Hagstad, H. V., Ochoa, R., McLure, J. and Barta, O. (1982) Clinical and epidemiologic characterization of a diarrheal syndrome in Basenji dogs. *Journal of the American Veterinary Medical Association*, **180**, 914–920.

26. Breuer, R. I., Buto, S. K., Christ, M. L. *et al.* (1991) Rectal irrigation with short chain fatty acids for distal ulcerative colitis. *Digestive Diseases and Science*, **36**, 185–187.

27. Brobst, D., Ferguson, A. B. and Carter, J. M. (1970) Evaluation of serum amylase and lipase activity in experimentally induced pancreatitis in the dog. *Journal of the American Veterinary Medical Association*, **157**, 1697–1702.

28. Burns, J. and Fox, S. M. (1986) The use of a barium meal to evaluate total gastric emptying time in the dog. *Veterinary Radiology*, **27**, 169–172.

29. Burrows, C. F. (1983) Metoclopramide. *Journal of the American Veterinary Medicine Association*, **183**, 1341–1343.

30. Burrows, C. F. (1986) Medical diseases of the colon. In *Canine and Feline Gastroenterology*. Eds. B. Jones and W. D. Liska. pp. 221–256. Philadelphia: W. B. Saunders.

31. Burrows, C. F. and Ignaszewski, L. (1990) Canine gastric dilatation–volvulus. *Journal of Small Animal Practice*, **31**, 495–501.

32. Burrows, C. F., Kronfeld, D. S., Banta, C. A. and Merritt, A. M. (1982) Effect of fibre on digestibility and transit time in dogs. *Journal of Nutrition*, **112**, 1726–1732.

33. Burrows, C. F. and Merritt, A. M. (1983) Influence of cellulose on myoelectric activity of proximal canine colon. *American Journal of Physiology*, **245**, G301–G306.

34. Burrows, C. F. and Merritt, A. M. (1992) Assessment of gastrointestinal function. In *Veterinary Gastroenterology*. Ed. N. V. Anderson. 2nd edn. pp. 16–42. Philadelphia: Lea and Febiger.

35. Burrows, C. F., Merritt, A. M. and Chiapella, A. M. (1980) Determination of fecal fat and trypsin output in the evaluation of chronic canine diarrhea. *Journal of the American Veterinary Medical Association*, **177**, 1128–1131.

36. Campbell, R. S. F., Brobst, D. and Bisgard, G. (1968) Intestinal lymphangiectasia in a dog. *Journal of the American Veterinary Medical Association*, **153**, 1050–1054.

37. Carman, R. J. and Lewis, J. C. M. (1983) Recurrent diarrhea in a dog associated with *Clostridium perfringens*, type A. *Veterinary Record*, **112**, 342–343.

38. Clemens, E. T. and Dobesh, G. D. (1988) Nutritional impact of the canine colonic microstructure and function. *Nutrition Research*, **8**, 625–633.

39. Cominelli, F., Zipser, R. D. and Dinarello, C. A. (1989) Sulphasalazine inhibits cytokine production in human mononuclear cells: a novel anti-inflammatory mechanism. *Gastroenterology*, **96** A96, (Abstr.).

40. Crow, S. E. (1985) Tumors of the alimentary tract. *Veterinary Clinics of North America*, **15**, 577–596.

41. Dennis, J. S., Kruger, J. M. and Mullaney, T. P. (1992) Lymphocytic/plasmacytic gastroenteritis in cats: 14 cases (1985–1990). *Journal of the American Veterinary Medical Association*, **200**, 1712–1718.

42. Dibartola, S. P., Rogers, W. A., Boyce, J. T. and Grimm, J. P. (1982) Regional enteritis in two dogs.

Journal of the American Veterinary Medical Association, **181**, 904–908.

43. Duffell, S. J. (1975) Some aspects of pancreatic disease in the cat. *Journal of Small Animal Practice,* **16**, 365–374.

44. Edmunds, C. J. and Pilcher, D. (1973) Electrical potential difference and sodium and potassium fluxes across rectal mucosa in ulcerative colitis. *Gut,* **14**, 784–789.

45. Eliakim, R., Karmeli, F., Razin, E. and Rachmite-witz, D. (1988) Role of platelet activating factor in ulcerative colitis: inhibition by sulphasalazine and prednisolone. *Gastroenterology,* **95**, 1167–1173.

46. Elson, C. O. (1988) The immunology of inflammatory bowel disease. In *Inflammatory Bowel Disease.* Eds. J. B. Kirsner and R. G. Shorter. pp. 97–164. Philadelphia: Lea and Febiger.

47. Ewing, G. O. and Gomez, J. A. (1973) Canine ulcerative colitis. *Journal of the American Animal Hospital Association,* **9**, 395–406.

48. Finco, D. R., Duncan, J. R., Schall, W. D., Hooper, B. E., Chandler, F. W. and Keating, K. A. (1973) Chronic enteric disease and hypopro-teinemia in 9 dogs. *Journal of the American Veterinary Medical Association,* **163**, 262–271.

49. Fiocchi, C. (1989) Lymphokines and the intestinal immune response. Role in inflammatory bowel disease. *Immunology Investigations,* **18**, 91–102.

50. Fioramonti, J., Fargeas, M. J. and Bueno, L. (1987) Stimulation of gastrointestinal motility by lopera-mide in dogs. *Digestive Diseases and Science,* **32**, 641–646.

51. Flesj, K. and Yri, T. (1977) Protein-losing enterop-athy in the Lundehund. *Journal of Small Animal Practice,* **18**, 11–23.

52. Fox, C., Moore, W. and Lichtenstein, L. (1991) Modulating of mediator release from human intestinal mast cells by sulfasalazine and 5-aminosalicylic acid. *Digestive Diseases and Science,* **36**, 179–184.

53. Fretland D. J., Widomski D., Bie-shung T. *et al.* (1990) Effect of a leukotriene B4 receptor antagonist SC-4190 on colonic inflammation in rat, guinea pig and rabbit. *Journal of Pharmacology and Experimental Therapeutics,* **255**, 572–576.

54. Freudiger, U. and Bigler, B. (1977) The diagnosis of chronic exocrine pancreatic insufficiency by the PABA test. *Kleintier praxis,* **22**, 73–79.

55. Gibbs, C. and Pearson, H. (1986) Localized tumours of the canine small intestine: a report of twenty cases. *Journal of Small Animal Practice,* **27**, 507–519.

56. Gionchetti, P., Guarneri, C., Campieri, M. *et al.* (1991) Scavenger effect of sulfasalazine, 5-aminosalicylic acid and olsalazine on superoxide radical generation. *Digestive Diseases and Science,* **36**, 174–178.

57. Goldman, P. and Peppercorn, M. A. (1975) Sulfasalazine. *New England Journal of Medicine,* **293**, 20–25.

58. Hall, E. J. and Batt, R. M. (1990) Enhanced intestinal permeability to ^{51}Cr-labeled EDTA in dogs with small intestinal disease. *Journal of the American Veterinary Medical Association,* **196**, 91–95.

59. Hall, E. J. and Batt, R. M. (1990) Development of a wheat-sensitive enteropathy in Irish Setters: morphologic changes. *American Journal of Veterinary Research,* **51**, 978–982.

60. Hall, E. J. and Batt, R. M. (1990) Development of a wheat-sensitive enteropathy in Irish Setters: biochemical changes. *American Journal of Veterinary Research,* **51**, 983–989.

61. Hall, E. J. and Batt, R. M. (1991) Abnormal permeability precedes the development of a gluten sensitive enteropathy in Irish Setter dogs. *Gut,* **32**, 749–753.

62. Hall, E. J. and Batt, R. M. (1991) Differential sugar absorption for the assessment of canine intestinal permeability: the cellobiose/mannitol test in gluten-sensitive enteropathy of Irish Setters. *Research in Veterinary Science,* **51**, 83–87.

63. Hall, E. J. and Batt, R. M. (1992) Dietary modulation of gluten sensitivity in a naturally occurring enteropathy of Irish Setter dogs. *Gut,* **33**, 198–205.

64. Hall, E. J., Bond, P. M., Mclean, C., Batt, R. M. and Mclean, L. (1991) A survey of the diagnosis and treatment of canine exocrine pancreatic insufficiency. *Journal of Small Animal Practice,* **32**, 613–619.

65. Harig, J. M. and Soergel, K. H. (1992) Short chain fatty acids in inflammatory bowel disease. *Progress in Inflammatory Bowel Disease,* **13**, 6–9.

66. Harig, J. M., Soergel, K. H., Komorowski, R. A. and Wood, C. M. (1989) Treatment of diversion colitis with short chain fatty acid irrigation. *New England Journal of Medicine,* **320**, 23–28.

67. Hawker, P. C., Mckay, J. S. and Turnberg, L. A. (1990) Electrolyte transport across colonic mucosa from patients with inflammatory bowel disease. *Gastroenterology,* **79**, 508–511.

68. Hayden, D. W. and Van Kruiningen, H. J. (1973) Eosinophilic gastroenteritis in German Shepherd dogs and its relationship to visceral larva migrans. *Journal of the American Veterinary Medical Association,* **162**, 379–384.

69. Hayden, D. W. and Van Kruiningen, H. J. (1982) Lymphocytic-plasmacytic enteritis in German Shepherd dogs. *Journal of the American Animal Hospital Association,* **18**, 89–96.

70. Hill, F. W. G. (1972) Malabsorption syndrome in

the dog: a study of thirty-eight cases. *Journal of Small Animal Practice,* **13**, 575–594.

71. Hill, F. W. G. and Kelly, D. F. (1974) Naturally occurring intestinal malabsorption in the dog. *American Journal of Digestive Disease,* **19**, 649–665.

72. Hill, F. W. G., Osborne, A. D. and Kidder, D. E. (1971) Pancreatic degenerative atrophy in dogs. *Journal of Comparative Pathology,* **81**, 321–330.

73. Hoult, J. R. S. (1986) Pharmacological and biochemical actions of sulfasalazine. *Dr.ugs,* **32** (Suppl. 7), 18–26.

74. Jacobs, G., Colins-Kelley, L., Lappin, M. *et al.* (1990) Lymphocytic–plasmacytic enteritis in 24 dogs. *Journal of Veterinary Internal Medicine,* **4**, 45–53.

75. Jacobs, R. M., Hall, R. L. and Rogers, W. A. (1982) Isoamylases in clinically normal and diseased dogs. *Veterinary Clinical Pathology,* **11**, 26–32.

76. Jacobs, R. M., Murtaugh, R. J. and Dehoff, W. D. (1985) Review of the clinicopathological findings of acute pancreatitis in the dog: use of an experimental model. *Journal of the American Animal Hospital Association,* **21**, 795–800.

77. Jergens, A. E. (1992) Feline inflammatory bowel disease. *Compendium of Continuing Education for the Practising Veterinarian,* **14**, 509–520.

78. Jergens, A. E., Moore, F. M., Haynes, J. S. and Miles, K. G. (1992) Idiopathic inflammatory bowel disease in dogs and cats: 84 cases (1987–1990). *Journal of the American Veterinary Medical Association,* **201**, 1603–1608.

79. Johnson, S. (1989) Loperamide: a novel antidiarrhoeal drug. *Compendium of Continuing Education for the Practising Veterinarian,* **11**, 1373–1375.

80. Kaneko, J. J., Moulton, J. E., Brodey, R. S. and Perryman, V. D. (1965) Malabsorption syndrome resembling nontropical sprue in dogs. *Journal of the American Veterinary Medical Association,* **146**, 463–473.

81. King, C. E. and Toskes, P. P. (1979) Small intestine bacterial overgrowth. *Gastroenterology,* **76**, 1035–1055.

82. Krook, A., Daniellson, D., Kiellander, J. and Jarnerot, G. (1981) The effect of metronidazole and sulfasalazine on the fecal flora in patients with Crohn's disease. *Scandinavian Journal of Gastroenterology,* **16**, 183–192.

83. Leib, M. S. (1991) Dietary management of chronic large bowel diarrhea in dogs. *Proceedings of the Eastern States Veterinary Conference,* **5**, 147–148.

84. Macdermott, P. R. and Stenson, W. F. (1988) Alterations in the immune system in ulcerative colitis and Crohn's disease. *Advances in Immunology,* **42**, 285–328.

85. Maclachlan, N. J., Breitschwerdt, E. B., Chambers, J. M., Argenzio, R. A. and De Buysscher, E. V. (1988) Gastroenteritis of Basenji dogs. *Veterinary Pathology,* **25**, 36–41.

86. Madara, J. L. (1988) γ-IFN enhances intestinal epithelial permeability by altering tight junctions. *Gastroenterology,* **94** (Abstr.), A276.

87. McIntyre, P. B., Powell–Tuck, J., Wood, S. R. *et al.* (1986) Controlled trial of bowel rest in the treatment of severe acute colitis. *Gut,* **27**, 481–485.

88. Meiderer, H. W. (1986) The gastric mucosal barrier. *Hepatogastroenterology,* **33**, 88–91.

89. Meyer, H., Schünemann, C., Elbers, H. and Junker, S. (1987) Precaecal and postileal protein digestion and intestinal urea conversion in dogs. In *Proceedings of the International Hanover Symposium.* Ed. A. T. B. Edney. pp. 27–30. Waltham Centre for Pet Nutrition.

90. Mia, A. S., Koger, H. D. and Tierney, M. M. (1978) Serum values of amylase and pancreatic lipase in healthy mature dogs and dogs with experimental pancreatitis. *American Journal of Veterinary Research,* **39**, 965–969.

91. Miller, R. I., Qualls, C. W. and Turnwald, G. H. (1983) Gastrointestinal phycomycosis in a dog. *Journal of the American Veterinary Medicine Association,* **182**, 1245–1248.

92. Moore, R. P. (1983) Feline eosinophilic enteritis. In *Current Veterinary Therapy VIII.* Ed. R. W. Kirk. pp. 791–793. Philadelphia: W. B. Saunders.

93. Murtaugh, R. J. and Jacobs, R. J. (1985) Serum amylase and isoamylase and their origins in healthy dogs and dogs with experimentally induced acute pancreatitis. *American Journal of Veterinary Research,* **46**, 743–747.

94. Neilson, O. H., Bukhane, K., Elmgreen, J. and Ahnfelt–Rhone, I. (1987) Inhibition of 5-lipoxygenase pathway of arachidonic acid metabolism in human neutrophils by sulfasalazine and 5-aminosalicylic acid. *Digestive Diseases and Science,* **32**, 577–582.

95. Nelson, R. W., Dimperio, M. E. and Long, G. G. (1984) Lymphocytic–plasmacytic colitis in the cat. *Journal of the American Veterinary Medical Association,* **184**, 1133–1135.

96. Nelson, R. W., Stookey, L. J. and Kazacos, E. (1988) Nutritional management of idiopathic chronic colitis in the dog. *Journal of Veterinary Internal Medicine,* **2**, 133–137.

97. Ochoa, R., Breitschwerdt, E. B. and Lincoln, K. L. (1984) Immunoproliferative small intestinal disease (IPSID) in Basenji dogs: morphological observations. *American Journal of Veterinary Research,* **45**, 482–490.

98. Odonkor, P., Mowat, C. and Hinal, H. S. (1981) Prevention of sepsis induced gastric lesions in dogs by cimetidine via inhibition of gastric secretion and by prostaglandin via cytoprotection. *Gastroenterology,* **80**, 375–379.

99. Parent, J. (1982) Effects of dexamethasone on pancreatic tissue and on serum amylase and lipase activities in dogs. *Journal of the American Veterinary Medical Association*, **180**, 743–746.

100. Pidgeon, G. (1982) Effect of diet on pancreatic insufficiency in dogs. *Journal of the American Veterinary Medical Association*, **181**, 232–235.

101. Pidgeon, G. (1987) Exocrine pancreatic disease in the dog and cat. Part 2: Exocrine pancreatic insufficiency. *Canine Practice*, **14**, 31–35.

102. Polzin, D. J., Osborne, C. A., Stevens, J. B. and Hayden, D. W. (1983) Serum amylase and lipase activities in dogs with chronic primary renal failure. *American Journal of Veterinary Research*, **44**, 404–410.

103. Porteous, J. R. *et al.* (1974) Induction of auto-allergic gastritis in dogs. *Journal of Pathology*, **112**, 138–146.

104. Quigley, P. J. and Henry, K. (1983) Eosinophilic enteritis in the dog: a case report with a brief review of the literature. *Journal of Comparative Pathology*, **91**, 387–392.

105. Rao, S. S. G. and Read, N. W. (1990) Gastrointestinal motility in patients with ulcerative colitis. *Scandinavian Journal of Gastroenterology*, **25** (Suppl. 172), 22–28.

106. Read, M., Read, N. W., Barber, C. D. *et al.* (1982) Effects of loperamide on anal sphincter function in patients complaining of chronic diarrhea with fecal incontinence and urgency. *Digestive Diseases and Science*, **27**, 807–814.

107. Richter, K. P. (1992) Lymphocytic plasmacytic enterocolitis in dogs. *Seminars in Veterinary Medicine and Surgery*, **7**, 134–144.

108. Ridgeway, M. (1984) Management of chronic colitis in the dog. *Journal of the American Veterinary Medical Association*, **185**, 804–806.

109. Rogers, W. A., Stradley, R. P., Sherding, R. G. *et al.* (1980) Simultaneous evaluation of pancreatic exocrine function and intestinal absorptive function in dogs with chronic diarrhea. *Journal of the American Veterinary Medical Association*, **177**, 1128–1131.

110. Roth, L., Leib, M. S., Davenport, D. J. and Monroe, W. E. (1990) Comparisons between endoscopic and histologic evaluation of the gastrointestinal tract in dogs and cats. *Journal of the American Veterinary Medical Association*, **196**, 635–638.

111. Roth, L., Walton, A. M., Leib, M. S. and Burrows, C. F. (1990) A grading system for lymphocytic plasmacytic colitis in dogs. *Journal of Veterinary Diagnostic Investigations*, **2**, 257–262.

112. Rutgers, H. C., Batt, R. M. and Kelly, D. F. (1988) Lymphocytic–plasmacytic enteritis asso-ciated with bacterial overgrowth in a dog. *Journal of the American Veterinary Medical Association*, **192**, 1739–1742.

113. Rutgers, H. C., Hall, E. J., Sørensen, S. H., Proud, J. and Batt, R. M. (1992) Differential sugar absorption for the assessment of diet-sensitive intestinal disease in dogs. *Proceedings of the Annual Congress of the British Small Animal Veterinary Association*. Birmingham, p. 184.

114. Sansom, J., Barnett, K. and Long, R. (1985) Keratoconjunctivitis sicca in the dog associated with the administration of salicylazosulfapyridine (sulfasala-zine). *Veterinary Record*, **116**, 391–393.

115. Sarna, S. K. (1991) Physiology and pathophysiology of colonic motor activity. *Digestive Diseases and Science* (Part one of two), **56**, 827–862.

116. Sarna, S. K. (1991) Physiology and pathophysiology of colonic motor activity. *Digestive Diseases and Science*, **56**, 998–1018 (Part two of two).

117. Sartor, R. B. (1991) Cytokines in inflammatory bowel disease. *Progress in Inflammatory Bowel Disease*, **12**, 5–8.

118. Schiller, L. R., Santa Ana, C., Morawski, S. G. and Fordtran, J. S. (1984) Mechanism of the antidiarrheal effect of loperamide. *Gastroenterology*, **86**, 1475–1480.

119. Sethi, A. K. and Sarna, S. K. (1991) Colonic motor activity in acute colitis in conscious dogs. *Gastroenterology*, **100**, 954–963.

120. Sethi, A. K. and Sarna S. K. (1991) Colonic motor response to a meal in acute colitis. *Gastroenterology*, **101**, 1537–1546.

121. Shaffer, J. L., Kershaw, A. and Houston, J. B. (1986) Disposition of metronidazole and its effect on sulphasalazine metabolism in patients with inflammatory bowel disease. *British Journal of Clinical Pharmacology*, **21**, 431–435.

122. Sherding, R. G. (1980) Canine large bowel diarrhoea. *Compendium of Continuing Education for the Practising Veterinarian*, **2**, 279–288.

123. Sherding, R. G. (1989) In *Textbook of Veterinary Internal Medicine*. Ed. S. J. Ettinger. 3rd edn. pp. 1323–1396. Philadelphia: W. B. Saunders.

124. Sheridan, V. (1975) Pancreatic deficiency in the cat. *Veterinary Record*, **96**, 229.

125. Simpson, J. W. and Doxey, D. L. (1988) Evaluation of faecal analysis as an aid to the detection of exocrine pancreatic insufficiency. *British Veterinary Journal*. **144(2)**, 174–178.

126. Simpson, K. W., Batt, R. M., McLean, L. and Morton, D. B. (1989) Circulating concentrations of trypsin-like immunoreactivity and activities of lipase and amylase after pancreatic duct ligation in dogs. *American Journal of Veterinary Research*, **50**, 629–632.

127. Simpson, K. W., Morton, D. B. and Batt, R. M. (1989) Effect of exocrine pancreatic insufficiency on cobalamin absorption in the dog. *American Journal of Veterinary Research,* **50**, 1233–1236.

128. Strombeck, D. R. (1978) New method for evaluation of chymotrypsin deficiency in dogs. *Journal of the American Veterinary Medical Association,* **173**, 1319–1323.

129. Strombeck, D. R., Farver, T. and Kaneko, J. J. (1981) Serum amylase and lipase activities in the diagnosis of pancreatitis in dogs. *American Journal of Veterinary Research,* **42**, 1966–1970.

130. Strombeck, D. R. and Guilford, W. G. (1990) Chronic gastritis, gastric retention, gastric neoplasms and gastric surgery. In *Small Animal Gastroenterology.* pp. 208–227. Davis, CA: Stonegate.

131. Strombeck, D. R. and Guilford, W. G. (1990) Infectious, parasitic and toxic gastroenteritis. In *Small Animal Gastroenterology.* pp. 320–337. Davis, CA: Stonegate.

132. Strombeck, D. R. and Hoenig, M. (1981) Maldigestion and malabsorption in a dog with chronic gastritis. *Journal of the American Veterinary Medical Association,* **179**, 801–805.

133. Strombeck, D. R., Wheeldon, E. and Harrold, D. (1984) Model of chronic pancreatitis in the dog. *American Journal of Veterinary Research,* **45**, 131–136.

134. Suter, M. M., Palmer, D. G. and Schenk, H. (1985) Primary intestinal lymphangiectasia in three dogs: a morphological and immunopathological investigation. *Veterinary Pathology,* **22**, 123–130.

135. Tams, T. R. (1986) Chronic feline inflammatory bowel disorders. Part 1. Idiopathic inflammatory bowel disease. *Compendium of Continuing Education for the Practising Veterinarian,* **8**, 371–378.

136. Tams, T. R. (1987) Chronic canine lymphocytic plasmacytic enteritis. *Compendium of Continuing Education for the Practising Veterinarian,* **9**, 1184–1191.

137. Teahon, K., Bjarnason, I. and Levi, A. J. (1988) The role of enteral and parenteral nutrition in Crohn's disease and ulcerative colitis. *Progress in Inflammatory Bowel Disease,* **12**, 1–4.

138. Turk, J., Fales, W., Miller, M. *et al.* (1992) Enteric *Clostridium perfringens* infection associated with parvoviral enteritis in dogs: (1987–1990). *Journal of the American Veterinary Medical Association,* **200**, 991–994.

139. Twedt, D. C. and Magne, M. L. (1986) Chronic gastritis. In *Current Veterinary Therapy IX.* Ed. R. W. Kirk. pp. 852–856. Philadelphia: W. B. Saunders.

140. Twedt, D. C. (1983) Differential diagnosis and therapy of vomiting. *Veterinary Clinics of North America,* **13**, 503–520.

141. Van Der Gaag, I. (1988) The histological appearance of large intestinal biopsies in dogs with clinical signs of large bowel disease. *Canadian Journal of Veterinary Research,* **52**, 75–82.

142. Van Der Gaag, I., Happé, R. P. and Wolvekamp, W. T. C. (1983) Eosinophilic enteritis complicated by partial ruptures and a perforation of the small intestine in a dog. *Journal of Small Animal Practice,* **24**, 575–581.

143. Van Kruiningen, H. J. (1972) Canine colitis comparable to regional enteritis and mucosal colitis of man. *Gastroenterology,* **62**, 1128–1142.

144. Van Kruiningen, H. J. (1976) Clinical efficacy of tylosin in canine inflammatory bowel disease. *Journal of the American Animal Hospital Association,* **12**, 498–501.

145. Vernon, D. F. (1962) Idiopathic sprue in a dog. *Journal of the American Veterinary Medical Association,* **140**, 1062–1067.

146. Walter, M. C., Goldschmidt, M. H., Stone, E. *et al.* (1985) Chronic hypertrophic gastropathy as a cause of pyloric obstruction in the dog. *Journal of the American Veterinary Medical Association,* **186**, 157–161.

147. Washabau, R. J. *et al.* (1986) Use of pulmonary hydrogen gas excretion to detect carbohydrate malabsorption in dogs. *Journal of the American Veterinary Medical Association,* **189**, 674–679.

148. Washabau, R. J., Strombeck, D. R., Buffington, C. A. and Harrold D. (1986) Evaluation of intestinal carbohydrate malabsorption in the dog by pulmonary hydrogen gas excretion. *American Journal of Veterinary Research,* **47**, 1402–1405.

149. Watson, A. D. J., Church, D. B., Middleton, D. H. *et al.* (1981) Weight loss in cats which eat well. *Journal of Small Animal Practice,* **22**, 473–482.

150. Westermarck, E. (1980) The hereditary nature of canine pancreatic degenerative atrophy in the German Shepherd dog. *Acta Veterinaria Scandinavica,* **21**, 389–394.

151. Westermarck, E. and Sandholm, M. (1980) Faecal hydrolase activity as determined by radial enzyme diffusion: a new method for detecting pancreatic dysfunction in the dog. *Research in Veterinary Science,* **28**, 341–346.

152. Westermarck, E., Wiberg, M. and Junttila, J. (1990) Role of feeding in the treatment of dogs with pancreatic degenerative atrophy. *Acta Veterinaria Scandinavica,* **31**, 325–331.

153. Whitbread, T. J., Batt, R. M. and Garthwaite, G. (1984) A relative deficiency of serum IgA in the German shepherd dog: a breed abnormality. *Research in Veterinary Science,* **37**, 350–352.

154. Wilcock, B. (1992) Endoscopic biopsy interpretation in canine or feline enterocolitis. *Seminars in Veterinary Medicine and Surgery,* **7**, 162–171.

155. Willard, M. D,. Dalley, J. B. and Trapp, A. L. (1985) Lymphocytic–plasmacytic enteritis in a cat. *Journal of the American Veterinary Medical Association,* **186**, 181–182.

156. Williams, D. A. (1985) Studies on the diagnosis and pathophysiology of canine exocrine pancreatic insufficiency. PhD Thesis, University of Liverpool.

157. Williams, D. A. (1992) Exocrine pancreatic insufficiency. In *Veterinary Gastroenterology.* Ed. V. Anderson. pp. 283–294. Philadelphia: Lea and Febiger.

158. Williams, D. A. and Batt, R. M. (1983) Diagnosis of canine exocrine pancreatic insufficiency by the assay of serum trypsin-like immunoreactivity. *Journal of Small Animal Practice,* **24**, 583–588.

159. Williams, D. A. and Batt, R. M. (1988) Sensitivity and specificity of radioimmunoassay of serum trypsin-like immunoreactivity for the diagnosis of canine exocrine pancreatic insufficiency. *Journal of the American Veterinary Medical Association,* **192**, 195–201.

160. Williams, D. A., Batt, R. M. and McLean, L. (1987) Bacterial overgrowth in the duodenum of dogs with exocrine pancreatic insufficiency. *Journal of the American Veterinary Medical Association,* **191**, 201–206.

161. Wolf, A. M. (1992) Feline lymphocytic plasmacytic enterocolitis. *Seminars in Veterinary Medicine and Surgery,* **7**, 128–133.

162. Yanoff, S. R., Willard, M. D., Boothe, H. W. and Walker, M. (1992) Short bowel syndrome in 4 dogs. *Veterinary Surgery,* **21**, 217–227.

163. Zeitz, M. (1990) Immunoregulatory abnormalities in inflammatory bowel disease. *European Journal of Gastroenterology and Hepatology,* **2**, 246–249.

CHAPTER 14

Feline Gastrointestinal Tract Disease

W. GRANT GUILFORD

Introduction

Gastrointestinal tract (GIT) dysfunction is common in the cat and may result from primary GIT disorders or diseases of other organ systems. An important consequence of GI dysfunction is loss of nutritional homeostasis. In addition the diet has a marked influence on the GIT (Fig. 14.1). It should therefore come as no surprise that successful management of GIT diseases usually requires a combination of nutritional and pharmacological therapy. Indeed, many GI diseases can and should be managed by dietary therapy only. Unfortunately pharmaceutical agents are often given inappropriate precedence. Drug therapy without appropriate nutritional management is likely to result in at best incomplete or delayed resolution of signs, and at worse exacerbation of the disorder.

The first part of this chapter reviews the diagnosis and initial therapy of clinical problems of the feline GIT. The second part is a synopsis of common gastroenteric diseases in which nutritional management is important.

INFLUENCE OF DIET ON THE GI TRACT	
Diet may alter	Cell renewal rate
	Motility
	Absorption
	Secretion of mucus, acid and enzymes
	Bacterial flora
	Luminal ammonia content
	Colonic volatile fatty acid content
Diet may contain	Toxic food additives
	Allergenic proteins
	Antigenic proteins
Diet may correct	Nutritional deficiencies

FIG. 14.1 Influence of the diet on the GI tract.

CAUSES OF VOMITING IN THE CAT

Acute vomiting

Food poisoning, intolerance or allergy
Toxicities (household plants, insecticides, drugs)
GIT obstruction (linear foreign bodies, intussusception, hair balls)
Toxemias (pyometra, peritonitis)
Viral gastroenteritis (panleukopenia, rotavirus, coronavirus)
Bacterial gastroenteritis (*Salmonella, Campylobacter?*, others)
Travel sickness

Chronic vomiting

Food intolerance/allergy
Inflammatory bowel disease (chronic gastritis, lymphoplasmacytic enteritis, eosinophilic gastroenteritis etc.)
GI obstruction (intussusception, pyloric stenosis, neoplasia)
Gastrointestinal neoplasia
Parasitism (*Ollulanus*, ascarids, *Physaloptera, Giardia*, heartworm)
Toxicities (lead)
Constipation
Chronic pancreatitis
Endocrine diseases (diabetes mellitus, hyperthyroidism)
Liver disease
Uremia
Cardiomyopathy
Neurological diseases (dysautonomia, CNS diseases)

FIG. 14.2 Causes of vomiting in the cat.

GI Problems

Vomiting

Introduction

Vomiting is a very common problem in the cat and can be due to a multitude of causes (Fig. 14.2). Great care must be taken to differentiate regurgitation from vomiting. The most reliable differentiating feature is the characteristic pronounced abdominal contractions of vomiting. The diagnostic procedures of value for the investigation of chronic vomiting are listed in Fig. 14.3 and a brief description follows.

History

Relevant history includes the likelihood of ingestion of foreign bodies, rubbish, organophosphates, household cleaners or ornamental plants. Aggravation of vomiting by a certain diet raises the possibility of food intolerance or allergy. Almost any drug can cause vomiting as an idiosyncratic reaction. Concurrent polyuria/polydipsia and vomiting suggests the presence of a systemic disease such as diabetes mellitus, renal failure or hyperthyroidism.

The frequency and chronicity of the vomiting, the volume and content of the vomitus, and the temporal relationship of the vomiting to the ingestion of food help to determine the cause of the problem. Vomiting within minutes to a few hours after ingestion supports gastric inflammation or irritation. Vomiting of food over 12 hr after ingestion is pathognomonic for delayed gastric emptying. Persistent vomiting of large volumes of liquid vomitus in spite of food restriction is highly suggestive of pyloric or upper intestinal obstruction. Occasional flecks of fresh blood in vomitus is not of concern but large amounts of digested blood ("coffee grounds") or melena suggest significant upper gastrointestinal hemorrhage. Yellow staining of vomitus is common whatever the cause of the vomiting. It is due to gastroduodenal reflux of bile and rules out complete pyloric obstruction.

| PROCEDURES FOR THE DIAGNOSIS OF CHRONIC VOMITING IN CATS* ||
Procedure	Usefulness
Dietary trials	Food sensitivity
CBC	Sepsis, toxemia, eosinophilic, lead, hydration
Serum chemistry profile	Diabetes mellitus, uremia, hypoproteinemia, liver disease, electrolyte concentrations
Urinalysis	Renal disease, liver disease, hydration, ketoacidosis
Blood gas	Acid–base status
Fecal flotation	Parasites
Fecal occult blood	Neoplasia, ulcers
Vomitus microscopic exam	*Ollulanus tricuspis*
Liver function test	Liver disease
Serum antibody/antigen tests	Heartworm, FeLV, FIV
T_4 level	Hyperthyroidism
Survey radiographs	Obstructions, foreign bodies, liver and kidney size, masses, pancreatitis, peritonitis
Ultrasonography	Masses, metastatic neoplasia, mural thickenings include pancreatitis
Contrast radiography	Gastric disorders, pyloric obstruction, gastric emptying, intestinal obstruction
Fluoroscopy	Gastrointestinal motility disorders
Endoscopy	Luminal and mucosal gastric, duodenal and large bowel disease
CSF tap	CNS diseases
Toxicology	Plasma cholinesterase, lead, serum osmolality (ethylene glycol)
Exploratory celiotomy	Full-thickness biopsy, gastrinoma, chronic pancreatitis

* Modified with permission from Strombeck, D.R. and Guilford, W.G. (1990) *Small Animal Gastroenterology*, 2nd edn. Davis: Stonegate Publishing.

FIG. 14.3 Procedures for the diagnosis of chronic vomiting in the cat.

Physical Examination

A complete physical examination is essential. Particular attention is paid to attitude, hydration and cardiovascular status, all of which indicate the severity of the vomiting. Close inspection should be made of the base of the tongue where linear foreign bodies can catch. Attentive palpation of the lymph nodes for lymphadenopathy and the neck for thyroid nodules should be performed. Detection of tachycardia, arrhythmias, murmurs and weak pulse suggests cardiomyopathy, a disorder that can result in vomiting. Both lymphosarcoma and hyperthyroidism are common causes of vomiting in cats. Methodical abdominal palpation is important. Distended gas-filled loops of bowel suggestive of obstruction may be apparent. A bowel infiltrated by inflammatory or neoplastic cells may feel thickened on palpation. Mesenteric lymphadenopathy or other abdominal masses may be detected.

Tabletop Assessment

By the conclusion of the clinical examination, the clinician should be able to differentiate expectoration, gagging, retching, regurgitation and vomiting, and should have assessed the severity of the condition and the fluid volume deficit (if any). The cause may be apparent or, more probably, the list of potential causes will have been narrowed.

Laboratory Data Base

A laboratory data base consisting of a complete blood count (CBC), serum chemistry panel, urinalysis and fecal flotation should be gathered (see Fig. 14.3). If available, a venous blood gas sample to determine acid–base status can be useful. Look for evidence on the data base of systemic diseases, such as kidney and liver insufficiency, toxemias (e.g. pyometra, peritonitis), diabetes mellitus and lead toxicity. If the data base points to any of these disorders, further diagnostic tests, such as organ biopsy, liver function tests and lead levels, may be required. Amylase and lipase have proved to be unreliable tests for chronic pancreatitis in cats.[4] The data base may detect significant or insignificant parasitism and aid in the identification of eosinophilic gastritis. Perusal of the packed cell volume (PCV), serum albumin, electrolyte panel and blood gas data helps tailor the fluid (crystalloid, plasma, blood) and electrolytes administered.

Survey Abdominal Radiographs

GIT foreign bodies and intestinal obstruction can usually be diagnosed by survey radiographs (see Fig. 14.3). Radiographs also assess gastric size, position and content, detect gross abnormalities in liver and kidney size and help detect abdominal fluid, abdominal masses, organ torsions, bowel perforation, peritonitis and pancreatitis. Radiographs showing tight clumping of the bowel gas pattern in midabdomen are characteristic of linear foreign bodies.

Controlled Diets

If no diagnosis is apparent after performing these initial diagnostic tests, and no warning signs of serious disease are apparent, it is good practice to discharge the cat on a controlled diet for 2–3 weeks to determine if the vomiting is food responsive. The controlled diet should initially consist of a meat to which the cat has not had frequent exposure in the last 6 months. Chicken is usually a good choice as it infrequently forms a significant proportion of a cat's diet and therefore is unlikely to be responsible for an allergic gastritis. If the controlled diet is used for more than 2–3 weeks it must be balanced (see dietary management of inflammatory bowel disease). Commercially available selected-protein diets are also useful for diagnostic purposes.

Additional Diagnostic Procedures

If the vomiting does not respond to the controlled diet, consideration should be given to the use of contrast radiography or, prefer-ably, referral of the cat to a specialist facility for endoscopy and/or abdominal ultrasound.

Endoscopy is a sensitive technique for the diagnosis of such common disorders as chronic gastritis and inflammatory bowel disease without the attendant risks and morbidity of full-thickness surgical biopsy. Exploratory laparotomy may be required for the diagnosis of mural diseases of the GIT, mucosal lesions out of reach of the endoscope, gastrinomas and low grade diseases of some organs such as chronic pancreatitis. Pyloric hypertrophy and large mass lesions of the gastric wall may be apparent on endoscopy, but definitive diagnosis may require full-thickness biopsy. Full-thickness biopsy of the intestine is disadvantaged by the increased likelihood of wound dehiscence: the biopsy sites of debilitated and hypoalbuminemic animals are particularly prone to dehiscence. In these patients, full-thickness intestinal biopsy should be performed only after great circumspection and preferably in association with nutritional support. Cerebrospinal fluid (CSF) taps or magnetic resonance imaging (MRI) may be required to detect central nervous system (CNS) lesions causing vomiting.

Symptomatic Therapy

Diet: Standard dietary recommendations for cats with vomiting include the provision of no food for 24–48 hr, followed by small quantities of a "bland" diet fed 3–4 times per day for 3–7 days.[16] These dietary recommendations have stood the test of time but are based more on common sense than scientific investigations.

Fasting the cat for a short period provides bowel "rest." This is traditionally considered of prime importance in the treatment of most GI problems although it has been recently challenged in the treatment of diarrhea (see below). In acute problems, bowel rest is accomplished by completely restricting the oral intake of food. Complete rest of the bowel for short periods (1–2 weeks) while still maintaining nutritional homeostasis can be achieved by total parenteral nutrition. The GIT rest can also be afforded by feeding highly digestible diets that are rapidly assimilated in the proximal small bowel.

By definition, a bland diet is one without coarse and spicy foods. Most commercial cat food could, therefore, be thought of as bland. One theoretical justification for *not* feeding a vomiting cat its usual diet is the observation in humans that acquired food allergies to proteins eaten during acute gastro-enteritis can delay recovery.[5,8] The same phenomenon may occur in the cat. Acquired food allergies to novel protein sources tempo-rarily introduced into a vomiting cat's diet (such as chicken) would not be as problematical as an acquired allergy to a dietary staple such as beef. Minimizing the protein content of the food fed to vomiting animals may reduce gastric acid secretion and dietary allergen-icity. Unfortunately, this is usually difficult to achieve with home-prepared diets. For delayed gastric emptying disorders, low fat diets are recommended, as food of high-energy density is emptied more slowly from the stomach.

Small frequent feedings have been recom-mended in order to limit the duration of acid secretion at each meal. Small volumes of food provoke less nausea during acute gastritis.

Fluid therapy: Lactated Ringer's spiked with an additional 5–15 mEq of KCl/500 ml is the preferred fluid for all vomiting cats with the exception of those vomiting due to an upper duodenal or pyloric obstruction or those with blood gas evidence of alkalosis. Cats with these latter problems should be rehydrated with 0.9% NaCl spiked with 10–20 mEq of KCl/500 ml. Avoid rehydration with potassium-free fluids.

Pharmacological therapy: Antiemetics are indicated only if vomiting is intractable. For a general purpose antiemetic use prochlor-perazine (0.1 mg/kg qid SC) or chlorpromazine (0.5 mg/kg tid SC). Anticholinergics, such as atropine, are contraindicated for routine use as antiemetics. H_2-blockers (cimetidine: 5–10 mg/kg tid PO, ranitidine: 3.5 mg/kg bid PO) are indicated if vomiting is associated with gastric erosions or ulcers (i.e. blood in the vomitus). Sucralfate (30–50 mg/kg tid PO) binds to areas of denuded mucosa and is particularly useful for the treatment of gastric ulcers. Bismuth salicylate has protective and antisecretory activities of value in the non-specific treatment of gastroenteritis (0.5–1 ml/ kg bid for 2–3 days). Antibiotics are primarily indicated if there is breach of the GI mucosal barrier as suggested by fever or blood in the vomitus. Amoxicillin (10–20 mg/kg bid SC) is usually satisfactory.

Symptomatic therapy of chronic vomiting with glucocorticoids is becoming widely prac-ticed because of the high incidence of inflam-matory bowel disease in cats. The use of these drugs without a diagnosis is fraught with dangers and as a general rule their symptomatic use is discouraged.

Diarrhea

Introduction

Diarrhea is an increase in the frequency, fluidity, or volume of bowel movements. Diarrhea can be due to primary disease of the GIT or may be secondary to a multitude of diseases of other organs (Fig. 14.4). The term "small bowel diarrhea" (malassimilation) refers to diarrhea as a consequence of small intestinal dysfunction resulting from diseases of the intestine itself (malabsorption), or from diseases of digestive organs such as the pancreas and liver (maldigestion) that inter-fere with the ability of the small intestine to absorb food. The term "large bowel diarrhea" refers to diarrhea resulting from diseases of the cecum, colon or rectum. In cats, chronic diarrhea is less commonly reported by owners than is chronic vomiting: in compari-son to dogs, cats less commonly develop diarrhea due to maldigestion or large bowel diseases.

History

The first goal of the history is to differentiate small bowel from large bowel diarrhea (Fig. 14.5). This is important because the diag-nostic and therapeutic approaches differ with each category. Additional goals of the history are to identify potential causes of the

CAUSES OF DIARRHEA IN THE CAT	
Acute small bowel-type diarrhea	Sudden change of diet Food poisoning, intolerance or allergy Viral enteritis (Panleukopenia, Coronavirus, Rotavirus, FeLV) Bacterial (*Campylobacter, Salmonella, Yersinia, Clostridium?, E. coli?*) Toxicosis (e.g. organophosphates) Toxemias (pyometra, abscess, peritonitis)
Chronic small bowel-type diarrhea	Bacterial overgrowth? (antibiotic misuse) Infectious enteritis (*Campylobacter*, FIV, FeLV) Parasites (*Giardia*, helminths) Food intolerance or allergy Inflammatory bowel disease (eosinophilic, lymphocytic-plasmacytic, other) Infiltrative neoplasia (lymphosarcoma, mastocytosis) Partial obstruction (neoplasia, strictures, intussusception) Exocrine pancreatic insufficiency (rare) Liver failure Uremia Hyperthyroidism
Acute large bowel-type diarrhea	Acute idiopathic colitis Infectious colitis (*Salmonella, Campylobacter, Clostridium*, FeLV) Abrasive colitis (bones, hair, wrapping material) Toxemias (pyometra, abscess, peritonitis) Toxicosis
Chronic large bowel-type diarrhea	Chronic colitis (lymphocytic–plasmacytic, eosinophilic, histiocytic) Infectious colitis (FIP, FeLV, *Campylobacter*, histoplasmosis) Foreign material (hair?) Parasites (*Cystisospora*) Partial obstruction (neoplasia, strictures, ileocolic intussusception) Neoplasia (adenocarcinoma, lymphosarcoma) Uremia

FIG. 14.4 Causes of diarrhea in the cat.

diarrhea, detect warning signs of serious disease and to identify any evidence of systemic disease (e.g. polydipsia/polyuria). With these aims in mind, questions to the client should include:

1. patient profile (age, breed, sex), parasite control program and vaccination status
2. duration of the complaint
3. whether the diarrhea ceases with fasting
4. historical associations (e.g. a particular food, environment, stressful situations)
5. defecation frequency
6. appearance of the feces (mucus, fresh blood, color, volume, smell)
7. presence of defecation urgency, tenesmus or dyschezia.

Young animals seem more prone to nutritional, microbial and parasitic causes of diarrhea. Adverse reactions to food are a prominent cause of diarrhea, and the history is a rapid way to identify responsible nutrients. For instance, chronic diarrhea in young cats is frequently caused by an intolerance to milk (Guilford, unpublished observations). Osmotic diarrhea usually results from malassimilation and is likely if the diarrhea ceases when the animal is fasted. Secretory diarrhea is usually due to infectious diseases and will often continue in the unfed animal. The frequency and appearance of bowel movements help differentiate large and small bowel diarrhea (Fig. 14.5). Rapid transit time is associated with yellow and green feces due to incompletely metabolized bilirubin. Normal cats commonly have black feces and it is important not to assume black feces are due to melena. Increased quantities of fecal fat (steatorrhea) impart a grey color and rancid smell to the feces. In the cat, steatorrhea is usually due

DIFFERENTIATION OF LARGE AND SMALL BOWEL DIARRHEA		
Clinical signs	**Small bowel**	**Large bowel**
Fecal volume	Large	Small
Fecal consistency	Loose	Loose to formed
Fecal mucus	Normal	Increased
Fecal fat	Increased	Normal
Fecal color	Variable	Usually brown
Fecal blood	Usually none Occasionally melena	Fresh blood frequent
Defecation frequency	Usually increased 2–4 times a day	Always increased 4–10 times a day
Defecation urgency	Rare	Frequent
Tenesmus	None	Frequent
Weight loss	Common	Infrequent
Exacerbating factors	Diet changes High fat diets Poorly digestible diets	Stress and psychological factors may be important

FIG. 14.5 Differentiation of large and small bowel diarrhea.

to inflammatory bowel disease or hyperthyroidism, both very common causes of chronic diarrhea.

Physical Examination

Careful attention should be paid to the cat's attitude, hydration and cardiovascular status and to examination of the oral cavity, lymph nodes, thyroid glands and abdomen. Weight loss as a result of malassimilation may be apparent. Loops of bowel filled with fluid and gas may be palpated in animals with enteritis or an obstruction. A bowel infiltrated by inflammatory or neoplastic cells may feel thickened on palpation. Detection of hepatomegaly raises the likelihood of hepatic disease. Palpation of the rectum may reveal fecal impaction or rectal masses. Examination of feces adherent to the thermometer may detect melena.

Tabletop Assessment

By the conclusion of the clinical examination, the clinician should be able to differentiate small bowel from large bowel diarrhea, and should have assessed the severity of the condition and the fluid volume deficit (if any). The cause of the diarrhea is rarely apparent but the list of potential causes will have been narrowed.

Diagnostic Evaluations

For complex or refractory cases of diarrhea the minimum laboratory data base should include a CBC, a serum chemistry profile with electrolyte concentrations, a urinalysis, a fecal flotation for parasite ova and a direct smear of saline-admixed fresh feces for protozoa. Cats with diarrhea in association with a peripheral eosinophilia may have eosinophilic gastroenteritis but parasitism (internal and external) and mast cell neoplasia are also important possibilities. Panhypoproteinemia (low serum albumin and globulin) characterizes protein-losing enteropathies. Panhypoproteinemia is uncommon in cats in comparison to dogs, but is occasionally seen with severe inflammatory bowel disease or severe subacute enteropathy secondary to nonfatal panleukopenia. Serological tests for feline leukemia (FeLV), feline immunodeficiency virus (FIV) and feline infectious peritonitis (FIP) may also be warranted. These diseases can be associated with chronic enteropathies. The FIP titer must be interpreted carefully due to cross-reactivity with enteric corona viruses. Serum thyroxine measurement is indicated in elderly cats with diarrhea particularly if steatorrhea is observed. In cats with small bowel diarrhea, a Sudan stain of fresh feces for fat can be worthwhile. Large quantities of undigested fats raise the unlikely

PROCEDURES FOR THE DIAGNOSIS OF CHRONIC DIARRHEA IN CATS*	
Procedure	**Usefulness**
CBC	Sepsis, toxemia, eosinophilic gastroenteritis, hydration
Serum chemistry profile	Uremia, protein-losing enteropathy, liver disease, electrolyte concentrations
Urinalysis	Renal disease, liver disease, hydration
Dietary trials	Food intolerance, food allergy
Blood gas	Acid–base status
Fecal flotation	Helminths, *Giardia, Cryptosporidia, Cystisospora*
Fecal direct smear	Giardiasis, other protozoans, neutrophils
Baerman technique	*Strongyloides* larvae
Fecal sudan stain	Steatorrhea
Fecal occult blood	Neoplasia, ulcers
Fecal digestion tests	Exocrine, pancreatic insufficiency
Fecal culture	*Salmonella, Campylobacter*
Serum antibody/ antigen tests	FeLV, FIV, FIP
Serum thyroxine level	Hyperthyroidism
Breath hydrogen test	Bacterial overgrowth?, carbohydrate malabsorption
Fecal total fat excretion	Fat malabsorption
Survey radiographs	Intussusception, foreign bodies, liver and kidney size, masses, pancreatic mass, peritonitis
Ultrasonography	Masses, metastatic, neoplasia, mural thickenings
Contrast radiography	Intestinal partial obstruction, neoplasia
Fluoroscopy	Gastrointestinal motility disorders
Endoscopy	Luminal and mucosal gastroduodenal, large bowel disease
Toxicology	Plasma cholinesterase, lead
ACTH stimulation test	Hypoadrenocorticism
Liver function test	Liver disease
Exploratory celiotomy	Full-thickness biopsy, chronic pancreatitis, phycomycosis, mycobacteria, focal neoplasia

* Modified with permission from Strombeck, D.R. and Guilford, W.G. (1990) *Small Animal Gastroenterology*, 2nd edn. Davis: Stonegate Publishing

FIG. 14.6 Procedures for the diagnosis of chronic diarrhea in cats.

possibility of exocrine pancreatic insufficiency that can be confirmed by azocasein hydrolysis tests.[18] If large bowel diarrhea is suspected a stained fecal smear and/or rectal scraping should be examined and may reveal inflammatory cells suggestive of colitis or fungi such as *Histoplasma*. Large amounts of neutrophils in a fecal smear or evidence of contagion are indications for fecal culture that is otherwise usually a low-yield procedure.

If no diagnosis is apparent after performance of these initial laboratory tests, and no warning signs of serious disease are apparent, it is customary to discharge the cat on a controlled diet for 2–3 weeks to determine if the diarrhea is food responsive. A similar controlled diet to that described for vomiting is appropriate.

If the diarrhea fails to respond to the controlled diet, further tests are indicated.

The high prevalence of inflammatory bowel disease means that upper GI biopsy is usually the next diagnostic procedure of choice. Where possible this should be performed by referral to an endoscopist to avoid the surgical morbidity of laparotomy. Full-thickness colon biopsies, in particular, should be avoided. Survey abdominal radiographs or contrast radiography rarely provide diagnostically useful information in cases of diarrhea but when they do, that information can be invaluable to patient well-being (e.g. partial obstruction). Miscellaneous other tests of occasional value are listed in Fig. 14.6.

Symptomatic Therapy

Diet and fluids: The traditional dietary and fluid therapy of diarrhea is similar to that described above for vomiting.[16] Lactated

Ringer's spiked with additional potassium chloride is the fluid of choice. Cats with diarrhea are usually starved for 24–48 hr and then offered a bland, low fat diet fed frequently and in small quantities for 3–7 days. Dietary fat is kept to a minimum because fat undergoes a complex digestion and absorption process that is easily disrupted by gastroenteritis. Furthermore, malabsorbed fatty acids and bile acids promote secretory diarrhea in the large bowel.[15]

In contrast to the traditional approach to dietary therapy of diarrhea, it has been suggested that cats with diarrhea tolerate high fat diets better than high carbohydrate diets. This suggestion needs further investigation before widespread application. Furthermore, the long-held trust in the value of bowel rest for the treatment of diarrhea has been challenged by the concept of "feeding-through" diarrhea. Recent studies of humans with acute diarrhea have shown that feeding through diarrhea maintains greater mucosal barrier integrity.[7] Furthermore, feeding highly digestible rice-based oral rehydration solutions reduces stool volume in comparison to glucose-based solutions.[9] In addition to the provision of energy, it has been demonstrated that the glucose and peptides in rice furnish organic substrates for electrolyte pumps that in turn drive water absorption.[1,3,10,12] One such rice-based rehydration solution used in humans is shown in Fig. 14.7. It is possible similar solutions, fed through nasogastric tubes, may be of value in the *short-term* symptomatic treatment of cats with acute diarrhea.

Pharmacological therapy: Motility modifiers such as opioids (loperamide 0.08 mg/kg tid PO) are indicated if diarrhea is intractable. Opioids are contraindicated if the diarrhea is due to invasive microorganisms and must be used with care in cats. Metoclopramide (0.2–0.4 mg/kg qid SC) is useful if diarrhea is associated with ileus. Anticholinergics are contraindicated in diarrhea. Activated charcoal (0.7–1.4 g/kg PO) is a useful adsorbent for the treatment of diarrhea due to toxins. Bismuth subsalicylate is indicated for nonspecific gastroenteritis (see above). Antibiotics are likely to delay recovery if injudiciously

RICE-BASED ORAL REHYDRATION SOLUTION	
Component	g/l water
Rice powder	50–80
Sodium chloride	3.5
Sodium bicarbonate	2.5
Potassium chloride	1.5

FIG. 14.7 Rice-based oral rehydration solution.

used and should only be utilized for the treatment of diarrhea if the diarrhea is hemorrhagic, a known pathogen has been cultured from the feces or there is evidence of bacterial overgrowth or sepsis. Fluoroquinolones are a good choice of antibiotic for the treatment of hemorrhagic diarrhea because they are very effective against enteric pathogens. Metronidazole (10–20 mg/kg bid for 5–7 days) is useful if giardiasis or anaerobic bacterial overgrowth is suspected. In young cats with acute large bowel diarrhea an oral sulfonamide is sometimes useful because of the possibility of coccidiosis. As with chronic vomiting, symptomatic therapy of chronic diarrhea with glucocorticoids is discouraged unless a diagnosis of inflammatory bowel disease has been made.

Dysphagia and Regurgitation

Diagnostic Approach

The term dysphagia refers to eating difficulties. Dysphagia should be differentiated from inappetence, a distinction that may escape the client. Dysphagia may result from disorders of the jaw, oral cavity, tongue, pharynx or esophagus. Cats with disorders of the oral cavity, tongue or jaws have difficulty lapping liquids and/or grasping and chewing food. Inability to open the mouth is occasionally seen and is often a result of a retrobulbar abscess or, less commonly, abnormalities of the temporomandibular joints. Common causes of prehension and mastication difficulty include stomatitis, oral neoplasia, mandibular fracture or subluxation and dental disorders.

Animals with dysphagia due to pharyngeal or esophageal disorders usually regurgitate. Regurgitation is the passive reflux of ingesta, usually from the esophagus or pharynx.

CAUSES OF OROPHARYNGEAL DYSPHAGIA AND REGURGITATION IN CATS	
Glossal disorders	Glossal neoplasia, trauma, ulceration
	Neuropathies of cranial nerves VII, IX, XII
	Hydrocephalus, brain stem lesions
Pharyngeal disorders	Pharyngitis
	Neoplasia, nasopharyngeal polyps
	Foreign bodies
	Retropharyngeal lymphadenopathy
	Rabies
	Cricopharyngeal achalasia (rare)
Palate disorders	Congenital defects
Esophageal disorders	Esophagitis (persistent vomiting, gastroesophageal reflux, idiopathic papillomatous)
	Foreign bodies
	Hiatal disorders (herniation, gastroesophageal intussusception)
	Megaesophagus (dysautonomia, lead poisoning, myasthenia gravis, idiopathic)
	Motility abnormalities

FIG. 14.8 Causes of oropharyngeal dysphagia and regurgitation in cats.

Unlike vomiting, it is not associated with signs of nausea or rhythmical abdominal contractions. Patients affected by oropharyngeal dysphagia make exaggerated swallowing movements and usually food will drop from the mouth within seconds of prehension. In contrast, esophageal dysphagia results in more delayed regurgitation and is usually not associated with exaggerated swallowing movements.

Regurgitation is an infrequent problem in the cat in comparison to the dog. It can occur as a result of any disease that obstructs the lumen of the esophagus or damages the components of the swallowing reflex. Important causes of oropharyngeal dysphagia and regurgitation are listed in Fig. 14.8. Diagnostic procedures of value in the investigation of cats with these problems are listed in Fig. 14.9.

PROCEDURES FOR THE DIAGNOSIS OF DYSPHAGIA AND REGURGITATION IN CATS*	
Procedure	**Usefulness**
Neurological examination	Cranial nerve disorders, CNS lesions, polyneuropathies
Oropharyngeal exam under GA	Masses, foreign bodies
Serum chemistry profile	Uremia
Blood lead	Lead toxicosis
Lymph node aspiration	Neoplasia
Survey radiographs	Dental and jaw disorders, megaesophagus, periesophageal masses, esophageal or pharyngeal foreign bodies
Swallowing study	Megaesophagus, motility disorders, stricture, esophageal masses, vascular ring anomalies, hiatal disorders, periesophageal masses
Ultrasound	Retropharyngeal disorders
Endoscopic exam of nasopharynx	Masses, foreign bodies
Esophagoscopy and biopsy	Esophagitis, obstructive diseases
Exploratory surgery	Periesophageal masses, hiatal disorders

* Modified with permission from Strombeck, D.R. and Guilford, W.G. (1990) *Small Animal Gastroenterology*, 2nd edn. Davis: Stonegate Publishing.

FIG. 14.9 Procedures for the diagnosis of dysphagia and regurgitation in cats.

Symptomatic Therapy

Therapy of regurgitation is best directed at the primary cause: unfortunately, this is frequently not possible. Symptomatic management of regurgitation due to esophageal disease consists of feeding and watering from an elevated container. Foods of differing consistency should be experimented with, as some individual animals are better suited to gruels, whereas others swallow better following stimulation of the swallowing reflex with coarse, dry foods. Gruels tend to be better tolerated by cats with esophageal strictures. Long-term nutritional homeostasis and esophageal rest can be achieved by gastrostomy tube placement.

Gagging, Retching and Expectoration

Gagging is a reflexive contraction of the constrictor muscles of the pharynx resulting from stimulation of the pharyngeal mucosa. Retching is an involuntary and ineffectual attempt at vomiting. Retching is produced by contractions of the diaphragm and abdominal muscles, the same motor events that cause vomiting. Expectoration is the ejection of airway and laryngopharyngeal mucus or discharges. Gagging and expectoration, alone or in combination, are common in cats. They are usually self-limiting and are commonly ascribed to accumulation of fur in the pharynx. Grass blades resting in the nasopharynx can produce similar signs. Diagnosis is usually made via history and oropharyngeal examination. Accumulation of fur can be minimized by regular combing.

Halitosis

Halitosis is a relatively frequent complaint in dogs, but is observed less in cats. Common causes of halitosis (listed Fig. 14.10) include dental disease, high protein diets and inflammatory or necrotizing oral or pharyngeal lesions such as neoplasia or ulcerative stomatitis. Highly proteinaceous diets result in halitosis via excretion into the breath of odiferous products of protein fermentation such as ammonia, hydrogen sulfide, indoles and skatoles.

Most cases of halitosis can be diagnosed through a history and a thorough physical examination of the oral cavity and pharynx. Occasionally, radiographs of the teeth, nasal cavity, pharynx or esophagus are required. Examination of the pharynx with a dental mirror or endoscope and the nasal cavity with a rhinoscope is sometimes helpful. Management

CAUSES OF HALITOSIS IN CATS*	
Dietary associated	Food remnants
	Highly proteinaceous diets
	Coprophagia
Disease of the lips	Cheilitis
Oral cavity disease	Inflammatory or necrotizing oral lesions
	Foreign bodies
Pharyngeal disease	Inflammatory or necrotizing pharyngeal lesions
	Pharyngeal foreign bodies
	Tonsillar crypt foreign bodies
Nasal cavity and sinus disease	Inflammatory diseases
	Necrotizing diseases
Dental disease	Periodontitis
	Gingivitis
	Tooth root abscesses
	Tartar
Esophageal disease	Megaesophagus
Malassimilation	IBD
Systemic disease	Uremia

* Modified with permission from Strombeck, D.R. and Guilford, W.G. (1990) *Small Animal Gastroenterology*, 2nd edn. Davis: Stonegate Publishing.

FIG. 14.10 Causes of halitosis in cats.

of halitosis depends on the primary cause but can include dental prophylaxis and a change to highly digestible lower protein diets.

Borborygmus and Flatulence

Borborygmus is a rumbling noise caused by the propulsion of gas through the GIT. Gas is produced by aerophagia or bacterial degradation of unabsorbed nutrients. Excessive gas usually results from dietary indiscretions, but on occasion can herald more serious GI disease such as malassimilation. Diets containing legumes are associated with gaseousness, as they contain large quantities of indigestible oligosaccharides. The management of these problems begins with a change to a highly digestible, low fiber diet, of moderate protein content. If dietary manipulation is not successful, consider investigation of the patient as described for the evaluation of small bowel diarrhea.

Tenesmus

Tenesmus refers to persistent or prolonged straining that is usually ineffectual and often painful. Owners often incorrectly equate tenesmus with constipation. Tenesmus is common in diseases of the lower GI, urinary or reproductive tracts. The relevant organ system can usually be identified from the history and by physical examination. Whether the tenesmus is associated with micturition or defecation can also aid differentiation. Obstructive disorders of the large bowel are more commonly associated with tenesmus before evacuation (e.g. constipation), whereas irritative disorders (colitis) are often associated with persistent tenesmus after evacuation. In addition to examination of the large bowel, anus and perineum, the clinician should carefully palpate the bladder, and closely examine the vagina or penis for evidence of pain, discharges, masses or calculi.

The diagnostic procedures are similar to those described for the evaluation of large bowel diarrhea. In cats the most common cause of tenesmus is constipation.

Constipation

Diagnostic Approach

Constipation is infrequent or absent defecation. Intractable constipation is called obstipation and has a guarded prognosis because irreversible degenerative changes in the colon may result. Constipation is a common presenting complaint and can result from many different causes (Fig. 14.11). In the cat the most common causes of constipation are ingestion of hair, tail-pull neuropathies, pelvic fractures and idiopathic megacolon and dehydration. Diagnosis is usually made by abdominal or rectal palpation. Neurological examination or pelvic radiography may reveal the cause.

Symptomatic Therapy

The first considerations in the treatment of constipation are (a) correction of any fluid and electrolyte derangements and (b) attention to the primary cause, if one can be identified. The initial approach to severely constipated patients should include anesthesia, the administration of a lubricating warm water enema and manual fragmentation and removal of the hardened stool. Precautions include adequate preparation of the patient for anesthesia and the use of endotracheal intubation to prevent aspiration of regurgitated GI content (which can be induced by the enema procedure). Phosphate enemas are contraindicated because of the likelihood of intoxication.

Animals with less severe constipation can be treated by regular warm water, lubricating enemas in association with softening agents such as dioctyl sodium sulfosuccinate (50 mg per cat bid–sid PO).

Prevention of subsequent constipation is best achieved by regular brushing of the cat's coat, use of bulk-forming laxatives such as psyllium fiber (1 tsp mixed with each meal) or bran (1–2 tblsp/400 g canned food), or the use of lubricant laxatives such as white petrolatum. Lactulose syrup (1–5 ml bid PO) may be useful in some cats. Mineral oil should be avoided.

CAUSES OF CONSTIPATION IN CATS	
Ingestion of indigestible material	Hair
Painful defecation	Pelvic or anorectal trauma
Mechanical obstruction	Pelvic fractures
	Intrapelvic masses
	Colonorectal mass
	Congenital anorectal lesions
Neurological disease	Sacral spinal cord deformity (Manx cats)
	Idiopathic megacolon
	Tail-pull neuropathy

FIG. 14.11 Causes of constipation in cats.

Fecal Incontinence

Fecal incontinence is the inability to retain feces until defecation is appropriate. It can be due to large bowel disorders, such as colitis, that interfere with the reservoir capability of the rectum. Fecal incontinence can also be due to failure of the sphincteric mechanisms, usually through neurological dysfunction. Treatment is dependent on the primary cause. Symptomatic therapy includes the feeding of low residue diets and the use of diphenoxylate (0.05–0.1 mg/kg tid PO) to increase sphincter tone.

Clinical Synopses of Common Gastroenteric Diseases

In the following section, the prevalent GI diseases of the cat in which nutritional management is important are reviewed.

Periodontal Disease

Inflammation of the tissues surrounding the teeth is a very common complaint that is usually caused by extension of gingivitis into the periodontal tissues. Treatment can include cleansing of the tooth crown and exposed root, gingivectomy to remove enlarged periodontal pockets and extraction of loose teeth or teeth affected by external root resorption. Recurrence of the disease may be reduced by feeding diets that require chewing (Guilford, personal observation).

Esophageal Disorders

Signs of esophageal disease include dysphagia, regurgitation and ptyalism. In the author's experience, the most frequent causes of esophageal disease in the cat are foreign bodies; esophagitis and/or strictures secondary to gastroesophageal reflux; and megaesophagus secondary to dysautonomia or lead poisoning. Other causes of esophageal disease are listed in Fig. 14.9.

Esophagitis is uncommon in the cat because of its fastidious eating habits. Gastroesophageal reflux due to chronic vomiting, anesthesia or hiatal disorders is the most common cause. The clinical signs are those of esophageal disease, however painful swallowing may be obvious. Diagnosis is usually made from the history, clinical signs, endoscopy and occasionally barium swallows. Nutritional management is difficult. High protein, low fat diets will increase gastroesophageal sphincter tone, theoretically minimizing gastroesophageal reflux. Unfortunately high protein diets have, however, the disadvantage of strongly stimulating gastric acid secretion. Pharmacological treatment can include metoclopramide (0.2–0.4 mg/kg tid PO), cimetidine (5 mg/kg bid PO) and sucralfate (30 mg/kg tid PO). Hiatal disorders may require surgical correction. Esophagitis can lead to esophageal stricture. This is best treated by repeated balloon dilation.

Esophageal motility abnormalities are uncommon. They may progress to esophageal dilatation (megaesophagus). Other clinical signs of dysautonomia may be apparent. The owners should be questioned about the access of the cat to lead that can result in megaesophagus. Idiopathic megaesophagus is particularly common in Siamese cats and is often associated with gastric emptying dysfunction.[11] It may also result from esophagitis. Diagnosis is reached by history, physical examination, radiography and fluoroscopy. Treatment is usually symptomatic and includes experimenting with diets of different consistency, use of elevated feeding bowls or forced feeding via a gastrostomy tube.

Acute Gastritis

Acute gastritis is common in cats. A history of recent exposure to other cats raises the likelihood of viral infections. Recent ingestion of foreign material, spoiled food or carrion may also be reported. Damage to the mucosa results in episodic discomfort and vomiting. Occasionally lethargy, depression, hematemesis, fever, dehydration, polydipsia and anterior abdominal pain may be present. Diagnosis is made on the basis of history and physical examination. The clinical signs are self-limiting and treatment is supportive (see discussion of vomiting above).

Chronic Gastritis

Chronic gastritis is characterized by chronic persistent (occasionally sporadic) vomiting. Affected cats are usually otherwise bright and alert. Vomitus may contain bile, partly digested food, small amounts of blood or only clear liquid. There is no specific relationship to the eating time. Specific types of chronic gastritis observed in cats include lymphocytic–plasmacytic gastritis (most common), eosinophilic gastritis and fibrosing gastritis. The latter disorder is probably an end-stage of the other forms of chronic gastritis and is also seen in association with *Ollulanus tricuspis*. The cause of chronic gastritis is usually not determined. Diagnosis is made on the basis of clinical signs, endoscopy and biopsy (via endoscopy). Contrast radiography is of little value. The gross appearance of the mucosa may be normal but histology reveals inflammatory cell infiltration in the lamina propria.

Treatment should be based on a bland hypoallergenic diet because of the possibility of associated or causal dietary sensitivity. As discussed above, commercial or home-prepared selected-protein diets based on a protein source not in the cat's usual diet make suitable hypoallergenic diets. Eliminate any other potential causative factors such as drugs or toxic household plants. *Ollulanus tricuspis* can be treated with fenbendazole (50 mg/kg daily for 3 days) and perhaps pyrantel (20 mg/kg). Cats with a confirmed diagnosis of lymphocytic–plasmacytic or eosinophilic gastritis usually also require immunosuppressive therapy as described for inflammatory bowel disease (see below).

Acute Gastroenteritis/Enteritis/Enterocolitis

Acute GI upsets are very common. They may affect just the stomach (gastritis), the stomach and small intestine (gastroenteritis) or the small and large intestine (enterocolitis). The clinical signs include a sudden onset of small bowel or a mixed small bowel/large bowel diarrhea that is usually watery but sometimes bloody (dysentery). There is usually associated vomiting. Some animals show depression, fever, abdominal pain and dehydration. Causes of acute diarrhea are listed in Fig. 14.4. The clinical signs of these different causes of gastroenteritis are similar and diagnosis is most often based on history (e.g. likelihood of exposure to a toxin; change of diet; contact with other animals etc.). Panleukopenia stands out from the other causes because it usually affects young unvaccinated cats and because of the severity of the depression, dehydration, diarrhea (voluminous and bloody) and leukopenia. Occasionally, FeLV or *Salmonella* spp. infections will produce a similar picture to panleukopenia. Other viruses such as rotavirus and enteric coronavirus may be responsible for outbreaks of acute diarrhea. A syndrome of transient third eyelid prolapse in association with self-limiting diarrhea, probably of viral origin, has also been described. *Salmonella*,

Campylobacter and *Clostridium* infections can produce acute diarrhea in cats as can *Yersinia pseudotuberculosis*, *Bacillus piliformis* and also possibly *Y. enterocolitica* and *Escherichia coli*. These infections often result in signs of enterocolitis rather than gastroenteritis. *Y. pseudotuberculosis* infections can be severe, resulting in pyogranulomatous infiltration and enlargement of the mesenteric nodes, liver and spleen. *Giardia* can produce acute to subacute or chronic diarrhea (see below). The coccidia, *Cystisospora felis* and *C. rivolta* are usually nonpathogenic but in some young cats appear to cause subacute large bowel diarrhea. Helminth parasites rarely cause significant diarrhea in cats. Two uncommon exceptions are *Strongyloides stercoralis* and *S. tumefaciens*.

Most causes of acute gastrointestinal upsets are self-limiting and cats will respond to supportive care regardless of the underlying cause. Therefore, few diagnostic procedures other than a fecal flotation and fresh smear for parasites such as coccidia and *Giardia* are usually performed. A white blood cell count assists the diagnosis of panleukopenia. Serum electrolyte levels will aid the symptomatic treatment of the animal by assisting the choice of fluid therapy (particularly the amount of potassium to be added to the fluids). Blood urea nitrogen level (by dipstick) and urine specific gravity are sometimes determined to rule out acute renal failure as a cause of the diarrhea. If the history is supportive of an infectious cause and particularly if clinical evidence of enterocolitis is apparent, fecal cultures are warranted.

Nutritional management of acute diarrhea is discussed above.

Panleukopenia requires broad spectrum antibiotic therapy in association with aggressive fluid and electrolyte therapy. Administration of plasma may also be helpful. Diarrhea due to *Campylobacter* rapidly responds to erythromycin (10 mg/kg tid PO). Most other bacterial diarrheas will resolve with symptomatic therapy with or without fluoroquinolones, ampicillin or trimethoprim–sulfa. Unfortunately, *Y. pseudotuberculosis* and *Bacillus piliformis* infections are often fatal. Diarrhea in association with coccidian parasites is usually self-limiting if nutrition and hygiene are attended to but resolution appears to be hastened by the use of trimethoprim–sulfonamide preparations (15–30 mg/kg bid–sid PO for 7 days). Most GI helminths can be safely eliminated with a pyrantel compound. *Strongyloides* is treated with fenbendazole (50 mg/kg PO sid for 5 days).

Inflammatory Bowel Diseases (Including Colitis)

Introduction and Diagnostic Approach

The idiopathic inflammatory bowel diseases (IBD) are a group of disorders characterized by persistent clinical signs of GI disease associated with histological evidence of inflammation of undetermined cause in the lamina propria of the small or large intestine. IBD are frequent causes of chronic vomiting and diarrhea in cats. The most frequent subclassifications recognized in cats are lymphocytic–plasmacytic enteritis and eosinophilic gastroenteritis. Other subclassifications of IBD seen less commonly in cats include lymphocytic–plasmacytic colitis, eosinophilic colitis and suppurative colitis. Rare cases of histiocytic and granulomatous IBD have been reported. The cause of IBD is unknown but appears to involve hypersensitivity to antigens (food, bacterial, mucus, epithelial) derived from the bowel lumen or mucosa. Diagnosis of IBD is made by eliminating known causes of intestinal inflammation and then acquiring a biopsy of the mucosa.

Dietary Therapy

The management of the different types of IBD is similar. Dietary therapy is a very important part of treatment. The ideal diet is highly digestible, low in lactose, moderately low in fat, hypoallergenic, not markedly hypertonic and contains generous doses of potassium, water-soluble and fat-soluble vitamins.[16] Good palatability, nutritional balance and ease of preparation are also required.

Diets of high digestibility provide bowel rest and reduce the allergenicity of the diet by decreasing the amount of dietary protein absorbed intact by the mucosa and the number of bacteria and bacterial products in the bowel. Both of these changes reduce the number of antigens presented to the gut-associated lymphoid tissue, minimizing the probability of hypersensitivity reactions. Disaccharides, especially lactose, should be avoided because the disaccharidases (lactase, maltase, isomaltose and sucrase) may be reduced. Moderate restriction of fat is theoretically advantageous because fat malassimilation is common in small bowel diarrhea. Hypoallergenicity minimizes the risks of acquired GI allergies secondary to mucosal barrier dysfunction. For this reason, the protein source is best chosen in relation to the animal's previous diet. It is best to avoid (if possible) any proteins the cat has been eating for the last 4–6 months. Hyperosmolar diets should be avoided because they predispose the animal to osmotic diarrhea.

A home-prepared diet that fulfills most of these ideal criteria is provided in Fig. 14.12. The diet is likely to be fed for many months and it must therefore be carefully balanced. The meat chosen should be diced into an easily absorbable carbohydrate source, such as baby rice cereal. The ratio of meat to carbohydrate should be between 1:1 to 1:2. Flavoring the rice with stocks or soups derived from the same animal species as the meat used, will often improve palatability. Vitamins and minerals should be added in sufficient quantities to balance the diet. The best way to do this is to use a human multivitamin and mineral supplement. Veterinary vitamin and mineral supplements labelled as "chewable" or "palatable" should be avoided in case they contain beef or lamb animal proteins as flavor enhancers.

The amount of dietary fiber is adjusted according to whether the small or large bowel is most affected by the inflammatory bowel disease. In small bowel disease, it is traditional to use low fiber diets because high dietary fiber content reduces the digestibility of many dietary components.[2,14,17] It should be pointed

DIET FOR THE TREATMENT OF CHRONIC DIARRHEA*†	
Rice baby cereal	8 oz
Diced chicken	6 oz
Vegetable oil	1 tsp
Dicalcium phosphate	1.5 tsp
Iodized 'lite' salt	0.5 tsp
Taurine	500 mg

* Balance with the addition of a suitable vitamin and mineral supplements. Diet provides approximately 475 kcal (1988 kJ) if necessary, palatability can be enhanced by adding small quantities of bacon grease or garlic powder or flavoring with chicken stock or chicken soup.
† For colitis, addition of fiber (psyllium: 1 tsp per meal) to the diet is useful.

FIG. 14.12 Diet for the treatment of chronic diarrhea.

out, however, that the binding and gelling properties of fiber are of potential benefit in the treatment of small bowel diarrhea. In contrast, if the predominant manifestation of the inflammatory bowel disease is colitis, high fiber diets are recommended. Fiber is quantitatively the most important nutrient reaching the large bowel and therefore, not surprisingly, can have significant effects on the large bowel function in healthy and diseased states. The type of fiber used markedly influences the nature of these effects. In the author's experience, soluble fibers (such as psyllium and fibrim) rather than insoluble fibers (such as bran) give better results. The beneficial effect of soluble fiber probably relates to the binding of irritant bile acids and the generation of volatile fatty acids, such as butyrate, that nourish the colonic epithelium and encourage growth of the normal bacterial flora (rather than pathogens).[6,17]

Pharmacological Therapy

In addition to the controlled diet, immunosuppressive therapy is often required. For small intestinal disorders use oral prednisone starting at 2 mg/kg bid for 2 weeks followed by a decreasing dose over 3 months. Intractable cats may require parenteral methylprednisolone injections (20 mg SC every 3–6 weeks). Refractory IBD may require metronidazole (15 mg/kg bid PO) or higher doses of prednisone (3 mg/kg bid PO) and the addition of azathioprine (0.3 mg/kg eod PO).

The latter drugs are often required when eosinophilic infiltrates involve other abdominal organs in addition to the bowel (hypereosinophilic syndrome). Regular CBC (including platelet counts) should be performed during azathioprine therapy because of the risk of bone marrow suppression. Sulfasalazine can be safely and effectively used for lymphocytic–plasmacytic colitis. Sulfasalazine is administered at 20–25 mg/kg bid for 1–2 weeks and then once daily for 2–5 weeks.

Suppurative colitis in cats may have a different pathogenesis to the other types of IBD. Mild forms of suppurative colitis are often associated with fur-impacted feces and it is possible that abrasion of the mucosa results in the suppurative response in some cats. Treatment includes lubricating laxatives such as white petrolatum, regular brushing of the coat and, if necessary, antibiotics such as ampicillin.

Idiopathic Megacolon

Idiopathic megacolon can affect cats of all ages, including kittens. The hallmarks of the disorder are a dilated colon with no evidence of physical or functional obstruction, and normal or near normal numbers of ganglion cells observed histologically in affected segments. Obstipation is the principal sign. Abdominal palpation and rectal examination are usually sufficient to identify an impacted colon and to rule out intestinal atresia and imperforate anus. Survey radiographs confirms megacolon and assesses the pelvis. After evacuation, a barium or air enema will help establish if a narrowed rectal or colonic segment is present. If a narrowed segment is detected proctoscopic evaluation with biopsy or surgical resection of the affected segment differentiates fibrous strictures, neoplastic proliferations and aganglionosis. Full thickness (usually excisional) biopsy is required to detect aganglionosis. The treatment of choice for idiopathic megacolon is subtotal colectomy which is very well tolerated by cats. Palliative medical therapies most often utilized are enemas, psyllium derivatives and various other laxatives such as lactulose liquid (1–5 ml bid PO).

References

1. Armstrong, W. M. (1987) Cellular mechanisms of ion transport in the small intestine. In *Physiology of the Gastrointestinal Tract*. Ed. L. R. Johnson. 2nd edn. pp. 1251–1265. New York: Raven Press.
2. Burrows, C. F., Kronfeld, D. S., Banta, C. A. and Merrit, A. M. (1982) Effects of fibre on digestibility and transit time in dogs. *Journal of Nutrition*, **112**, 1726–1732.
3. Carpenter, C. C. J., Greenough, W. B. and Pierce, N. F. (1988) Oral-rehydration therapy—the role of polymeric substrates. *New England Journal of Medicine*, **319**, 1346–1348.
4. Duffell, S. J. (1975) Some aspects of pancreatic disease in the cat. *Journal of Small Animal Practice*, **16**, 365–374.
5. Gryboski, J. D. (1991) Gastrointestinal aspects of cow's milk protein intolerance and allergy. *Immunology and Allergy Clinics of North America*, **11**, 733–797.
6. Eastwood, M. A. (1992) The physiological effect of dietary fibre: an update. *Annual Reviews of Nutrition*, **12**, 19–35.
7. Isolauri, E., Juntunen, M. and Wiren, S. (1989) Intestinal permeability changes in acute gastroenteritis: effects of clinical factors and nutritional management. *Journal of Pediatric Gastroenterology and Nutrition*, **8**, 466–473.
8. Iyngkaran, N., Robinson, M. J. and Sumithran, E. (1978) Cow's milk protein sensitive enteropathy: an important factor prolonging diarrhea of acute infectious enteritis in early infancy. *Archives of Diseases of Children*, **53**, 1503–1512.
9. Molla, A. M., Molla, A., Rhode, J. and Greenough, W. B. (1989) Turning off the diarrhoea, the role of food and ORS. *Journal of Pediatric Gastroenterology and Nutrition*, **8**, 81–84.
10. Patra, F. C., Mahalanabis, D., Jalan, K. N., Sen, A. and Banerjee, P. (1982) Is oral rice electrolyte solution superior to glucose electrolyte solution in the infantile diarrhea? *Archives of Diseases of Children*, **57**, 910–912.
11. Pearson, H., Gaskell, C. J., Gibbs, C. and Waterman, A. (1974) Pyloric and oesophageal dysfunction in the cat. *Journal of Small Animal Practice*, **15**, 487–501.
12. Powell, D. W. (1987) Intestinal water and electrolyte transport. In *Physiology of the Gastrointestinal Tract*. Ed. L. R. Johnson. 2nd edn. pp. 1267–1305. New York: Raven Press.
13. Simpson, K. W., Shiroma, J. T., Biller, D. S., Wicks, J., Johnson, S. E., Dimski, D. and Chew, D. Diagnosis of feline pancreatitis: clinical, ultrasonographic and pathologic findings in four cats. *Journal of Small Animal Practice* (in press).

14. Stock–Damge, C., Aprahamian, M., Raul, F., Humbert, W. and Bouchet, P. (1984) Effects of wheat bran on the exocrine pancreas and the small intestinal mucosa in the dog. *Journal of Nutrition,* **114**, 1076–1082.

15. Strombeck, D. R. and Guilford, W. G. (1990) Maldigestion, malabsorption, bacterial overgrowth, and protein-losing enteropathy. In *Small Animal Gastroenterology*. 2nd edn. pp. 296–319. Davis, CA: Stonegate Publishing.

16. Strombeck, D. R. and Guilford, W. G. (1990b) Nutritional management of gastrointestinal diseases. In *Small Animal Gastroenterology*. 2nd edn. pp. 690–709. Davis, CA: Stonegate Publishing.

17. Vahouny, G. V. (1987) Effects of dietary fibre on digestion and absorption. In *Physiology of the Gastrointestinal Tract*. Ed. L. R. Johnson. pp. 1623–1648. New York: Raven Press.

18. Williams, D. A. and Reed, S. D. (1989) Comparison of two methods for evaluation of pancreatic function in cats by assay of fecal proteolytic activity. *Proceedings of the American College of Veterinary Internal Medicine* (Abstr.), 1047.

CHAPTER 15

Hepatic Disease

H. CAROLIEN RUTGERS and JOHN G. HARTE

Introduction

Hepatic disease is quite common in the dog and cat, but can be difficult to recognize. Clinical signs and physical findings are often nonspecific and biochemical recognition is hindered by the fact that test results may also be influenced by drugs or diseases that involve the liver secondarily.[119] The incidence of hepatic lesions is relatively high and has been quoted in one study as between 10 and 12 cases per 1000, based upon biopsy or post-mortem results.[167] Not all these lesions, however, are due to primary liver disease. Degenerative or reactive secondary hepatopathies and chronic hepatitis are the histological lesions encountered most commonly,[167,186] but in many cases these changes are secondary to a primary disease elsewhere. Such changes usually resolve with treatment of the underlying disease. In the diagnostic evaluation of an animal with elevated liver enzymes it is therefore important to rule out primary non-hepatic disease prior to embarking upon invasive tests such as a liver biopsy.

The clinical spectrum of primary liver disease may range from occult chronic disease to fulminant acute liver failure. The absence of clinical signs does not rule out liver disease. The liver has a large functional reserve capacity and development of clinical signs implies marked diffuse impairment of liver function. With the increasing use of bio-chemical screening, liver disease is now frequently recognized earlier. This is important because timely diagnosis and treatment may in many cases prevent or reduce irreversible progression of the liver changes.

Pathophysiology of Liver Disease

The liver is the most metabolically active organ in the body and performs a key role in the metabolism of protein, carbohydrate, fat, minerals and vitamins (Fig. 15.1). The liver also has important regulatory and circulatory functions and plays a major role in the metabolic regulation of both endogenous (such as steroid hormones) and exogenous compounds (such as drugs). The hepatic reticuloendothelial system (RES) further removes waste products and potentially harmful substances, especially from the portal venous blood, and thus functions as a sieve for blood coming from the intestine before it reaches the systemic circulation.

Liver disease may cause a wide variety of metabolic derangements (Fig. 15.2). Fasting hypoglycemia may occur due to glycogen

IMPORTANT NUTRITIONAL AND METABOLIC FUNCTIONS OF THE HEALTHY LIVER	
Nutrient	**Functional role of the liver**
Protein	Storage of newly synthesized amino acids
	Protein synthesis
	Deamination/transamination of amino acids
	Protein reserve
	Purine and pyrimidine synthesis
Carbohydrate	Blood glucose regulation
	Glycogen reserve
	Intermediary metabolism
Lipid	Phospholipid synthesis
	Cholesterol synthesis
	Bile salt synthesis/bile secretion
	Fatty acid oxidation
	Lipoprotein synthesis
Vitamins	Storage of vitamins A, D, E, K, B$_{12}$, C
	Vitamin D hydrolysis for renal activation
	Utilization of B vitamins as cofactors
Minerals	Storage of iron, copper, zinc, manganese

FIG. 15.1 Important nutritional and metabolic functions of the healthy liver.

depletion and inadequate gluconeogenesis but is uncommon because of the large hepatic reserve for maintaining euglycemia; it is seen with severe acute liver failure, large neoplasms (especially hepatoma) or congenital porto-systemic shunts (particularly in toy breeds).[14] The liver is the major site of synthesis for most plasma proteins (except γ-globulins).

The reduced production of coagulation factors is one of the factors contributing to the increased bleeding seen in liver disease and reduced synthesis of albumin may result in hypoalbuminemia in chronic liver disease. Albumin production, however, is also strongly dependent upon the availability of amino acids (especially tryptophan) and is reduced

PATHOPHYSIOLOGY OF LIVER DISEASE		
Function		**Clinical effects**
Nutrient metabolism	Carbohydrate	Hypo/hyperglycemia
	Protein	Hypoalbuminemia
	Lipid	Hyperammonemia
		Decreased bile salt formation
		Abnormal cholesterol/lipid levels
		Decreased lipoproteins and clotting factors
	Vitamins	Coenzyme deficiencies
Catabolism	Bile pigment metabolism	Hyperbilirubinemia
	Bile acid metabolism	Malabsorption
	Steroid metabolism	Hyperadrenocorticism
	Drug metabolism	Hyperestrogenism
		Impaired drug detoxification
Circulatory	Portal blood pressure homeostasis	Increased portal blood pressure (ascites)
		Hyperammonemia where portal circulation is by-passed e.g. shunts

FIG. 15.2 Pathophysiology of liver disease.

by anorexia and malnutrition. The serum albumin concentration may be further lowered as a consequence of intestinal or renal protein loss or may appear reduced if the volume of distribution is increased (as in portal hypertension with ascites) and it is therefore an insensitive index of hepatic function. However, monitoring of serum albumin is important in the dietary management of an animal with hepatic disease because it also reflects the animal's nutritional status. Serum globulin concentrations are usually normal in liver disease because γ-globulins are synthesized outside the liver; hyperglobulinemia may occur if there is antigen-mediated hepatic disease or decreased clearance of portal antigens.

Hyperammonemia is one of the hallmarks of hepatic encephalopathy (HE), which is a clinical syndrome characterized by an altered state of consciousness, abnormal mental status and impaired neurological function.[35,65,77,160,170] Ammonia is generated from the catabolism of proteins, nucleic acids and urea. Most exogenous ammonia is produced in the colon through the degradation of dietary amines and the action of bacterial ureases on urea, after which it is absorbed into the portal blood and converted into urea in the liver by the Krebs–Henseleit urea cycle. In patients with abnormal blood flow to the liver (extra- or intrahepatic shunting of blood) or, less commonly, with severe parenchymal damage, ammonia is not detoxified and enters the systemic circulation.[174] Ammonia is neurotoxic, but its exact role in the pathogenesis of HE is controversial.[2,123,160,163,170] Increases or imbalances of other metabolites such as mercaptans, free fatty acids and further breakdown products of protein metabolism probably also play a role in the pathogenesis of HE (Fig. 15.3)[2,123,160,163,170]; however, blood ammonia is currently the only parameter easily measurable in the laboratory that may indicate the presence of HE. Abnormal amino acid concentrations, particularly a low ratio of branched chain (BCAA) to aromatic amino acids (AAA), have also been implicated, but treatments aimed at normalizing these imbalances (such as infusions with BCAA)

SUBSTANCES BELIEVED TO BE INVOLVED IN THE DEVELOPMENT OF HEPATIC ENCEPHALOPATHY

Ammonia
Mercaptans (e.g. methanethiol)
Octanoic acid (short chain fatty acid)
False neurotransmitters:
  Tyramine (precursor of octopamine)
  Octopamine
  Tryptophan (aromatic amino acid; metabolized to serotonin)
Decreased branched chain to aromatic amino acid ratio
Gammaaminobutyric-acid (GABA) and GABA-like agents
Endogenous benzodiazepine receptor ligands

FIG. 15.3 Substances believed to be involved in the development of hepatic encephalopathy.

have given inconsistent results.[66,89,123,131] Very rarely, hyperammonemic encephalopathy has been reported in disorders not associated with acquired hepatocellular disease or portosystemic shunts, such as congenital hepatic urea cycle enzyme deficiencies,[65,66,168] organic acidemia, urinary tract obstruction and infection with urea-splitting bacteria.[72] Blood urea may also be low when ammonia metabolism is altered, but this is not a consistent finding due to the many variables influencing urea levels.

The consequences of abnormal lipid metabolism are still largely unknown. Total serum cholesterol may be elevated by extrahepatic bile duct obstruction (increased synthesis and decreased excretion) while it may be decreased in severe hepatocellular failure or portosystemic shunting of blood (decreased synthesis, decreased absorption from the gut if fat absorption is impaired and possibly increased conversion into bile acids). Abnormalities in lipoprotein metabolism may play a role in the etiopathogenesis of hepatic lipidosis in the cat.[15,17,41]

Diagnosis of Liver Disease

History and Physical Findings

Clinical signs (Fig. 15.4) are rarely evident until liver disease is advanced, but even then

HISTORICAL AND CLINICAL FINDINGS IN LIVER DISEASE

Depression
Anorexia
Weight loss
Vomiting
Diarrhea
Polyuria/polydipsia
Grey (acholic) stools
Pigmented urine
Bleeding tendency (skin, gut, urine)
Jaundice
Ascites
Drug intolerance
Encephalopathy
(depression and dementia, pacing, circling,
head-pressing, hypersalivation (cat), amaurosis
(blindness), seizures, coma)

FIG. 15.4 Historical and clinical findings in liver disease.

DIFFERENTIAL DIAGNOSIS OF HEPATOMEGALY

Infiltrative liver disease (neoplasia, amyloidosis)
Severe congestion (congestive heart failure,
posthepatic venous obstruction)
Diffuse inflammation (cat–acute and chronic;
dog–acute only)
Excessive storage (lipidosis, glycogen)
Nodular hyperplasia
Work hypertrophy (severe anemia)

FIG. 15.6 Differential diagnosis of hepatomegaly.

they are often vague. None of the signs are diagnostic for liver disease and even the more specific signs, such as jaundice and ascites, can also be found with other disorders such as hemolysis or congestive heart failure. Signs may be acute or chronic and in the latter case they are frequently intermittent. Chronic hepatic disease can, however, be associated with acute onset of clinical signs once the liver's large reserve capacity has been surpassed and this can initially seem to be an acute disease.

Findings on physical examination are usually nonspecific (Fig. 15.5). Jaundice, hepatomegaly and ascites are the only abnormalities that may point toward an underlying problem. Hepatomegaly (Fig. 15.6) is common in feline liver disease, whereas canine liver diseases are often associated with a reduction in liver

size, especially when chronic. Low-protein ascites due to portal hypertension may occur in dogs with chronic hepatitis, cirrhosis and fibrosis but is rarely found in the cat. In cats with suspected liver disease the thyroid gland should be palpated carefully because hyperthyroidism can result in clinical and biochemical changes resembling liver disease. Ocular fundic examination in the cat may also reveal chorioretinitis in animals with systemic disease such as feline infectious peritonitis (FIP), feline leukemia virus (FeLV) infection or toxoplasmosis.

Laboratory Evaluation

Baseline Laboratory Evaluation

Baseline evaluation is essential for diagnostic evaluation and to rule out other metabolic illnesses. It should at least include hematology, a serum biochemical profile, urinalysis and in cats a FeLV/FIV test. A T_4 determination should be done in older cats because hyperthyroidism may cause nonspecific elevation of liver enzymes. In a nonjaundiced animal this may be followed up with liver function tests if needed. An algorithm for the diagnostic approach can be found in Fig. 15.7.

Hematology is often normal but occasionally a mild to moderate anemia, consistent with anemia of chronic disease, is seen. Marked anemia suggests hemolysis or blood loss due to a coagulopathy or a bleeding gastrointestinal ulcer. Acanthocytes, or poikilocytes, are abnormally shaped erythrocytes that may be seen in cats with portosystemic shunts.[41,42,158] Microcytosis (MCV < 65 fl),

PHYSICAL FINDINGS IN LIVER DISEASE

Weight loss
Stunted growth
Jaundice
Ascites
Hepatomegaly
Painful liver
Hemorrhage

FIG. 15.5 Physical findings in liver disease.

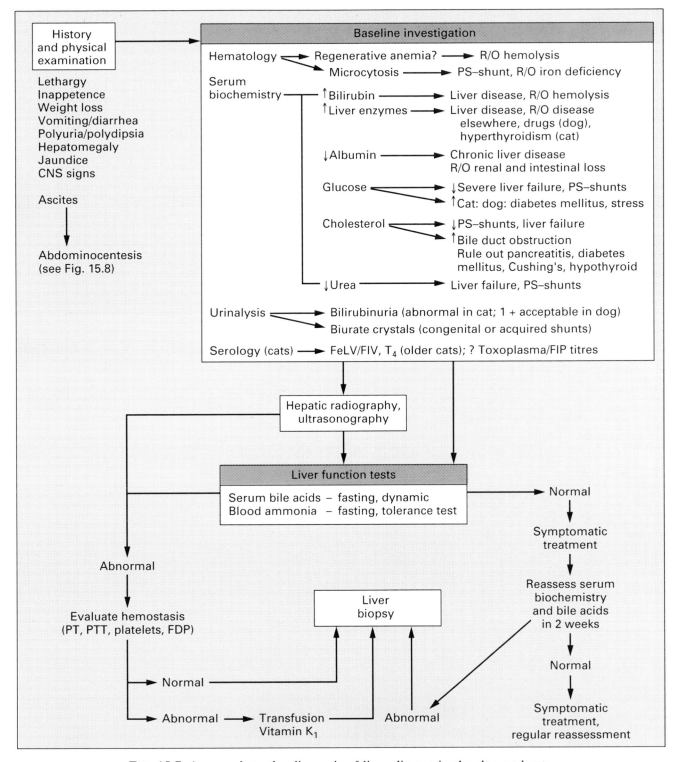

FIG. 15.7 Approach to the diagnosis of liver disease in the dog and cat.

the cause of which is unknown, occurs in up to 70% of dogs with portosystemic shunts.[46,71]

Serum enzymes used as biochemical "markers" of hepatobiliary disease are alanine aminotransferase (ALT) and aspartate amino-transferase (AST), enzymes that reflect hepatocellular injury, and alkaline phosphatase (ALP) and γ-glutamyl transferase (GGT), which increase as a result of increased production due to cholestasis or drugs. ALT occurs in the soluble component of the hepatocellular cytoplasm, while AST is located in both the cytoplasm and mitochondria; increased serum AST activity therefore implies greater hepatic damage. Elevations in ALT are associated with any type of hepatocellular membrane damage, degeneration or necrosis, and the degree of elevation generally reflects the number of hepatocytes damaged. The half-life of this enzyme is short, but levels may remain elevated for several weeks while the liver regenerates after the original insult.

Increased serum activities of ALP and GGT reflect increased synthesis by epithelial cells lining the bile canaliculi, which generally is the result of reduced bile flow and cholestasis. In the dog, anticonvulsants and glucocorticoids (either endogenous or exogenous) can increase serum activities of ALP and GGT by inducing enzyme production; glucocorticoids can in addition stimulate the production of a novel steroid-induced isoenzyme that can result in dramatic increases in serum levels.[9,148] Glucocorticoids can also induce hepatocyte production and release of ALT but this is usually only mild. The various isoenzymes of ALP (steroid-induced isoenzyme, bone isoenzyme and liver isoenzyme) can be separated using several biochemical methods, but this is of limited diagnostic usefulness.[197] It is important to appreciate that primary hepatocellular disease may cause ALP elevation if there is significant intrahepatic cholestasis, either due to bile duct compression by swollen hepatocytes or in association with fibrosis and cirrhosis.[49] Periportal lesions cause greater increases than do centrilobular lesions. In cats, ALP has a very short half-life (6 hr in the cat, vs. 3 days in the dog); therefore minor increases are significant. GGT is more reliable than ALP as a marker of cholestasis in the cat and is also less influenced by nonhepatic disease or drugs in the dog.[49]

Measurement of total bilirubin is a relatively insensitive test of hepatic function. Division into unconjugated and conjugated bilirubin using the Van den Bergh test has been used historically to distinguish between hemolytic (predominantly unconjugated hyperbilirubinemia), posthepatic (predominantly conjugated) and hepatic jaundice (varying combinations), but this is now considered too inaccurate.[190] Recently, it has been shown that there is a form of conjugated bilirubin (biliprotein) that is covalently bound to albumin and thus has a long half-life of approximately 10 days.[190] The slow degradation of this biliprotein can therefore prolong measured hyperbilirubinemia long beyond resolution of the original insult.

Bilirubinuria is a sensitive indicator of hyperbilirubinemia. Dogs have a low renal threshold for bilirubin and mild bilirubinuria may be normal in a male dog. The cat has a high threshold and any bilirubinuria is therefore abnormal.[41] Urobilinogen should be absent in complete biliary obstruction, but this is an unreliable test and of little real use. Ammonium biurate crystals are seen with chronic hyperammonemia; they are usually seen in congenital portosystemic shunts but occasionally also in acquired shunts.[155]

Liver Function Tests

These are used to assess the functional integrity of the liver if serum bilirubin is normal. Liver function tests of value in the assessment of nonhyperbilirubinemic liver disease include the sulfobromophthalein (BSP) retention test, indocyanine green (ICG) clearance, ammonia tolerance and serum bile acid analysis.[42,44,46,118,119]

Serum bile acids: Because of its sensitivity and convenience, the preferred liver function test is the measurement of serum bile acid levels during fasting and 2 hr postprandially.[47] Serum bile acids are formed in the liver as metabolites of cholesterol, excreted via bile into the intestine, reabsorbed from the ileum and removed from the portal blood by the liver (enterohepatic circulation). Normal fasting levels in the dog have been reported as less than 5 or 15 μmol/l.[47,55,97] Normal postprandial levels (2 hr after feeding a standard meal) are less than 20 μmol/l in dogs.[47,55] This test is especially useful for the diagnosis of portosystemic shunts and can also be used to support the need for a liver biopsy.[47,121]

Blood ammonia: Blood ammonia may be

increased especially in portosystemic shunting of blood, but also if there is severe reduction in functional hepatic mass. Although common, hyperammonemia is not an invariable finding in hepatic encephalopathy. If fasting ammonia is normal an oral ammonia tolerance test may be performed. Plasma ammonia is measured before and 30 min after a dilute solution of ammonium chloride (NH_4Cl 100 mg/kg) is given via a stomach tube; no increase indicates a normal tolerance.[118] This test is most useful for the diagnosis of portosystemic shunts and in documenting HE in patients with primary signs of central nervous system (CNS) dysfunction. Its usefulness is limited by the lability of ammonia; blood should be collected on crushed ice and analyzed within 2 hr. Plasma may be stored at $-20°C$ for 48 hr and still give reliable results in cats but not in dogs.[87,130]

Coagulation tests: These are indicated in animals with a bleeding tendency and prior to blind biopsy techniques. The liver synthesizes most coagulation factors and decreased synthesis and storage in severe diffuse liver disease may cause clinical or subclinical coagulopathies.[8,10] The prothrombin time (PT) is usually prolonged first because factor VII has the shortest half-life and this is later followed by a prolonged activated partial thromboplastin time (PTT). Severe diffuse liver disease may also result in disseminated intravascular coagulation (DIC) and in that case thrombocytopenia, prolonged PT and PTT and elevated fibrin degradation products may be found. Animals in DIC usually have clinical bleeding (gastrointestinal bleeding, skin ecchymoses) and prognosis in these patients is grave. In extrahepatic bile duct obstruction, reduced micellation of fats can result in the reduced uptake of fat-soluble vitamins and consequently a vitamin K-dependent coagulopathy. Coagulation studies before and after the administration of vitamin K can distinguish between these two coagulopathies: correction of the coagulopathy supports vitamin K deficiency.

Peritoneal fluid cytology: Abdominocentesis is performed in animals with ascites. A modified transudate (protein < 25 g/l, few cells) is common in dogs with chronic liver disease. Cats with liver disease usually do not have ascites, with the exception of cats with lymphocytic cholangitis, half of which have a high protein effusion.[107] The fluid should always be examined grossly as well as for total protein and cell types in order to distinguish between neoplasia, FIP infection, hemorrhage, chylous effusion and ruptured gall bladder (Fig. 15.8).

Liver Imaging

Radiography: Survey abdominal radiographs give an idea about liver size (hepatomegaly, microhepatia) and shape (hepatic mass). In dogs reduced hepatic size may be an indicator of chronicity and is also a common finding in animals with congenital portosystemic shunts. Renomegaly is common in animals with congenital portosystemic shunts, but concomitant renal or cystic calculi are often radiolucent and only observed on contrast studies.

Ultrasonography: Ultrasonographic examination of the hepatobiliary system can provide valuable information about the internal structure of the liver and the biliary tract (Fig. 15.9).[103] It allows visualization of dilated bile ducts suggestive of extrahepatic obstruction and is much more sensitive than radiography in detecting choleliths. Intrahepatic masses can be visualized (neoplasia, granulomas, abscesses), and a diffuse increase or decrease in echogenicity may indicate generalized disease (hepatitis, lymphosarcoma, cirrhosis).[22] It is also sometimes helpful in the localization of portosystemic vascular anomalies and can be used to guide the needle during percutaneous liver biopsy, especially useful when biopsying focal lesions.

Contrast radiography: Cholecystography is rarely done in the dog, but angiography (usually jejunal vein portography) is commonly used for the identification of congenital vascular anomalies. This technique involves laparotomy and cannulation of a mesenteric vein using a catheter of the largest possible diameter (e.g. 19 gauge, 8 in.), after which 1 ml/kg of an aqueous contrast medium is

CHARACTERISTICS OF PERITONEAL EFFUSIONS			
	Transudate	**Modified transudate**	**Exudate**
Parameter			
Total protein (g/l)	< 25	25–50	> 30
Specific gravity	< 1.018	1.010–1.030	>> 1.018
Number of cells (x10^9/l)	<< 0.5	0.5–5.0	> 5.0
Clinical correlate	Renal protein loss Intestinal protein loss Liver disease Arteriovenous shunting	Liver disease Heart disease Vena caval disease Neoplasia	Peritonitis FIP Pancreatitis Urinary tract rupture Biliary tract rupture Hemorrhage Chylous effusion

FIG. 15.8 Characteristics of peritoneal effusions.

manually injected as rapidly as possible and a radiograph taken at the end of injection.[24,25] This is repeated for the ventrodorsal exposure. Portal pressure can be measured simultaneously (normal 8–13 cm H_2O).

Scintigraphy: Nuclear hepatic scintigraphy is a noninvasive test for measurement of relative hepatic blood flow that has been used to assess liver size and as an aid in the diagnosis of neoplasia and portosystemic shunts.[69,88] Quantitative hepatobiliary scintigraphy has recently also been described as a measure of bile flow in dogs with cholestasis.[151] The need for radioactive materials limits the use of these techniques at the moment to research and/or teaching institutions.

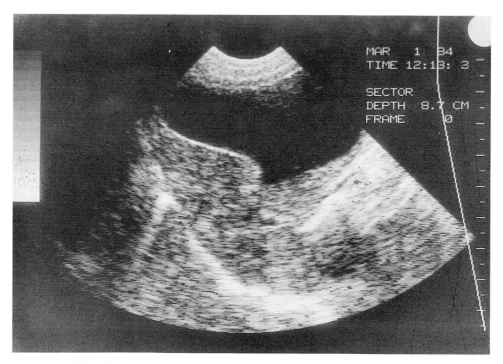

FIG. 15.9 Ultrasonogram of a 3-year-old Scottish Terrier with marked cholestatic jaundice demonstrating enlargement of the gallbladder and multiple hypoechoic areas in the hepatic parenchyma. Histological diagnosis was lymphosarcoma. The neoplastic foci in this case caused intrahepatic obstruction to bile flow.

Liver Cytology and Biopsy

Histological examination of liver tissue is often necessary to verify hepatic disease, to define whether the problem is primary or secondary and to clarify the cause of abnormal liver tests and/or size (Figs. 15.10 and 15.11).[100,132] Liver aspirates for cytology can be obtained relatively easily by needle aspiration when there is hepatomegaly; this is particularly useful in the cat and for the diagnosis of diffuse diseases such as lymphosarcoma.

REASONS FOR PERFORMING A LIVER BIOPSY

- Explain abnormal liver tests
  Persistent elevations of serum liver enzymes
  Abnormal liver function tests
- Abnormal liver size (hepatomegaly or microhepatia)
- Progression of liver disease
- Evaluate response to treatment

FIG. 15.10 Reasons for performing a liver biopsy.

The animal's coagulation status should be assessed prior to a blind percutaneous biopsy. A full coagulation profile includes a platelet count, measurement of prothrombin and partial thromboplastin time and of fibrin degradation products (FDP). In practice, normal whole blood clotting time (fresh blood in a glass tube should clot within 6 min) and platelet numbers (at least 12–15 platelets/$1000\times$ field) are generally sufficient for a liver biopsy to proceed.[167]

The advantages of percutaneous biopsy are that it is minimally invasive, does not require general anesthesia and can be repeated.[64,132] The major disadvantage is the potential for sampling error and it is therefore best used in diffuse disease. Ultrasound-guided biopsy does allow some selection of area to be

---

LIVER BIOPSY TECHNIQUES

- Percutaneous
  Needle aspirate for cytology
  Blind biopsy with Menghini or Tru-Cut needle
  Ultrasound-guided biopsy
- Keyhole (Tru-Cut needle)
- Laparoscopy
- Laparotomy

FIG. 15.11 Liver biopsy techniques.

---

biopsied. Laparotomy allows complete visualization and can also be therapeutic (e.g. in biliary obstruction), but it requires anesthesia (preferably isoflurane) and is maximally invasive. Biopsy in an animal with abnormal coagulation tests should be addressed on an individual basis.[119] Vitamin K responsive coagulopathies should respond within 24–48 hr to subcutaneous injection of vitamin K_1. If the patient does not respond, the coagulopathy is probably due to hepatocellular insufficiency and a transfusion with fresh plasma or whole blood to replenish clotting factors is required prior to biopsy.

Primary Versus Secondary Liver Disease

Due to its central role in metabolism, the liver may be affected by a variety of other disease conditions ("innocent bystander") associated with episodes of anoxia, toxin release, nutritional imbalances, hormone and metabolic changes and infective organisms. Often this is only manifested by a mild rise in liver enzymes; occasionally, there are some nonspecific changes on liver biopsy that resolve following the resolution of the primary disease. Examples of disorders that secondarily lead to liver disease are acute pancreatitis, severe acute or chronic small intestinal disease, severe extrahepatic bacterial infection, shock, anemia, congestive heart failure, trauma and, in the cat, hyperthyroidism.

Cholestatic Syndromes

These are characterized by increased plasma concentrations of substances normally excreted in the bile (e.g. bilirubin, bile acids and cholesterol) or increased activities of liver enzymes associated with the biliary epithelium (ALP, GGT). Cholestasis may be intrahepatic, resulting from functional or diffuse mechanical alteration of hepatocyte, canaliculus or intrahepatic ductules, or it may be extrahepatic, associated with mechanical obstruction of the extrahepatic biliary tract. Extrahepatic cholestasis tends to be more severe, with marked elevation of serum bilirubin and bile acid levels as well as of cholesterol and liver enzyme activities.[156]

Sometimes the two syndromes can only be distinguished via liver biopsy.[120,122]

Approach to the Jaundiced Animal

Jaundice, or icterus, is characterized by hyperbilirubinemia and deposition of bile pigment in tissues, clinically evident when total serum bilirubin exceeds 30 μmol/l (Fig. 15.12). Prehepatic jaundice occurs through diseases causing accelerated red blood cell destruction; bilirubin levels rise later in the

DIFFERENTIAL DIAGNOSIS OF ICTERUS
Hemolytic

Immune-mediated
  Autoimmune hemolytic anemia (AIHA)
  Toxic (onions, zinc, drug reaction)
  Incompatible blood transfusion
  FeLV-related (cat)
Infectious
  Babesiosis
  Hemobartonellosis
  Ehrlichia
Toxic
  Acetaminophen
  Lead
  Copper
  Methylene blue (cat)

Hepatocellular

Toxic
  Idiopathic hepatic necrosis (toxin?)
  Bacterial endotoxin
  Drugs (arsenic, mebendazole, trimethoprim-sulpha, acetaminophen, especially
    the cat)
  Hepatic copper accumulation
Neoplasia
  Primary or metastatic
  FeLV-related (cat — lymphosarcoma, myeloproliferative)
Degenerative
  Hepatic fibrosis/cirrhosis
  Lipidosis (idiopathic feline, diabetes mellitus, secondary)
Infectious
  Bacterial (leptospirosis, septicemia)
  Viral (ICH, FIP)
Inflammatory of unknown origin
  Chronic hepatitis (dog)
  Cholangiohepatitis (cat)
Congenital defects (not yet reported in the dog or cat)
  Impaired uptake, conjugation or excretion of bilirubin

Obstructive

Intrahepatic
  Cholangitis/cholangiohepatitis
  Neoplasia
  Any severe hepatopathy associated with cell swelling and/or inflammatory
    infiltrates in the periportal area (especially dogs)
Extrahepatic
  Acute pancreatitis — bile duct compression (especially in dog!)
  Neoplasm compressing bile duct (biliary tree, pancreas, duodenum)
  Cholecystitis/cholelithiasis
  Stricture of bile duct
  Traumatic rupture of biliary tree
  Sludged bile syndrome with cholangiohepatitis (cat)

FIG. 15.12 Differential diagnosis of icterus.

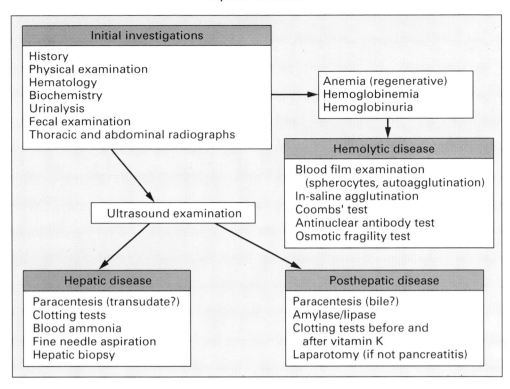

FIG. 15.13 Diagnostic approach to the jaundiced dog and cat.

course of the disease when superimposed hepatic damage occurs. Hepatic jaundice is not a sensitive indicator of disease and its presence depends on the extent and location of the lesion. Liver diseases that are associated with marked cell swelling and/or inflammatory infiltrates in the periportal region usually cause more severe cholestasis and jaundice. Posthepatic jaundice is due to biliary obstruction and is usually associated with the most marked cholestasis; however, numerous intrahepatic disorders can clinically mimic posthepatic jaundice, especially in the cat. The diagnostic approach to patients with jaundice is summarized in Fig. 15.13.

Canine Liver Diseases

Primary liver disease in the dog can be classified histologically into inflammatory and noninflammatory diseases (Fig. 15.14). Noninfectious inflammatory diseases (chronic hepatitis and cirrhosis) are by far the largest category, followed by hepatic neoplasia, toxic hepatopathy and portosystemic shunts.[185]

Infectious Liver Diseases

Leptospirosis (caused by *Leptospira icterohemorrhagiae* and *L. canicola*) and infectious canine hepatitis (ICH, caused by canine adenovirus 1) are rare today because most dogs are vaccinated against them. Leptospirosis, however, should still be suspected in dogs with acute liver failure, especially when

CLASSIFICATION OF LIVER DISEASE IN THE DOG	
Inflammatory	
Infectious	Bacterial (leptospirosis, ascending from gut) Viral (ICH, herpes)
Noninfectious	Chronic hepatitis Cirrhosis/fibrosis Toxic and drug-induced
Noninflammatory	
Portosystemic shunts Neoplasia Steroid-induced hepatopathy Lipidosis (obesity, endocrine) Surgical (trauma, torsion, entrapment in hernia)	

FIG. 15.14 Classification of liver disease in the dog.

ETIOLOGY OF CANINE CHRONIC HEPATITIS	
Hepatic copper accumulation	Copper-storage hepatitis in Bedlington Terriers and in West Highland White Terriers
	Copper-associated hepatitis in Doberman Pinschers and in Skye Terriers
Putative infectious causes	Leptospirosis-associated chronic hepatitis
	Infectious canine-adenovirus-associated chronic hepatitis
	Canine acidophil cell hepatitis
Drug-induced	Anticonvulsants (primidone, phenytoin)
Idiopathic	Canine chronic active hepatitis
	Lobular dissecting hepatitis
	Idiopathic chronic hepatitis

FIG. 15.15 Etiology of canine chronic hepatitis.

accompanied by acute renal failure.[11] Non-specific bacterial invasion of the liver may occasionally occur due to ascending biliary infections (often *Escherichia coli*) or as a result of trauma and/or hypoxia allowing proliferation of bacteria normally present in the liver (such as clostridia).[51,105]

Chronic Hepatitis

Chronic hepatitis encompasses a range of inflammatory liver diseases resulting from multiple causes,[75,76] most of which can end in cirrhosis. The etiology is often never determined; metabolic, infectious, toxic and autoimmune factors have been implicated. Copper-associated chronic hepatitis has been documented in a number of breeds as an inherited etiology, and the increased incidence of specific types of chronic hepatitis in certain breeds suggests a genetic predisposition.[5]

Copper accumulation: Copper accumulation associated with chronic hepatitis may result from either a primary defect of copper metabolism (as in Bedlington and West Highland White Terriers) or from secondary

FIG. 15.16 Micronodular cirrhosis in a 5-year-old Bedlington Terrier with chronic copper hepatotoxicosis.

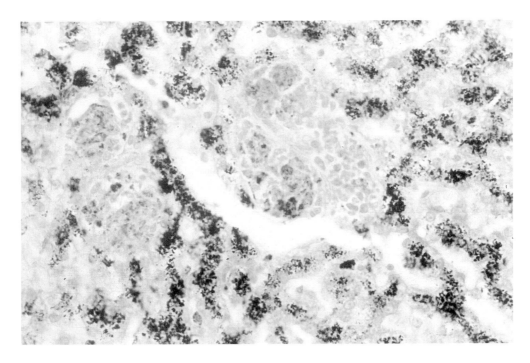

FIG. 15.17 A copper stain (rubeanic acid) identifies copper as dark-staining granules in the liver of a 5-year-old Bedlington Terrier with copper hepatotoxicosis.

copper retention resulting from altered normal biliary copper excretion. Copper-storage hepatitis in the Bedlington Terrier is inherited in an autosomal recessive manner and has a high incidence.[83,84,95,172] Clinical signs may range from those of an acute fulminant hepatitis, more common in young-adult dogs, to those of chronic hepatitis and cirrhosis (Fig. 15.16), usually seen in middle-aged to older dogs. Some affected animals are clinically asymptomatic but hepatitis may be detected on the basis of a persistent elevation of serum ALT activity.[84,187] In early stages of the disease, the liver may appear histologically normal, but increased copper can be demonstrated cytochemically (Fig. 15.17) using rubeanic acid or rhodamine stains[94,179] and quantitative analysis demonstrates increased liver copper values (normal values are less than 200 µg/g dry weight[187]). A similar hereditary copper-induced liver disease has also been described in West Highland White Terriers in the U.S.A.[183] It differs from the disease in Bedlington Terriers in that copper accumulation is not age-related in West Highland White Terriers while the magnitude of copper accumulation is lower. Hepatic copper accumulation has further been reported in a form of chronic hepatitis seen in Dobermann Pinschers.[56,61,96] Affected animals usually are presented with advanced liver disease and fulminant clinical signs. The predominantly periportal distribution of copper suggests that increased hepatic copper in these dogs is a consequence of cholestasis and reduced biliary excretion. Chronic hepatitis and cirrhosis associated with hepatic copper accumulation also occurs in related Skye Terriers; it has been speculated that this is caused by a disorder of intracellular bile metabolism resulting in disturbed bile secretion and copper accumulation.[78] Clinical signs frequently are intermittent, with ascites as a prominent feature, but acute and fatal liver failure can occur. In addition to the above breeds, familial liver disease associated with high liver copper levels has also been suggested to occur in Cocker Spaniels and Labrador Retrievers[181] and elevated liver copper levels have been reported in other breeds as well as mixed breeds.[177,181,182,185]

Final diagnosis of copper toxicosis relies on

wedge biopsies in which copper content is assessed quantitatively and histologically.[179] Cytological detection of copper following fine needle aspiration of the liver has been described[175] but this may occasionally result in false negative results.[79] Measurement of ^{64}Cu excreted in the stool during 48 hr after an intravenous dose of isotope has recently been described as an aid in the early diagnosis of copper toxicosis in dogs.[31] This technique was associated with few false negatives but many (up to 30%) false positives, so positive dogs still need a liver biopsy to verify the diagnosis.

Infectious: Persistent infection with canine adenovirus-1 (the cause of infectious canine hepatitis) may play a role in the induction and maintenance of chronic hepatocellular damage in some dogs,[68,140] but this is unlikely to be a common cause. The same applies to infection with leptospires that has been related to an outbreak of chronic active hepatitis in five American Foxhounds from a kennel.[27] An as yet unidentified agent (virus?) has recently also been suggested as the cause of chronic hepatitis and cirrhosis in dogs in the Glasgow area of the U.K.[91,92]

Drug-induced: Some drugs may cause a chronic hepatitis and specifically the anticonvulsant drugs (notably primidone) have been implicated.[36-38,139] Dogs treated with anticonvulsant drugs often develop moderate increases in serum liver enzyme activities due to enzyme induction,[119] but up to 15% of dogs receiving extended anticonvulsant drug therapy may develop significant hepatotoxicity, initially manifested by persistent marked abnormalities in hepatic function tests and later by clinical signs of liver disease.[36] By the time clinical signs are noticed the liver disease is frequently advanced[38]; hence, in dogs given anticonvulsant therapy, liver tests should be monitored periodically. Marked enzyme elevations may necessitate a change in anticonvulsant drugs or a decrease in dosage.[139]

Immune-mediated: Immune-mediated mechanisms may play a role in many cases of chronic hepatitis. It is thought that non-specific damage to hepatocytes results in release of liver antigens whereafter antibodies are formed against these antigens; these antibodies can then damage intact hepatocytes. Primary immune-mediated disease similar to chronic active hepatitis (CAH) in

FIG. 15.18 Cirrhotic appearance of the liver of a 7-year-old female Labrador Retriever with chronic active hepatitis.

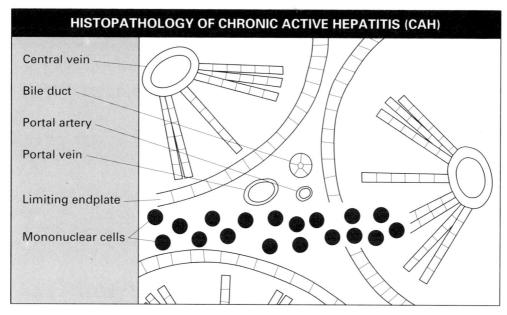

HISTOPATHOLOGY OF CHRONIC ACTIVE HEPATITIS (CAH)

Central vein
Bile duct
Portal artery
Portal vein
Limiting endplate
Mononuclear cells

FIG. 15.19 Schematic diagram of a liver lobule, demonstrating the pathology of chronic active hepatitis. Mononuclear inflammatory cells (lymphocytes and plasma cells) infiltrate into the portal area and destroy the limiting endplate; the inflammatory process can then extend into the hepatic parenchyma resulting in diffuse inflammation, fibrosis and cirrhosis.

humans may also occur in a small number of dogs (Fig. 15.18).[13,61,165,178,179] Diagnosis of CAH in the dog generally is made on the basis of liver biopsy changes compatible with human criteria for CAH,[30,80] such as peri-portal inflammation (Fig. 15.19), piecemeal necrosis and bridging necrosis.[13,61,117,180] Human patients with autoimmune CAH often have evidence of immune system dysfunction, characterized by hypergammaglobulinemia and circulating autoantibodies against smooth muscle, liver membrane and nuclear protein. In a recent study of dogs with various forms of liver disease, antinuclear and anti-liver membrane autoantibodies were found in low and varying levels and it was suggested that, in contrast to man, these autoantibodies might occur secondarily to the liver damage and be of minor importance in the pathogenesis and diagnosis of liver disease.[6]

Idiopathic: There are other forms of chronic hepatitis wherein the etiology is unknown, whereas hepatic histological findings do not resemble human CAH. Sometimes these are breed-related; an example is lobular dissecting hepatitis prevalent in young black Standard Poodles in the U.K.[19]

It is likely that there are as yet unascertained metabolic abnormalities underlying such diseases.

Idiopathic chronic hepatitis is more common in female dogs between 2 and 10 years-of-age. In the early stages clinical signs are usually absent or minimal, but when the disease progresses signs of liver disease develop. Signs such as ascites, jaundice and hepatic encephalopathy usually imply advanced disease and poor prognosis. Laboratory findings typically include elevated liver enzymes (especially ALT). Liver biopsy is required for definitive diagnosis and in all cases copper levels should be determined to rule out copper hepatotoxicosis. Early diagnosis is important so that appropriate therapy may be instituted. Unfortunately, many cases are only diagnosed when clinical signs appear and the disease is quite advanced.

Treatment of chronic hepatitis consists of removing the inciting cause, if identified, and antiinflammatory therapy. Decoppering agents are indicated in animals with copper-storage hepatitis; they are less likely to be beneficial in dogs with copper accumulation secondary to cholestasis. Corticosteroids are often prescribed although their use is controversial.[153,165,167] Although prednisolone therapy may prolong survival[165] it probably does not prevent progression into cirrhosis. Antiinflammatory drugs of unknown efficacy in the dog that can be tried in cases that do not respond to conventional treatment are azathioprine, colchicine and ursodeoxycholic acid.[153,185] Prognosis in dogs with chronic hepatitis is always guarded, even more so when the animal has developed clinical signs or if biopsy indicates cirrhosis.

Cirrhosis and Fibrosis

Cirrhosis is the end-stage of many liver diseases and is characterized by disruption of normal architecture by regenerative nodules and fibrosis. Animals generally have clinical and biochemical evidence of chronic liver failure. Biopsy is essential to differentiate it from potentially reversible liver diseases. Treatment is symptomatic only and prognosis is poor. Fibrosis usually occurs as a component of chronic inflammatory disease and/or cirrhosis to repair hepatic injury. Idiopathic hepatic fibrosis has recently been reported in young dogs (German Shepherd dogs were predisposed) with evidence of chronic hepatic failure, portal hypertension and hepatic encephalopathy.[154,155] Intensive treatment of hepatic encephalopathy results in long-term survival in a number of cases, although in general the prognosis for recovery is guarded.[155]

Toxic Liver Disease

Many chemicals, drugs, anesthetics or biological toxins can damage the liver (for examples see Fig. 15.20), although the actual cause is often never identified.[99] Hypoxemia due to severe anemia, acute circulatory failure, shock or thromboembolism may also cause hepatotoxic changes. Toxicity may be predictable (intrinsic toxins) or idiosyncratic.[134] Hepatic damage may vary from mild degenerative changes to full-blown acute hepatic necrosis. Clinical onset is often acute. Elevated serum transaminases and bilirubin are common laboratory findings. Liver biopsy is useful to rule out other diseases and/or to support a diagnosis of hepatotoxic injury; the latter may be suggested by a characteristic pattern of zonal necrosis. Prognosis for full recovery depends upon the severity of the original insult and is poor for massive hepatic necrosis. Animals that do not die from acute liver failure often recover with supportive care.

Congenital Portosystemic Shunts

Portosystemic shunts (PSS) are abnormal vascular connections between the portal and systemic venous systems that occur due to incomplete postnatal closure of venous shunts

HEPATOTOXINS IN THE DOG AND CAT	
Chemicals	Organic solvents (CCl$_4$)
	Pesticides (dieldrin)
	Heavy metals (copper, iron, lead, mercury)
	Industrial chemicals (chlorinated biphenyls, naphthalenes)
Drugs	Analgesics (acetaminophen, phenylbutazone)
	Anthelmintics (mebendazole, oxibendazole)
	Anticonvulsants (primidone, phenytoin)
	Antineoplastics (methotrexate)
	Ketoconazole
	Sulfonamides (sulfadiazine/trimethoprim)
	Thiacetarsamide
Anesthetics	Halothane, methoxyflurane
Biological toxins	Algatoxins (Rutland Water blue–green algae)
	Mycotoxins (aflatoxins)
	Bacterial endotoxin

FIG. 15.20 Hepatotoxins in the dog and cat.

normally present during fetal life.[174] The shunts divert portal blood away from the liver into the systemic circulation, thereby allowing toxins that are normally cleared by the liver to remain in the systemic circulation, resulting in hepatic encephalopathy and other signs of liver dysfunction.[93,162,174] Progressive liver atrophy occurs due to lack of hepatotrophic factors. Congenital shunts are usually single and may be extra- or intrahepatic. They must be distinguished from multiple acquired shunts that develop secondarily to portal hypertension caused by diffuse chronic liver disease. Extrahepatic shunts are most common and occur usually in small breeds of dogs or in cats. Dogs of large breeds are more likely to develop intrahepatic shunts such as a patent ductus venosus.[46,93] There is an increased incidence in Yorkshire Terriers, miniature Schnauzers, Old English Sheepdogs and Irish Wolfhounds, and in the latter breed PSS has been shown to be inherited.[116] Clinical signs

are usually recognized when the animal is less than 1 year-of-age, but congenital PSS have also been reported in dogs as old as 8–10 years-of-age.[98] The presenting history and clinical findings can be highly variable with the exception of stunted growth and failure to gain weight. Signs may be related to the CNS, gastrointestinal tract or urinary tract and are frequently episodic in nature. Neurological signs due to hepatic encephalopathy are noted in over 90% of cases.[93,152] These signs tend to wax and wane and may be particularly prominent following ingestion of a high protein meal. Urolithiasis (urate calculi) may occur because of increased urinary excretion of ammonia and uric acid and is sometimes the main presenting complaint (Fig. 15.21).

Dynamic bile acid testing and blood ammonia measurement are the main laboratory tests indicating shunting of blood[44,71,118,121]; abdominal radiography usually reveals reduced

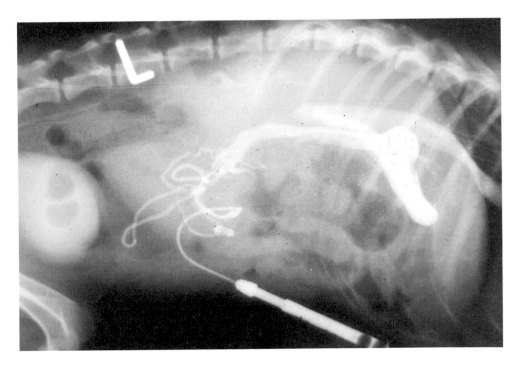

FIG. 15.21 Portogram of a 10-month-old male Chihuahua with an extrahepatic portosystemic shunt. There is a catheter in a jejunal vein through which contrast material has been injected. Contrast medium from a previous injection has been recirculated and excreted through the kidneys into the bladder; note the two filling defects in the bladder, which proved to be radiolucent urate stones. From Rutgers, H. C., 1993. Diagnosis and management of portosystemic shunts. *In Practice*, **15**, 175–181. Reprinted with permission.

liver size and renomegaly. Positive contrast portography (Fig. 15.21) is required for definitive diagnosis[26,113] and liver biopsy is essential to rule out primary or secondary liver disease.

The treatment of choice is partial or total ligation of the shunt, which is more feasible in extrahepatic shunts.[24,25,33,34,113] Ligation of multiple acquired shunts is contraindicated because life-threatening portal hypertension may result. Most animals improve dramatically following recovery from anesthesia and surgery. Medical management of hepatic encephalopathy, using low protein diets, oral antibiotics and/or lactulose, is important pre-operatively and should also continue post-operatively because regeneration of liver tissue requires several months.[160,173] If the animal does not improve adequately portography should be repeated; if this demonstrates continued shunting, further ligation may be necessary. Long-term medical management is only indicated when surgical intervention is not feasible or when there is evidence for portal hypertension. It is primarily directed at control of hepatic encephalopathy and does not reverse the progressive hepatic atrophy and ensuing metabolic alterations; long-term prognosis for these animals is usually poor.

It recently has been recognized that there is a category of dogs in which clinicopatho-logical and hepatic histological findings suggest a congenital PSS, yet this cannot be confirmed at portography. A microvascular hepatic dysplasia has been suggested as the under-lying cause, resulting in shunting at the level of the hepatic lobule.[137] Cairn Terriers may have an increased incidence. Symptomatic treatment of these cases using a protein-restricted diet has been more successful than for dogs with macroscopic PSS.

Neoplasia

Metastatic hepatic neoplasia is quite common, especially lymphosarcoma. Metastases can also come from carcinomas or sarcomas originating from the pancreas, spleen, gastro-intestinal tract, bone marrow, mammary gland and thyroid gland. Primary hepatic neoplasia is uncommon (hepatoma, hepato-cellular carcinoma, rarely neuroendocrine tumors[135,136,196]) and is more likely to occur in older dogs; signs are vague and nonspecific but there may be associated hepatomegaly or hepatic masses.[114,135] Hypoglycemia may occur in dogs with hepatoma and is thought to be due to excessive glucose consumption by the tumor and perhaps release of insulin-like factors by the tumor. Hepatoma may be amenable to surgical resection and has a better prognosis than the others.

Steroid-Induced Hepatopathy

This may be due to either exogenous steroid administration or endogenous hyperadreno-corticism.[9,148] Hepatomegaly occurs due to glycogen accumulation in hepatocytes. Pro-longed, high-dose treatment with glucocorti-costeroids will cause steroid hepatopathy in most dogs, but some dogs are very sensitive to even small amounts given for a short time. Clinical signs are those of hypercortisolism and include polyuria/polydipsia, abdominal distension and haircoat changes. Increased serum ALP activity is the most consistent biochemical abnormality and may be the steroid-induced isoenzyme of ALP as well as the hepatic enzyme. Biopsy may be necessary to rule out other causes of liver disease and shows typical centrilobular vacuolization of hepatocytes.[9] Steroid hepatopathy is reversible but this may take months. It is important to recognize it as such and not to confuse it with primary liver disease.

Feline Liver Diseases

Cats have certain unique features of hepatic metabolism and anatomy that are important for the understanding of feline liver disorders. Metabolic features are a relative deficiency of glucuronyl transferase, affecting the liver's ability to metabolize drugs and chemicals, and an inability to synthesize arginine, an important part of the hepatic urea cycle, which predis-poses cats to hyperammonemia during periods of inadequate food intake or anorexia. An anatomical feature is that the major pancreatic duct joins the common bile duct before its entry into the duodenum, which may explain the frequent coexistence of pancreatic and biliary tract disease in the cat.

Chronic hepatitis and hepatotoxicity are not as common in the cat as in the dog; the latter reflects the cat's more fastidious nature. Hepatic necrosis in the cat is more likely to be due to hepatic hypoxia, neoplastic disease and infections, rather than to hepatotoxins.[41,167] Several systemic infectious and metabolic illnesses, such as septicemia, FIP, diabetes mellitus, hyperthyroidism, myeloproliferative disorders and hypoxia, may also lead to secondary hepatic involvement and abnormal liver tests in the cat.[18,41] As for the dog, thorough clinical and laboratory evaluations are necessary to help determine the primary cause of hepatic changes (Fig. 15.22).

Cholangitis/Cholangiohepatitis Complex

Cholangiohepatitis is a disease complex consisting of cholangitis, cholangiohepatitis and biliary cirrhosis. Lymphocytic cholangitis is the most common form and is thought to have an immune-mediated etiology, possibly subsequent to suppurative inflammation resulting in immune-mediated self-perpetuating inflammation. Suppurative cholangitis probably results from bacterial invasion of the bile duct by enteric bacteria.[86] There may be concurrent interstitial pancreatitis.[101] Biliary cirrhosis, characterized by extensive fibrosis and chronic inflammation, is the uncommon final stage.[41] Clinical signs are vague and recurrent in the early stages and may include intermittent anorexia, fever, lethargy, vomiting and jaundice. Cats with lymphocytic cholangitis may be polyphagic in the early stages and are frequently presented because of ascites and/or jaundice.[106]

Biochemical testing shows enzyme elevation and in later stages hyperbilirubinemia and hyperglobulinemia. Definitive diagnosis of cholangiohepatitis must be based upon liver biopsy and bile culture. For the suppurative form, systemic antibiotics (e.g. ampicillin,

CLASSIFICATION OF LIVER DISEASE IN THE CAT
Inflammatory
Infectious • Viral (FIP) • Protozoal (toxoplasmosis) • Bacterial (sepsis, *Bacillus pisiformis*) • Parasitic (liver flukes) Cholangitis/cholangiohepatitis Toxic and drug-induced
Noninflammatory
Neoplastic Lipidosis • Idiopathic • Obesity, endocrine (diabetes mellitus) Portosystemic shunts Extrahepatic bile duct obstruction

FIG. 15.22 Classification of liver disease in the cat.

amoxycillin) should be administered for at least 1–2 months, and corticosteroids may be given later to prevent progression into chronic lymphocytic cholangiohepatitis. The lymphocytic form requires immunosuppressive treatment with prednisolone. Sludging and inspissation of bile is a common complication and may lead to choleliths and biliary obstruction requiring surgery. Prognosis is very variable and the aim of treatment is more to control than to cure the disease. Intermittent therapy may be necessary as the disease activity waxes and wanes. Successful treatment of suppurative cholangiohepatitis using cholecystojejunostomy and prolonged antibiotic therapy has been reported.[162]

Feline Hepatic Lipidosis

Fatty infiltration of the liver may occur with obesity, endocrine disorders such as diabetes mellitus, various toxins and hypoxia. Idiopathic hepatic lipidosis is a unique syndrome of cats and is characterized by extreme accumulation of triglycerides within hepatocytes due to a disruption in hepatic lipid metabolism, resulting in severe liver dysfunction.[12,90,146,184] This syndrome is very common in cats in North America but not elsewhere. The etiology is unknown, but obesity (insulin-resistant), acute starvation, rapid weight loss, protein–calorie imbalance, deficiencies of specific nutrients

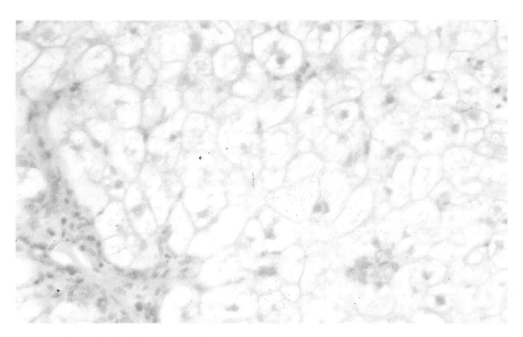

FIG. 15.23 Liver biopsy in a 4-year-old DSH cat with hepatic lipidosis before treatment, showing severe distension of hepatocytes with lipid. (H & E)

(e.g. carnitine, arginine, taurine), toxic metabolites (orotic acid), bacterial toxins and abnormalities of lipoprotein formation and secretion have been suggested.[195]

Common clinical signs are anorexia of several weeks duration, progressive weight loss, jaundice, hepatomegaly and, less frequently, hepatic encephalopathy. Obesity is a predisposing factor. Many cats have a history of vomiting. There is often a history that the cat was previously obese, but a stressful event caused sudden loss of appetite from which the cat never recovered. Liver biochemistries are characterized by hyperbilirubinemia and a marked elevation of ALP, consistent with severe intrahepatic cholestasis. Serum GGT activity, however, usually remains within the normal range or it becomes only mildly elevated, in contrast to other feline liver disorders wherein the magnitude of increase in GGT parallels that of ALP. Definitive diagnosis requires a liver aspirate or biopsy (Fig. 15.23) demonstrating obvious hepatocellular cytosolic vacuolation. Supportive care and long-term force-feeding of a balanced cat food diet are the most important aspects of therapy (Fig. 15.24) and it is frequently necessary to place a nasogastric or gastrostomy tube to facilitate force-feeding. The nasogastric tube is less suitable than a gastrostomy tube as it is less comfortable and is too narrow for feeding puréed commercial diets. The latter is therefore used for long-term (3–6 weeks) nutritional management at home.[59] Prognosis used to be regarded as poor, but recently a 65% survival rate was described following weeks to months of intensive nutritional support via a surgically placed gastrostomy tube.[90]

Neoplasia

Primary hepatic tumors are uncommon in the cat. Hepatocellular carcinoma, bile duct carcinoma and carcinoids have been reported, causing progressive weight loss, jaundice, intermittent vomiting and sometimes a palpable abdominal mass.[1,63] More commonly, the liver is the site of metastatic neoplasia or it is involved in systemic neoplasia; lymphosarcoma, myeloproliferative disorders and disseminated mastocytosis are most common. Although primary hepatic lymphosarcoma is rare, the liver is often involved in multicentric

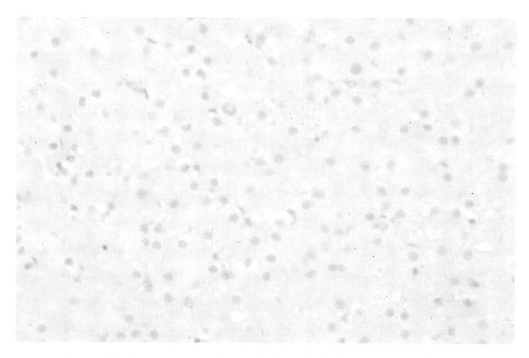

FIG. 15.24 Liver biopsy of the cat in FIG. 15.23 after 1 month of intensive nutritional management. Note the marked resolution of the fatty infiltration. (H & E)

lymphosarcoma along with the spleen and lymph nodes. Combination chemotherapy is possible for lymphosarcoma and mastocytosis.

Congenital PSS

Congenital PSS in cats are becoming more frequently recognized.[20,28,152,158] Most cats with congenital shunts demonstrate neuro-behavioral signs from a young age, with the first signs appearing around the age of 3 months. Clinical signs are mostly related to hepatic encephalopathy. Hypersalivation is an especially prominent finding in cats, probably being a manifestation of both nausea and encephalopathy. Common findings on physical examination are small body size, prominent kidneys and a typical copper-colored iris.[41] Elevated blood ammonia and bile acid concentrations are the most consistent laboratory findings. Urinary ammonium biurate crystals occur less frequently than in the dog. Liver size is often normal on radiographs. Mesenteric venography is usually performed to confirm the diagnosis and locate the anomalous vessel. Most congenital PSS in cats are single and extrahepatic. Surgical correction by partial or complete ligation of the anomalous vessel is the treatment of choice,[24] although this appears to be less successful in the cat than in the dog.[25] In some cases, recurrence of neurological signs following surgery was associated with recanalization of the shunt.[28] If surgical ligation is not possible, conservative treatment may be continued; prognosis for these cases has been variable.

Bile Duct Obstruction

Extrahepatic bile duct obstruction may be due to sludged bile associated with the cholangiohepatitis complex, bile pigment stones, neoplasia, compression of the common bile duct by pancreatitis, or bile duct strictures. Clinical signs vary dependent upon the underlying cause but usually include anorexia, vomiting, fever and weight loss. Jaundice is seen with complete biliary obstruction. Serum bilirubin, cholesterol and ALP activity and bile acid concentrations are generally markedly increased, while transaminase activities vary. With complete bile duct obstruction, stools may become acholic and there may be bleeding tendencies that are responsive to vitamin K administration. Choleliths, which are generally composed of cholesterol, bilirubin and calcium, are usually radiolucent and do not show up on plain abdominal radiographs but can be identified on ultrasound.[81] Surgical intervention will be necessary in the majority of the cases and may involve a cholecystectomy and bile culture in case of strictures or neoplasia. Cholelithiasis is often associated with cholangitis, necessitating long-term antibiotic therapy postoperatively.

Miscellaneous Liver Diseases

Infection with liver flukes (*Platynosonum* spp.) has been described in cats in North America and tropical regions.[21,104] Signs are often mild and diagnosis is difficult because the eggs are only sporadically found on fecal examination.

Biliary cysts are occasionally found and may be associated with pancreatic or renal cysts. They rarely give clinical problems, although large cysts may cause progressive abdominal enlargement and discomfort. Congenital biliary atresia with associated multiple large biliary cysts has been reported in a kitten with signs of jaundice and hepatomegaly.[73]

Treatment of Liver Disease

Objectives of therapy are: to eliminate causative agents, if known; to stop inflammation and minimize fibrosis; to provide optimum conditions for hepatic regeneration; and to control complications such as secondary bacterial infection, ascites and hepatoencephalopathy.

Nutritional Management

General Considerations

The liver is central to the metabolism of many nutrients and a range of metabolic disturbances may result from liver disease (Fig. 15.2). The reduction of dietary intake caused by vomiting and/or inappetence secondary to liver disease also renders the animal more dependent on catabolic lipid oxidation, glycogen breakdown and gluconeogenesis. The caloric and nutritional needs of the patient should be individually assessed depending on age, activity, ideal body weight and body condition. Reduced efficiency and regulation of lipid, protein and carbohydrate metabolism secondary to liver disease (e.g. abnormal synthesis of amino acids, proteins, purines and pyrimidines, abnormal synthesis of cholesterol, phospholipids and bile salts) may contribute to clotting factor deficiencies, portal hypertension, hypoproteinemia and hypoalbuminemia. Failure to address the patient's needs for high-quality nutrition may retard recuperation by further depleting metabolic reserves e.g. glycogen and hepatic protein reserve depletion.[131]

Liver disease may disrupt hepatic detoxification and excretory functions. Hepatic encephalopathy, a syndrome of reduced hepatic ability to detoxify waste products, may be seen in congenital anomalies of the portovascular system, acute fulminant hepatic failure or chronic progressive hepatic disorders that lead to end stage liver failure. Dietary modification to decrease the quantity of waste products handled by the diseased liver may reduce the severity of clinical signs and retard deterioration. Dietary management must aim to reduce the detoxification and excretion workload of the liver yet still provide adequate high quality nutrition to prevent deleterious catabolic changes.

The goals of nutritional therapy are to supply adequate nutrition to meet the requirements of the recuperating patient, replenish associated deficits and promote hepatic regeneration.[14,16,74] Prior to instituting dietary modification, it is essential that the patient's fluid balance is evaluated, corrected and maintained.[70,160] The specific liver disease affecting the patient should ideally be determined prior to instituting therapy. Dietary management and total medical management of the patient will depend on the severity of clinical signs and the animal's response to therapy. Regular monitoring of parameters including body weight, general body condition, total protein and serum albumin concentrations is recommended. Nutrition is the most easily manipulated variable[144] in the management of the patient with liver disease, facilitating restoration and maintenance of the normal range of permissive and stimulatory metabolic factors required by the recuperating patient (including insulin, glucagon, triiodothyronine, steroids, products of hepatic protein synthesis and growth hormone).

Palatability

Liver disease is often complicated by inappetence that promotes catabolic body tissue losses. Dietary management of hepatic disease requires a highly palatable diet that animals with reduced appetites will find acceptable despite the inherent protein restriction. Practical measures such as the use of highly odorous food, the warming of food prior to feeding, the maceration of food to enhance its texture and prompting learned feeding behavior by petting and stroking can stimulate eating. The smell of food is vital in soliciting and maintaining the patient's interest: strong meat, fish or cheese odors are often favored.

Cases of total anorexia may respond temporarily to appetite stimulants.[110] Benzodiazepine tranquilizers, being γ-amino butyric acid analogues,[124,129] may promote encephalopathy and are contraindicated in liver disease. Where pharmacological appetite stimulants are used, careful attention must be paid to discrepancies between the nutritional requirements and actual intakes in these patients and, where clinically appropriate, more aggressive therapy such as tube feeding should be considered.

Metabolic Requirements in Liver Disease

Protein–calorie malnutrition (PCM; Fig. 15.25) may occur in acute fulminant and severe chronic liver failure, complicating the patient's catabolic status and reducing visceral protein synthesis, cellular immunity[131] and total lymphocyte count. The stress of hospitalization, concurrent medical conditions and recuperation indicate that the liver patient requires a highly digestible source of energy. A range of factors directly impact on the nutritional requirements of patients with liver dysfunction (Figs 15.2 and 15.25). Cirrhotic human patients show reduced hepatic glucose production, decreased peripheral glucose utilization and decreased hepatic glycogen stores despite total body resting energy expenditures not being significantly altered in compensated chronic liver disease.[40,89,131,171] As a result cirrhotic patients recruit alternative fuels rapidly and have metabolic profiles after an overnight fast that resemble normal individuals undergoing more prolonged fasts.[89] Hepatic disease resembling human cirrhosis is, however, uncommon in companion animals[167,169] except in terminal end stage liver disease and the metabolic requirements of dogs and cats with liver disease are not known. However, frequent feeding with a highly digestible diet

---

PROTEIN – CALORIE MALNUTRITION IN LIVER DISEASE

Decreased food intake
- Anorexia, nausea, vomiting
- Poor quality diet

Malabsorption
- Potential deficiency of pancreatic exocrine secretions and bile salts
- Possible enteropathy and malabsorption of D-xylose, thiamine, folic acid

Alterations in protein metabolism

Medical complications
- Gastrointestinal bleeding
- Altered neurological status
- Purgation, neomycin/lactose therapy

Stress

FIG. 15.25 Clinical signs of protein–calorie malnutrition in liver disease.

is recommended to minimise catabolism of body tissue, reduce the release of potentially neurotoxic fatty acids from adipose tissue and moderate hypoglycemia. These general recommendations complement dietary modifications of specific nutritional components (e.g. lipids, proteins, carbohydrates).

Dietary Therapy

Nitrogen Metabolism

Skeletal muscle wasting and abnormal plasma amino acid profiles may accompany liver disease.[193] These abnormalities demonstrate the vital role of the liver in protein metabolism and underline the need for adequate high-quality protein to reduce catabolic losses. Dietary therapy must aim to provide adequate protein to meet the requirements for hepatic regeneration and repair, yet minimize nitrogenous waste products of protein catabolism. Hence, the type and quantity of protein must be modified in patients with severe liver disease to avoid encephalopathy.

HE is a clinical syndrome characterized by an altered state of consciousness, abnormal mental status and impaired neurological function associated with exposure of the CNS to predominantly nitrogenous, neurotoxic waste products that are normally metabolized by the liver (Fig. 15.3).[35,65,66,77,160,170,174] HE is of great concern in severe liver dysfunction and PSS. An increase in the plasma concentration of almost all amino acids is observed with the notable exception of BCAA and arginine.[129,141,145,192] Abnormal liver function contributes to these changes in amino acid concentrations.

The role of AAA in HE is controversial.[150] AAA are thought to favor binding of "fake" neurotransmitters, which resemble dopamine and norepinephrine, to neuronal receptors resulting in reduced neural excitation and ineffective transmission. AAA transport across the blood–brain barrier is thought to be facilitated by hyperammonemia.[2]

BCAA, i.e. valine, leucine, isoleucine, are essential in the diet and are metabolized outside the liver, predominantly in the muscle. Unusually low levels of BCAA are seen in two forms of protein intolerance: chronic renal failure and HE.[2,62] Subsequently it has been proposed that BCAA supplementation may have therapeutic benefits in the management of HE by competitively inhibiting transport of aromatic and sulfur-containing amino acids across the blood–brain barrier. Keto acid analogues of amino acids may reduce both protein depletion and hyperammonemia by reacting with amino group donors to provide a source of amino acids for protein synthesis[62] and to reduce protein catabolism.[149] Human clinical data imply that ornithine salts of α-keto acid analogues may improve the neurological status and could be useful in dietary management of HE.[2,62,129,141,149,150] Currently, little veterinary clinical evidence exists to support this theory. Restoration of the normal BCAA: AAA ratio and the use of dopamine-like drugs does not consistently reverse signs of HE in human clinical patients.[66,89,123,131]

Where HE is mild, an initial approach of frequent feeding (4–6 small meals daily) of a highly digestible, nutritionally complete diet may aid management. Therapeutic rationales aim to reduce the toxic waste products that must be handled by the compromized liver while providing adequate protein for regeneration, repair and maintenance. In more severe cases a highly digestible, protein-restricted, high biological value (>75) protein diet is recommended e.g. cottage cheese. Dairy proteins have a high biological value and appropriate BCAA:AAA ratio but are generally low in arginine. A dietary source of arginine, an important component of the urea cycle, is essential in the treatment of HE, especially in cats. Supplementation with an arginine source, such as egg protein, should be considered. Egg protein has a high biological value, is highly digestible, rich in arginine, but has a high methionine content. Fish meal protein is generally a high biological value protein source that is highly palatable, especially to cats, but may be high in purine, potentially leading to increased uric acid production, further predisposing patients to

FOOD VALUES OF SOME COMMON INGREDIENTS USED IN HOME-MADE DIETS (per g)			
Food value	Creamed cottage cheese	Whole egg	Cooked rice
Calories (kcal)	1.1	1.6	1.1
Protein (mg)	140	130	20
Carbohydrate (mg)	29	8	244
Fat (mg)	40	115	1
Amino acids			
Tryptophan	1.5	2.1	0.2
Phenylalanine	7.3	7.5	1.0
Leucine	14.6	11.4	1.7
Isoleucine	7.9	8.5	0.9
Lysine	11.4	8.3	0.8
Valine	7.0	9.6	1.4
Methionine	3.8	4.0	0.4
Threonine	6.4	6.5	0.8
BCAA/AAA	3.5	3.1	3.3

Adapted from Pennington, J.A. and Church, N.N.: Bowes and Church's, *Food Values of Portions Commonly Used*, 13th edn. New York: Harper and Row, 1980.

FIG. 15.26 Food values of some common ingredients used in home-made diets.

urate stone formation. Alternative protein sources include vegetable proteins (soy flour, corn grits, rice) that have a favorable amino acid profile, a low content of both methionine and mercaptans. Ideally the diet should be based on a highly digestible carbohydrate (e.g. rice, pasta), be adequate in vitamin content and highly palatable to combat inappetence often associated with hepatic disease. Dietary management should initially use a highly palatable diet with favorable amino acid profile and reduced protein content (~17 g/400 kcal metabolizable energy, ME). On the basis of the clinical response, and in conjunction with concurrent medical management, the protein content of the diet may be progressively increased or the protein sources amended.[14,173] The protein level recommended may not be appropriate in all cases and substitution of ingredients may be performed.[14,16,173] Ingredient substitution should take into account the protein and amino acid contents of the ingredients (Fig. 15.26). Dietary management aims to ensure that adequate, highly digestible, high quality dietary protein is supplied to meet the animal's metabolic needs without exceeding the detoxification capacity of the compromised liver.

Lipid Requirements

Cholestatic liver diseases may be associated with reduced bile acid production[169] but steatorrhea and intestinal malabsorption are usually mild with liver disease.[176] Bile salts are not essential for *in vitro* pancreatic lipase activity or long chain fatty acid absorption.[176] Fat restriction is indicated where steatorrhea is clinically evident. Dietary lipids provide a highly palatable, concentrated source of energy, so mild restriction of lipids, which retains the palatability enhancing effects of animal fats, may be a positive asset. Fatty acids may aggravate HE by the direct action of short and medium chain length fatty acids on the CNS[157] and indirectly by reducing the conversion of ammonia to urea, aggravating postprandial hyperammonemia and providing a further indication for the use of a reduced fat diet.[124] Medium chain triglycerides have been associated with deterioration in HE, but the extreme levels of up to 50% of calories in experimental diets bear little resemblance to commercially available products. It can be hypothesized that a low fat, low protein, high carbohydrate diet may be of benefit in the management of HE due to its stimulation of insulin production, resulting in a fall in fatty acid levels, but veterinary clinical data substantiating fatty acid modification in the management of liver disease are currently unavailable.

Carbohydrate

Highly digestible carbohydrates such as rice and pasta are absorbed in the proximal gastrointestinal tract and provide a non-encephalogenic energy source for animals with liver disease.

The role of fiber in the management of hepatic encephalopathy remains unclear. Fiber may increase bacterial fermentation, decrease palatability and increase nitrogenous losses due to the abrasive desquamation of epithelial cells. In humans, oral administration of purified soluble pectin fiber increased fecal nitrogen excretion and reduced ammonia production within the intestine (suggesting a primary inhibition of ureolysis), without altering the fecal pH.[82,191,194] Further clinical studies are required before a firm veterinary recommendation can be supported on this issue.

Vitamin Requirements

Mineral and vitamin deficiencies associated with hepatic disease may occur due to a combination of poor dietary intake, reduced intestinal absorption and increased demands due to catabolism and regeneration.[58,188] Synthesis, storage and conversion of vitamins to metabolically active metabolites may all be reduced by liver disease. Water-soluble vitamin supplementation with B vitamins should ensure that nutritional regimens meet maintenance requirements. However, ascorbic acid supplementation should be at levels of up to 25 mg/day to compensate for decreased synthesis.[171] Reduced bile salt production secondary to acute or chronic parenchymal liver damage

or biliary obstruction may cause decreased absorption of fat-soluble vitamins.[175] Dietary supplementation with vitamin E (500 mg/day for dogs, 100 mg/day for cats) provides an important hepatocellular protective effect against copper toxicity and lipid peroxidation injuries. Vitamin A supplementation is unwarranted and dangerous (maximum canine and feline dietary levels 40,000 IU/400 kcal) due to the risk of synergism between vitamin A and cytotoxins in provoking hepatocyte damage.

Minerals

Copper: Certain breeds, most notably the Bedlington and West Highland White Terriers are prone to disorders of hepatic storage of copper leading to hepatotoxicity.[83,84,95,96,181–183,186,187] Dietary therapy may aid the medical management of this condition via restricted levels of dietary copper and the use of zinc supplementation (2 mg/kg/day). Dietary copper restriction to a level close to the NRC minimum (0.8 mg/1000 kcal ME)[128] may complement concurrent medical therapy in severe cases. Dietary copper restriction is best obtained by avoiding offal (e.g. liver and sweetbreads). Zinc may reduce intestinal absorption of copper by divalent ion competition and affects the synthesis of metallothein, the copper-binding protein that binds to intestinal copper excreted in the bile and reduces its absorption.[85,107,108]

Zinc: Zinc supplementation may also assist the management of HE. Zinc deficiency may alter neurological status by altering levels of brain neurotransmitters, increasing muscle synthesis of ammonia from aspartate and reducing urea synthesis capacity. Human patients with HE appear to be more profoundly zinc depleted than patients without HE.[7,142]

Sodium: Portal hypertension secondary to liver disease may be associated with hypovolemia, ascites and altered renal retention of sodium and water. The exact interrelationship and sequence of these events remains controversial.[161,167] The overall regulation of hepatic blood flow essential for regulated clearance of metabolites necessitates a complex interaction of flow through the hepatic artery and portal circulation. Sodium retention is a very early response in liver disease associated with exhaustion of normal compensatory mechanisms. Sodium restriction has been advocated to reduce portal hypertension.[161] Extreme dietary sodium restriction may, however, cause hyponatremia; thus electrolyte concentrations should be monitored carefully if sodium-restricted diets are fed. Sodium restriction may also reduce palatability and sodium-restricted diets may be difficult to formulate. A diet containing approximately 0.30–0.40 g/1000 kcal of sodium is recommended to provide a moderate, yet practical, level of dietary sodium restriction.

Nutrition in Feline Liver Disease

Cats are obligate carnivores with different nutritional requirements to dogs.[109,128,147] Normal feline diets must provide adequate highly digestible, high biological value protein to meet the unique feline requirements for protein and essential amino acids. Diets must provide adequate protein to meet the nutritional requirements of the cat. (The minimum maintenance level is 120 g crude protein/kg of diet.)[127] Feline acceptance of meat free diets is generally poor.

Diets should be highly palatable, containing adequate quantities of arginine and taurine and be supplemented with B vitamins because these become rapidly depleted in feline hepatic disease.[188] Arginine deficiency in the cat results in a rapidly developing inability to detoxify nitrogenous compounds via the urea cycle. Hyperammonemia, HE and rapid clinical deterioration are associated with arginine deficiency due to the inability of the cat to synthesize the amino acid ornithine, an essential intermediate compound in the urea cycle.[39,125,147]

Hepatic Lipidosis

Feline hepatic lipidosis is a life-threatening hepatobiliary disease affecting the cat in which restoration of nutritional intake is essential for the patient's survival. Hepatic lipidosis may result from defects in the assembly of very low density lipoproteins, phospholipids or apoproteins. A range of metabolic defects including diabetes mellitus, fructose intolerance and galactosemia and toxic reactions to oxytetracycline, puromycin and carbon tetrachloride have been implicated in the etiology of hepatic lipidosis.[50,57] In cases of hepatic lipidosis due to diabetes mellitus, calorie restriction appropriate to the management of diabetes combined with protein supplementation will promote adequate lipoprotein synthesis to permit lipid transport from the liver.

Idiopathic hepatic lipidosis is characterized by the lack of any underlying medical cause and is most commonly associated with overweight cats with a history of anorexia.[4,23,41,48,90,195] Nutritional therapy should aim to provide a readily digestible source of energy that is adequate in protein content, yet does not accentuate hepatic encephalopathy. Considering the low plasma taurine levels noted with this condition and feline requirements for a dietary source of arginine, supplementation with taurine and arginine is recommended. Carnitine is an important intermediary in hepatic fatty acid metabolism and carnitine supplementation has been proposed as being useful in the management of lipidotic cats.[14,15,48] Highly digestible carbohydrate may worsen the effects of glucose intolerance, which frequently accompanies idiopathic hepatic lipidosis and will complicate diabetes mellitus associated lipidosis.[50,57] Nasoesophageal and gastrostomy tubes may provide the most efficient, least stressful means of supplying nutrition.[58,59,67] Ideally a liquid diet for management of hepatic lipidosis should contain 200–300 kcal of ME daily and have an osmolality of 225–300 mOsm/l to avoid secondary osmotic diarrhea. Protein-restricted diets are only indicated where lipidotic cats show HE.

Supportive and Symptomatic Therapy

This is summarized in Fig. 15.27. Fluid therapy is often indicated in the initial treatment

DRUGS USED IN THE TREATMENT OF HEPATOBILIARY DISEASE

Drug	Dose	Administration	Comments
Antibacterials			
Ampicillin	20 mg/kg	tid PO, SC	Concentrated in bile
Amoxycillin	22 mg/kg	bid PO	Concentrated in bile
Cephalexin	15 mg/kg	tid PO, SC, IM	Concentrated in bile
Enrofloxacin	2.5 mg/kg	bid PO	Concentrated in bile
Metronidazole	7.5 mg/kg	bid–tid, PO	Reduces encephalopathogenic bacteria; anti-anaerobes
Neomycin	10–20 mg/kg	tid PO	Reduces encephalopathogenic bacteria; not absorbed
Do not use: tetracyclines, sulfonamides, trimethoprim–sulfonamide, erythromycin (hepatotoxic or dependent on hepatic handling)			
Immunosuppressive drugs			
Prednisolone	1–2 mg/kg	sid PO, SC	High dose until clinical remission, then taper down to 0.5–1.0 mg/kg every other day
Azathioprine	1–2 mg/kg	sid or eod, PO	Use in conjunction with prednisolone; monitor WBC
Fluid therapy			
Saline or Hartmann's	40–60 ml/kg/day	IV	Use 0.45% NaCl for maintenance
KCl	10–20 mmol/500 ml maintenance fluids; monitor serum potassium and adjust as needed		
Dextrose	Add to maintenance fluids to make 2.5–5.0% solution		
Decoppering agents			
D-Penicillamine	10–15 mg/kg	bid PO	Give on empty stomach; may cause vomiting and anorexia
Trientene	10–15 mg/kg	bid PO	If intolerant of D-penicillamine
Zinc chloride or acetate	5–10 mg/kg	bid PO	More effective in the prevention of copper accumulation than in actual decoppering; give on an empty stomach
Antifibrotic drugs			
Glucocorticosteroids	1–2 mg/kg	sip PO	If fibrosis is associated with inflammation
D-Penicillamine	10–15 mg/kg	bid PO	May cause vomiting
Colchicine	0.03 mg/kg	sid PO	
Treatment for gastroduodenal ulceration			
Cimetidine	5–10 mg/kg	tid PO, IV	Interferes with hepatic p450 system; drug interactions
Ranitidine	2–4 mg/kg	bid PO	Interferes with hepatic p450 system, but less than cimetidine
Famotidine	0.5 mg/kg	sid PO	No hepatic interaction
Sucralfate	(D) 0.5–1.0 g (C) 0.25 g	tid–qid PO	Local protection; not absorbed
Alteration of intestinal flora			
Lactulose	(D) 5–20 ml (C) 0.25–0.5 ml	tid–qid PO sid–tid PO	Titrate dose to 2–3 soft motions per day; excess causes diarrhea
Lactulose retention enema	(D) 20–70 ml (C) 10–15 ml	bid–tid	Dilute 1:2 with water to 50–200 ml total; retain 20–30 min
Diuretics			
Spironolactone	1–2 mg/kg	bid PO	
Frusemide	1–2 mg/kg	bid PO	

Table 15.27 continued

FIG. 15.27 Drugs used in the treatment of hepatobiliary disease.

DRUGS USED IN THE TREATMENT OF HEPATOBILIARY DISEASE (continued)			
Drug	Dose	Administration	Comments
Treatment of coagulopathy			
Vitamin K$_1$	(D) 2 mg/kg (C) 5 mg	bid IM, SC	
Fresh plasma	6–10 ml/kg	IV	sid–bid as needed
Heparin	50–100 IU/kg	bid–tid SC	For DIC, use with plasma
D — dog; C — cat.			

FIG. 15.27 continued.

of animals with decompensated liver disease. Supplementation with dextrose and potassium is important in animals with severe acute or chronic liver failure to prevent hypoglycemia and hypokalemia that may precipitate HE. Animals with liver disease are more susceptible to gut-borne infections and septicemia because hepatic reticuloendothelial function is reduced; consequently, antibiotics (broad-spectrum penicillins) are often given either for prophylaxis or to treat an existing infection.[60]

Gastrointestinal ulceration is a common complication in dogs with liver disease and may be due to gastric acid hypersecretion, impaired gastric mucosal blood flow secondary to portal hypertension and reduced epithelial cell turnover.[164] Because gastrointestinal bleeding may precipitate HE, early treatment with H$_2$ blockers and/or sucralfate is advocated. Ranitidine may be preferable to cimetidine because it causes less inhibition of hepatic microsomal enzymes. Sucralfate, a locally acting gastric mucosal protecting agent, is safe and effective and is useful when given either on its own or concurrently with other medication.

Treatment of hepatoencephalopathy is directed toward reducing the formation and uptake of toxic substances (e.g. ammonia) from the intestinal tract and consists of dietary protein restriction, alteration of the intestinal flora using neomycin or metronidazole (which can be given together because they act synergistically)[160,173] and lactulose. Lactulose is a nondigestible disaccharide that is metabolized by colonic bacteria resulting in colonic acidification and ammonia trapping. It also alters colonic bacterial flora and causes osmotic diarrhea, thus reducing the time available for both production and absorption of ammonia and other toxins. The dose should be adjusted to avoid diarrhea and achieve 2–3 soft stools a day. Its sweet taste may make it unpalatable especially for cats and it may then be replaced by powdered lactitol, which is less sweet. Lactulose and neomycin can also be given as retention enemas in the treatment of hepatic coma after an initial cleansing enema. Hepatic encephalopathic crises are often precipitated by other factors (metabolic imbalances, gastrointestinal bleeding, infections and other hypercatabolic states) and it is important to identify and control these as well. In the treatment of ascites with diuretics care should be taken to prevent hypokalemia and alkalosis from occurring because both factors can exacerbate encephalopathy. Spironolactone is in this respect safer because it is potassium-sparing, but it is not as effective as frusemide. For this reason they are often used in combination.

Coagulopathies associated with liver disease signify severe disease, usually due to either DIC and/or decreased production of clotting factors. Prognosis in these cases is usually poor despite treatment with plasma (to replenish coagulation factors) and/or heparin. They can be differentiated from vitamin K deficiency due to severe cholestasis by lack of response to treatment with vitamin K$_1$.

Drug Therapy

In liver disease all drugs should be used with caution. The liver is the major site of drug metabolism and impaired liver function may result in major alterations to the expected metabolism. Drugs that are primarily dependent upon the liver for inactivation or excretion (e.g. chloramphenicol, erythromycin, tranquilizers) should not be used or should be used at reduced dosages. Potentially hepatotoxic drugs (e.g. tetracycline, anticonvulsants, anthelmintics) should be avoided.[133,134]

Antibiotics are specifically indicated in the treatment of bacterial hepatitis, cholangiohepatitis and/or cholecystitis and hepatic abscesses. Antibiotics that are excreted in an active form and at therapeutic concentrations in bile, without being hepatotoxic, are most suitable for the treatment of hepatobiliary infections (Fig. 15.27). Antibiotics are also useful in the short-term management of HE and prophylactically in some patients with acute liver failure, chronic hepatitis and cirrhosis because of the increased risk of septicemia and endotoxemia associated with reduced hepatic reticuloendothelial activity and to prevent invasion by intestinal bacteria.[60]

Corticosteroids are used in the treatment of chronic hepatitis in the dog and lymphocytic cholangitis/cholangiohepatitis in the cat. In the dog, they are most useful in the treatment of idiopathic chronic active hepatitis but they may be indicated in some other cases because of their antiinflammatory and antifibrotic effects.[112,165] There are no controlled studies documenting their clinical efficacy in dogs with chronic hepatitis. Corticosteroids have the disadvantage of being catabolic and immunosuppressive, contraindicating their use in animals with hepatoencephalopathy or infectious hepatitis; high dosages may cause a reversible hepatopathy.[148] They are recommended in dogs with chronic hepatitis only when there is persistent clinical and biochemical evidence of hepatic dysfunction, chronic hepatitis is confirmed histologically and no known cause can be identified.[76,185] Tapering prednisolone is preferred and long-term alternating day treatment is often necessary. Follow-up biopsies may be required to assess efficacy of treatment because biochemical reevaluation is hampered by steroid-induced enzyme induction. For patients that fail to

respond to corticosteroid therapy alone, or to decrease the severity of side-effects, prednisolone may be combined at a reduced dose with the immunosuppressive drug azathioprine.[76,112]

The bile acid ursodeoxycholic acid is currently receiving considerable clinical attention for the treatment of chronic liver disease in humans and may offer a safe and novel alternative in the management of chronic cholestatic intrahepatic disease. Its mechanisms of action are not well defined but may include a combination of increased bile flow, displacement of cytotoxic bile acids from the enterohepatic circulation, direct cellular protection and immunomodulation. A dose of 10–15 mg/kg PO sid, extrapolated from dosages in humans, has proved safe in normal dogs and cats and has been associated with transient clinical improvement in a single dog with chronic liver disease. Further studies of this drug appear warranted in dogs and cats with liver disease, especially when given early in the course of the disease.[117]

Copper chelation therapy is indicated in Bedlington or West Highland White Terriers with copper hepatotoxicosis. D-Penicillamine is used most commonly; it reduces liver copper content by forming a chelate with serum copper that is subsequently excreted in the urine. It is best given on an empty stomach because food interferes with its absorption. Decoppering may take months or years of chronic therapy.[181,186] Common side-effects are vomiting and diarrhea. These signs can be ameliorated by reducing the dose followed by slowly increasing over time, or by giving the drug with some food. Dogs on chelator therapy should have repeat liver biopsies, which can be taken percutaneously, to assess the efficacy of the treatment. Trientine is a copper chelator that can be used in dogs that do not tolerate D-penicillamine; this drug has a similar potency to D-penicillamine but has few or no side-effects.[3,186] 2,3,2-Tetramine is related to trientine but is more potent; however, it is not commercially available at this moment. Treatment with zinc gluconate or acetate is currently attracting a lot of clinical interest as a possible treatment for copper hepatotoxicosis in both humans and dogs. Zinc induces the production of an intestinal cell metallothein that binds to intestinal copper and prohibits its absorption; copper is then lost in the stool when the intestinal cell sloughs. Zinc treatment in dogs may be of most value in the prevention of copper reaccumulating following earlier copper chelator therapy and it also is promising as treatment for presymptomatic affected dogs. However, in a recent study, zinc acetate was also found to be useful in lowering elevated liver copper levels, presumably through an indirect process involving release from liver copper stores to compensate for reduced copper absorption.[32] Zinc should be given 1 hr before feeding, as food interferes with its absorption. A dose of 5–10 mg/kg bid has been recommended; alternatively, a Bedlington Terrier may be given an induction dose of 100 mg bid for 3 months, followed by a maintenance dose of 50 mg bid.[32] Dogs on zinc should have blood zinc levels monitored periodically so that these do not exceed 200 μg/dl.[32] Pending further studies it is recommended to initiate treatment with a copper chelator (usually D-penicillamine); when decoppering is judged sufficient based upon liver copper level determination, maintenance treatment with either zinc or intermittent chelator therapy can be started.

The efficacy of drugs that inhibit hepatic synthesis of collagen or promote its removal in the treatment of hepatic fibrosis is uncertain. Corticosteroids have a mild antifibrotic action and may be of use in patients in which hepatic fibrosis is associated with significant underlying inflammation. Colchicine has a more specific effect and has been reported as causing clinical and histological improvement in both a dog with inflammatory hepatic fibrosis and in a dog with idiopathic hepatoportal fibrosis.[29,154] Further studies are needed to see whether this drug, which currently is not licensed for use in the dog, will be useful in the treatment of hepatic fibrosis in dogs. Side-effects of colchicine in humans are numerous (vomiting, diarrhea, bone marrow dyscrasias, neuropathies) but were not noticed in these two dogs. D-Penicillamine also has some antifibrotic effects in addition to its chelator properties, but its use is limited by the frequency of its side-effects.

References

1. Alexander, R. W. and Kock, R. A. (1982) Primary hepatic carcinoid (APUD cell carcinoma) in the cat. *Journal of Small Animal Practice,* **23**, 767.
2. Alexander, W. F., Spindel, E., Harty, R. F. *et al.* (1989) The usefulness of branched chain amino acids in patients with acute or chronic hepatic encephalopathy. *American Journal of Gastroenterology,* **84**, 91–96.
3. Allen, K. G. D., Twedt, D. C. and Hunsaker, M. P. (1987) Tetramine cupruretic agents: a comparison in dogs. *American Journal of Veterinary Research,* **48**, 28–30.
4. Alpers, D. H. and Sabesin, S. M. (1985) Fatty liver: biochemical and clinical aspects. In *Diseases of the Liver.* Eds. L. Schiff and E. R. Schiff. 6th edn. Philadelphia: Lippincott.
5. Andersson, M. and Sevelius, E. (1991) Breed, sex and age distribution in dogs with chronic liver disease: a demographic study. *Journal of Small Animal Practice,* **32**, 1–5.
6. Andersson, M. and Severlius, E. (1992) Circulating autoantibodies in dogs with chronic liver disease. *Journal of Small Animal Practice,* **33**, 389–394.
7. Antoniello, S., Auletta, M., Cerini, R. and Cepesso, A. M. (1986) Zinc deficiency and hepatic encephalopathy. *Italian Journal of Gastroenterology,* **18**, 17–21.
8. Badylak, S. E. (1988) Coagulation disorders and liver disease. *Veterinary Clinics of North America: Small Animal Practice,* **18**, 87–93.
9. Badylak, S. F. and Van Fleet, J. F. (1981) Sequential morphologic and clinicopathologic alterations in dogs with experimentally induced glucocorticoid hepatopathy. *American Journal of Veterinary Research,* **42**, 1310–1315.
10. Badylak, S. E. and Van Fleet, J. F. (1981) Alterations of prothrombin time and activated partial thromboplastin time in dogs with hepatic disease. *American Journal of Veterinary Research,* **42**, 2053–2056.
11. Baldwin, C. J. and Atkins, C. E. (1987) Leptospirosis in dogs. *Compendium of Continuing Education,* **9**, 499–507.
12. Barsanti, J. A., Jones, B. D., Spano, J. S. and Taylow, H. W. (1977) Prolonged anorexia associated with hepatic lipidosis in three cats. *Feline Practice,* **7**, 52–57.
13. Barton, C. (1977) Chronic active hepatic disease with cirrhosis in a dog. *Missouri Veterinarians,* **28**, 17–20.

14. Bauer, J. E. (1986) Nutrition and liver function: nutrient metabolism in health and disease. *Compendium of Continuing Education*, **8**, 923–931.

15. Bauer, J. E. (1988) Feline lipid metabolism and hepatic lipidosis. Kal Kan symposium for the treatment of small animal disease. *Feline Medicine*, October, 56–59.

16. Bauer, J. E. and Schenck, K. L. (1988) Nutritional management of hepatic disease. *Veterinary Clinics of North America*, **19**, 513–527.

17. Bauer, J. E., Meyer D. J., Goring, R. L. *et al.* (1988) Lipoprotein cholesterol distribution in experimentally produced canine cholestasis. In *Nutrition of the Dog and Cat*. Eds. I. Burger and J. Rivers. Cambridge: Cambridge University Press.

18. Bennett, A. M. (1977) Tyzzer's disease in cats experimentally infected with feline leukaemia virus. *Veterinary Microbiology*, **2**, 49.

19. Bennett, A. M., Davies, J. D., Gaskell, C. J. G. and Lucke, V. M. (1983) Lobular dissecting hepatitis in the dog. *Veterinary Pathology*, **20**, 179–188.

20. Berger, B., Whiting, P. G., Breznock, E. M., Bruhl–Day, R. and Moore, P. F. (1986) Congenital feline portosystemic shunts. *Journal of the American Veterinary Medical Association*, **188**, 517–520.

21. Bielsa, L. M. and Greiner, E. C. (1984) Liver flukes (*Platynosonum conicinnum*) in cats. *Journal of the American Animal Hospital Association*, **21**, 269–274.

22. Biller, D. S., Kantrowitz, B. and Miyabashi, T. (1992) Ultrasonography of diffuse liver disease: a review. *Journal of Veterinary Internal Medicine*, **6**, 71–76.

23. Biourge, V. (1991) Sequential findings in cats with hepatic lipidosis. *Proceedings of the Ninth ACVIM Forum*, pp. 189–191.

24. Birchard, S. J. (1984) Surgical management of portosystemic shunts in dogs and cats. *Compendium of Continuing Education*, **6**, 795–800.

25. Birchard, S. J. and Sherding, R. G. (1992) Feline portosystemic shunts. *Compendium of Continuing Education*, **14**, 1295–1300.

26. Birchard, S. J., Biller, D. S. and Johnson, S. E. (1989) Differentiation of intrahepatic versus extrahepatic portosystemic shunts in dogs using positive-contrast portography. *Journal of the American Animal Hospital Association*, **25**, 13–17.

27. Bishop, L., Strandberg, J. D., Adams, R. J., Brownstein, D. G. and Patterson, R. (1979) Chronic active hepatitis in dogs associated with leptospires. *American Journal of Veterinary Research*, **40**, 839–844.

28. Blaxter, A. C., Holt, P. E., Pearson, G. R., Gibbs, C. and Gruffydd–Jones, T. E. (1988) Congenital portosystemic shunt in the cat: a report of 9 cases. *Journal of Small Animal Practice*, **29**, 631–645.

29. Boer, H., Nelson, R. W. and Long, G. G. (1984) Colchicine therapy for hepatic fibrosis in a dog. *Journal of the American Veterinary Medical Association*, **185**, 303–305.

30. Boyer, J. L. and Miller, D. J. (1982) Chronic hepatitis. In *Diseases of the Liver*. Eds. L. Schiff and E. R. Schiff. 5th edn. pp. 771–811. Philadelphia: Lippincott.

31. Brewer, G. J., Schall, W., Dick, R., Yuzbasiyan–Gurkan, V., Thomas, M. and Padgett, G. (1992) Use of [64]copper measurements to diagnose canine copper toxicosis. *Journal of Veterinary Internal Medicine*, **6**, 41–43.

32. Brewer, G. J., Dick, R. D., Schall, W., Yuzbasiyan–Gurkan, V., Mullanaey, T. P., Pace, C., Lindgren, J., Thomas, M. and Padgett, G. (1992) Use of zinc acetate to treat copper toxicosis in dogs. *Journal of the American Veterinary Medical Association*, **201**, 564–568.

33. Breznock, E. M. (1979) Surgical manipulation of portosystemic shunts in dogs. *Journal of the American Veterinary Medical Association*, **174**, 819–826.

34. Breznock, E. M. (1983) Surgical manipulation of intrahepatic portocaval shunts in dogs. *Journal of the American Veterinary Medical Association*, **182**, 788–793.

35. Bunch, S. E. (1991) Hepatic encephalopathy. *Progress in Veterinary Neurology*, **2**, 207–296.

36. Bunch, S. E., Baldwin, B. H., Hornbuckle, W. E. and Tennant, B. C. (1984) Compromised hepatic function in dogs associated with anticonvulsant drugs. *Journal of the American Veterinary Medical Association*, **184**, 444–448.

37. Bunch, S. E., Castleman, W. L., Baldwin, B. H., Hornbuckle, W. E. and Tennant, B. C. (1985) Effects of long-term primidone and phenytoin administration on canine hepatic function and morphology. *American Journal of Veterinary Research*, **46**, 105–115.

38. Bunch, S. E., Castleman, W. L., Baldwin, B. H., Hornbuckle, W. E. and Tennant, B. C. (1982) Hepatic cirrhosis associated with long-term anticonvulsant drug therapy in dogs. *Journal of the American Veterinary Medical Association*, **181**, 357–362.

39. Burger, I. H. (1988) A basic guide to nutrient requirements. In *The Waltham Book of Dog and Cat Nutrition*. Ed. A. T. Edney. 2nd edn. Oxford: Pergamon Press.

40. Cabre, E., Gonzalez–Huiz, F., Abad Lacruz, Z. *et al.* (1990) Effect of total enteral nutrition on the short term outcome of severely malnourished cirrhotics. *Gastroenterology*, **98**, 715–720.

41. Center, S. A. (1986) Feline liver disorders and their management. *Compendium of Continuing Education,* **8**, 889–902.
42. Center, S. A., Baldwin, B. H. and Erb, H. (1986) Bile acid concentrations in the diagnosis of hepatobiliary disease in the cat. *Journal of the American Veterinary Medical Association,* **189**, 891–896.
43. Center, S. A., Baldwin, B. H., Erb, H. and Tennant, B. C. (1985) Bile acid concentrations in the diagnosis of hepatobiliary disease in the dog. *Journal of the American Veterinary Medical Association,* **187**, 935–940.
44. Center, S. A., Baldwin, B. H., De Lahunta, A., Dietze, A. E. and Tennant, B. C. (1985) Evaluation of serum bile acid concentrations for the diagnosis of portosystemic venous anomalies in the dog and cat. *Journal of the American Veterinary Medical Association,* **186**, 1090–1094.
45. Center, S. A., Bunch, S. E., Baldwin, B. H., Hornbuckle, W. E. and Tennant, B. C. (1983) Comparison of sulfobromophthalein and indocyanine green clearances in the cat. *American Journal of Veterinary Research,* **44**, 727–730.
46. Center, S. A. and Magne, M. L. (1990) History, physical examination and clinicopathologic features of portosystemic vascular anomalies in the dog and cat. *Seminars in Veterinary Medicine and Surgery (Small Animals),* **5**, 83–93.
47. Center, S. A., Manwarren, T., Slater, M. R. and Wilentz, E. (1991) Evaluation of twelve-hour preprandial and two-hour post-prandial serum bile acid concentrations for diagnosis of hepatobiliary disease in dogs. *Journal of the American Veterinary Medical Association,* **199**, 217–226.
48. Center, S. A., Thompson, M., Wood, P. A., Millington, D. S. and Chace, D. H. (1991) Hepatic ultrastructural and metabolic derangements in cats with severe hepatic lipidosis. *Proceedings of the Ninth ACVIM Forum,* pp. 193–196.
49. Center, S. A., Slater, M. R., Manwarren, T. and Prymak, K. (1992) Diagnostic efficacy of serum alkaline phosphatase and gamma-glutamyltransferase in dogs with histologically confirmed hepatobiliary disease: 270 cases. *Journal of the American Veterinary Medical Association,* **201**, 1258–1264.
50. Chang, S. and Silvis, S. E. (1974) Fatty liver produced by hyperalimentation of rats. *American Journal of Gastroenterology,* **62**, 410–418.
51. Cobb, L. M. and Mackay, K. A. (1962) A bacteriologic study of the liver of the normal dog. *Journal of Comparative Pathology,* **72**, 92.
52. Conn, H. O. (1979) Ammonia tolerance in liver disease. *Journal of Laboratory and Clinical Medicine,* **55**, 855.
53. Conn, H. O. (1982) Ammonia tolerance in assessing the potency of portocaval anastomoses. *Archives of Internal Medicine,* **131**, 221.
54. Cornelius, L. M., Thrall, D. E., Halliwell, W. H. Frank, G. M., Kern, A. J. and Woods, C. B. (1975) Anomalous portosystemic anastomoses associated with chronic hepatic insufficiency in six young dogs. *Journal of the American Veterinary Medical Association,* **167**, 220–228.
55. Counsell, L. J. and Lumsden, J. H. (1988) Serum bile acids: reference values in healthy dogs and comparison of two kit methods. *Veterinary Clinical Pathology,* **17**, 71–74.
56. Crawford, M. A., Schall, W. D., Jensen, R. K. and Tasker, J. B. (1985) Chronic active hepatitis in 26 Doberman Pinschers. *Journal of the American Veterinary Medical Association,* **187**, 1343–1350.
57. Creutzfeldt, W., Frerichs, H. and Sickinger, K. (1970) Liver disease and diabetes mellitus. In *Progress in Liver Diseases.* Vol. 3. Eds. H. Popper and F. Schaffner. pp. 371–407.
58. Crowe, D. T. (1990) Nutritional support for the hospitalized patient: an introduction to tube feeding. *Compendium of Continuing Education,* **12**, 1711–1720.
59. Crowe, D. T. (1986) Enteral nutrition for critically ill or injured patients—Part II. *Compendium of Continuing Education,* **8**, 719–726.
60. Davenport, D. J. (1990) Antimicrobial therapy for gastrointestinal, pancreatic and hepatic disorders. *Problems in Veterinary Medicine,* **2**, 374–393.
61. Doige, C. E. and Lester (1981) Chronic active hepatitis in dogs—a review of 14 cases. *Journal of the American Animal Hospital Association,* **17**, 725–730.
62. Eriksson, L. S. and Conn, H. O. (1989) Branched-chain amino acids in the management of hepatic encephalopathy: an analysis of variant. *Hepatology,* **10**, 228–246.
63. Feldman, B. F., Strafuss, A. C. and Gabbert, N. (1976) Bile duct carcinoma in the cat. Three case reports. *Feline Practice,* **6**, 33.
64. Feldman, E. C. and Ettinger, S. J. (1976) Percutaneous transthoracic liver biopsy in the dog. *Journal of the American Veterinary Medical Association,* **169**, 805–810.
65. Flannery, D. B., Hsia, Y. E. and Wolf, B. (1982) Current status of hyperammonemic syndromes. *Hepatology,* **2**, 495–506.
66. Gammal, S. H. and Jones, E. A. (1989) Hepatic encephalopathy. *Medical Clinics of North America,* **73**, 793–813.
67. Garvey, M. S. (1986) Feline liver disease. *Feline Medicine I, Annual Kal Kan Symposium, Eastern States Veterinary Conference,* pp. 19–26.
68. Gocke, D. J., Presig, R., Morris, T. Q., McKay,

D. G. and Bradley, S. E. (1967) Experimental viral hepatitis in the dog. Production of persistent disease in partially immune animals. *Journal of Clinical Investigation,* **46**, 1505–1517.

69. Godshalk, C. P., Twardock, A. R. and Kneller, S. K. (1989) Nuclear scintigraphic assessment of liver size in clinically normal dogs. *American Journal of Veterinary Research,* **50**, 645–650.

70. Grauer, G. F. and Nichols, C. E. R. (1985) Ascites, renal abnormalities and electrolyte and acid base disorders associated with liver disease. *Veterinary Clinics of North America,* **15**, 197–203.

71. Griffiths, G. L., Lumsden, J. H. and Valli, V. E. O. (1981) Haematologic and biochemical changes in dogs with portosystemic shunts. *Journal of the American Animal Hospital Association,* **17**, 705–710.

72. Hall. J. A., Allen, T. A. and Fettman, M. J. (1987) Hyperammonemia associated with urethral obstruction in a dog. *Journal of the American Veterinary Medical Association,* **191**, 1116–1118.

73. Hampson, E. C. G. M., Filippich, L. J., Kelly, W. R. and Evans, K. (1987) Congenital biliary atresia in a cat: a case report. *Journal of Small Animal Practice,* **28**, 39–48.

74. Hand, M. S., Crane, S. W. and Buffington, C. A. Surgical nutrition. In *Manual of Small Animal Surgical Therapeutics.* Eds. C. W. Betts and S. W. Crane. pp. 91–115. New York: Churchill Livingstone.

75. Hardy, R. M. (1985) Chronic hepatitis: an emerging syndrome in dogs. *Veterinary Clinics of North America: Small Animal Practice,* **15**, 135–150.

76. Hardy, R. M. (1986) Chronic hepatitis in dogs: a syndrome. *Compendium of Continuing Education,* **8**, 904–914.

77. Hardy, R. M., (1990) Pathophysiology of hepatic encephalopathy. *Seminars in Veterinary Medicine and Surgery (Small Animal),* **5**(2), 100–106.

78. Haywood, S., Rutgers, H. C. and Christian, M. K. (1988) Hepatitis and copper accumulation in Skye Terriers. *Veterinary Pathology,* **25**, 408–414.

79. Haywood, S. H. and Hall, E. H. (1992) Copper toxicosis in Bedlington Terriers (letter to the editor). *Veterinary Record,* **19**, 272.

80. Hazzi, C. H. (1986) Diagnosis and management of chronic hepatitis. *American Journal of Gastroenterology,* **81**, 85–90.

81. Heidner, G. L. and Campbell, K. L. (1985) Cholelithiasis in a cat. *Journal of the American Veterinary Medical Association,* **186**, 176–177.

82. Herrman, R., Shakoor, T. and Weber, F. L. (1987) Beneficial effects of pectin in chronic hepatic encephalopathy. *Gastroenterology,* **92**, 1795 (Abstr.).

83. Herrtage, M. E., Seymour, C. A., Jefferies, A. R., Blakemore, W. F. and Palmer, A. C. (1987) Inherited copper toxicosis in the Bedlington Terrier: a report of two clinical cases. *Journal of Small Animal Practice,* **28**, 1127–1140.

84. Herrtage, M. E., Seymour, C. A., White, R. A. S., Small, G. M. and Wight, D. G. D. (1987) Inherited copper toxicosis in the Bedlington Terrier: the prevalence in asymptomatic dogs. *Journal of Small Animal Practice,* **28**, 1141–1151.

85. Hill, G. M., Brewer, C. J., Pvasod, A. S. et al. (1986) Oral zinc therapy for Wilson's disease patients. *Clinical Research,* **31**, 466A,

86. Hirsch, V. M. and Doige, C. E. (1983) Suppurative cholangitis in cats. *Journal of the American Veterinary Medical Association,* **182**, 1223–1226.

87. Hitt, M. E. and Jones, B. D. (1986) Effect of storage temperature and time on canine plasma ammonia concentrations. *American Journal of Veterinary Research,* **47**, 363–364.

88. Hornoff, W. J., Koblik, P. D. and Breznock, E. M. (1983) Radiocolloid scintigraphy as an aid to the diagnosis of congenital portocaval anomalies in the dog. *Journal of the American Veterinary Medical Association,* **182**, 44–46.

89. Hunout, D., Aicardi, V., Hirsch, S. et al. (1989) Nutritional support in hospitalized patients with alcoholic liver disease. *European Journal of Clinical Nutrition,* **43**, 615–621.

90. Jacobs, G., Cornelius, L., Allen, S. and Greene, C. (1989) Treatment of idiopathic hepatic lipidosis in cats: 11 cases (1986–1987). *Journal of the American Veterinary Medical Association,* **195**, 635–639.

91. Jarrett, W. F. H. and O'Neill, B. W. (1985) A new transmissible agent causing acute hepatitis, chronic hepatitis and cirrhosis in dogs. *Veterinary Record,* **116**, 629–635.

92. Jarrett, W. F. H., O'Neill, B. W. and Lindholm, I. (1987) Persistent hepatitis and chronic fibrosis induced by canine acidophil cell hepatitis virus. *Veterinary Record,* **120**, 234–235.

93. Johnson, C. A., Armstrong, P. J. and Hauptman, J. G. (1987) Congenital portosystemic shunts in dogs: 46 cases (1979–1986). *Journal of the American Veterinary Medical Association,* **191**, 1478–1483.

94. Johnson, G. F., Gilbertson, S. R., Goldfischer, S., Grushof, P. S. and Sternlieb, I. (1984) Cytochemical detection of inherited copper toxicosis of Bedlington Terriers. *Veterinary Pathology,* **21**, 57–60.

95. Johnson, G. F., Sternlieb, I., Twedt, D. C., Grushoff, P. S. and Scheinberg, I. H. (1980) Inheritance of copper toxicosis in Bedlington Terriers. *American Journal of Veterinary Research,* **41**, 1865.

96. Johnson, G. F., Zawie, D. A., Gilbertson, S. R. and Sternlieb, I. (1982) Chronic active hepatitis in Doberman Pinschers. *Journal of the American Veterinary Medical Association,* **180**, 1438–1442.

97. Johnson, S. E., Rogers, W. A., Bonagura, J. D. and Caldwell, J. H. (1985) Determination of serum bile acids in fasting dogs with hepatobiliary disease. *American Journal of Veterinary Research,* **46**, 2048–2053.

98. Johnson, S. E., Crisp, S. M., Smeak, D. D. and Fingeroth, J. M. (1989) Hepatic encephalopathy in two aged dogs secondary to a presumed congenital portal-azygous shunt. *Journal of the American Animal Hospital Association,* **25**, 129–132.

99. Johnson, S. E. (1992) Liver and biliary tract. In *Veterinary Gastroenterology.* Eds. N. V. Anderson, R. G. Sherding, A. M. Merritt and R. H. Whitlock. pp. 504–569. London: Lea and Febiger.

100. Jones, B. D., Hitt, M. and Hurst, T. (1985) Hepatic biopsy. *Veterinary Clinics of North America* **15**, 39–66.

101. Kelly, D. F., Baggott, D. G. and Gaskell, C. J. (1975) Jaundice in the cat associated with inflammation of the biliary tract and pancreas. *Journal of Small Animal Practice,* **16**, 163.

102. Laflamme, D. P. (1989) Hepatoencephalopathy associated with multiple portal systemic shunts in a dog. *Journal of the American Animal Hospital Association,* **25**, 199–202.

103. Lamb, C. R. (1990) Abdominal ultrasonography in small animals: examination of the liver, spleen and pancreas. *Journal of Small Animal Practice,* **31**, 6–15.

104. Lewis, D. T., Malone, J. B., Taboada, J., Hribernik, T. N., Pechman, R. D. and Dean, P. W. (1991) Cholangiohepatitis and choledochectasia associated with *Amphimerus pseudofelineus* in a cat. *Journal of the American Animal Hospital Association,* **27**, 156–161.

105. Lord, P. F., Carb, A., Halliwell, W. H. and Prueter, J. C. (1982) Emphysematous hepatic abscess associated with trauma, necrotic hepatic nodular hyperplasia and adenoma in a dog. *Veterinary Radiology,* **23**, 46.

106. Lucke, V. M. and Davies, J. D. (1984) Progressive lymphocytic cholangitis in the cat. *Journal of Small Animal Practice,* **25**, 249–260.

107. Lucke, V. M. and Herrtage, M. E. (1987) Copper associated liver disease in the dog. *Veterinary Annual,* **27**, 264.

108. Ludwig, J., Owen, C. A., Barham, S. S. *et al.* (1982) The liver in inherited copper disease of Bedlington Terriers. *Laboratory Investigation,* **106**, 432.

109. MacDonald, M. L., Rogers, Q. R. and Morris, J. G. (1984) Nutrition of the domestic cat, a mammalian carnivore. *Annual Review of Nutrition,* **4**, 521–562.

110. Macy, D. W. and Ralston, S. L. (1989) Cause and control of decreased appetite. In *Current Veterinary Therapy. X, Small Animal Practice.* Ed. R. W. Kirk. pp. 18–24. Philadelphia: W. B. Saunders.

111. Maddison J. E. (1981) Portosystemic encephalopathy in two young dogs: some additional diagnostic and therapeutic considerations. *Journal of Small Animal Practice,* **22**, 731–739.

112. Magne, M. L. and Chiapella, A. M. (1986) Medical management of canine chronic hepatitis. *Compendium of Continuing Education,* **8**, 915–921.

113. Martin, R. A. and Freeman, L. E. (1987) Identification and surgical management of portosystemic shunts in the dog and cat. *Seminars of Veterinary Medicine and Surgery (Small Animal Practice),* **2**, 302–306.

114. McConnell, M. F. and Lumsden, J. H. (1983) Biochemical evaluation of metastatic liver disease in the dog. *Journal of the American Animal Hospital Association,* **19**, 173–176.

115. Merino, C. E., Thomas, T. J., Doizaki, W. and Najarian, J. S. (1975) Methionine induced hepatic coma in dogs. *American Journal of Surgery,* **130**, 41–46.

116. Meyer, H. P. and Rothuizen, J. (1991) Congenital portosystemic shunts in dogs are a genetic disorder. *Tijdschrift voor Diergeneeskunde,* **116**, 80S.

117. Meyer, D. J. and Thompson, M. B. (1993). Bile acids—beyond their value as a liver function test. *Proceedings of the Eleventh ACVIM Forum.* pp. 210–212. Washington.

118. Meyer, D. J., Strombeck, D. R., Stone, A. E., Zenoble, R. D. and Buss, D. D. (1978) Ammonia tolerance test in clinically normal dogs and in dogs with portosystemic shunts. *Journal of the American Veterinary Medical Association,* **173**, 377–379.

119. Meyer, D. J. and Noonan, N. E. (1981) Liver tests in dogs receiving anticonvulsant drugs. *Journal of the American Animal Hospital Association,* **17**, 261–264.

120. Meyer, D. J. and Center, S. A. (1986) Approach to the diagnosis of liver disorders in the dog and cat. *Compendium of Continuing Education,* **8**, 880–888.

121. Meyer, D. J. (1986) Liver functions tests in dogs with portosystemic shunts: measurement of serum bile acid concentrations. *Journal of the American Veterinary Medical Association,* **188**, 167–199.

122. Meyer, D. E., Iverson, W. O. and Terrell, T. G. (1980) Obstructive jaundice associated with chronic active hepatitis in a dog. *Journal of the American Veterinary Medical Association,* **176**, 41–44.

123. Michel, H., Bories, P., Aubin, J. P. *et al.* (1985) Treatment of acute hepatic encephalopathy in cirrhotics with a branched chain amino-acids enriched versus a conventional amino acids enriched mixture. *Liver,* **5**, 282–289.

124. Morgan, H. M., Bolton, C. H., Morris, J. S. and Read, A. E. (1974). Medium chain triglycerides and hepatic encephalopathy. *Gut,* **15**, 180–184.
125. Morris, J. J. and Rogers, Q. R. (1978) Arginine: an essential amino acid for the cat. *Journal of Nutrition,* **18**, 1944–1978.
126. Mullen, K. D. and Weber, F. L. (1991) Role of nutrition in hepatic encephalopathy. *Seminars in Liver Disease,* **11**, 292–305.
127. National Research Council (1978) *Nutrient Requirements of Cats, No. 13. Nutrient Requirements of Domestic Animals.* p. 27. Washington DC: National Academy of Sciences.
128. National Research Council (1985) *Nutrient Requirements of Dogs, Nutrient Requirements of Domestic Animals.* Washington DC: National Academy of Sciences.
129. Naylor, C., O'Rourke, K., Detsky, A. and Baker, J. (1989) Parenteral nutrition with branched chain amino-acids in hepatic encephalopathy. *Gastroenterology,* **97**, 1033–1042.
130. Ogilvie, G. K., Engelking, L. R. and Anwer, S. (1985) Effect of plasma sample storage on blood ammonia, bilirubin and urea nitrogen concentration. *American Journal of Veterinary Research,* **46**, 2619–2622.
131. O'Keefe, S. J. D., El-Zayadi, A., Carraher, T. *et al.* (1980) Malnutrition and immuno-competence in patients with liver disease. *Lancet,* **ii**, 615–617.
132. Osborne, C. A., Hardy, R. M., Stevens, J. B. and Perman, V. (1974) Liver biopsy. *Veterinary Clinics of North America,* **4**, 333–350.
133. Osborne, C. A., Hardy, R. M. and Davis, L. S. (1982) Use and misuse of drugs in hepatic disorders. *Journal of the American Animal Hospital Association,* **18**, 23–31.
134. Papich, M. G. and Davis, L. E. (1985) Dr.ugs and the liver. *Veterinary Clinics of North America,* **15**, 77–96.
135. Patnaik, A. K., Hurvitz, A. I. and Lieberman, P. H. (1980) Canine hepatic neoplasms: a clinico-pathologic study. *Veterinary Pathology,* **17**, 553–564.
136. Patnaik, A. K., Lieberman, P. H., Hurvitz, A. I. and Johnson, G. F. (1981) Canine hepatic carcinoids. *Veterinary Pathology,* **18**, 445–453.
137. Phillips, L., Tappe, J. and Lyman, R. (1993) Hepatic microvascular dysplasia without demonstrable macroscopic shunts. *Proceedings of the Eleventh ACVIM Forum,* pp. 438–439. Washington.
138. Pion, P. D., Kittleson, M. D., Rogers, Q. R. and Morris, J. G. (1987) Myocardial failure in cats associated with low plasma taurine: a reversible cardiomyopathy. *Science,* **237**, 764.
139. Poffenbarger, E. M. and Hardy, R. M. (1985) Hepatic cirrhosis associated with long term pri-midone therapy in a dog. *Journal of the American Veterinary Medical Association,* **187**, 978–980.
140. Rakich, P. M., Rogers, K. W., Lukert, P. D. and Cornelius, L. M. (1986) Immunohistochemical detection of canine adenovirus in paraffin sections of liver. *Veterinary Pathology,* **23**, 478–484.
141. Record, C. O., Buxton, B., Chase, R. A. *et al.* (1976) Plasma and brain amino-acids in fulminant hepatic failure and their relationship to hepatic encephalopathy. *European Journal of Clinical Investigation,* **72**, 483–487.
142. Reding, P., Duchateau, J. and Bataille, C. (1983) Oral zinc supplementation improves hepatic encephalopathy: results of a randomised controlled trial. *Lancet,* **ii**, 493–495.
143. Ritland, S. and Berger, A. (1975) Plasma concentration of lipoprotein X(LP-X) in experimental bile duct obstruction. *Scandinavian Journal of Gastroenterology,* **10**(17).
144. Robinson, C. H. and Lawler, M. R. (1977) Diet in disturbances of the liver, gallbladder and pancreas. In *Normal and Therapeutic Nutrition.* 15th edn. pp. 467–469. New York: Macmillan.
145. Rocchi, E., Cassanelli, M., Gilbertini, P. *et al.* (1985) Standard or branched chain amino-acid infusions as short-term nutritional support in liver cirrhosis. *Journal of Parenteral and Enteral Nutrition,* **9**, 447–451.
146. Rogers, K. S. and Cornelius, L. M. (1985) Feline icterus. *Compendium of Continuing Education,* **7**, 391.
147. Rogers, Q. R. and Morris, J. G. (1983) Protein and amino acid nutrition of the cat. *American Animal Hospital Association Proceedings,* p. 333.
148. Rogers, W. A. and Ruebner, B. H. (1977) A retrospective study of probable glucocorticoid-induced hepatopathy in dogs. *Journal of the American Veterinary Medical Association,* **170**, 603–606.
149. Rosen, H. M., Yoshimura, N., Hodeman, J. M. *et al.* (1977) Plasma amino acid patterns in hepatic encephalopathy of differing etiology. *Gastroenterology,* **72**, 483–487.
150. Rossi–Fanelli, F., Freund, H., Krause, R. *et al.* (1982) Induction of coma in normal dogs by the infusion of aromatic amino acids and its prevention by the addition of branched-chain amino acids. *Gastroenterology,* **83**, 664–671.
151. Rothuizen, J. and Van den Brom, W. E. (1990) Quantitive hepatobiliary scintigraphy as a measure of bile flow in dogs with cholestatic disease. *American Journal of Veterinary Research,* **51**, 253–256.
152. Rothuizen, J., Van den Ingh, T. S. G. A. M., Voorhout, G., Van der luer, R. J. T. and Wouda,

W. (1982) Congenital portosystemic shunts in sixteen dogs and three cats. *Journal of Small Animal Practice*, **23**, 67–81.

153. Rutgers, H. C. and Haywood, S. (1988) Chronic hepatitis in the dog. *Journal of Small Animal Practice*, **29**, 679–690.

154. Rutgers, H. C., Haywood, S. and Batt, R. M. (1990) Colchicine treatment in a dog with hepatic fibrosis. *Journal of Small Animal Practice*, **31**, 97–101.

155. Rutgers, H. C., Haywood, S. and Kelly, D. F. (1993) Idiopathic hepatic fibrosis in fifteen dogs. *Veterinary Record*, **133**, 115–118.

156. Rutgers, H. C., Stradley, R. P. and Johnson, S. E. (1988) Serum bile acid analysis in dogs with experimentally induced cholestatic jaundice. *American Journal of Veterinary Research*, **49**, 317–320.

157. Samson, F. E., Dahl, N. and Dahl, D. (1956) A study of the narcotic action of the short chain fatty acids. *Journal of Clinical Investigation*, **35**, 1291–1298.

158. Scavelli, T. D., Hornbuckle, W. E., Roth, L., Rendano, V. T., de Lahunta, A., Center, S. A., French, T. W. and Zimmer, J. F. (1986) Portosystemic shunts in cats: seven cases (1976–1984). *Journal of the American Veterinary Medical Association*, **189**, 317–325.

159. Sherding, R. G. (1985) Acute hepatic failure. *Veterinary Clinics of North America*, **15**, 119.

160. Sherding, R. G., (1979) Hepatic encephalopathy in the dog. *Compendium of Continuing Education*, **1**, 55–63.

161. Schrier, R. W., Arroyo, V., Bernadi M., Epstein, M., Hendricksen J. H. and Rodes, J. (1988) Peripheral arterial vasodilation hypothesis: a proposal for the initiation of renal sodium and water retention in cirrhosis. *Hepatology*, **8**, 1151–1157.

162. Shaker, E. H., Zawie, D. A., Garvey, M. S. and Gilbertson, S. R. (1991) Suppurative cholangiohepatitis in a cat. *Journal of the American Animal Hospital Association*, **27**, 148–150.

163. Smith, A. R., Rossi–Fanelli, F., Ziparo, V., James, J. H., Perelle, B. A. and Fischer, J. E. (1978) Alterations in plasma and CSF amino acids, amines and metabolites in hepatic coma. *Annals of Surgery*, **187**, 343–350.

164. Stanton, M. E. and Bright, R. M. (1989) Gastroduodenal ulceration in dogs. *Journal of Veterinary Internal Medicine*, **3**, 238–242.

165. Strombeck, D. R. and Gribble, D. G. (1978) Chronic active hepatitis in the dog. *Journal of the American Veterinary Medical Association*, **173**, 380–386.

166. Strombeck, D. R. (1978) Clinicopathologic features of primary and metastatic neoplastic disease of the liver. *Journal of the American Veterinary Medical Association*, **173**, 267–269.

167. Strombeck, D. R. and Guildford, W. G. (1991) *Small Animal Gastroenterology*. 2nd edn. London: Wolfe.

168. Strombeck, D. R., Meyer, W. and Freedland, R. A. (1975) Hyperammonemia due to a urea cycle enzyme deficiency in two dogs. *Journal of the American Veterinary Medical Association*, **166**, 1109–1111.

169. Strombeck, D. R., Schaffer, M. L. and Rogers, Q. R. (1983) Dietary therapy for dogs with chronic hepatic insufficiency. In *Current Veterinary Therapy VIII*. Ed. R. W. Kirk. pp. 817–821. Philadelphia: W. B. Saunders.

170. Strombeck, D. R., Weiser, M. G. and Kaneko, J. J. (1975) Hyperammonemia and hepatic encephalopathy in the dog. *Journal of the American Veterinary Medical Association*, **166**, 1105–1108.

171. Strombeck, D. R., Harrold, D. Rogers, Q. *et al.* (1983) Plasma amino acid, glucagon and insulin concentrations in dogs with nitrosamine-induced-hepatic disease. *American Journal of Veterinary Research*, **44**, 2028–2036.

172. Su, L.-C., Owen, C. A., Zollman, P. E. and Hardy, R. M. (1982) A defect of biliary excretion of copper-laden Bedlington Terriers. *American Journal of Physiology*, **243**, G231–G236.

173. Tabaoda, J. (1990) Medical management of animals with portosystemic shunts. *Seminars in Veterinary Medicine and Surgery: Small Animal Practice*, **5**, 107–119.

174. Tams, T. R. (1985) Hepatic encephalopathy. *Veterinary Clinics of North America: Small Animal Practice*, **15**, 177–195.

175. Teske, E., Brinkhuis, B. G. A. M., Bode, P., Van den Ingh, T. S. G. A. M. and Rothuizen, J. (1992) Cytological detection of copper for the diagnosis of inherited copper toxicosis in Bedlington Terriers. *Veterinary Record*, **131**, 30–32.

176. Thompson, A. B. R. (1973) Intestinal absorption of lipids: influence of the unstained water layer and bile acid micelle. In *Disturbances in Lipid and Lipoprotein Metabolism*. Eds. J. M. Dietschy, A. M. Goth and J. A. Ontko. pp. 29–55. Baltimore: American Physiological Society.

177. Thornburg, L. P. and Rottinghaus, G. (1985) What is the significance of hepatic copper values in dogs with cirrhosis? *Veterinary Medicine*, **80**, 50–54.

178. Thornburg, L. P. (1982) Chronic active hepatitis —what is it and does it occur in dogs? *Journal of the American Animal Hospital Association*, **18**, 21–22.

179. Thornburg, L. P., Beissenherz, M., Dolan, M. and Raisbeck, M. F. (1985) Histochemical demonstration of copper and copper-associated protein in the canine liver. *Veterinary Pathology*, **22**, 327–332.

180. Thornburg, L. P., Moxley, R. A. and Jones, B. D. (1981) An unusual case of chronic hepatitis in a Kerry Blue. *Veterinary Medicine/Small Animal Clinics*, **76**, 363–364.

181. Thornburg, L. P., Polley, D. and Dimmitt, R. (1984) The diagnosis and treatment of copper toxicosis in dogs. *Canadian Practice*, **11**, 36–39.

182. Thornburg, L. P., Rottinghaus, G. and Gage, H. (1986) Chronic liver disease associated with high hepatic copper concentration in a dog. *Journal of the American Veterinary Medical Association*, **188**, 1190–1191.

183. Thornburg, L. P., Shaw, D., Dolan, M., Raisbeck, M., Crawford, S., Dennis, G. L. and Olwin, D. B. (1986) Hereditary copper toxicosis in West Highland White Terriers. *Veterinary Pathology*, **23**, 148–154.

184. Thornburg, L. P., Simpson, S. and Diglio, K. (1982) Fatty liver syndrome in cats. *Journal of the American Animal Hospital Association*, **18**, 397.

185. Twedt, D. C. (1992) The clinical approach to canine liver disease. *Proceedings of the Tenth ACVIM Forum*. pp. 113–118. San Diego.

186. Twedt, D. C. and Whitney, E. L. (1989) Management of copper hepatotoxicosis in dogs. In *Current Veterinary Therapy, X*. Ed. R. W. Kirk. pp. 891–896. Philadelphia: W. B. Saunders.

187. Twedt, D. C., Sternlieb, I. and Gilbertson, S. R. (1979) Clinical, morphologic and chemical studies on copper toxicosis of Bedlington Terriers. *Journal of the American Veterinary Medical Association*, **175**, 269–275.

188. Vaden, S. L., Wood, P. A., Ledley, F. D. *et al.* (in press) Cobalamin deficiency associated with methyl malonic acidemia in a cat. *Journal of the American Veterinary Medical Association*.

189. Van den Ingh, T. S. G. A. M. and Rothuizen, J. (1982) Hepatoportal fibrosis in three young dogs. *Veterinary Record*, **110**, 575–578.

190. Van den Ingh, T. S. G. A. M., Rothuizen, J. and Van den Brom, W. E. (1986) Extrahepatic cholestasis in the dog and the differentiation of extrahepatic and intrahepatic cholestasis. *Veterinary Quarterly*, **8**, 150–157.

191. Vince, A. J., Burridge, A. M., Park, M. and O'Grady, F. (1973) Ammonia production by intestinal bacteria. *Gut*, **14**, 171–177.

192. Walser, M. (1984) Therapeutic aspects of branched chain and keto acids. *Clinical Science*, **66**, 1–15.

193. Watanabe, A., Tetsuyu, S., Okita, M. *et al.* (1983) Effect of a branched chain amino-acid enriched nutritional supplement on the pathophysiology of the liver and the nutritional state of patients with liver cirrhosis. *Acta Medica Okayama*, **37(4)**, 21–33.

194. Weber, F. L., Stephen, A. M., Karagiannis and E. M. (1985) Effects of dietary fiber on nitrogen metabolism in cirrhotic patients. *Gastroenterology*, **88**, 1704 (Abstr.).

195. Weingand, K. W. (1991) Pathogenic mechanisms of hepatic lipidosis. *Proceedings of the Ninth ACVIM Forum*. pp. 185–187.

196. Willard, M. D., Dunstan, R. W. and Faulkner, J. (1988) Neuroendocrine carcinoma of the gallbladder in a dog. *Journal of the American Veterinary Medical Association*, **192**, 926–928.

197. Wilson, S. M. and Feldman, E. C. (1992) Diagnostic value of the steroid-induced isoenzyme of alkaline phosphatase in the dog. *Journal of the American Animal Hospital Association*, **28**, 245–250.

198. Zieve, L., Doizaki, W. and Zieve, F. (1974) Synergism between mercaptans and ammonia or fatty acids in the production of coma: a possible role for mercaptans in the pathogenesis of hepatic coma. *Journal of Laboratory and Clinical Medicine*, **83**, 16–28.

CHAPTER 16

Feline Renal Disease

ROBERT MORAILLON and ROGER WOLTER

Acute Kidney Failure

Acute renal failure is associated with a rapid diminution of renal function that leads to an oliguric/anuric syndrome, with a resultant accumulation of nitrogenous wastes and disruptions in the fluid electrolyte and acid–base balance.

Acute kidney failure may be of prerenal, renal or postrenal origin.

Clinical examination and anamnesis enable the clinician to quickly distinguish prerenal and postrenal uremia. Prerenal acute kidney failure is the consequence of a decrease in glomerular blood flow that occurs in hypotension secondary to hypovolemia, e.g. infectious gastroenteritis, states of shock and during the progression of heart failure. In the cat, postrenal acute kidney failure is usually a consequence of urethral obstruction by magnesium ammonium phosphate uroliths or mucus plugs. Obstruction of the urinary system may also be caused by compression with abdominal or pelvic masses. Rupture of the urinary tract, usually associated with abdominal trauma, also causes postrenal acute kidney failure. Therefore in the prerenal uremia syndrome, hypovolemia and a state of shock are dominant, whereas in the postrenal form, existence of a distended bladder (vesical globe) and urethral obstruction are characteristic.

Parenchymatous renal failure is diagnosed by urine analysis (inappropriately low urine specific gravity, significant proteinuria, granular/epithelial/red cell urinary casts). The study of the urine creatinine:plasma creatinine ratio (<10 in parenchymatous renal failure) and urinary sodium concentration (>40 mmol/l in parenchymatous renal failure) may also be helpful for the diagnosis of parenchymatous renal failure.

The kidney lesions responsible for acute parenchymatous renal failure are either tubular lesions of toxic origin (aminoglycoside antibiotics, antiinflammatory drugs, ethylene glycol)[27,56] or interstitial lesions observed during septicemia or as a complication of chronic infections (pyometritis, cutaneous abscess, pyothorax).

Pathophysiology

The major cause of prerenal acute kidney failure is a decrease in the glomerular filtration rate due to a fall in renal perfusion pressure. The reduction in the glomerular filtration rate results in oliguria due to an increased reabsorption of sodium chloride by the proximal convoluted tubules and activation of the renin–angiotensin system. A larger quantity of water is also reabsorbed in the collecting tubules.

In parenchymatous acute kidney failure oliguria results from failure of filtration, obstruction of renal tubules by casts and cellular debris and cellular and interstitial edema. The pathophysiological mechanisms resulting in oliguria–anuria are responsible for the disturbances in the water–electrolyte balance (overhydration, hyperkalemia, hyperphosphatemia), metabolic acidosis and the accumulation of nitrogen waste products.

Depending on the etiology of the renal injury the lesions may be reversible. In such cases an extensive diuresis may occur during recovery due to reduced proximal tubular reabsorption of sodium chloride, and a lack of responsiveness of the collecting tubules to antidiuretic hormone (ADH).

Clinical Signs and Clinicopathological Findings

Acute parenchymal kidney failure is characterized by oliguria–anuria combined with gastrointestinal signs (vomiting and diarrhea) and the rapid development of shock.

The biochemical consequences of acute renal failure in the cat are similar to those observed in other species: elevation of blood urea and creatinine (azotemia) and fluid–electrolyte imbalances, especially hyperkalemia. Hyperkalemia may be accentuated by metabolic acidosis and can lead to cardiac arrhythmias resulting in death.

Analysis of the urine reveals an inappropriately low specific gravity (<1.035) and the urine sediment may contain casts (the nature and number of which depend on the renal lesions) and glucosuria may be detected.

In postrenal acute kidney failure, the signs occur in two phases.

The first phase is associated with obstruction of the urethra and is characterized by oliguria–anuria, stranguria, dysuria and possibly hematuria. Uremia develops after approximately 24 hr. On abdominal palpation the bladder is found to be distended. The biochemical changes are dominated by

an elevation in blood urea, creatinine, phosphate and potassium and the development of metabolic acidosis.

When the obstruction is removed, the second phase is characterized by a profuse diuresis that may cause hypokalemia with anorexia, muscle weakness and intestine and bladder atony.

Diagnosis

Kidney failure is diagnosed by measuring blood urea and creatinine values and by changes in certain urinary parameters (specific gravity and sediment examination).

To distinguish between acute prerenal kidney failure and acute renal failure requires an assessment of the case history, the urine specific gravity and the urine creatinine/plasma creatinine ratio, which is high with prerenal acute kidney failure and low with acute renal failure.

Acute postrenal kidney failure is characterized by the inability to catheterize the urethra and by finding a large distended bladder on abdominal palpation, radiography or ultrasonography. Other causes of postrenal kidney failure that must also be considered are urethral or bladder rupture, particularly in road traffic accident cats.

More definitive diagnosis of parenchymatous renal failure is achieved by percutaneous biopsy through the abdominal wall. The kidney is localized by palpation and the biopsy procedure can be assisted by ultrasonographic guidance of the biopsy needle.

Treatment

Acute kidney failure requires immediate symptomatic treatment, the aims of which are: rehydration in prerenal failure, the elimination of toxic substances or treatment for suppurative foci in parenchymatous kidney failure and the removal of the urethral obstruction in postrenal kidney failure.

Treatment aimed at correcting water–electrolyte imbalances includes slow intravenous infusion of lactated Ringer's solution or isotonic saline solution. The volume of fluid given depends on the percentage dehydration and on the weight of the animal.[14,36,55]

$$\text{Volume in mls} = \%\ \text{dehydration} \times \text{body weight (kg)} \times 1{,}000.$$

In order to avoid any risk of fluid overload, the infusion should be administered over a 6-hr period. Maintenance requirements must also be given using an infusion rate of 60 ml/kg/day. Fluid input and uric output should be monitored carefully. Sequential assessment of packed cell volume, total protein and body weight enables fluid therapy to be monitored.

Correction of metabolic acidosis can be undertaken whenever the alkaline reserve is less than 15 mEq/l or pH < 7.1.

The amount of bicarbonate to be injected may be determined by applying the following equation:

$$NaHCO_3\ (mEq) = \text{body weight (kg)} \times 0.3 \times (15 - HCO_3\ \text{measured in mEq}).$$

Bicarbonate must be injected carefully—½ dose slow IV, then add ½ to IV fluids. Further amounts are given after checking blood pH and HCO_3^-. If rehydration is not sufficient to create a diuresis, either forced diuresis with furosemide or mannitol dialysis should be used to eliminate nitrogenous waste products. The effectiveness of intravenous furosemide at a dosage of 2–6 mg/kg can be checked by measuring the urine output, which should be about 1–4 ml/min. If furosemide is unsuccessful, mannitol may be used at a dosage of 0.5 g/kg. An excessive diuresis can have two adverse consequences that must be anticipated: dehydration and hypokalemia.[22] Urinary losses are compensated for systemically, and if hypokalemia occurs potassium

chloride is added to the infusion (Fig. 16.1). When these treatments fail, peritoneal dialysis is indicated and may be undertaken in some centers.[10,16] Peritoneal dialysis is not often carried out and constitutes injections into the peritoneal cavity, using a special catheter, of a dialysis liquid heated to 38–39°C, left for around 30 min and then siphoned off. The administration rate should not exceed 0.5 mEq/kg/hr.

Chronic Kidney Failure

Etiology

The lesions responsible for chronic kidney failure are shown in Fig. 16.2.[14] Distinction between chronic and acute renal failure cannot be made on the basis of azotemia or hyperphosphatemia. The diagnosis of chronic renal failure is indicated when the duration of clinical signs is greater than 2 weeks and by the presence of anemia, weight loss and small kidneys.

Chronic tubulointerstitial nephritis is the most common lesion associated with chronic kidney failure in the cat.

Glomerulonephritis can have multiple causes: infections (feline leukemia virus; feline infectious peritonitis, mycoplasmosis); inflammatory disorders (disseminated lupus erythematosus; pancreatitis); or toxic injury (mercury derivatives).

Renal amyloidosis is characterized by the deposition of amyloid, a specific fibrillar protein, in the medullary interstitial tissues, and to a lesser extent in the glomeruli.

Amyloidosis in the cat has a breed and familial incidence, being more common in the Abyssinian.[12]

Amyloid, a glycoprotein, is an inert substance, however it can damage tissues in which it is deposited. In the cat, deposition of amyloid in the medulla leads to compression of the vasa recta, reducing blood flow in this area. This may cause papillary necrosis or chronic interstitial fibrosis characterized by a moderate inflammatory response. The glomerular lesions are less common and occur later in the disease, which is why cats suffering

| TREATMENT OF HYPOKALEMIA ||
Blood potassium (mEq/liter)	Potassium chloride added to 250 ml perfusion solution (mEq)
2.0–2.5	20
2.6–3.0	10
3.1–3.5	7
The administration rate should not exceed 0.5 mEq/kg/hr.	

FIG. 16.1 Treatment of hypokalemia.

KIDNEY LESIONS RESPONSIBLE FOR CHRONIC KIDNEY FAILURE IN THE CAT	
Acquired lesions	Chronic tubulointerstitial nephritis
	Chronic glomerulonephritis
	Amyloidosis
	Renal lymphosarcoma
	Chronic pyelonephritis
	Hydronephrosis
	Pyogranulomatous nephritis of feline infectious peritonitis
	Polycystic renal disease of adults
	Hypokalemic nethropathy
	Periarteritis nodosa
Congenital lesions	Agenesis of the renal cortex
	Renal polycystitis
	Hydronephrosis
	Amyloidosis in Abysinnians

FIG. 16.2 Kidney lesions responsible for chronic kidney failure in the cat.

from amyloidosis have only slight proteinuria. Other tissues in which amyloid may be deposited include the adrenal glands, the thyroid glands, the digestive tract, the spleen, the pancreas and the tongue.[19,20]

Amyloidosis in cats occurs secondary to other diseases.[21] It is most commonly observed with chronic diseases especially those involving chronic suppurative foci, collagen disorders, autoimmune diseases and malignant tumors. In the majority of cases, no underlying lesion is identified.

Diagnosis of amyloidosis requires kidney biopsy and special staining of tissues.

Other diseases responsible for kidney failure in the cat include renal lymphosarcoma, renal carcinoma (which is much rarer than lymphosarcoma), periarteritis, chronic pyelonephritis and pyogranulomatous nephritis due to infectious peritonitis.

Kidney failure can also result from congenital lesions including, agenesis of the renal cortex and renal polycystitis that is usually found in long-haired cats and may be associated with intrahepatic cysts.[37,54]

Pathophysiology

In the cat and the dog, chronic kidney failure is associated with a reduction in the number of functional nephrons, and an increase in the filtration rate of the remaining healthy nephrons (hyperfiltration) in order to maintain a normal glomerular filtration rate.

Glomerular hyperfiltration occurs due to glomerular capillary hypertension, which is independent of the systemic arterial pressure. Renal hemodynamics may also be influenced by dietary protein intake. In rats high protein intake is associated with glomerular capillary hypertension and hyperfiltration and promotes glomerular sclerosis.[6] Restricting dietary protein intake reduces glomerular hyperfiltration in rats by a complex process involving glucagon, growth hormone, renal prostaglandins, the renin–angiotensin system, catecholamines and glomerulopressin, a hepatic derived renal vasodilator, as shown in humans.[43]

Although glomerular hyperfiltration is important for maintaining excretory function in the failing kidney there is evidence to suggest that this compensatory mechanism may damage surviving nephrons by inducing glomerular sclerosis.

The specific compensatory adaptations of the diseased kidney in the cat are dependent on the substance to be excreted.

The excretion of urea and creatinine is dependent on the glomerular filtration rate. In the failing kidney compensation initially occurs due to hyperfiltration by the remaining healthy nephrons. With continuing deterioration of kidney function, elevation in plasma levels of nitrogen waste products occurs and results in systemic dysfunction.[15]

In the dog with chronic kidney failure there is a progressive increase in the fraction of filtered water. This is due to the osmotic diuresis resulting from the elimination of nitrogen waste products. With chronic kidney disease in the cat this phenomenon is less pronounced and the ability to concentrate urine is prolonged.

As the glomerular filtration rate begins to fall the residual functioning nephrons increase the fractional excretion of sodium. Urinary excretion of sodium is maintained by osmotic diuresis and an increasing concentration of atrial natriuretic peptide.

The kidneys maintain potassium homeostasis by increasing the fractional excretion of potassium. This is achieved by increased tubular secretion of potassium through the surviving nephrons. Hypokalemia may, however, occur in some cases because of enhanced renal losses induced by the osmotic diuresis. K^+ depletion may be exacerbated by metabolic acidosis.

With the reduction in the glomerular filtration rate there is an increased fractional excretion and reduced tubular reabsorption of phosphates as a consequence of hyperparathyroidism induced by increasing blood phosphate concentrations. Hyperphosphatemia occurs when the glomerular flow rate is less than 30% of its normal value. A consequence of the secondary hyperparathyroidism is the development of bone disorders characterized by osteofibrosis.[33]

Although hyperphosphatemia is common during the course of chronic kidney failure, calcium concentrations are only affected in the terminal stages of the disease. Reduced renal synthesis of active vitamin D (1,25-dihydrocholecalciferol) is associated with a decrease in intestinal absorption of calcium. This mechanism is compensated for by secondary hyperparathyroidism releasing skeletal calcium.

The acid–base status in the cat is maintained until chronic kidney failure is very advanced. The kidney adjusts the elimination of fixed acids by increasing the reabsorption of filtered bicarbonate through surviving healthy nephrons and by excreting H^+ ions in the form of titratable acid and ammonium. Addition of bicarbonate to a nonacidifying diet is felt unnecessary. The use of an acidifying diet is not recommended.

A nonregenerative anemia is common in chronic kidney failure. Production of erythropoietin by the diseased kidneys is not adequate to compensate for blood losses associated with hemolysis and gastrointestinal hemorrhages and premature senescence of red cells.

Clinical Signs

Due to the functional reserve of the kidney in the cat,[44] polyuria–polydipsia rarely occurs. Similarly, vomiting is uncommon in cats because, in the authors' experience, gastric ulceration is rare in this species.

The most common clinical signs in cats with chronic kidney failure include anorexia, weight loss and a general deterioration of condition. Examination of the mouth often reveals pallor and ulceration of the mucosal membranes.[18] Ocular signs are sometimes reported, notably atrophy of the optic fundus or retinal detachment.[41] Systemic hypertension is considered to cause extravasation through the retinal vessels, edema and retinal detachment.

Physical examination of the cat may reveal an increase in kidney size due to hydronephrosis, polycystosis, lymphosarcoma or granulomatous nephritis, or a decrease in kidney size due to chronic tubulointerstitial nephritis, chronic glomerulonephritis or pyelonephritis.[15]

The main biochemical changes are an increase in the blood concentrations of urea, creatinine and phosphate, and a decrease in the alkaline reserve (metabolic acidosis). Hyperproteinemia may also occur due to dehydration, except in the nephrotic syndrome (see below), which is characterized by hypoproteinemia and hypoalbuminemia due to an abundant proteinuria. In some polyuric cats hyponatremia may also occur. Prolonged anorexia combined with polyuria may also induce hypokalemia, precipitating hypokalemic polymyopathy. Hypercholesterolemia may also occur, and is more common in the cat than the dog.

Analysis of the urine may reveal a decreased urine specific gravity, increased protein concentration and the presence of casts. The degree and nature of these changes depend on the lesions responsible for the disease.[17]

The diagnosis of chronic kidney failure is confirmed in a similar way to acute renal failure by an elevation of the blood urea, creatinine levels and decreased urine specific gravity. Precise determination of the renal lesions is achieved either by ultrasonography,[59,60,66] which can give extremely useful information, or by percutaneous needle biopsy.[42,43] Stereotactic needle biopsy guided by ultrasound will certainly become more popular in the future. Azotemia and hyperphosphatemia are not sufficient to distinguish chronic from acute renal failure. Chronicity is confirmed by anemia, weight loss and small kidneys.

Dietary Therapy

Dietary therapy is important in reducing uremia and perhaps simultaneously exerting a protective effect on surviving nephrons.

Cats with chronic kidney failure require specific modifications in their dietary levels of protein, lipid, minerals and vitamins. These modifications are significantly different to those required by dogs with chronic kidney failure.

Water and Food Intake

Water intake should be encouraged. Milk and broths may be given to encourage maximum fluid intake.

Dietary energy is important for providing energy and for the maintenance of nitrogen metabolism. Nitrogen waste products may increase in chronic kidney failure due to metabolic disturbances associated with uremia (elevated insulin and glucagon concentrations). In formulating specific diets it is important that provision is made for adequate nonprotein calories. However, the diet must be palatable and a compromise between the ideal diet and one that is acceptable to the cat may be necessary. Many of the suggested modifications (decreased protein and salt etc.) reduce palatability. To increase the palatability of these diets it is necessary to flavor the food with small amounts of animal fat or other flavor enhancers.

While encouraging increased dietary intake it is important to be aware of the possible adverse digestive effects associated with uremia such as buccal, mucosal and gastrointestinal ulceration. *Ad libitum* feeding may also promote glomerular sclerosis and it may be beneficial to feed only once or twice daily.[49,50] In some uremic patients, however, feeding 3 or 4 times a day may be necessary to stimulate appetite, facilitate digestion and reduce postprandial nausea. Palatability may be enhanced by warming the food first.

During periods of severe uremia it may be necessary to use assisted feeding for the patient (tubes). See Chapters 4 and 5 for information on critical care nutrition and techniques of enteral nutrition support.

Nitrogen Intake

The dietary protein intake should be of high biological value and at a level that prevents excessive nitrogen catabolism both within the digestive tract as well as body tissues. Sufficient protein is required to maintain lean body weight, digestive secretions, hormone production, hair growth and the immune system etc.

The following reference values are available for the adjustment of the dietary nitrogen intake.[31,35] Dietary protein adjustment is not necessary when:

blood urea concentration < 0.6 g/l and
serum protein concentration ≥ 55 g/l
or
serum albumin ≥ 23 g/l.

Quantitative Requirements

The protein requirements for a cat with chronic kidney disease have not been clearly established. Rationales exist for both an increase in amino acid requirements, and also a decrease.

Intestinal losses are increased because of impaired intestinal absorption and possible diffusion of plasma proteins into the intestinal tract. Albuminuria also increases protein losses, the significance of which is dependent upon the degree of proteinuria.

In humans with chronic kidney failure the recommended protein intake is 1 g/kg body weight (BW). This is twice the normal value for maintenance.

In animals with chronic kidney failure high protein intake may adversely affect renal hemodynamics resulting in damage to the surviving nephrons. Experimental studies using partially nephrectomized rats fed high protein diets have demonstrated glomerular hypertension and hyperfiltration (hyperfiltration theory) with progressive glomerular sclerosis and destruction of surviving nephrons. Loss of kidney function results in an accumulation of toxic nitrogen waste products, such as ammonia and guanidine derivatives. Increased levels of ammonia are toxic to renal tissue and can perpetuate injury to the kidneys. Supplementation with sodium bicarbonate has been shown to lower tissue ammonia concentrations.

Inadequate protein nutrition also has significant adverse effects including: a reduction in appetite, loss of muscle mass, a progressive reduction in serum albumin concentrations and increased protein catabolism with accumulation of nitrogen waste products. Insufficient protein also reduces other renal functions,[32] adversely effects erythropoiesis, hormone synthesis and the immune system and urinary excretion of magnesium. Urinary excretion of magnesium is important as it plays a part in preventing the precipitation of calcium phosphate in the kidneys.[51] Dogs fed a diet with a protein level of 9.5% develop protein malnutrition characterized by hypoalbuminemia.[49,50]

In formulating the dietary protein requirements the goal should be to provide the optimum amount of protein and amino acids to minimize progressive renal damage while maintaining adequate nutrition. Protein of high biological value should be used with the emphasis on quality rather than quantity. An overall protein requirement

RECOMMENDED PROTEIN LEVEL RELATIVE TO THE BLOOD CREATININE VALUE	
Blood creatinine (mg/l)	Dietary protein (% DM)
35	25
40	19
45	14
50	10
55	7

FIG. 16.3 Recommended protein level relative to the blood creatinine value.

(crude protein × digestibility × biological value) of 10% can be achieved with a crude protein level of 18% with 75% digestibility and a biological value of 75%, or by using 12% crude protein with 90% digestibility and 90% biological value.

In a study by Polzin and coworkers,[47] dogs fed a diet containing 17% protein did better than those fed a diet containing 44 or 8% protein. The diet containing 44% protein increased proteinuria and uremia. From a practical aspect, complete dry diets generally contain more moderate levels of protein than moist foods that have not been specifically formulated.[25] In another study by Bovée,[4] diets with 57% protein were *not* detrimental to renal function. Conversely, nitrogen restriction did not improve urinary excretion of either phosphorus or sodium. Other studies[62] have also demonstrated the benefits of using a diet containing 32% protein rather than 15% protein in the recovery phase of acute kidney failure. Brown *et al.*[7] demonstrated that dietary protein below 32% did not prevent glomerular hyperfiltration and hypertension in partially nephrectomized dogs.

Clinical observations have led to recommendations for the level of dietary protein relative to the blood creatinine value (Fig. 16.3).

Studies in the cat[49,50] have demonstrated adverse effects with diets containing 52% protein. Diets containing a protein level less than 18%, however, adversely affect appetite and result in protein malnutrition and clinical and biochemical signs of uremia. The higher protein requirements for cats mean that low protein diets specifically formulated for dogs are unsuitable.

It is currently recommended that cats with chronic kidney failure are fed diets containing approximately 20% of their calories as protein, equivalent to about 3.5 g/kg BW of high biological value protein. In practice this corresponds to a protein level of 20–22% in a dry diet and 8% in a moist diet.[45] Palatability is enhanced by using appetizing protein sources, such as liver, fish and meat, and by the addition of small quantities of animal fat.

Other benefits from reducing the dietary protein content include a reduction in phosphorus, sulfur and magnesium intake; however, this effect is dependent on the use of high quality proteins.

Qualitative Requirements

Protein quality: It is important when formulating therapeutic diets for chronic kidney failure that a suitable balance between protein and nonprotein calories exists. This is measured using the protein–calorie ratio (PCR = g crude protein per megacalorie of metabolizable energy). For maintenance in the healthy adult cat this ratio should be at least 65, and normal sources of protein can be used. In cats with chronic kidney failure this value should be reduced to about 55 during the early stages of the disease and decreased further to about 45 as the severity of the disease progresses. While decreasing the quantity of protein in the diet its quality should be increased.

Trimmed meat, red offal, fish, milk derivatives, eggs, yeasts and soya protein concentrates have a high digestibility (>85%) and provide high levels of essential amino acids, in particular, lysine and tryptophan.

Lysine supplementation may be used to enhance the biological value of the diet. The arginine content should also be evaluated as this amino acid is important in the detoxification of ammonia and may be a limiting factor in low protein diets.

In contrast the diet should contain limited amounts of methionine and phenylalanine. These amino acids tend to be increased in rats with chronic kidney failure due to associated hepatic dysfunction, which may result in

clinical signs of hepatic encephalopathy.[3] To minimize these adverse effects it may be appropriate to increase the dietary intake of valine, leucine and isoleucine using milk proteins.

The use of α-keto acids for protein synthesis may also be beneficial in reducing ammonium levels in the cat.

Altered protein synthesis: Anabolic steroids can be employed to stimulate protein synthesis. The proposed mechanism includes a more efficient recycling of nitrogen waste products. With a reduction in the degree of uremia, appetite is improved and body weight is maintained. They are also recommended in the treatment of anemia associated with chronic kidney failure. Nandrolone decanoate and testosterone enanthate are considered to be among the most effective; however, results in practice have not always been convincing.[23]

Ammoniagenesis and absorption: In order to reduce the release or absorption of intestinal ammonia several measures may be taken:

- an increase in dietary fiber content to reduce transit time of digesta;
- the acidification of colonic contents by dietary components that ferment in the large intestine. These substances include soluble fibers, crude starch, lactose, lactulose, polydextrose and other polyalcohols. In an acid environment ammonia is converted to ammonium that is less diffusible and remains within the colon, being excreted rather than absorbed. The use of inositol, however, should be avoided as it may cause neurotoxicity in animals with chronic kidney failure;
- the addition of clays (zeolite, attapulgite, clinoptilolite etc.), which adsorb ammonia and increase its fecal excretion, to the diet;
- the use of probiotics. Lactose fermenting, nonurease containing bacteria, such as *Lactobacillus*, are fed in an attempt to repopulate the colon. This requires supplementing the diet with large continuous quantities of *Lactobacillus*-containing drugs or yogurts;
- the use of antibiotic therapy to suppress urea-splitting intestinal flora that contribute

significantly to blood ammonia concentrations. Other beneficial effects of oral antibiotics result from a reduction in bacterial deamination of amino acids, reducing the production of aromatic amino acids and other potentially toxic products produced by gut bacteria.

Lipid Requirements

A relatively high dietary lipid intake increases palatability when feeding reduced protein diets. Lipids also increase the energy concentration of the diet, which is beneficial when food intake is reduced due to inappetence.

The composition of dietary lipids can adversely affect systemic blood pressure, platelet aggregation, blood viscosity, erythropoiesis, the immune system and the skin. The mechanisms responsible for these effects may be related in part to disturbances in prostaglandin formation. Long-chain saturated fatty acids predispose to hyperlipidemia and may increase the rate of development of glomerulosclerosis.[32] Lipids can therefore adversely affect renal blood flow and accelerate deterioration of renal function. In formulating the dietary lipid content an appropriate balance between saturated and unsaturated fatty acids must be maintained, paying special attention to supplementation with essential fatty acids of the ω6 and especially ω3 series. In some cases it may also be preferable to supplement the diet with carnitine due to reduced renal synthesis of this amino acid.

For practical purposes the lipid content should constitute 13–15% of a dry diet. Vegetable oils, especially soya and rapeseed, and fish oils are preferred.

Mineral and Vitamin Requirements

The diet can be supplemented with zinc and copper, as well as vitamins A, E and B complex, together with other antioxidants. Hyperphosphatemia induces precipitation of calcium phosphate in the kidneys resulting in nephrocalcinosis, proteinuria and further deterioration of renal function. A reduction in the level of phosphate, either through dietary restriction or reduced intestinal absorption is important in limiting renal injury. Consideration should also be given to other electrolyte disturbances and deficiencies of several trace elements and vitamins.

Phosphorus and calcium: The level of dietary phosphate should be restricted in order to maintain a plasma phosphate concentration of between 20 and 30 mg/l (0.65–1 mmol/l).

Lewis and colleagues[35] recommend a dietary phosphate intake of about 1 mg/kcal (0.35% for a dry diet containing 3500 kcal/kg). Leibetseder and Neufeld[34] obtained good results with a diet containing 0.38% phosphorus per 1.28% calcium (21.5% protein). Finco *et al.*[24] improved glomerular filtration in dogs with kidney failure by using a diet containing 0.4% phosphorus and 0.6% calcium (32% protein), compared to a diet containing 1.4% phosphorus and 1.9% calcium. It is important that the calcium/phosphorus ratio is maintained at a level (Ca/P = 1.2–1.4:1) that prevents excessive assimilation of phosphates. Calcium carbonate is an effective phosphate binding agent and has additional benefits because of its alkalinizing properties. Besides maintaining the calcium/phosphorus ratio it is also important to evaluate the quantity of each mineral in the diet. High dietary levels of calcium and phosphorus result in reduced intestinal absorption of calcium and a proportionally increased intestinal absorption of phosphorus causing hyperphosphatemia. Foods high in phosphorus, such as wheat bran, should be avoided. Excessive supplementation with vitamin D should also be avoided due to the risk of inducing hypercalcemia in patients predisposed to hyperphosphatemia. This would increase the potential for renal mineralization and further deterioration of renal function.

Other intestinal phosphate binding agents include aluminum carbonate or hydroxide. These agents render ingested phosphates unabsorbable as well as acting as antacid compounds both in the dog[8] and in the cat.[2] Aluminum contained in phosphate binding antacids may also be absorbed from the intestinal tract and accumulate in various body tissues, such as bone and brain. Encephalopathies and osteomalacia related to aluminum toxicity have been extensively reported in humans.[31] The accumulation of aluminum in the body can be reduced by the onerous treatment with desferrioxamine.[11] Because of these adverse effects it is probably safer to include clays such as zeolite and kaolinite in the diet. These substances reduce the intestinal absorption of phosphorus, act as intestinal adsorbents for ammonia, amines and other toxins and may reduce gastrointestinal irritation decreasing the incidence of vomiting and diarrhea.[65]

Sodium, potassium and magnesium: Urinary excretion of sodium is reduced in chronic kidney failure despite an increasing diuresis. Sodium retention contributes to systemic hypertension which is a relatively common complicating factor with chronic kidney failure. This potentially exacerbates existing renal hyperfiltration and promotes further deterioration of surviving nephrons.

Dietary sodium should therefore be limited to 0.20–0.25% of the dry matter, which is equivalent to 0.5–0.6% salt (NaCl) and corresponds to the minimum requirements established by the National Research Council (NRC, 1986) for a healthy cat. If ascites, edema or congestive heart failure occur, the intake of sodium may need to be reduced further.

Metabolic acidosis unresponsive to treatment with calcium carbonate may require treatment with sodium bicarbonate. In order to treat the acidosis without increasing the proportion of sodium in the diet it is necessary to partially substitute the salt content of the diet with the sodium bicarbonate (1.4 parts of bicarbonate per 1 part of chloride).

Disturbances of the potassium balance predispose toward hyperkalemia in dogs, and hypokalemia in cats with chronic kidney failure.

Hypokalemia may increase damage to surviving nephrons by increasing systemic blood pressure,[57] and by a direct adverse affect on renal metabolism. It also predisposes toward muscle ischemia causing muscle weakness and rhabdomyolysis—the syndrome of hypokalemic polymyopathy.

Hypokalemia may be induced by excessive dietary sodium. It is therefore recommended that the dietary potassium/sodium ratio is ≥2. For practical purposes dietary potassium should be at least 0.5–6%.

Metabolic acidosis and urinary acidifiers, such as ammonium chloride and methionine used in the treatment of the feline urological syndrome (FUS), may also induce hypokalemia. Urinary loss of potassium, together with reduced levels of taurine also increase the risk of dilated cardiomyopathy and retinal degeneration.

An increased retention of magnesium can occur in chronic kidney failure. A reduction in dietary intake has therefore been recommended. However, hypomagnesemia increases urinary loss of potassium, predisposing to hypokalemia and increasing renal damage. Together with calcium and potassium, magnesium is important in preventing arterial hypertension. Urinary excretion of magnesium also reduces the precipitation of calcium phosphates within the kidneys.

It is therefore recommended to keep dietary magnesium to a level of 0.10–0.12%.

Trace Elements and Vitamins

In chronic kidney failure there is reduced intestinal absorption of trace elements and increased urine losses. Iron and zinc deficiencies may have significant consequences. Iron is important in the treatment of anemia associated with reduced erythropoietin production and for ceruloplasmin efficiency. Zinc is important in taste sensation, maintenance of muscle mass, wound healing and the immune system.

Uremic patients tend to develop vitamin deficiencies except for vitamin A. Supplementation of the diet with vitamins may be necessary.

Supplementation of the diet with vitamin D should be moderately increased because of impaired renal conversion of vitamin D precursors to their most active form (1,25-vitamin D) and to prevent hyperphosphatemia. Hypervitaminosis D, however, must be prevented as hypercalcemia may result in calcification of renal tubular cells and further deterioration of renal function.[46,53]

MAIN NUTRITIONAL REQUIREMENTS FOR CATS WITH CHRONIC KIDNEY FAILURE (per kg of DM)			
	NRC (1986)	Present suggestions	
	(minimum values)	Early stages of chronic kidney failure	Later stages of chronic kidney failure
Energy kcal ME	4000	≥ 4,200	≥ 4,400
Proteins (g)	240	220-230	200
RPC= gProt mega Cal	65	55	45
Lipids (g)	90	120-150	150-180
Calcium (g)	8	8	7
Phosphorus (g)	6	4	3
Ca/P	1.33	2	2.3
Sodium (g)	2	2-2.5	2
Potassium (g)	4	6	6
Magnesium (g)	0.4	1.1-1.3	1.1-1.3
Trace elements	–	↑	
Vitamins A(IU)	3300	=	↓
D(IU)	500	↑	↑
B	–	↑	↑
C	0	↑	↑
Taurine (mg)	0.1-0.2	↑	↑
Carnitine	0	↑	↑

– specific levels have not been recommended; ↑ Increased supplementation suggested; ↓ decreased supplementation suggested.

FIG. 16.4 Main nutritional requirements for cats with chronic kidney failure (per kg of dry matter).

If vitamin D supplementation is necessary it is preferable to use hydroxylated precursors, especially 1-OH-vitamin D, with hydroxylation in the 25-position being performed by the liver and intestines.[13]

Supplementation with vitamin K may also be important. This is because of reduced dietary absorption in the small intestine, reduced microbial synthesis of vitamin K in the large intestine and an increased incidence of coagulation disorders. Gastrointestinal hemorrhage is the most common complication.

Supplementation with B complex vitamins needs to be greatly increased due to reduced dietary intake, altered enzyme and microbial function in the intestinal tract and increased loss of water-soluble vitamins in the urine. Increasing the dietary intake of B complex vitamins may also increase appetite.[35] Similarly vitamin C should also be supplemented.

Dietary carnitine may also need to be supplemented due to reduced production by the kidneys. Additional taurine may also be necessary especially in cases with metabolic acidosis and hypokalemia.

Figure 16.4 compares the main nutritional requirements for cats with chronic kidney failure to those nutritional requirements for normal healthy cats (NRC recommendations) (Fig. 16.4).

Medical Therapy

In most cases dietary therapy needs to be complemented by medical treatment.[45] If anemia is present anabolic products such as nandrolone (1 mg/kg IM once a week) or stanozolol (2–4 mg/animal/day PO) may be used. Vomiting may require antiemetic treatment using antacids such as cimetidine (15 mg/kg/day PO tid) or ranitidine (4 mg/kg/day PO bid) or antiemetics such as metoclopramide (0.2–0.6 mg/kg/day PO or parenterally). With hyperphosphatemia, aluminum hydroxide may be given (30–90 mg/kg/day PO tid). In animals with poor appetite, benzodiazepines such as diazepam (0.2–0.6 mg/kg/day IM) or oxazepam (0.2–0.4 mg/kg/day PO) may be used as appetite stimulants.

If there are signs of arterial hypertension (retinal hemorrhage or detachment), antihypertensive agents such as propranolol (5–15 mg/animal/day) may be used in combination with a diuretic. Other antihypertensive agents include angiotensin converting inhibitors (ACE), but because these may aggravate kidney failure they should be used with extreme caution and the blood urea and creatinine levels should be checked 4–5 days after the commencement of therapy.

Nephrotic Syndrome

Nephrotic syndrome is defined by a clinical combination of massive proteinuria, hypoproteinemia and edema. This condition is rare in the cat.

Etiology

In most cases the nephrotic syndrome is a consequence of membranous glomerulonephritis. These glomerulonephropathies have an immunological origin that is characterized by the deposition of immune complexes on the glomerular membrane or subepithelial region. These deposits are composed of IgG and complement (C_3), and less frequently IgA or IgM. The prognosis is more serious if the immune complexes consist of several different types of immunoglobulin (Ig).

In many cases the antigen responsible for the formation of this immune complex is unknown, although viral antigens have sometimes been identified (feline leukemia virus, infectious peritonitis virus).[1,28–30]

Amyloidosis is not responsible for the nephrotic syndrome in the cat because amyloid is preferentially deposited in the medullary interstitial tissue in this species.

Clinical Signs

The edema characteristic of the nephrotic syndrome is found in the lower extremities of the limbs, face and thoracoabdominal wall. Fluid may also be located in the body cavities, as ascites or hydrothorax.

The clinical signs of the nephrotic syndrome are often complicated at an early stage by kidney failure.

Laboratory tests demonstrate an abundant proteinuria, hypoalbuminemia and hypercholesterolemia. Increased blood urea, creatinine, protein/urinary creatinine ratio values frequently occur.

The diagnosis of nephrotic syndrome is relatively easy. However, determination of the exact cause requires biopsy puncture of the kidney and the use of histoimmunological techniques using immunofluorescence to demonstrate immune complexes.[65]

Treatment

When the nephrotic syndrome is not complicated by kidney failure, treatment is aimed at controlling the edema using diuretics such as furosemide (2–4 mg/kg/day PO bid) or bumetamide (0.25 mg/kg sid). Corticosteroids such as prednisolone (2–4 mg/kg/day PO bid) may be used to stop the progression of the lesions.

The dietary therapy used depends on whether kidney failure is also present. When kidney failure is not detectable the diet must contain increased protein in order to compensate for urinary protein losses.

If the nephrotic syndrome is complicated by kidney failure, the diet must be protein restricted and only diuretics are used. In these cases the prognosis is poor irrespective of the treatment or diet given.

References

1. Anderson, L. J. and Jarrett, W. F. H. (1971) Membranous glomerulonephritis associated with leukemia in cats. *Research in Veterinary Science*, **12**, 179.
2. Barsanti, J. A., Gitter, M. L. and Crowell, W. A. (1981) Long-term management of chronic renal failure in a cat. *Feline Practice*, **11(1)**, 10–20.
3. Bocock, M. A. and Zlotkin, S. H. (1990) Hepatic sulfur amino acid metabolism in rats with chronic renal failure. *Journal of Nutrition*, **120**, 691–699.

4. Bovée, K. C. (1991) Influence of dietary protein on renal function in dogs. *Journal of Nutrition,* **120**, S128–S139.

5. Boyce, J. T., DiBartola, S. P., Chew, D. J. and Gasper, P. W. (1984) Familial renal amyloidosis in Abyssinian cats. *Veterinary Pathology,* **21**, 33.

6. Brazy, P. C., Stead, W. W. and Fitzwilliam, J. F. (1989) Progression of renal insufficiency: role of blood pressure. *Kidney International,* **35**, 670–674.

7. Brown, S. A., Finco, D. R., Crowell, W. A. and Navar, L. G. (1991) Dietary protein intake and the glomerular adaptations to partial nephrectomy in dogs. *Journal of Nutrition,* **121**, S125–S127.

8. Bush, B. M. (1971) Essential feature of the treatment of chronic renal failure in the dog. *Journal of Small Animal Practice,* **12**, 569–774.

9. Cartee, R. E., Selcer, B. A. and Patton, C. S. (1980) Ultrasonographic diagnosis of renal disease in small animals. *Journal of the American Veterinary Medical Association,* **176**, 426.

10. Carter, L. S., Wingfield, W. E. and Allen, T. A. (1989) Clinical experience with peritoneal dialysis in small animals. *Compendium of Continuing Education for the Practising Veterinarian,* **11**, 1335–1343.

11. Charhon, S. P., Chavassieux, P., Chapuy, M. C., Accominotti, M., Traeger, J. and Meunier, P. J. (1986) Traitement par la ferrioxamine de l'ostéomalacie par intoxication à l'aluminum. *Presse Médicale,* **15(2)**, 55–59.

12. Chew, D. J., DiBartola, S. P., Boyce, J. T. and Gasper, P. W. (1982) Renal amyloidosis in related Abyssinian cats. *Journal of the American Veterinary Medical Association,* **181**, 139.

13. Coen, G., Messa, F., Massimetti, C., Mazzaferro, S., Manganiello, M., Donato, G., Finistauri, D., Giuliano, G. and Cinotti, G. A. (1984) 1–Alpha-OH-cholecalciferol (1–alpha-OHD$_3$) and low phosphate diet in predialysis chronic renal failure: effects on renal function and on secondary hyperparathyroidism. *Acta Vitaminology Enzymology,* **6(2)**, 129–135.

14. Cotard, J. P. (1989) Traitement de l'insuffisance rénale aiguä chez le chien et le chat. *Pratique Médicale et Chirurgicle de l'Animal de Compagnie,* **24**, No. Spécial Néphrologie, 387.

15. Cotard, J. P. (1991) L''insuffisance rénale chronique chez le chien et le chat. *Pratique Médicale et Chirurgicale de l'Animal de Compagnie,* **26**, 507–523.

16. Crisp, M. S., Chew, D. S., DiBartola, S. P. *et al.* (1989) Peritoneal dialysis in dogs and cats: 27 cases (1976–1987). *Journal of the American Veterinary Medical Association,* **95**, 467–474.

17. DiBartola, S. P. (1980) Acute renal failure: pathophysiology and management. *Compendium of Continuing Education for the Practising Veterinarian,* **11**, 952–958.

18. DiBartola, H., Rutgers, H. C., Zack, P. M. and Tarr, M. J. (1987) Clinicopathologic findings associated with chronic renal disease in cats: 74 cases (1973–1984). *Journal of the American Veterinary Medical Association,* **190**, 1196.

19. DiBartola, S. P. and Tarr, M. J. (1986) Tissue distribution of amyloid deposits in Abyssinian cats with familial amyloidosis. *Journal of Comparative Pathology,* **96**, 387.

20. DiBartola, S. P., Spaulding, G. L., Chew, D. J. and Lewis, R. M. (1980) Urinary protein excretion and immunopathologic findings in dogs with glomerular disease. *Journal of American Veterinary Medical Association,* **177**, 6.

21. DiBartola, S. P., Reiter, J. A., Cornacoff, J. B., Kociba, G. J. and Bensons, M. D. (1989) Serum amyloid A protein concentration measured by radial immunodiffusion in Abyssinian and non Abyssinian cats. *American Journal of Veterinary Research,* **50**, 1414.

22. Fettman, M. J. (1989) Feline kaliopenic polymyopathy/nephropathy syndrome. *Veterinary Clinics of North America (Small Animal Practice),* **19**, 415.

23. Finco, D. R. and Barsanti, J. A. (1984) Effects of an anabolic steroid on acute uremia in the dog. *American Journal of Veterinary Research,* **45 (11)**, 2285–2288.

24. Finco, D. R., Brown, S. A., Crowell, W. A., Groves, C. A., Duncan, J. R. and Barsanti, J. A. (1992) Effects of phosphorus/calcium-restricted and phosphorus/calcium-replete 32% protein diets in dogs with chronic renal failure.

25. Finco, D. R., Crowell, W. A., Barsanti, J. A. (1985) Effects of three diets on dogs with induced chronic renal failure. *American Journal of Veterinary Research,* **46**, 646–653.

26. Finco, D. R. (1980) Nutrition during uremic crisis. In *Current Veterinary Therapy VII (Small Animal Practice).* Ed. R. W. Kirk. pp. 1104–1106. Philadelphia: W. B. Saunders.

27. Grauer, G. F. and Thrall, M. A. (1982) Ethylene glycol (antifreeze) poisoning in the dog and the cat. *Journal of the American Animal Hospital Association,* **18**, 492.

28. Hayashi, T., Ishida, T. and Fujiwara, K. (1982) Glomerulonephritis associated with feline infectious peritonitis. *Japanese Journal of Veterinary Science,* **44**, 909.

29. Jacobs–Geel, H. E. L. *et al.* (1982) Isolation and characterization of feline C3 and evidence for immune complex pathogenesis of feline infectious

peritonitis. *Japanese Journal of Veterinary Science,* **44**, 909.

30. Jeraj, K. P., Hardy, R., O'Leary, T. P., Vernier, R. L. and Michael, A. F. (1985) Immune complex glomerulonephritis in a cat with renal lymphosarcoma. *Veterinary Pathology,* **22**, 287.

31. Kopple, J. D. (1981) Nutritional therapy in kidney failure. *Nutrition Reviews,* **39(5)**, 103–205.

32. Kopple, J. D. (1991) Role of diet in the progression of chronic renal failure: experience with human studies and proposed mechanisms by which nutrients may retard progression. *Journal of Nutrition,* **121**, S124.

33. Lau, K. (1989) Phosphates excess and progressive renal failure: the precipitation calcification hypothesis. *Kidney International,* **36**, 918.

34. Leibetseder, J. L. and Neufeld, K. W. (1991) Effects of medium protein diets in dogs with chronic renal failure. *Journal of Nutrition,* **121**, S145–S149.

35. Lewis, L. D. and Morris, M. L. (1984) *Small Animal Clinical Nutrition,* 2nd edn. Topeka, KS: Mark Morris Associates, 544 p.

36. Linder, A. *et al.* (1979) Synergism of dopamine plus furosemide in preventing acute renal failure in the dog. *Kidney International,* **16**, 158.

37. Lulich, J. P., Osborne, C. A., Walter, P. A. and O'Brien, T. D. (1988) Feline idiopathic polycystic kidney disease. *Compendium of Continuing Education for the Practising Veterinarian,* **10**, 1029.

38. Moraillon, R. and Cotard, J. P. (1991) L'insuffisance rénale chez le chat. Monographie, Cambridge: Waltham.

39. Morgan, R. (1986) Systemic Hypertension in four cats; ocular and medical findings. *Journal of the American Animal Hospital Association,* **22**, 615.

40. Nash, A. S., Boyd, J. S., Minto, A. W. and Wright, N. J. (1983) Renal biopsy in the normal cat: the effects of a single needle biopsy. *Research in Veterinary Science,* **34**, 347.

41. Nash, A. S., Boyd, J. S., Minto, A. W. and Wright, N. J. (1986) Renal biopsy in the normal cat: an examination of the effects of repeated needle biopsy. *Research in Veterinary Science,* **40**, 112.

42. NRC (1985) Nutrient requirements of dogs. Subcommittee on dog nutrition. Committee of animal nutrition. Washington DC: National Academy Press.

43. O'Donnel, M. P., Kasiske, B. L. and Keane, W. F. (1987) Nonhemodynamic factors contribute to accelerated glomerular injury in nephrectomized rats fed a high protein diet. *Kidney International,* **31**, 390.

44. Osbaldiston, G. W. and Fuhrman, W. (1970) The clearance of creatinine, insulin para-aminohippurate and phenolsulphophthalein in the cat. *Canadian Journal of Comparative Medicine,* **34**, 138.

45. Osborne, C. A., Polzin, D. J., Abdullahi, S., Klausner, J. S. and Rogers, Q. R. (1982) Role of diet in management of feline chronic polyuric renal failure: current status. *Journal of the American Animal Hospital Association,* **18(1)**, 11–20.

46. Pages, J. P. and Trouillet, J. L. (1984) Néphropathie hypercalcémique: à propos de quatre intoxications aiguës par la vitamine D chez le chien. *Pratique Médicale et Chirurgicale de l'Animal de Compagnie,* **19(4)**, 293–300.

47. Polzin, D. J. and Osborne, C. A. (1986) Update— conservative medical management of chronic renal failure. In *Current Veterinary Therapy IX (Small Animal Practice).* Ed. R. W. Kirk. pp. 1167–1173. Philadelphia: W. B. Saunders. pp. 1167–1173.

48. Polzin, D. J., Leininger, J. L. and Osborne, C. A. (1988) Chronic, progressive renal failure: can progression be modified? In *Renal Diseases in Dogs and Cats. Comparative and Clinical Aspects.* Oxford: Blackwell Scientific Publications.

49. Polzin, D. J., Osborne, C. A. and Lulich, J. P. (1991) Effects of dietary protein/phosphate restriction in normal dogs and dogs with chronic renal failure. *Journal of Small Animal Practice,* **32**, 289–295.

50. Polzin, D. J., Osborne, C. A., Larry, G. and Adams (1991) Effect of modified protein diets in dogs and cats with chronic renal failure: current status. *Journal of Nutrition,* **121**, S140–S144.

51. Ritskes–Hoitinga, J., Lemmens, A. G., Danse, L. H. J. C. and Beunen, A. C. (1989) Phosphorus-induced nephrocalcinosis and kidney function in female rats. *Journal of Nutrition,* **119**, 1423–1431.

52. Russo, E. A., Lees, G. E. and Hightower, D. (1986) Evaluation of renal function in cats using quantitative urinalysis. *American Journal of Veterinary Research,* **47**, 1308.

53. Spangler, L., Gribble, D. H. and Lee, T. C. (1979) Vitamin D intoxication and the pathogenesis of vitamin D nephropathy in the dog. *American Journal of Veterinary Research,* **40(1)**, 73–83.

54. Stebbins, K. E. (1989) Polycystic disease of the kidney and liver in an adult persian cat. *Journal of Comparative Pathology,* **100**, 327.

55. Thornhill, J. A. (1981) Peritoneal dialysis in the dog and cat. An update. *Compendium of Continuing Education for the Practising Veterinarian,* **3**, 20.

56. Thrall, M. A., Grauer, G. F. and Mero, K. N. (1984) Clinicopathologic findings in dogs and cats with ethylene glycol intoxication. *Journal of the American Veterinary Medical Association,* **184**, 37.

57. Tobian, L. (1988) Potassium and hypertension. *Nutrition Reviews,* **46**, 8.

58. Vollman, A. M., Reusch, C., Kraft, W. *et al.* (1991) Atrial natriuretic peptide concentration in dogs with congestive heart failure, chronic renal failure and hyperadrenocorticism. *Americal Journal of Veterinary Research,* **52**, 1831–1834.

59. Walter, P. A., Johnston, G. R., Feeney, D. A. and O'Brien, T. D. (1988) Application of ultrasonography in the diagnosis of parenchymal kidney disease in cats: 24 cases (1961–1986). *Journal of the American Veterinary Medical Association,* **192**, 92.

60. Walter, P. A., Johnston, G. R. and Feeney, D. A. (1988) Renal ultrasonography in healthy cats. *American Journal of Veterinary Research,* **48**, 600.

61. Walter, M. and Mitch, W. E. (1977) Dietary management of renal failure. *Kidney,* **10**, 13–17.

62. White, J. V., Finco, D. R., Crowell, W. A., Brown, S. A. and Hirakawa, D. A. (1991) Effect of dietary protein on functional, morphologic, and histologic changes of kidney during compensatory renal growth in dogs. *American Journal of Veterinary Research,* **52(8)**, 1357–1365.

63. Wolfe, M. J., Smith, C. A. and Lewis, R. M. (1982) Immunofluorescent staining of cutaneous and renal biopsy specimens: a comparison of preservation by quick-freezing with or without storage in transport medium of Michel. *Journal of the American Animal Hospital Association,* **18**, 444.

64. Wolter, R. and Dunoyer, C. (1990) La kaolinite du Bassin des Charentes en alimentation animale. Innocuité et efficacité zootechnique. *Recueil de Médecine Vétérinaire,* **166(5)**, 487–499.

65. Wolter, R., Dunoyer, C., Henry, N. and Seegmuller, S. (1990) Les argiles en alimentation animale: intérêt général. *Rec. Méd. Vét.,* **116(1)**, 21–27.

66. Yeager, A. B. and Anderson, W. I. (1989) Study of association between histologic features and echogenicity of architecturally normal cat kidneys. *American Journal of Veterinary Research,* **50(6)**, 860.

CHAPTER 17

Feline Lower Urinary Tract Disease

PETER J. MARKWELL and C. TONY BUFFINGTON

Introduction

Clinical problems associated with the urinary tract of the cat are not recent phenomena. Evidence of nephrolithiasis was found in 0.22% of cats examined at the Dr.esden pathological institute between 1862 and 1897,[36] and many reports of problems associated with the lower urinary tract have been published over the years. For example, "retention of urine" was described by Kirk as a "very common" condition in cats; the commonest cause of this problem was obstruction of the urethra by a sabulous material.[42] Less frequent causes included cystic or urethral calculi. Blount noted that seven different types of urinary calculi could occur in cats, and that triple phosphates were present in the majority of calculi deposited in alkaline urine.[8] Milks recorded only one urethral calculus from a cat in his own studies, but suggested that there was evidence indicating them to be fairly common in cats.[58] This is in contrast to the observations of Krabbe who noted no examples of "real stone formation" in a series of over 1000 cats seen at the Royal Veterinary and Agricultural College in Copenhagen throughout the 1930s and 1940s.[45]

Sedimentation of the urine was, however, reported in approximately 1% of cases; this was believed to be associated with infection. More recently Stansbury and Truesdail suggested that the incidence of calculi formation in cats ranged from 0.6 to 10%, the average being 2.3%.[78] They were, however, probably estimating proportional morbidity, as information from small animal hospitals was included.

Although these reports demonstrate that uroliths and urethral plugs have afflicted cats for many years, it is difficult to compare these observations with more recent data. The observation by Krabbe of an approximate 1% incidence, however, is strikingly similar to estimates of 0.64 and 0.85% reported recently.[47,80] More recent data on the incidence of lower urinary tract disease (LUTD) in cats are lacking.

When reviewing epidemiological data, care must be taken to differentiate between estimates of incidence in the population and proportional morbidity rates, which may be much higher. The proportional morbidity is the ratio of cases of LUTD to all cases seen in a clinic or hospital, and thus its denominator is biased toward sick animals, those owned by people willing and able to bring their pets to veterinarians and the interests of the veterinarians in the study. Estimates of proportional morbidity due to LUTD range from 1 to 6%.[81]

The term feline urological syndrome (FUS) was coined by Osbaldiston and Taussig in 1970 to describe, "The feline disease syndrome characterized by dysuria, urethral obstruction, urolithiasis and hematuria . . .".[61] Study of 46 cats with FUS led them to conclude that, "The need for further investigation of FUS is emphasized by observations in this study which indicate that the condition may not be a single disease entity, but rather a group of separate urologic problems." Subsequent epidemiological studies identified many risk factors associated with FUS.[81] Proposed dietary influences, results of many diet-related studies and the fact that struvite (the stone most commonly associated with FUS) is composed of magnesium, ammonium and phosphorus, led to the conclusion that most cases of FUS were diet-induced and investigation of other potential causes was unnecessary.

In 1984 Osborne *et al.* proposed that FUS be used as a synonym for lower urinary tract disorders in cats (Osbaldiston and Taussig's original meaning), reiterating that FUS could be induced by single or multiple causes, and that specific treatment would depend on the underlying cause(s), some of which are unknown.[63] Barsanti and Finco proposed a definition of exclusion, which effectively restricted FUS to idiopathic LUTD.[7] However, although this definition appears to apply to a majority of cases,[68] it may become a "moving definition" in that idiopathic cases will presumably decrease in frequency as research continues and more causes of the signs of LUTD are elucidated.

We propose that the term FUS be abandoned altogether because of the confusion that surrounds it. We recommend that signs of lower urinary tract disease *in the absence of a specific diagnosis* be called simply idiopathic LUTD. Where a specific cause of disease is identified, the appropriate descriptive term should be used. This definition helps to emphasize the need for approaching cases on an individual basis, and acknowledges that we still know very little about the etiology of the signs in most cases.

The most detailed study of specific causes of feline LUTD was reported by Osborne *et al.* and Kruger *et al.*[46,68] In this prospective study of 143 cases of hematuria and dysuria, urethral plugs were present in 32 cases, urolithiasis without urinary tract infection (UTI) in 30 cases, UTI alone in two cases and uroliths with UTI in two cases. Seventy-seven cases were classified as idiopathic, based on absence of evidence of infection (bacteria, ureaplasma, mycoplasma or virus), urolithiasis, neoplasia or congenital abnormality. These data were collected from cases between 1982 and 1985, and thus may not be representative of the current profile of causes of LUTD. The clinical impression of the authors is that true uroliths more recently have become less common than urethral plugs.

Quantitative analysis of feline uroliths by polarized light microscopy and X-ray diffraction methods has revealed a variety of chemical types (Fig. 17.1).[65] As indicated in this figure, 64.5% of the uroliths in this cumulative series were composed of 70–100% struvite. Calcium oxalate was found in 20.1% of uroliths. It is interesting to compare these findings with data from the same center published 3 years previously, which reported 70.3% struvite and only 10.3% oxalate from a total of 1180 uroliths.[68] These findings suggest that although struvite is still the most important type of feline urolith, its importance is decreasing and that of oxalate increasing. At the Ohio State University Veterinary Hospital, only 4 of 15 calculi surgically removed from cats over the period from 1989 to 1991 were struvite (10 were oxalate, one was cystine), again indicating the growing importance of nonstruvite calculi in cats.

Most urethral plugs also contain a mineral component, although a proportion may just be matrix. Of the minerals found, struvite is again the most common (Fig. 17.2). However, whether the relative importance of struvite in urethral plugs is still as great as suggested by these data is unknown, as again they were

MINERAL COMPOSITION OF FELINE UROLITHS*		
Predominant mineral type	**Number of uroliths**	**%**
Magnesium ammonium phosphate $6H_2O$	1296	64.5
100%	(955)	(47.5)
70–99%†	(341)	(17.0)
Magnesium hydrogen phosphate $3H_2O$	5	0.2
70–99%†	(5)	(0.2)
Calcium oxalate	405	20.1
Calcium oxalate monohydrate		
100%	(135)	(6.7)
70–99%†	(144)	(7.2)
Calcium oxalate dihydrate		
100%	(30)	(1.5)
70–99%†	(42)	(2.1)
Calcium oxalate monohydrate and dihydrate		
100%	(37)	(1.8)
70–99%†	(17)	(0.8)
Calcium phosphate	43	2.1
Calcium phosphate		
100%	(18)	(0.9)
70–99%†	(12)	(0.6)
Calcium hydrogen phosphate $6H_2O$		
100%	(5)	(0.2)
70–99%†	(6)	(0.3)
Tricalcium phosphate		
100%	(1)	(<0.1)
70–99%†	(1)	(<0.1)
Uric acid and urates	123	6.1
Ammonium acid urate		
100%	(85)	(4.2)
70–99%†	(31)	(1.5)
Sodium urate		
70–99%†	(1)	(<0.1)
Uric acid		
100%	(3)	(0.1)
70–99%†	(3)	(0.1)
Cystine	2	0.1
Silica	0	0
Mixed‡	68	3.4
Compound§	22	1.1
Matrix	46	2.3
Total	2010	100%

* Uroliths analyzed by polarized light microscopy and X-ray diffraction methods
† Uroliths composed of 77–99% of mineral type listed; no nucleus and shell detected.
‡ Uroliths did not contain at least 70% of mineral type listed; no nucleus or shell detected.
§ Uroliths contained an identifiable nucleus and one or more surrounding layers of a different mineral type.
From Osborne *et al* (1992) *Journal of Small Animal Practice* **33**, 172–177. Reproduced with permission.

FIG. 17.1 Mineral composition of feline uroliths evaluated by quantitative methods.

largely collected during the mid 1980s (C. A. Osborne, personal communication).

It is therefore important to approach cases of LUTD on an individual basis, as would occur with other species. This will enable, at least in some cases, identification of specific causes of disease and in some of these the institution of specific management programs aimed at cure and/or prevention of further recurrence of disease.

Pathophysiology

Urolithiasis may be defined as the presence of macroscopic mineralization within the

Peter J. Markwell and C. Tony Buffington

MINERAL COMPOSITION OF FELINE URETHRAL PLUGS*		
Predominant mineral type	**Number of uroliths**	**%**
Magnesium ammonium phosphate 6H₂O	496	81.4
100%	(400)	(65.7)
70–99%†	(96)	(15.8)
Calcium oxalate	14	2.3
Calcium oxalate monohydrate		
100%	(4)	(0.7)
70–99%†	(5)	(0.8)
Calcium oxalate dihydrate		
100%	(2)	(0.3)
70–99%†	(2)	(0.3)
Calcium oxalate monohydrate and dihydrate	(1)	(0.2)
Calcium phosphate	16	2.6
Calcium phosphate		
100%	(6)	(1.0)
70–99%†	(7)	(1.1)
Calcium hydrogen phosphate 6H₂O		
100%	(2)	(0.3)
70–99%†	(1)	(0.2)
Ammonium acid urate	6	1.0
100%	(6)	(1.0)
Xanthine	1	(0.2)
Sulfadiazine	1	0.2
Mixed‡	17	2.8
Matrix	58	7.4
Total	609	100%

* Urethral plugs examined by polarizing light microscopy and X-ray diffraction methods.
† Uroliths composed of 70–99% of mineral type listed; no nucleus and shell detected.
‡ Uroliths did not contain at least 70% of mineral type listed; no nucleus or shell detected.
From Osborne *et al* (1992) *Journal of Small Animal Practice* **33**, 172–177. Reproduced with permission.

FIG. 17.2 Mineral composition of feline urethral plugs analyzed by quantitative methods.

urinary system, from sand to radiographically obvious urocystoliths. UTI is not considered to play an important part in most cases of struvite urolithiasis in cats, in contrast to the situation in dogs. It has been estimated that 75–80% of struvite uroliths observed in cats are sterile, whereas 66–77% of naturally occurring canine struvite calculi are associated with infection.[10,68]

Diet contributes to the etiology and prevention of struvite formation because dietary ingredients and feeding patterns influence the pH, solute concentration and volume of urine. Nutritional research has played a key role in the development of programs for the management of struvite-associated LUTD in cats, although most researchers thought they were studying FUS at the time that their studies were published. The aspects of diet that have been particularly studied with regard to struvite formation include mineral content and the effect(s) on urinary pH and water turnover.

The effects of the mineral content of the diet originally were ascribed to the ash content. Ash is the total noncombustible material of the diet, determined by burning an aliquot of the diet for 2 hr at 600°C and weighing the residue. In 1956, Dickinson and Scott reported their failure to produce urinary calculi in kittens by the addition of bone meal to a basal diet.[24] Ash concentrations in this study were 8, 22 and 30%. Thus, although components of ash may be significant with regard to struvite formation, the ash content of the diet *per se* is of no relevance. Specific dietary minerals that have been studied include magnesium, calcium, phosphorus and sodium. Several studies have shown that urolithiasis and UTI can be induced by feeding diets containing high levels of magnesium (Fig. 17.3). A number

MAGNESIUM IN FELINE LOWER URINARY TRACT DISEASE		
Author	Dietary magnesium concentration (%)	Cats obstructing/ cats observed
Rich *et al.*[71]	0.1	0/4
	0.25	0/8
	0.5	0/8
	0.75	2/8
	1.0	5/6
Hamar *et al.*[33]	0.75–1.0	6/8
Lewis *et al.*[50]	0.75–1.0	17/24
Kallfelz *et al.*[40]	1.0	7/16

FIG. 17.3 Studies of the role of magnesium in feline lower urinary tract disease.

of questions have to be asked before these types of studies can be clearly related to the "naturally" occurring disease. First, in some studies (e.g. Rich *et al.*[71]), the crystals found were reported to be magnesium phosphate, not magnesium ammonium phosphate. Second, the magnesium content used in these studies was far in excess of that contained within commercial foods, or that likely to be encountered by the cat in its natural diet (i.e. 1 vs 0.1%). Third, the choice of magnesium salt used to increase the dietary magnesium content may have influenced the results through its concurrent effect on urine pH. Some of these issues have been addressed in other studies.

One exception to the large quantities of magnesium generally fed was reported in a study by Finco *et al.*[30] In this study one cat of a group of four obstructed with struvite crystals, when fed a diet containing 0.17% dry matter (DM) magnesium, although at the time of obstruction this cat had an alkaline urine. These investigators found that, within diet groups, urine magnesium concentration was not higher in cats with urethral obstruction than in those that did not obstruct. They concluded that factor(s) other than urine magnesium concentration were important in urethral obstruction.

One of these factors is the pH of the urine. Taton *et al.* showed that radiographically visible bladder calculi actually dissolved in cats consuming a diet containing 0.37% magnesium when ammonium chloride was

added, which resulted in a urine pH of less than 6.0.[79] In another study, a cat receiving a semipurified diet containing only 0.05% magnesium obstructed with struvite, but the average urine pH of its group was 8.2.[18] These results both emphasize that the urine pH is much more important than the magnesium content of the diet with respect to struvite formation.

The effects of the calcium and phosphorus content of the diet on urolithiasis have also been studied.[22,50] These studies found that dietary phosphorus had no influence on calculi production in diets containing large quantities of magnesium (>0.75%) when the calcium content was low (0.85%). High phosphorus (1.6%) in the presence of high calcium (2.0%) appeared to exacerbate the adverse effects of magnesium. In these studies, however, increasing dietary calcium and magnesium also increased the urine pH. Although increasing the diet phosphate content tended

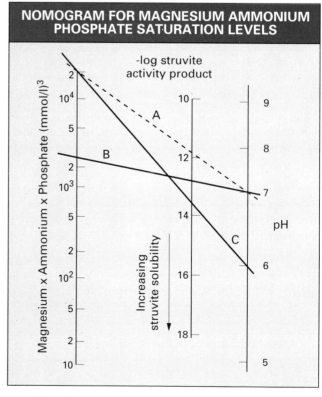

FIG. 17.4 Nomogram for magnesium ammonium phosphate saturation levels. From Marshall and Robertson, 1976. *Clinica Chimica Acta,* **72**, 255–260. With permission of Elsevier.

to decrease the urine pH, it was still alkaline in the presence of the high calcium and/or magnesium. Thus it is possible that the calculogenic effect of calcium and magnesium addition resulted from effects on urine pH.

The theoretical basis for the relative importance of urine pH versus the urine concentration of the components of struvite, magnesium, ammonium or phosphate, is demonstrated by the nomogram shown in Fig. 17.4.[57] The activity product (solute activity is the concentration that is free to react with other solutes in a solution), which is the ultimate determinant of crystal formation, is expressed in the nomogram as its negative logarithm, just as pH is the negative logarithm of hydrogen ion concentration. Changes in pSAP are directly proportional to struvite solubility; as pSAP increases, struvite solubility increases (and the risk of struvite formation declines). The nomogram shows that if pH is kept constant and magnesium, ammonium or phosphate concentrations are reduced by a factor of 10, then the activity product decreases by approximately one log unit (i.e. approximately a 10-fold change; lines A and B). However, if pH is changed by one unit (i.e. a 10-fold change in hydrogen ion concentration) the activity product changes by approximately 1.8 units (approximately a 60-fold change; line C). Thus, while reduction of magnesium, ammonium or phosphate concentrations in urine will reduce the activity product, changes in urinary pH have a sixfold greater effect. Increasing urine volume, which decreases the activity of all crystalloids simultaneously, has an intermediate effect. The key effect of urine pH is that it determines the proportion of total urine phosphate present as the divalent anion, the form of phosphate necessary for struvite formation. Increasing pH increases the concentration of the divalent anion as the monovalent anion is deprotonated (Fig. 17.5). During crystallization of struvite a further proton is shed resulting in a lowering of the pH of the solution in which the reaction is taking place.[9]

The rationale for the management of struvite-associated LUTD is thus based on reducing the struvite activity product in

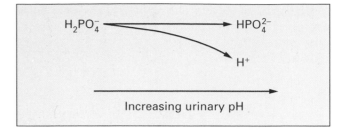

FIG. 17.5 The effect of increasing pH on phosphate ions.

urine to below the solubility product (K_{sp}), creating urine that is undersaturated with regard to the components of struvite; under these circumstances crystals will not form and preformed struvite will dissolve (Fig. 17.6).[15] As discussed above, undersaturation may be promoted through modification of the diet to increase urine volume and to reduce mineral excretion and pH of the urine. Of these factors, the cornerstone is assurance that consumption of the diet does not result in persistently alkaline urine.

Epidemiological studies of LUTD conducted in the 1970s implicated dry cat foods as a risk factor.[70,80,81] Explanations offered for this observation were that cats might not repair a water deficit as well as dogs, and that cats fed dry foods take in less water. It has been suggested that the cats' descent from desert living ancestors was responsible for the relative insensitivity of its homeostatic responses to heat and dehydration.[3,4] These interpretations require further discussion, however.

Adolph found, in studies of five cats and three dogs, that both incurred similar water deficits in 48°C environments when water was available.[2] After heat exposure without available water, dogs replaced moderate, but not severe, deficits more rapidly than did cats. Both species, however, drank proportionately more than did humans when dehydration exceeded 5%. These results suggest that differences in response to dehydration between dogs and cats are relatively small, and probably not clinically relevant as a risk factor for urolithiasis. Moreover, available data suggest that urolithiasis occurs

ZONES OF URINE SATURATION	
HIGH	Zone of oversaturation – spontaneous crystallization – rapid growth of crystals
Formation of product ⟶ Product of $[Mg^{2+}] \times [NH_4^+] \times [PO_4^{3-}]$ Solubility product ⟶	Metastable zone of supersaturation – no dissolution – possible growth of preformed crystals
LOW	Zone of undersaturation – no crystallization – preformed crystals dissolve

FIG. 17.6 Zones of urine saturation with magnesium ammonium phosphate. (From Buffington.[15])

more commonly in dogs than it does in cats.[16]

It is also doubtful that cats reduce water intake in some unusual way when fed dry foods. Burger *et al.* reported that dogs fed a variety of diets of differing moisture content drank similarly, whereas cats fed diets containing differing amounts of moisture drank quite different amounts of water.[19] Unfortunately, the results were not corrected for the potential renal solute load (PRSL) of the diets fed. The PRSL of a diet is the amount of solute, i.e. minerals and nitrogen, that must be excreted in the urine.[43,82] The PRSL has been estimated as the urea (mg N ÷ 28) plus twice the sum of the sodium and potassium content of the diet (mg N/28 + 2(Na + K)).[60] Although Burger *et al.* did not report the potassium content of their diets, a crude estimate of the PRSL can be obtained by summing the DM protein and ash content of a diet. When water intakes reported by Burger *et al.* were regressed against this estimate of diet PRSL, 84% of the decrease in water intake was explained by the lower solute load of the dry diets. In other studies, more than 90% of diet-induced changes in water intake[39,41] or urine formation[73] could be explained by solute load (protein + ash content) of the diet. Gaskell reported similar data from cats fed a single diet to which increasing amounts of water were added.[32] When the water content of the food was 10 or 45%, total water intake and urine volume and specific gravity, were not different between groups. When the water content of the food was increased to 75%, however, total water intake and urine volume increased, and urine specific gravity decreased. Because the diet was the same in all cases, water intake (from food) probably increased as a consequence of increased food intake to meet energy needs from the water-diluted diet. Similar responses have been reported by others.[20] Gaskell also found that food (and PRSL) intake decreased when water availability was limited.[31] The urine specific gravity of the water restricted cats was the same (1.045) as that found in cats fed the same dry diet with unrestricted access to water. If cats really are more tolerant of dehydration than other species, one might have expected them to continue to eat to meet their energy needs and let urine specific gravity rise, or to become dehydrated. The observed results argue against these explanations. Thus the differences in urine volumes and specific gravities observed in some of the studies discussed may be more of a reflection of differences in PRSL and/or energy content of dry and wet diets, rather than moisture content *per se*. Therefore, if a cat is changed from a dry to a wet diet as part of a management program for LUTD, it is important to ensure that the specific gravity of the urine is checked to determine that it actually decreases.

Finally, although changes in urine specific gravity may have contributed to the increased risk associated with dry foods noted in the epidemiological studies, other factors may also have been involved. Data presented in the early 1980s suggested that at least some of the dry diets available at that time could produce substantial alkaline tides when meal fed.[51] Thus the effect of dry diets on urine pH may have been one of these additional factors.

The pathophysiology of other types of urolith is, in general, much less well understood than that of struvite. Oxalate urolithiasis has been considered to be the sequela of a group of disorders disturbing the balance between urine concentrations of calculogenic components (calcium and oxalic acid) and crystallization inhibitors (such as citrate, phosphorus and magnesium). Thus hypercalciuria and hyperoxaluria are both etiological risk factors for the condition.[54] Dietary acidification can have an impact on urinary calcium excretion, and hence the risk of oxalate urolithiasis.

Fettman et al. recently showed that the type of acidifying ingredient was not an important determinant of the effect of acidification.[28] They found that cat foods acidified by two methods had similar adverse effects on the acid–base, mineral and bone homeostases. One of the diets was "naturally" acidified, whereas the other was acidified by the addition of phosphoric acid. Both diets induced uncompensated chronic metabolic acidosis, significantly decreased whole-body balance of calcium, phosphorus, potassium and magnesium and decreased bone formation.

Fettman et al. also found significant increases in both urine concentration and fractional excretion of calcium, presumably caused by diet-induced acidosis.[28] The calciuria produced by ingestion of acidic diets, or the low urine magnesium concentrations resulting from consumption of magnesium-restricted diets, may increase the risk of calcium oxalate urolithiasis.[74] To evaluate the effect of diet on calcium oxalate solubility, a computer program[1] is available to estimate (from published values[17]) the relative supersaturation (RSS) of calcium oxalate (and struvite) in urine of cats fed a variety of diets. These calculations are presented in Fig. 17.7. It is noteworthy that the diet (MgO supplemented) that was most saturated with respect to struvite was least saturated with respect to calcium oxalate, whereas the two commercial diets, while only slightly saturated with respect to struvite, were the most highly supersaturated with calcium oxalate. The observation that the ($MgCl_2$-supplemented) diet producing the lowest urine pH and the greatest calciuria did not cause the greatest saturation with calcium oxalate, suggests the importance of urine magnesium in inhibiting calcium oxalate formation.

Hyperoxaluria has been detected less frequently in dogs with calcium oxalate uroliths than hypercalciuria.[55] Its importance in cats is unknown, although it has been observed in kittens fed diets deficient in vitamin B_6.[5]

Little is known of the causes of urate urolithiasis in cats, although it has been recommended that animals developing ammonium acid urate calculi should be evaluated for

EFFECTS OF DIET ON URINE pH AND MINERAL CONCENTRATIONS, AND STRUVITE AND CALCIUM OXALATE RELATIVE SUPERSATURATION

	Basal	MgCl$_2$	MgO	Dry diet	Canned diet
pH	7.2±0.3	5.8±0.1	7.9±0.3	6.0±0.2	6.7±0.4
Magnesium (mM)	7.3±2.8	53.1±16.3	49.1±14.1	6.9±1.2	2.7±0.6
Calcium (mM)	4.7±1.5	15.5±8.2	8.1±3.6	6.0±2.8	4.3±1.4
Relative supersaturation					
Struvite	24.7	0.7	87.1	1.9	4.8
Calcium oxalate	41.3	12.8	8.6	49.1	47.8

FIG. 17.7 Effects of diet on urine pH and mineral concentrations, and struvite and calcium oxalate relative supersaturation. The higher the relative supersaturation, the greater the probability of stone formation. (Data from Buffington et al.)[17]

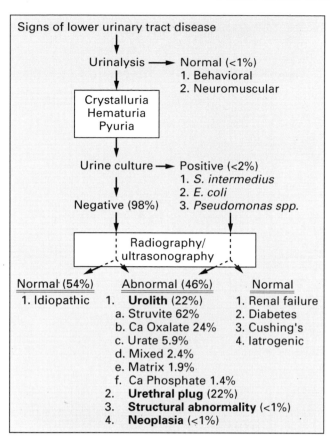

Signs of lower urinary tract disease

↓

Urinalysis ⟶ Normal (<1%)
          1. Behavioral
          2. Neuromuscular

↓

Crystalluria
Hematuria
Pyuria

↓

Urine culture ⟶ Positive (<2%)
          1. *S. intermedius*
          2. *E. coli*
Negative (98%)   3. *Pseudomonas spp.*

↓

Radiography/
ultrasonography

Normal (54%)   Abnormal (46%)   Normal
1. Idiopathic   1. **Urolith** (22%)   1. Renal failure
             a. Struvite 62%   2. Diabetes
             b. Ca Oxalate 24%   3. Cushing's
             c. Urate 5.9%   4. Iatrogenic
             d. Mixed 2.4%
             e. Matrix 1.9%
             f. Ca Phosphate 1.4%
          2. **Urethral plug** (22%)
          3. **Structural abnormality** (<1%)
          4. **Neoplasia** (<1%)

FIG. 17.8 Diagnostic algorithm for feline lower urinary tract disease.

portosystemic shunts and liver disease.[75] In one case a renal tubular reabsorption defect was proposed as a cause.[38]

Diagnosis

A flow chart for the diagnosis of LUTD in cats is given in Fig. 17.8. Percentages refer to data from 43 female and 98 male cats presented by Kruger *et al.*[46] The percentage of cases with crystalluria, hematuria and pyuria varied with the eventual diagnosis, and 46% of 26 normal controls had crystalluria and cystocentesis induced hematuria. No cause for signs of LUTD were found in the majority of cases. Of diagnosable causes, struvite urolithiasis and urethral plugs were the most common.

To assist further with differential diagnosis of crystal types, photomicrographs of the common types of crystals likely to be encountered

in feline urine samples are shown in Figs. 17.9–17.12.

Therapeutics

The therapeutic approach to any case depends on identification of the underlying cause of disease if possible, and subsequent implementation of an appropriate management strategy. Treatment recommendations presented are for obstructed cats, non-obstructed cats with urolithiasis and cats with idiopathic disease.

The initial management of obstructed cats is similar regardless of the cause of obstruction. However, follow-up management will depend on the initiating cause of the problem and may involve long-term diet therapy, particularly if the cause appears to be associated with the formation of struvite in the lower urinary tract.

Obstructed Cats

The first consideration for cats with urethral obstruction of less than 24–48 hr duration is relief of the obstruction. This will normally require anesthesia because cats with this duration of urethral obstruction usually are not systemically ill. The distal penis should be examined first, and an attempt made to dislodge manually any obstructing material present. If the obstruction is not located at the tip of the penis, the next step is aseptic passage of a well lubricated urinary catheter. Open-end, closed-ended and special purpose catheters have been advocated.[67] The penis should be directed dorsally and caudally so that its long axis parallels the vertebral column to eliminate the natural curvature of the urethra.[7] The catheter is advanced to the site of obstruction, and fluids syringed in to try to dislodge the obstructing material while attempting to advance the catheter into the bladder. A sterile, physiological solution such as 0.9% saline or lactated Ringer's at room temperature is recommended for the purpose of flushing. An acetate buffer (pH 4.5) has

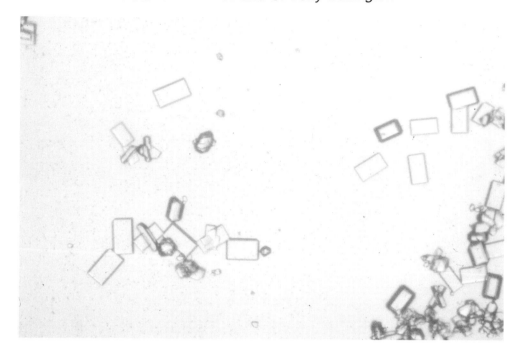

FIG. 17.9 Struvite crystals.

been described as an irrigating solution,[37] but is not recommended because of concerns about its effect on the systemic acid–base balance if absorbed through the damaged bladder mucosa of a cat likely to have metabolic acidosis. Careful attention to aseptic technique and a gentle approach at this stage may prevent later complications (e.g. bacterial UTI, recurrent obstruction due to extraluminal swelling).

If a catheter cannot be passed, the distal urethra is occluded and flushing is continued to increase intraluminal hydrostatic pressure and attempt to force the obstructing material

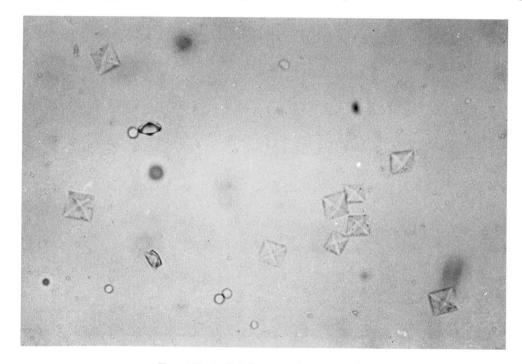

FIG. 17.10 Calcium oxalate crystals.

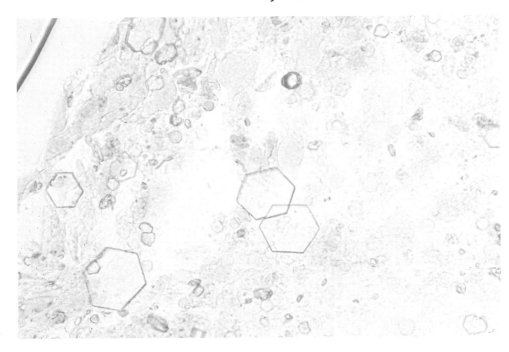

FIG. 17.11 Cystine crystals.

back into the bladder as the catheter is advanced. If hydropulsion is successful, the urethra and bladder are irrigated with sterile 0.9% saline or lactated Ringer's solution until the returning fluid is free of blood and crystalline debris. If hydropulsion fails, cystocentesis is employed to decompress the bladder and

retrograde flushing is attempted again. If this is unsuccessful, emergency cystotomy with placement of a Foley catheter or perineal urethrostomy must be considered.

If obstruction is relieved by retrograde flushing, the clinician must decide if an indwelling urinary catheter should be placed.[48]

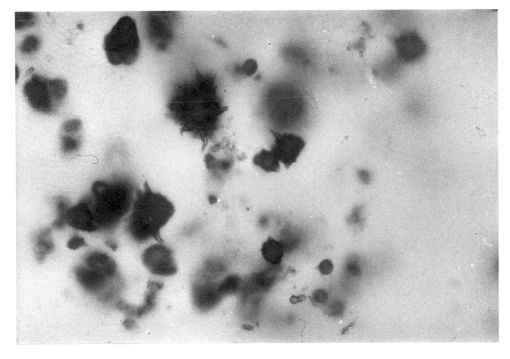

FIG. 17.12 Urate crystals.

Placing a catheter will prevent recurrence of obstruction, keep the bladder decompressed and facilitate monitoring of urine output. On the other hand, indwelling catheterization contributes to continued urethral trauma and ascending bacterial UTI. If a catheter is placed, it should be maintained for the minimum length of time. Ascending bacterial UTI frequently occurs and the additional trauma to the urethra and inflammation it causes might contribute to recurrence of obstruction after removal of the catheter. Treatment with ampicillin during indwelling catheterization reduced the frequency of bacterial UTI, but resulted in the occurrence of highly resistant bacteria in four out of 12 cats that developed UTI despite its use.[49] An Elizabethan collar usually is required to prevent the cat from removing the catheter. After the catheter is removed, urine culture should be performed and appropriate antibiotic treatment instituted for bacterial UTI, if present.

Subcutaneous administration of a balanced electrolyte solution such as lactated Ringer's solution at approximately 90–120 ml/kg/day is used to promote diuresis and prevent early recurrence of obstruction. The bladder should be palpated periodically to detect recurrence of obstruction or failure of adequate emptying (i.e. bladder hypotony). If the bladder returns to normal size soon after relief of the obstruction, bladder hypotony is unlikely. If not, the clinician must ensure that the bladder is kept empty by repeated manual compression because some degree of bladder hypotony is likely to be present. Cats resent repeated manual compression of the bladder and can generate marked abdominal muscle tone to resist the clinician's attempts to empty the bladder.

If urinary tract obstruction has been present for more than 24–48 hr, case management will be complicated by the effects of postrenal renal failure. Therapy is aimed at correcting these effects as well as relieving urinary tract obstruction. Thus the additional therapeutic goals are to correct hyperkalemia, acid–base imbalance, dehydration and azotemia with appropriate fluid therapy. A

number of reports have described the emergency management of cats with urethral obstruction.[6,7,34,44,66]

Following relief of the obstruction (and stabilization of the patient with prolonged obstructions), consideration must be given to the correction of the primary cause of disease. If the obstruction resulted from urolithiasis, attempts should be made to identify the type of urolith involved (see differential diagnosis). If composed of struvite, dietary modification should be considered to try to achieve dissolution (see below).

Long-term management of cats subsequent to relief of obstruction caused by urethral plugs has not been well studied, although it seems reasonable to assume that measures effective in dissolving or preventing struvite urolith formation will also be effective in preventing recurrence of struvite formation in urethral plugs, if this is the mineral component. Whether control of struvite formation will bring about prevention of recurrence of disease is not clear, although a recent study provides some evidence in support of this.[29] In this clinical study of cats with a history of obstructive (11 cases) or nonobstructive (10 cases) LUTD, obstructing material was characterized as sabulous plugs in three cats, a matrix plug in one cat and "microcalculi" in another. It was not characterized in the remainder of the obstructed group. Most of the cats had shown more than one bout of urinary obstruction in the previous 18 months. Cats were placed on a veterinary diet designed for the management of struvite associated disease and monitored for 48 weeks, during which no signs of recurrence of obstruction were noted. Considering the previous history of the cats, and the high probability of recurrence as noted in older epidemiological studies,[81] these data provide encouragement that diet may be of benefit in the long-term prevention of recurrence of obstruction.

Nonobstructed Cats with Urolithiasis

The key to management of this group of cats is to identify the type of urolith involved and to adopt an appropriate management program. For struvite uroliths this will involve dietary modification, but for other types of urolith surgical intervention is currently the most reliable method of management. The following comments thus apply specifically to the management of struvite uroliths and assume that bacterial UTI is not present. The management of bacterial UTI has been described in detail elsewhere.[64] Based on the experimental data presented on the formation of struvite, the following factors should be considered in dietary management: acidification of the cat's urine, enhancement of urine volume and reduction of specific gravity and restriction of crystalloid intake.

The ideal urine pH for the dissolution of struvite has not been established in experimental studies. Struvite crystals are, however, rarely found at pH values of less than 6.5,[72] and thus this value may be assumed to equate approximately to the struvite solubility product.[16] It follows, therefore, that urine pH values below this are likely to be associated with undersaturation with struvite and in time should bring about dissolution of preformed crystalline material. It has been shown in clinical studies that a urine pH of ~6, maintained by a canned diet with restricted magnesium, can successfully bring about dissolution of struvite uroliths in periods ranging from 14 to 141 days.[68]

An acidic urine pH may be maintained by diet and/or urine acidifiers. It is probably easier to feed a prepared diet that promotes an acid urine pH, but methionine and ammonium chloride also acidify urine and can be mixed into a patient's food. The dosage of these acidifiers must be adjusted carefully for each patient and will depend on both the diet fed and the metabolism of the individual patient. A starting dose of 100–200 mg/kg/day of either acidifier should be incorporated into food, and increased cautiously to attain the desired urine pH. It should be remembered that any change to the composition of the patient's diet may affect the dose of acidifier required. Acidifying agents should not be added to diets already supplemented with these compounds, nor to diets that normally result in a urine pH of 6.4

or less within 4 hr of feeding. Acidifiers should be used with particular caution in animals less than 1 year-of-age and in animals with documented hepatic or renal disease. Regardless of the method of acidification, the patient's urine pH should be 6.0–6.5 on a urinalysis dipstick 4–6 hr after ingestion of a meal. After treatment for several weeks, a venous blood gas determination or biochemical profile that includes total CO_2 content should be obtained to be certain that the serum bicarbonate or total CO_2 exceeds 17 mmol/l.

Acidification of the urine is not without potential toxicity, however.[23] Methionine has been recommended as a urinary acidifier at a dosage of 1500 mg/cat/day.[52] For a 3-kg cat eating 60 g DM/day, this would constitute 2.5% of DM. When fed experimentally to cats at a dosage of 2.8–5.5% of diet DM (560–1100 mg/kg/day), methionine caused hemolytic anemia, methemoglobinemia and Heinz body formation. This amount of methionine also decreased food intake and weight gain in kittens.[27] Also, kittens consuming very large amounts of a combination product containing ammonium chloride and D,L-methionine developed weight loss and neurological abnormalities and subsequently died.[12]

Ammonium chloride also can be toxic. Some cats given 1000 mg/day of ammonium chloride developed anorexia, vomiting and diarrhea.[53] Diets containing 1.5% ammonium chloride provide a cat with approximately 800 mg ammonium chloride/day. A study of this dosage given as powder or tablet to meal-fed cats showed that 4 hr postprandial urine pH was less than 6.0.[77] In this study, venous blood pH and bicarbonate concentrations were decreased to values at the lower end of the range reported for normal cats. A combination product supplying 580 mg each of ammonium chloride and methionine (equivalent to 0.8–1.5% of each compound in the diet) given to meal-fed cats resulted in urine pH of 6, 4 hr after eating.[76] This product reduced venous blood pH and bicarbonate concentrations more than ingestion of 800 mg ammonium chloride alone, but the values still fell within the reported normal range. The investigators did not report observing adverse gastrointestinal signs in either of these studies.

Chronic acidification can have detrimental effects on renal function and bone development and can cause potassium depletion[24] that can contribute to renal dysfunction. Feeding a diet marginally replete in potassium that also contains an excessive amount of acidifying potential may cause chronic metabolic acidosis and depletion of body potassium stores. Potassium depletion and hypokalemia in turn may lead to chronic tubulointerstitial nephritis and increased urinary fractional excretion of potassium, further aggravating potassium depletion. Dow *et al.* reported a clinical syndrome of hypokalemic nephropathy in cats fed an acidifying diet low in potassium.[25] Several of the cats had metabolic acidosis.

When acid intake exceeds output, acid–base homeostasis reestablishes itself at a decreased blood bicarbonate concentration by recruiting bone buffers (carbonate and phosphate) to buffer the chronic acid load. Chronic acid feeding has been reported to affect bone metabolism in young cats. Diets containing 3% ammonium chloride slowed growth, decreased blood pH and bicarbonate concentrations, lowered urine pH, increased urine calcium excretion and promoted bone demineralization of caudal vertebrae of young cats.[14,15] Similar changes in blood and urine were found in adult cats fed diets containing 1.5% ammonium chloride.[21] Similar effects would be expected with chronic excessive intake of other acidifying agents.

To enhance urine volume, water intake should be maximized and fecal water loss minimized. For this latter reason diets with a low digestibility should be avoided because fecal water loss may be much greater than with highly digestible foods.[56] Water intake may be increased by adding it to canned or dry diets, or by feeding relatively energy dilute canned diets. Specific targets for urine specific gravity have not been defined; however, struvite stones in cat bladders have dissolved despite urine specific gravities greater than 1.050. Notwithstanding this, it seems reasonable to try to achieve a more dilute urine and specific gravities less than 1.035 are a reasonable target. Augmenting water turnover will not create problems for most animals or owners, and may help to hasten stone dissolution. Increasing water intake also increases urination frequency, providing less time for crystal growth. Holme noted that hematuria in a group of cats receiving a high magnesium, low moisture content diet, was almost completely abolished when water intake was increased by feeding the same diet ground up as a slurry containing 80% moisture.[35]

Restriction of dietary crystalloid intake is also a desirable goal, although it is probably not essential if urine pH can be maintained below 6.6. A reasonable target is restriction of dietary magnesium to approximately 20–40 mg and phosphorus to 125–250 mg/100 kcal. Struvite urolithiasis has never been reported in cats fed diets containing less than 40 mg magnesium/100 kcal when urine pH was less than 7. When the urine pH was 8.2, however, urolithiasis occurred in kittens fed a purified diet containing less than 20 mg magnesium/100 kcal.[18] Restriction of phosphorus to 125 mg/100 kcal is consistent with the minimum recommendations for growth,[59] adult maintenance[26] and studies of struvite urolithiasis in cats.[50] In this latter study, however, serum phosphorus concentrations averaged 3.4 ± 1.5 mg/dl, near the lower limit of normal. Moreover, consumption of acidified diets has been shown to decrease the phosphorus balance in adult cats. Lewis *et al.* also showed that both magnesium and phosphorus absorption were influenced by the dietary content of other minerals.[50] Due to interactions with parameters such as other minerals and the acidification potential of the diet, the appropriate dietary content of magnesium and phosphorus will be that which maintains normal homeostasis without increasing the risk of struvite urolithiasis.

It should be reemphasized that these measures are specific for the management of struvite urolithiasis and may (with the exception of enhancing urine volume) be quite inappropriate for the management of other types of urolith in cats. Medical protocols for dissolution of other types of urolith in cats are not well developed,[68] although dietary modification may have a role to play in reducing the risk of recurrence of some types. However, published studies in support of this in cats are currently lacking. Treatment will thus require surgical intervention; discussion of methods for cystotomy are outside the scope of this chapter, but may be found in standard surgical texts. Recommendations to help prevent recurrence of calcium oxalate urolithiasis in dogs are for a diet moderately restricted in protein, calcium, oxalate and sodium, and not restricted or supplemented with phosphorus and magnesium.[54] These recommendations may also be of value in cats, although published studies supporting this suggestion are lacking. Application to cats of a modified protocol for the prevention of ammonium acid urate uroliths in dogs has also been advocated.[75]

Cats with Idiopathic Disease

In the series of 143 cats with hematuria and dysuria already discussed, 77 (54%) were classified as idiopathic.[68] This represented 58% of the nonobstructed female, 79% of the nonobstructed male and 28% of the obstructed male cats. If this series of cases is reasonably representative of the general pattern of disease, then it is likely that the clinician will be unable to identify a specific cause, and thus formulate an appropriate management program in the majority of cats with LUTD. Of the group of 10 non-obstructed cats studied by Filippich,[29] only one showed significant crystalluria and only two had positive bacterial cultures (although the author considered these to be secondary infections, as urine samples from previous bouts were sterile). Thus most of this group of cats could be considered to have idiopathic disease; in addition, most had had multiple bouts of disease prior to the study. Recurrence of disease was noted in only three of the cats during the study; one cat showed three bouts of dysuria and hematuria and

thus did not apparently respond to dietary change. The other two showed three bouts between them, and two of these were known to have been associated with dietary change. Two of these cats also showed a recurrence of disease when returned to a normal diet at the end of the study. Thus, it appears that the use of a canned, restricted magnesium, acidifying diet may have played a role in preventing recurrence of disease in most of these apparently idiopathic cases. What is not clear is which aspect(s) of the diet was/were likely to have been most important, or whether the stress of the diet change influenced reoccurrence. The author noted that the mean urine pH at the start of the trial in this non-obstructed group was 6.25 ± 0.52 as a result of the fact that many of the cats were already on some form of urinary acidification at the time of referral. Average urine pH values measured in the whole group of cats remained between 6 and 6.25 through the remainder of the study. These data suggest that the apparent response of the cats was probably not mediated through urinary pH, and thus disease was probably not struvite related. Urine specific gravity (SG) of this group of cats averaged 1.048 at the start of the study (average of all cats 1.046). During the study the SG averaged between 1.030 and 1.034 (all cats), thus suggesting that the dietary regimen brought about a dilution of the urine. It seems most logical to hypothesize that this may have been an important factor in the clinical response of the cats.

Although idiopathic disease remains an important challenge both to researchers rying to investigate the cause or causes of the condition and to clinicians trying to manage cases, this information does provide at least a tentative basis for making recommendations that may be of value in some cases.

These are:

1. To change the cat to a food containing 20–40 mg magnesium and 125–250 mg phosphorus/100 kcal.
2. To monitor urine pH and to add an acidifier if necessary to maintain urine pH between 6.0 and 6.5 4–6 hr after feeding.

3. To monitor urine SG and to add water to the cat's food if it is consistently above 1.035.

It must be emphasized that these measures are not based on an understanding of the underlying disease process and thus may be ineffective in some cases. However, they may be regarded as conservative measures and are relatively easy to implement, although the foregoing comments on risks of acidification should always be considered.

Summary

The diet influences the parameters that determine the probability of struvite formation in a variety of ways. Moreover, dietary constituents have different effects depending on their concentration, form and interactions with other ingredients in the food. The complexity of these interactions make prediction of the effects of a particular diet on struvite formation impossible. Fortunately, reputable petfood manufacturers attempt to formulate diets that will not promote struvite formation. Their success is evident in the declining numbers of struvite uroliths submitted for analysis in recent years. The increase in calcium oxalate may suggest the need for more fine tuning in the future, or may be the "unmasking" of a more naturally occurring pattern of urolithiasis by diminishing the incidence of diet-related urolithiasis.

References

1. Ackermann, D., Brown, C., Dunthorn, M. *et al.* (1989) Use of the computer program EQUIL to estimate pH in model solutions and human urine. *Urology Research,* **17,** 157.
2. Adolph, E. F. (1947) Tolerance to heat and dehydration in several species of mammals. *American Journal of Physiology,* **151,** 564–575.
3. Anderson, R. S. (1982) Water balance in the dog and cat. *Journal of Small Animal Practice,* **23,** 588–598.
4. Anderson, R. S. (1983) Fluid balance and diet. *Proceedings of the Seventh Kal Kan Symposium.* pp. 19–24. Vernon, CA: Kal Kan.
5. Bai, S. G., Sampson, D. A., Morris, J. G. and Rogers, Q. R. (1989) Vitamin B6 requirement of growing kittens. *Journal of Nutrition,* **119,** 1020–1027.
6. Barsanti, J. A. and Finco, D. R. (1984) Management of postrenal uremia. *Veterinary Clinics of North America,* **14,** 609–616.
7. Barsanti, J. A. and Finco, D. R. (1986) Feline urologic syndrome. In *Nephrology and Urology.* Ed. E. W. Breitschwerdt. pp. 43–74. New York: Churchill Livingstone.
8. Blount, W. P. (1931) Urinary calculi. *Veterinary Journal,* **87,** 561–576.
9. Boistelle, R., Abbona, F. and Madsen, H. E. L. (1983) On the transformation of struvite into newberyite in aqueous systems, *Physics and Chemistry of Minerals,* **9,** 216–222.
10. Bovee, K. C. (1984) Urolithiasis. In *Canine Nephrology.* Ed. K. C. Bovee. pp. 355–379. Philadelphia: Harwel.
12. Brown, J. E. and Fox, L. M. (1984) Ammonium chloride/methionine toxicity in kittens. *Feline Practice Journal,* **14,** 16–19.
13. Buffington, C. A. (1988) Effects of diet on the feline struvite urolithiasis syndrome. Ph.D. Thesis. University of California, Davis.
14. Buffington, C. A. (1988) Effects of age and food deprivation on urinary pH in cats. *Proceedings of the Third Annual Symposium of the ESVNU.* pp. 60–72. Barcelona: Intercongress.
15. Buffington, C. A. (1988) Feline struvite urolithiasis: effect of diet. *Proceedings of the Third Annual Symposium of the ESVNU.* pp. 73–112. Barcelona: Intercongress.
16. Buffington, C. A. (1992) Nutritional aspects of struvite urolithiasis in dogs and cats. *Proceedings of the Sixteenth Waltham/OSU Symposium for the Treatment of Small Animal Diseases.* pp. 51– 57. Vernon, CA: Kal Kan.
17. Buffington, C. A., Rogers, Q. R. and Morris, J.G. (1990) Effect of diet on struvite activity product in feline urine. *American Journal of Veterinary Research,* **51,** 2025–2029.
18. Buffington, C. A., Rogers, Q. A., Morris, J. G. and Cook, N. E. (1985) Feline struvite urolithiasis: magnesium effect depends on urinary pH. *Feline Practice Journal,* **15,** 29–33.
19. Burger, I. H., Anderson, R. S. and Holme. D. W. (1980) Nutritional factors affecting water balance in the dog. In *Nutrition of the Dog and Cat.* Ed. R. S. Anderson. pp. 145–156. Oxford: Pergamon.
20. Castonguay, T. W. (1981) Dietary dilution and intake in the cat. *Physiology and Behaviour,* **27,** 547–549.
21. Ching, S. V., Fettman, M. J. and Hamar, D. W.

(1989) The effect of chronic dietary acidification using ammonium chloride on acid–base and mineral metabolism in the adult cat. *Journal of Nutrition,* **119**, 902–915.

22. Chow, F. H. C. (1977) Dietary mineral effects on feline urolithiasis. *Proceedings of the Kal Kan Symposium for Treatment of Dog and Cat Diseases.* pp. 36–39. Vernon, CA: Kal Kan.

23. Dibartola, S. P. and Buffington, C. A. (in press) Feline urologic syndrome. In *Textbook of Small Animal Surgery.* Ed. D. J. Slatter. 2nd edn. Philadelphia: W. B. Saunders.

24. Dickinson, C. D. and Scott, P. P. (1956) Failure to produce urinary calculi in kittens by the addition of mineral salts, derived from bone meal, to the diet. *Veterinary Record,* **68**, 858–859.

25. Dow, S. W., Fettman, M. J., LeCouteur, R. S. and Hamar, D. W. (1987) Potassium depletion in cats: renal and dietary influences. *Journal of the American Veterinary Medical Association,* **191**, 1569–1575.

26. Dzanis, D. A., Corbin, J. E., Czarnecki–Mauldin, G. L., Hirakawa, D. A., Kallfelz, F. A., Morris, M. L. Jr., Rogers, Q. R., Sheffy, B. E. and Thompson, A. (1992) *AAFCO Nutrient Profiles for Cat Foods.* Report of the feline nutrition expert committee. Rockville, MD: U.S. Food and Dr.ug Administration.

27. Fau, D., Smalley, D. A., Rogers, Q. R. and Morris, J. G. (1983) Effects of excess methionine in the kitten. *Federation Proceedings,* **42**, 542 (Abstr. 1469).

28. Fettman, M. J., Coble, J. M., Hamar, D. W., Norrdin, R. W., Seim, H. B., Kealy, R. D., Rogers, Q. R., McCrea, K. and Moffat, K. (1992) Effect of dietary phosphoric acid supplementation on acid–base balance and mineral and bone metabolism in adult cats. *American Journal of Veterinary Research,* **53**, 2125–2135.

29. Filippich, L. (in press) Feline lower urinary tract disease: Clinical dietary study. *Australian Veterinary Practitioner.*

30. Finco, D. R., Barsanti, J. A. and Crowell, W. A. (1985) Characterization of magnesium-induced urinary disease in the cat and comparison with feline urologic syndrome. *American Journal of Veterinary Research,* **46**, 391–400.

31. Gaskell, C. J. (1979) Studies on the feline urolithiasis syndrome. Ph.D. Thesis. University of Bristol.

32. Gaskell, C. J. (1985) Feline urological syndrome: a United Kingdom perspective. *Feline Medicine: Proceedings, Eastern States Veterinary Conference, Orlando.* pp. 27–32. New Jersey: Veterinary Learning Systems.

33. Hamar, D., Chow, F. H. C., Dysart, M. I. and Rich, L. J. (1976) Effect of sodium chloride in prevention of experimentally produced phosphate uroliths in male cats. *Journal of the American Animal Hospital Association,* **12**, 514–517.

34. Hause, W. R. (1984) Management of acute illness in cats. *Modern Veterinary Practice,* **65**, 359–362.

35. Holme, D. W. (1977) Research into the feline urological syndrome. *Proceedings of the Kal Kan Symposium for Treatment of Dog and Cat Diseases.* pp. 40–45. Vernon, CA: Kal Kan.

36. Hutyra, F., Marek, J. and Manninger, R. (1938) *Special Pathology and Therapeutics of the Diseases of Domestic Animals. Vol. III.* p. 47. London: Balliére, Tindall and Cox.

37. Jackson, O. F. (1971) The treatment and subsequent prevention of struvite urolithiasis in cats. *Journal of Small Animal Practice,* **12**, 555–568.

38. Jackson, O. F. and Sutor, D. J. (1970) Ammonium acid urate calculus in a cat with a high uric acid excretion possibly due to a renal tubular reabsorption defect. *Veterinary Record,* **86**, 335.

39. Jackson, O. F. and Tovey, J. D. (1977) Water balance studies in domestic cats. *Feline Practice,* **7(4)**, 30–33.

40. Kallfelz, F. A., Bressett, J. D. and Wallace, R. J. (1980) Urethral obstruction in random source and SPF male cats induced by high levels of dietary magnesium and phosphorus. *Feline Practice Journal,* **10**, 25–35.

41. Kane, E., Rogers, Q. R., Morris, J. G. and Leung, P. M. B. (1981) Feeding behaviour of the cat fed laboratory and commercial diets. *Nutrition Research,* **1**, 499–507.

42. Kirk, H. (1925) *The Diseases of the Cat.* p. 261. London: Balliére, Tindall and Cox.

43. Kohn, C. W. and DiBartola, S. P. (1992) Composition and distribution of body fluids in dogs and cats. In *Fluid Therapy in Small Animal Practice.* Ed. S. P. DiBartola. pp. 1–34. Philadelphia: W. B. Saunders.

44. Kolata, R. J. (1984) Emergency treatment of urethral obstruction in male cats. *Modern Veterinary Practice,* **65**, 517–521.

45. Krabbe, A. (1949) Urolithiasis in dogs and cats. *Veterinary Record,* **61**, 751–759.

46. Kruger, J. M., Osborne, C. A., Goyal, S. M., Wickstrom, S. L., Johnston, G. R., Fletcher, T. F. and Brown, P. A. (1991) Clinical evaluation of cats with lower urinary tract disease. *Journal of the American Veterinary Medical Association,* **199**, 211–216.

47. Lawler, D. F., Sjolin, D. W. and Collins, J. E. (1985) Incidence rates of feline lower urinary tract disease in the United States. *Feline Practice Journal,* **15**, 13–16.

48. Lees, G.E. and Osborne, C. A. (1984) Use and misuse of indwelling urinary catheters in cats. *Veterinary Clinics of North America,* **14**, 599–608.

49. Lees, G. E., Osborne, C. A., Stevens, J. B. and Ward, G. E. (1981) Adverse effects of open indwelling urethral catheterization in clinically normal male cats. *American Journal of Veterinary Research,* **42**, 825–833.

50. Lewis, L. D., Chow, F. H. C., Taton, L. F. and Hamar, D. W. (1978) Effect of various dietary mineral concentrations on the occurrence of feline urolithiasis. *Journal of the American Veterinary Medical Association,* **172**, 559–563.

51. Lewis, L. D. and Morris, M. (1983) *Small Animal Clinical Nutrition.* pp. 8–24. Topeka, KS: Mark Morris Associates.

52. Lewis, L. D. and Morris, M. L. Jr. (1984) Feline urologic syndrome: causes and clinical management. *Veterinary Medicine/Small Animal Clinician,* **79**, 323–337.

53. Lloyd, W. E. and Sullivan, D. J. (1984) Effects of orally administered ammonium chloride and methionine on feline urinary acidity. *Veterinary Medicine,* **79**, 773–778.

54. Lulich, J. P., Osborne, C. A., Felice, L. J., Polzin, D. J., Thumchai, R. and Sanderson, S. (1992) Calcium oxalate urolithiasis. *Nephrology and Urology. The Sixteenth Annual Waltham/OSU Symposium.* pp. 69–74. Vernon, CA: Kal Kan.

55. Lulich, J. P., Osborne, C. A., Nagode, L. A. *et al.* (1991) Evaluation of urine and serum metabolites in minature Schnauzers with calcium oxalate urolithiasis. *American Journal of Veterianary Research,* **52**, 1583–1590.

56. Markwell, P. J. and Gaskell, C. J. (1991) Progress in understanding feline lower urinary tract disease. *Waltham International Focus,* **1**, 22–29.

57. Marshall, W. and Robertson, W. G. (1976) Nomograms for the estimation of the saturation of urine with calcium oxalate, calcium phosphate, magnesium ammonium phosphate, uric acid, sodium acid urate, ammonium acid urate and cystine. *Clinica Chimica Acta,* **72**, 253–260.

58. Milks, H. J. (1935) Urinary calculi. *Cornell Veterinarian,* **25**, 153–161.

59. National Research Council (1986) *Nutrient Requirements of Cats.* Washington: National Academy Press.

60. O'Conner, W. J. and Potts, D. J. (1969) The external water exchanges of normal laboratory dogs. *Quarterly Journal of Experimental Physiology,* **54**, 244–265.

61. Osbaldiston, G. W. and Taussig, R. A. (1970) Clinical report on 46 cases of feline urological syndrome. *Veterinary Medicine/Small Animal Clinician,* **65**, 461–468.

62. Osborne, C. A., Kruger, J. M., Lulich, J. P., Beauclair, K. D., Unger, L. K., Bird, K. A., Koehler, L. A., Bartges, J. W., Felice, L. J., Molitor, T., Polzin, D. J., Hall, C. L. and O'Brien, T. D. (1992) Feline lower urinary tract diseases: state of the science. *Proceedings of the Waltham/OSU Symposium for Treatment of Small Animal Diseases: Nephrology and Urology.* pp. 89–98. Vernon, CA: Kal Kan.

63. Osborne, C. A., Johnston, G. R., Polzin, D. J., Kruger, J. M., Bell, F. W., Poffenbarger, E. M., Feeney, D. A., Stevens, J. B. and McMenomy, M. F. (1984) Feline urologic syndrome: a heterogenous phenomenon? *Journal of the American Animal Hospital Association,* **20**, 17–32.

64. Osborne, C. A., Kruger, J. M., Johnston, G. R. and Polzin, D. J. (1989) Feline lower urinary tract disorders. In *Textbook of Veterinary Internal Medicine.* Ed. S. J. Ettinger. 3rd edn. pp. 2057–2082. Philadelphia: W. B. Saunders.

65. Osborne, C. A., Kruger, J. P., Lulich, J. P., Bartges, J. W., Polzin, D. J., Molitor, T., Beauclair, K. D. and Onffroy, J. (1992) Feline matrix-crystalline plugs: a unifying hypothesis of causes. *Journal of Small Animal Practice,* **33**, 172–177.

66. Osborne, C. A., Lees, G. E., Polzin D. J. and Kruger, J. M. (1984) Immediate relief of feline urethral obstruction. *Veterinary Clinics of North America,* **14**, 585–598.

67. Osborne, C. A., Polzin, D. J., Johnston, G. R. and Kruger, J. M. (1986) Medical management of feline urologic syndrome. In *Current Veterinary Therapy IX.* Ed. R. W. Kirk. pp. 1196–1206. Philadelphia: W. B. Saunders.

68. Osborne, C. A., Polzin, D. J., Kruger, J. M., Lulich, J. P., Johnston, G. R. and O'Brien, T. D. (1989a) Relationship of nutritional factors to the cause, dissolution, and prevention of feline uroliths and urethral plugs. *Veterinary Clinics of North America,* **19**, 561–581.

69. Osborne, C. A., Sanna, J. J., Unger, L. K., Clinton, C. W. and Davenport, M. P. (1989) Mineral composition of 4500 uroliths from dogs, cats, horses, cattle, sheep, goats, and pigs. *Veterinary Medicine,* **84**, 750–764.

70. Reif, J. S., Bovee, K., Gaskell, C. J., Batt, R. M. and Maguire, T. G. (1977) Feline urethral obstruction: a case control study. *Journal of the American Veterinary Medical Association,* **170**, 1320–1324.

71. Rich, L. J., Dysart, I., Chow, F. C. and Hamar D. (1974) Urethral obstruction in male cats: experimental production by addition of magnesium and phosphate to diet. *Feline Practice Journal,* **4**, 44–47.

72. Rich, L. J. and Kirk, R. W. (1969) The relationship of struvite crystals to urethral obstruction in cats. *Journal of the American Veterinary Medical Association,* **154**, 153–157.

73. Sauer, L. S., Hamar, D. and Lewis, L. D. (1985) Effect of diet composition on water intake and excretion by the cat. *Feline Practice Journal,* **15,** 16–21.

74. Schwille, P. O. and Herrmann, U. (1992) Environmental factors in the pathophysiology of recurrent idiopathic calcium urolithiasis (RCU), with emphasis on nutrition. *Urology Research,* **20,** 72.

75. Senior, D. F. (1992) Urate urolithiasis. *Nephrology and Urology. The Sixteenth Annual Waltham/OSU Symposium.* pp. 59–68. Vernon, CA: Kal Kan.

76. Senior, D. F., Sundstrom, D. A. and Wolfson, B. B. (1986) Testing the effects of ammonium chloride and D,L-methionine on the urinary pH of cats. *Veterinary Medicine,* **81,** 88–93.

77. Senior, D. F., Sundstrom, D. A. and Wolfson, B. B. (1986) Effectiveness of ammonium chloride as a urinary acidifier in cats fed a popular brand of canned cat food. *Feline Practice Journal,* **16,** 24–26.

78. Stansbury, R. L. and Truesdail, R. W. (1955) Occurrence of vesical calculi in cats receiving different diets. *The North American Veterinarian,* **36,** 841–845.

79. Taton, G. F., Hamar, D. W. and Lewis, L. D. (1984) Urinary acidification in the prevention and treatment of feline struvite urolithiasis. *Journal of the American Veterinary Medical Association,* **184,** 437–443.

80. Walker, A. D., Weaver, A. D., Anderson, R. S., Crighton, G. W., Fennel, C., Gaskell, C. J. and Wilkinson, G. T. (1977) An epidemiological survey of the feline urological syndrome. *Journal of Small Animal Practice,* **18,** 283–301.

81. Willeberg, P. (1984) Epidemiology of naturally occurring feline urologic syndrome. *Veterinary Clinics of North America,* **14,** 455–469.

82. Ziegler, E. E. and Fomon, S. J. (1989) Potential renal solute load of infants formulas. *Journal of Nutrition,* **119(12S),** 1785–1788.

CHAPTER 18

Canine Renal Disease

SCOTT A. BROWN

Renal Disease

Incidence/Prevalence

Renal disease is a frequent cause of illness and death in dogs, affecting approximately 1% of all dogs.[64] The prevalence increases with advancing age, with chronic renal failure affecting 10% of all dogs over 15 years-of-age.[78] Renal failure sufficient to cause uremia is associated with a high rate of mortality and tremendous financial cost of further therapy, i.e. renal transplantation and/or dialysis. For dogs with renal failure, the general goals of dietary management are to maximize the quality and quantity of life. To accomplish these goals, the veterinary practitioner must strive to reduce the prevalence and/or severity of uremia, maintain adequate nutrient and energy intake, reduce mortality from the complications of uremia and arrest progression of renal injury.

Clinical Syndromes

The kidney is an organ of homeostasis. Renal failure occurs when a disease prevents the kidney from performing a vital homeostatic function. Clinically, this is usually defined as an accumulation of nitrogenous wastes, referred to as azotemia, and is due to inadequate renal excretory function. Azotemic renal failure is recognized in clinical patients by detecting elevations in the serum concentrations of creatinine and urea nitrogen. Azotemia only occurs after renal disease has destroyed about three-quarters of functional renal tissue.[13] Other functions of the kidney, such as the ability to concentrate, may well be disrupted earlier in renal disease in dogs. However, renal failure at this early, nonazotemic stage is not often recognized in dogs. Consequently, the clinician attempting to manage a dog with renal failure is generally faced with either an azotemic patient or one in which clinical signs, such as vomiting and lethargy, are associated with azotemia, i.e. uremia.

Renal failure may be of recent onset (acute renal failure) or of long-term duration (chronic renal failure). The temporal distinction between acute and chronic renal failure is somewhat arbitrary; accordingly some have suggested that renal failure of more than 2-weeks duration be referred to as chronic renal failure.

Some animals with renal disease have generalized glomerular injury of a type that interferes with the selectivity of the glomerular barrier to plasma proteins. Consequently, plasma proteins normally prevented from entering the glomerular filtrate traverse the filtration barrier and proteinuria ensues. Severely proteinuric animals may develop the nephrotic syndrome, which is characterized by hypoalbuminemia, hypercholesterolemia and edema. Although dogs with the nephrotic syndrome uniformly exhibit marked proteinuria, not all will be azotemic. Proteinuric animals may develop a deficiency of antithrombin III that, coupled with the presence of hypoalbuminemia, predisposes them to the development of thromboses.[81,82,98]

A wide variety of other renal diseases, including obstruction, nephrolithiasis, dysuria/hematuria/pollakiuria, urinary tract infections and abnormal micturition are important in the clinical practice of urology in dogs. Unfortunately, our knowledge of the dietary therapy in dogs with diseases of the urinary tract is limited to those with renal failure and those with urolithiasis. The latter clinical problem affects predominantly the lower urinary tract in dogs and will be considered in Chapter 19. This chapter will address issues related to the nutritional management of renal failure in dogs.

Acute Renal Failure

Definition

Acute renal failure is a sudden reduction in renal function that occurs from intrarenal mechanisms, leading to a disruption of one or more of the renal homeostatic functions. Because renal function is most often assessed by evaluation of excretory function, acute renal failure is generally synonymous with a sudden fall in the glomerular filtration rate (GFR) due to a renal disease. This definition excludes causes of reduced GFR that are either prerenal in origin (e.g. dehydration and systemic hypotension) or postrenal in origin (e.g. urinary tract obstruction and rupture). Although this definition of acute renal failure does not require the presence of azotemia, clinically identified acute renal failure is generally defined as a sudden fall in GFR leading to the retention of nitrogenous wastes.

Pathophysiology

Factors leading to the development of acute renal failure in dogs are many. Generally, these include a variety of nephrotoxins, such as aminoglycosides,[12] ethylene glycol,[44] amphotericin B[87] and nonsteroidal antiinflammatory drugs.[7] In contrast, primary ischemic renal failure is not commonly recognized in dogs.

Although several investigators have distinguished between factors responsible for the initiation and maintenance of acute renal failure, the critical stage in the nutritional therapy of acute renal failure is that time during which recovery of renal function occurs. During this time, compensatory processes lead to an enhancement of renal function. If life is to be sustained, this recovery phase must be encouraged to allow GFR to reach levels capable of sustaining life.

Recovery of renal function following an acute renal insult requires a variable, but prolonged, period of time. In dogs with gentamicin-associated acute renal failure this increase in GFR, evidenced by a fall in serum creatinine concentrations, will not begin until 7–10 days after acute renal failure is initially recognized.[12] Increases in renal function may continue for several months, although most (about 85%) of the compensatory renal response in dogs occurs within the first 60 days following a sudden reduction in GFR.[15] The key to therapy during renal recovery in these patients is to provide adequate nutrient intake to allow renal repair and prevent the development of a catabolic state that will worsen uremia.

Diagnosis

The diagnosis of acute renal failure requires the identification of a sudden onset of azotemia (or reduced GFR) and the exclusion of prerenal and postrenal causes of azotemia.

Prognosis

The prognosis for acute renal failure is dependent upon the extent of renal injury, the cause of the primary insult and the extent of uremic complications. The presence of oliguria or anuria is a poor prognostic sign.[12] Acute renal failure from some causes seems to carry a better prognosis than others. For example, animals suffering from aminoglycoside nephrotoxicity appear to have a better prognosis than those with renal failure due to ethylene glycol ingestion. The severity of azotemia is inversely correlated with the prognosis, though severely azotemic patients may recover, particularly those showing a dramatic response to rehydration with parenteral fluid administration. Therapy should be attempted for a minimum of 7 days to ascertain the course of acute renal failure in any individual animal.

Therapy

The goals of therapy in these patients are to minimize the extent of uremia, provide adequate nutrient intake and enhance the recovery of renal function. To achieve these goals, clinical problems must be identified and appropriate therapy instituted (Fig. 18.1). Ideally, therapy will allow renal recovery from an acute insult to occur while the animal is in a state of adequate nutrient intake and without the metabolic abnormalities of the uremic syndrome. Unfortunately, this frequently requires the use of dialytic therapy. Generally, these dogs should be managed with fluid and electrolyte therapy based upon measurements of the serum chemistry. The fluid and electrolyte therapy must be individualized, as some animals will be polyuric and thus generally hypokalemic (e.g. aminoglycoside nephrotoxicity) while others will be oliguric ($\leq$0.5 ml/kg/hr) and thus frequently hyperkalemic (e.g. ethylene glycol toxicity).

Oliguric patients may benefit from attempts to enhance urine formation. Initially, the animal should be rehydrated and have electrolyte abnormalities corrected by conventional fluid therapy. Subsequently, a test dose of an osmotic diuretic such as glucose (0.5–1.0 ml/kg of 10% dextrose) or mannitol (0.5–1.0 g/kg IV over 30 min) should be administered through a central venous line. If this is ineffective, a loop diuretic such as furosemide

(2–8 mg/kg IV bolus, repeated in 30 min if no response is noted) and/or dopamine (1–5 μg/kg/min given by constant IV infusion in an isotonic solution) may be administered. If there is no response to the combined administration of dopamine plus furosemide, then other measures to encourage urine formation generally prove inadequate. At this time, conservative fluid therapy, dialysis, or transplantation are the remaining alternatives. Some dogs with oliguric acute renal failure will respond to conservative fluid therapy, become polyuric and develop adequate GFR to sustain life.

Animals with acute renal failure require adequate energy intake. Human beings with acute renal failure exhibit a marked increase in basal metabolic rate.[69] Affected dogs are likely to have a similar increase in energy requirements. Acute renal failure is characterized by a catabolic state with high rates of protein turnover. Unless adequate energy intake is achieved in these patients, protein catabolism leads to generation of nitrogenous wastes and worsening of the uremic state. It is not reasonable to expect adequate regeneration of renal tissue to occur when a severely uremic environment is superimposed upon malnutrition. Consequently, the most important nutritional principle is to provide sufficient energy, which will reduce the degree of uremia and enhance the return of renal function.[34,63,103]

Enteral nutritional support may be necessary in anorexic dogs. A nasogastric tube may be placed, without general anesthesia, to allow the administration of electrolytes, dextrose solutions or liquid diets. A gastrostomy tube is undesirable as the additional irritation of the tube and its placement may induce vomiting in a markedly azotemic dog not otherwise showing clinical signs. Alternatively, in vomiting dogs, a jejunostomy tube may be placed utilizing a limited laparotomy.

Gastrointestinal abnormalities associated with the uremic syndrome mean that the initial energy support therapy may need to be parenteral. Various studies have evaluated the use of intravenously administered glucose and/or amino acid solutions in acute renal

DIAGNOSIS AND THERAPY OF COMPLICATIONS OF RENAL FAILURE

Clinical problem	Observed in acute (A) or chronic (C) renal failure	Diagnostic parameter	Therapy
Malnutrition	AC	Daily energy intake Body weight	Increase palatability of diets Energy supplements Enteral nutritional support Parenteral nutritional support
Uremia	AC	BUN Clinical signs	Dietary protein restriction Antiemetics Transplantation or dialysis
Oliguria	AC	Urine volume	Rehydration Osmotic diuresis Loop diuresis Dopamine ± loop diuresis Conservative medical therapy
Hypokalemia Hyperkalemia	AC AC	Plasma K^+ Plasma K^+	Dietary K^+ supplementation Bicarbonate therapy Insulin plus glucose K^+ exchange resins
Metabolic acidosis	AC	Plasma HCO_3^- or plasma total CO_2	Dietary alkalinization Vegetable source protein
Hyperphosphatemia	AC	Plasma PO_4^{2-}	Dietary phosphate restriction Intestinal PO_4^{2-} binders
Systemic hypertension Anemia	C C	Blood pressure Retinal examination Hematocrit	Dietary sodium restriction Antihypertensive agents Adequate energy intake Anabolic steroids(?) Erythropoietin
Renal osteodystrophy	C	Plasma PTH Plasma PO_4^{2-} Fractional clearance of PO_4^{2-}	Dietary phosphate restriction Calcitriol
Progressive renal disease	C	Sequential serum Cr	Dietary PO_4^{2-} restriction (modification of dietary content of lipids, sodium, acid, and/or protein?)

FIG. 18.1 Diagnosis and therapy of complications of renal failure.

failure. Treatment of rats with renal disease enhanced protein synthesis in the damaged kidney.[94,95] These results are supported by some, but not all, studies in human beings and by a study in dogs with experimental renal disease.[1,2,60,102] Some possible side-effects of amino acid solutions include the production of metabolic acidosis and hyperammonemia. As noted above, hypertonic glucose solutions have been used routinely and recommended in clinical practice. A frequently used dose is 20–30 ml of 10% dextrose/kg body weight (BW) given over 2 hr and repeated tid–bid. This dose provides 10–30% of daily energy needs. Trials to evaluate the effectiveness of either glucose or amino acid solutions and appropriate dosages for amino acid solutions have yet to be performed in dogs.

Although the cause–effect relationship has not been clearly established, irreversible renal injury and nephrocalcinosis may occur in an animal with renal disease and hyperphosphatemia.[15] For this reason, it is desirable to reduce the extent of hyperphosphatemia in dogs with acute renal failure. Dietary restriction of phosphorus in conjunction with the administration of intestinal phosphorus binders should be aggressively employed in an attempt to normalize the serum phosphorus concentration, using the same principles as outlined for therapy of chronic renal failure (see below).

It has been argued that the toxicity of hyperphosphatemia is, in part, mediated by renal secondary hyperparathyroidism.[66–68,93] However, it has never been established that parathyroid hormone is a nephrotoxin. Therefore, on the basis of currently available evidence, efforts designed solely to reduce the extent of hyperparathyroidism in the recovery phase of acute renal failure, such as vitamin D administration,[21] are unwarranted.

Dietary protein restriction in animals with renal failure has been advocated for two different reasons. First, dietary protein restriction will reduce the production of uremic toxins. In uremic animals, diets restricted in protein that supply 2.0–3.0 g protein/kg BW/day should be considered. However, the most critical nutritional concern in these patients is adequate energy intake and if affected animals ingest an inadequate amount of a low protein diet, they will catabolize body protein stores for energy, thereby worsening uremia. The second rationale for dietary protein restriction is to slow the rate of progression of renal disease (see below) by minimizing glomerular hypertension,[11,52] metabolic acidosis[73] and/or cellular oxidant or metabolic stress.[49] However, in dogs that are nonazotemic or are only mildly azotemic following an acute renal insult, there is currently little justification for dietary protein restriction. In these animals, acidosis can be treated by dietary alkalinizing agents, such as potassium citrate or sodium bicarbonate. Long-term studies have clearly established that glomerular hypertension by itself does not produce rapid decrements in renal function in dogs with renal failure[19,20,37,77,84] and the reversal of glomerular hypertension does not seem to be a reasonable sole rationale for dietary protein restriction in dogs immediately after an acute renal insult. In contrast, results of a recent study support the use of a higher protein diet for several months after an acute renal injury to enhance renal recovery.[100]

Chronic Renal Failure

Definition

Chronic renal failure is defined as azotemia (or reduced GFR) of a chronic duration (>2 weeks) with the exclusion of prerenal and postrenal causes.

Pathophysiology

Glomerular Hyperfiltration

Frequently, the diagnosis of chronic renal failure is established in a dog that is azotemic, but with renal function remaining adequate to prevent the development of the uremic syndrome. These dogs frequently have a good quality of life for weeks, months or even years. However, these dogs often suffer progressive decrements of renal function over time and are destroyed. In certain inbred strains of rats, this progression from early to end-stage renal disease is an inherent property of their kidney.[11,52] In these strains of rats, the progression occurs regardless of the inciting cause and will continue even if the inciting cause is no longer present. For example, following removal of seven-eighths of functional renal tissue, the remaining eighth is initially normal and renal function is adequate to sustain life. However, the rats will suffer progressive decrements of renal function and develop uremia.[80] At postmortem, the renal tissue, although it was initially normal, now appears scarred and a marked expansion of mesangial matrix (glomerulosclerosis) is generally observed. An analogy has been drawn between progressive renal disease in pets and that observed in these rodents.[11,52] The development of endstage renal disease requires costly therapeutic intervention, using renal transplantation and/or dialytic therapy; therefore attempts to ameliorate this progressive nephropathy have received considerable attention. In veterinary medicine, where the availability of renal transplantation and dialysis is limited for technical, scientific and financial reasons, therapies aimed at limiting the progression of renal disease have great potential for benefit.

It has been proposed that the same factors produce progressive renal disease in all animals.[11,52] Proponents of this theory argue that once a critical mass of renal tissue is removed or destroyed by disease, progression to end-stage renal disease is inevitable.[11] These investigators have argued that the compensatory response of the kidney to injury is, paradoxically, responsible for the progressive nature of renal disease (Fig. 18.2). This compensatory response of the kidney to disease is universal, regardless of the animal species or the primary cause of the injury. Specifically, as nephrons are destroyed, the surviving or remnant nephrons enlarge and have enhanced function.[55] Micropuncture studies verified that this response is associated with an increase in single nephron blood flow, single nephron filtration rate and elevated pressure within the glomerular capillary bed in rats[32,52] and dogs.[13,16] Subsequent studies have indicated that elevations of blood pressure within the glomerular capillaries, referred to as glomerular hypertension, and increases in glomerular volume, referred to as glomerular hypertrophy, are the causative factors in glomerular injury in rodent models.[11,52,104] Any stimulus that leads to renal hypertrophy and hyperfunction within nephrons could thereby produce progressive renal injury in rodents. This has been termed the hyperfiltration theory of progressive renal disease (Fig. 18.2).[11,52]

Until further studies are completed, it seems plausible to assume that dogs with spontaneous renal disease do exhibit glomerular hypertension and hypertrophy. Given this presumption, the question remains whether or not these changes are injurious to canine nephrons. If injurious, dogs with glomerular hypertension and hypertrophy should suffer progressive decrements of renal function. However, the course of both experimental and spontaneous chronic renal disease in dogs is variable and is not progressive in all cases.[15,19,20,37,77,84]

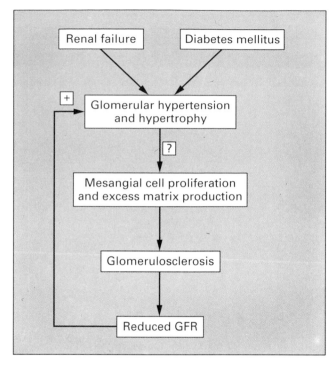

FIG. 18.2 The hyperfiltration theory of progressive chronic renal failure. Proponents argue that adaptations to renal disease lead to elevations of glomerular capillary pressure (glomerular hypertension) and glomerular enlargement (glomerular hypertrophy). It is argued that these changes are maladaptive and injure the glomerulus, ultimately leading to glomerulosclerosis and progressive renal dysfunction.

While some dogs with early chronic renal disease suffer progressive decrements in renal function and exhibit a linear decline of renal function over time,[4] others have stable, improving or episodically declining GFR.[9,15] Unfortunately, we do not yet know why some dogs exhibit a progressive pattern of declining renal function while others do not. One possible explanation is that the degree of glomerular hypertension and hypertrophy may vary between dogs and that those animals exhibiting the most marked adaptive changes will undergo progressive renal dysfunction. However, this and several alternative hypotheses remain untested.

One may hypothesize that dogs that exhibit a pattern of progressively or episodically declining renal function will benefit from measures designed to limit glomerular hypertension and hypertrophy. To this end, it has been proposed that dietary protein restriction will reduce or prevent the development of glomerular hypertension and hypertrophy. Importantly, the link between dietary protein restriction and normalization of glomerular size and pressure in remnant nephrons has yet to be established. One study on the effects of lowering glomerular hypertension in dogs with renal disease was conducted in uninephrectomized dogs with diabetes mellitus.[18] Treatment of dogs with an angiotensin-converting enzyme inhibitor (lisinopril) lowered glomerular capillary pressure, while a calcium antagonist related to diltiazem (TA-3090) did not. Although both agents reduced the extent of proteinuria and mesangial expansion, only the angiotensin-converting enzyme inhibitor limited the extent of glomerulosclerosis. Because the diabetic state may alter the response of the renal glomerulus to hypertension, it is tenuous to extrapolate these results to all dogs with chronic renal failure.

Several studies have examined the possibility that variations in dietary protein intake will modify the course of chronic renal disease. The effects of three levels of dietary protein intake were studied in 35 female dogs for 48 months following three-quarters or one-half nephrectomy plus the induction of pyelonephritis and there was no evidence of progressive decline of renal function.[20,84] While a relationship between protein intake and glomerular lesions was demonstrated in these dogs by light microscopy, this relationship was not confirmed by ultrastructural studies. Another study employed three groups of young male Beagle dogs fed three diets varying in protein, fat and mineral content.[77] During the 40-week study period, the high protein, high phosphorus diet was associated with a higher mortality rate attributed to uremia due to high protein intake rather than to progressive nephron destruction. A study of $^{11}/_{12}$-nephrectomized male Beagles documented the presence of glomerular lesions but did not establish a relationship between protein intake and renal dysfunction.[79] A recent study similarly failed to find any evidence that high protein intake results in progressive decrements of renal function in markedly azotemic dogs with experimental renal dysfunction.[37] In short, there is currently no clear evidence that high protein is injurious to canine kidneys or that reduction of protein intake in dogs with renal dysfunction results in a preservation of either renal structure or function.

Some of the components of the hyperfiltration theory have been tested in dogs. Recently, micropuncture and morphometric studies have verified the existence of glomerular hypertension and hypertrophy in remnant nephrons of dogs with experimental renal disease.[13,16] An important difference between dogs and rats in their adaptive response is that the increase in glomerular size in dogs is associated with an increase in the glomerular capillary ultrafiltration coefficient (K_f).[13,16] A principal determinant of the ultrafiltration coefficient is the surface area of the capillary wall. Thus the enlarging canine glomerulus appears to provide a beneficial increase in surface area that allows GFR to increase. Consequently, glomerular enlargement in dogs with renal disease has some apparent benefits to the animal. This is not the case in rats where remnant glomeruli enlarge without

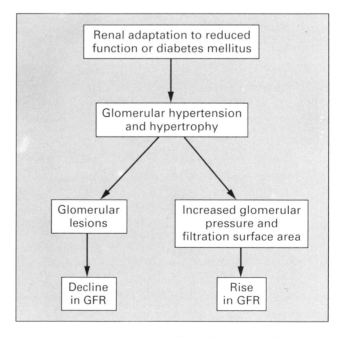

FIG. 18.3 Although some studies indicate that glomerular lesions develop in dogs with renal disease,[18–20,36,37,77,79,84] there are beneficial effects to the increase in glomerular pressure and size and these changes will tend to increase GFR.

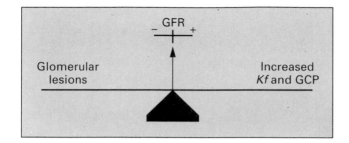

FIG. 18.4 Stable chronic renal disease. According to this hypothesis, there is a balance between the GFR lowering effect of glomerular morphological lesions and the GFR raising effects of glomerular hypertension (an increase in glomerular capillary pressure, GCP) and hypertrophy (an increase in the glomerular ultrafiltration coefficient, K_f).

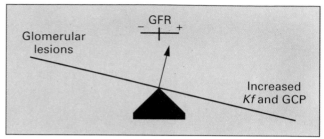

FIG. 18.5 Increasing GFR in a dog with chronic renal disease. In some dogs, the GFR raising effects of glomerular hypertension (an increase in glomerular capillary pressure, GCP) and hypertrophy (an increase in the glomerular utlrafiltration coefficient, K_f) predominate and GFR increases over time. This scenario predominates early in chronic renal failure and may be observed in some dogs for several years.[15]

exhibiting an increase in filtration surface area.

In the diseased dog kidney, two counteracting processes are present (Fig. 18.3). First, the enlarging glomerulus is subject to local hypertension and the associated accumulation of mesangial matrix (glomerulosclerosis). Although no causal link has yet been established in dogs, it may be that these glomerular hemodynamic and structural changes are self-injurious in dogs. Consistent with this hypothesis, studies have demonstrated that dogs with reduced renal function do exhibit glomerular lesions.[15,36,37,79] These morphological changes in the glomerulus should reduce GFR. However, this undesirable effect of glomerular adaptation to renal disease is initially offset by the counteracting effect of glomerular hypertension (increase glomerular capillary pressure, GCP) and hypertrophy (increased glomerular ultrafiltration coefficient, K_f) to enhance GFR. According to this hypothesis, a balance exists between these two processes (Fig. 18.4). In those dogs with chronic renal failure in which enhanced filtration surface area and increased GCP predominate, whole kidney GFR will be stable or increase (Fig. 18.5). In other dogs with chronic renal failure, hypertensive injury predominates and progressive decrements of whole kidney GFR ensue (Fig. 18.6). Because the ever enlarging, hypertensive glomerulus will be increasingly susceptible to renal injury and worsening morphological changes, the

enhancement of filtration surface area and GCP may reach a point of no return. Consequently, although the process of GFR enhancement may predominate early, the process of GFR decline may be observed in the latter stages of chronic renal failure. This tendency for GFR to be stable (or actually increase) for several months following a marked reduction in renal mass, followed by a pattern of decreasing GFR, has been observed in dogs with experimental chronic renal failure.[15,36,37,79] Dietary protein restriction has not been shown to effect the relationship between the factors that increase GFR and those that decrease GFR; however, intake of other nutrients may alter this relationship. Specifically, dietary phosphorus restriction will prolong the duration of GFR stability.[15,37]

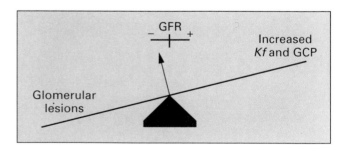

FIG. 18.6 Progressive chronic renal disease. In some dogs, the GFR lowering effects of glomerular morphological lesions predominate. This scenario may be observed late in chronic renal failure.[15,36,37]

Disorders of Phosphorus Homeostasis

Excessive dietary phosphorus or calcium intake can lead to renal mineralization in normal animals.[42,65] In rats with renal failure, restriction of dietary phosphorus below normal dietary levels reduces the extent of proteinuria and nephrocalcinosis and preserves renal function.[3,41,54] There are several factors that play a critical role in determining the effects of dietary phosphorus on renal function in rats. First, the nephrotoxicity of phosphorus is dependent upon the level of renal function.[50] Thus, animals with less renal function develop nephrocalcinosis at lower levels of dietary phosphorus intake, reflecting the dependence of phosphorus excretion upon the level of renal function (i.e. GFR). Second, the nephrotoxicity of phosphorus is directly related to the ratio of dietary calcium to dietary phosphorus.[23,51] The extent of phosphorus-associated nephrocalcinosis in rats is affected by both gender and age, as female[27] and young[65] rats are clearly more susceptible. Finally, low protein diets and a low dietary magnesium/phosphorus ratio may further exacerbate the nephrotoxicity of phosphorus.[43,96]

The mechanism of phosphorus toxicity is not clear. Some studies have suggested that hyperphosphatemia directly interacts with ionized calcium to produce calcium phosphorus salts within the renal parenchyma, nephrocalcinosis. In support of this hypothesis, inhibitors of calcium phosphorus crystal growth have provided protection in rats with experimental renal disease.[41] An alternate explanation is that high dietary intake of phosphorus leads to increased renal handling of phosphorus, producing an energy/oxidant stress on the renal tubular cells. This has been referred to as renal hypermetabolism and has been supported by several experimental studies in rats.[49] Another hypothesis, based upon the fact that hyperphosphatemia contributes to the genesis of renal secondary hyperparathyroidism, proposes that parathyroid hormone directly mediates renal injury. This hypothesis is supported by one published study.[90] Alternatively, parathyroid hormone may indirectly contribute to renal injury by enhancing metabolic acidosis.[6] Recent studies in rodents have indicated that heightened renal ammonia generation occurs during metabolic acidosis and ammonia can activate the alternate pathway for complement, thereby producing complement-mediated renal injury.[73]

Two recent studies in dogs have demonstrated that a balanced reduction of dietary phosphorus and calcium intake preserves renal function and prolongs survival in markedly azotemic dogs.[15,36,37] These studies clearly establish a rationale for dietary phosphorus restriction in dogs with chronic renal failure and the results of one of these studies indicate that female dogs may be more susceptible to the renal toxicity of phosphorus, especially in the setting of a low protein diet, than males.[15]

Systemic Hypertension

Hypertension is the presence of abnormally elevated blood pressure within the circulatory system. Hypertension may be local or generalized in the systemic arteries. For example, hypertension is present locally within the glomerular capillaries of dogs with renal disease.[13,16] When the elevated hydrostatic pressures are observed within the arteries of the systemic circulatory system, this is referred to as systemic hypertension. Mild elevations of systemic blood pressure have been reported in dogs with experimental renal failure.[15,26] Surveys of dogs with spontaneous renal disease also suggest that systemic hypertension is present in many affected animals[28,99] and may be more severe and prevalent than in experimental animals.

A review of the variety of factors that result in the genesis and maintenance of systemic hypertension are beyond the scope of this chapter. The consequences of systemic hypertension do, however, have important implications for dogs with renal disease. First, because a glomerulus within a diseased canine kidney is already subject to high intraglomerular pressure, a further elevation of systemic arterial blood pressure could worsen the extent of glomerular capillary hypertension, thereby predisposing the kidney to hypertensive injury. Preliminary results of a major clinical study of dietary intervention in people with renal disease indicate that the most critical factor in the progression of renal disease is the degree of systemic hypertension.[58]

The only study to evaluate the effects of systemic hypertension on dogs with renal disease was conducted in unnephrectomized dogs with diabetes mellitus. Treatment of dogs with an angiotensin-converting enzyme inhibitor (lisinopril) or a calcium antagonist related to diltiazem (TA-3090) or both reduced systemic blood pressure and proteinuria. In those diabetic Beagles, combination therapy or treatment with the angiotensin-converting enzyme inhibitor alone was of most benefit.[18] However, the degree of systemic hypertension observed in untreated diabetic dogs was relatively mild and the protective effects of lisinopril may relate to the ability of this angiotensin-converting enzyme inhibitor to preferentially lower glomerular capillary pressure.

The consequences of systemic hypertension depend upon both the duration and the severity of the hypertension. Ruptures of blood vessels leading to infarcts are rarely reported in dogs. The most commonly recognized side-effect of systemic hypertension is blindness, and uncontrolled severe hypertension can lead to retinal hemorrhage and/or detachment and permanent blindness.[28,62,83,86] A possible side-effect of chronic, uncontrolled systemic hypertension is damage to blood vessels and the development of arteriosclerosis. Again, this is not commonly observed in dogs. In dogs with preexistent renal disease, however, the presence of systemic hypertension should be interpreted as a risk factor for progressive loss of renal function and effective management of systemic hypertension is indicated.

Dietary Lipids

Humans with renal disease exhibit a wide variety of abnormalities of lipid metabolism, with an increase in low and very low density plasma lipoprotein concentrations and an associated hypercholesterolemia and hyper-triglyceridemia.[72] Although the cause of these abnormalities in lipid metabolism are only now being unravelled, there is abundant experimental support for a toxic effect of elevated plasma lipid concentrations on the kidneys[38,45,72] and for an effect of dietary lipids on renal disease.[8,88] One recent study established a link between hyperlipidemia and progression of renal disease in dogs.[15] Although this study did not demonstrate a cause–effect relationship, these results do provide a rationale for further study of the role of dietary lipids in renal disease.

There are at least two effects of dietary lipids that may alter renal function. First, diets high in cholesterol and saturated fat content may increase plasma concentrations of tri-glycerides and cholesterol,[38,45,72,88] whereas diets high in polyunsaturated fatty acid content, such as some fish or vegetable oils, tend to lower plasma lipid concentrations. Indeed diets rich in polyunsaturated fatty acids (e.g. n3 fatty acids) lower plasma cholesterol concentrations in dogs with renal failure.[17] A second effect of changes in the type of dietary fat ingested is to alter renal production of eicosanoids (prostaglandins, thromboxanes, leukotrienes and prostacyclin), which are usually derived from arachidonic acid. The production of eicosanoids by the kidney is apparently enhanced in chronic renal disease and appears to play a critical role in the genesis of glomerular hypertension.[8,88] Because fatty acids within the plasma membrane are often derived from dietary fat, prostaglandin metabolism can be affected by alterations in the type of dietary fatty acid ingested. Vegetable oils contain a large amount of linoleic acid which is referred to as an $\omega 6$ (n6) fatty acid and is readily converted to arachidonic acid, the immediate precursor of the eicosanoids. Menhaden and other oils from fish or animals feeding on plankton are rich in fatty acids referred to as $\omega 3$ fatty acids (n3). These n3 fatty acids serve as poor substrates for the generation of eicosanoids. Consequently, relatively few eicosanoids are formed when an animal eats a diet in which the fat is predominantly composed of n3 fatty acids. In addition, the eicosanoids produced from n3 fatty acids are less effective in promoting vasoconstriction and platelet aggregation. There is, therefore, a theoretical basis for the hypothesis that diets supplemented with n3 fatty acids will benefit animals with chronic renal failure.

Metabolic Acidosis

The injurious effect of enhanced ammonium generation in chronic renal failure was emphasized in a study of tubulointerstitial disease in rodents with induced renal disease.[73] The authors proposed that ammonium activated complement by the alternate pathway and they provided evidence that reduction of renal ammonium generation by dietary bicarbonate supplementation led to reduced tubulointerstitial disease and less complement deposition. A role for acidosis in spontaneous renal disease in cats has also been suggested.[33] Although a role for acidosis in the progression of renal disease in dogs has not been thoroughly evaluated, tubulointerstitial lesions are prevalent in dogs with experimental[15,36,37] and spontaneous[25,70] chronic renal disease, providing a rationale for dietary alkalinization (e.g. use of alkalinizing agents, substitution of vegetable-source for animal-source proteins and/or dietary protein restriction).

Diagnosis

The diagnosis of chronic renal failure requires the identification of a chronic (>2 weeks) onset of azotemia (or reduced GFR) and the exclusion of prerenal and postrenal causes of azotemia. Other findings that confirm chronicity include the presence of a normocytic, normochromic nonregenerative anemia and evidence of renal osteodystrophy.

Prognosis

The prognosis for chronic renal failure is dependent upon the extent of renal injury, the remaining reserve of renal compensatory processes, the extent of uremic complications and whether or not the primary renal disease can be identified and successfully treated. The presence of oliguria or anuria is a poor prognostic sign. In general, because nephrons of dogs with chronic renal failure have already undergone maximal compensatory changes, the prognosis for dogs with chronic renal failure is poorer than for those with acute renal failure.

General Therapy

It should be noted that as a result of the increased emphasis upon the progressive nature of renal disease in people and animals, therapy directed at slowing the rate of progression of renal disease has received a great deal of emphasis in recent years. This is not, however, the sole or even the most important consideration for nutritional therapy in these animals. The general goals of therapy of dogs with chronic renal failure may be summarized as efforts designed to maximize the quality and the quantity of life through:

1. maintenance of adequate nutrient and energy intake,
2. reduction of the prevalence and severity of uremia,
3. reduction of mortality from the complications of uremia and
4. arresting progressive renal injury.

Careful attention to detail is required in the medical management of dogs with chronic renal failure and clinical problems should be identified, appropriate therapy instituted and the progress of clinical problems carefully and regularly monitored (Fig. 18.1). Unfortunately, little is known about the dietary needs of dogs with renal failure. Generally, it is assumed that nutrient requirements of dogs with renal failure are similar to those of normal dogs. However, people with acute renal failure exhibit enhanced energy requirements.[69] It is likely that dogs with chronic renal failure may also have decreased requirements for some nutrients e.g. dogs with experimental renal failure have an apparent daily requirement for dietary phosphorus intake that is considerably less than the U.S. National Research Council's recommended minimum daily intake (89 mg P/kg BW sid).[74] In short, it is likely that dogs with chronic renal failure have increased requirements for some nutrients (e.g. energy) and reduced requirements for others (e.g. phosphorus).

A thorough consideration of other therapies for chronic renal failure, such as the administration of recombinant erythropoietin in anemic animals[29,40] or calcitriol in animals with complications from renal secondary hyperparathyroidism[21] is beyond the scope of this chapter.

Nutritional Therapy

Dietary Phosphorus Restriction

Restriction of dietary phosphorus intake may prevent the development of renal secondary hyperparathyroidism and protect renal function.[15,36,37] Dietary phosphorus should be restricted in proportion to the degree of reduction of GFR. Because azotemia is generally not recognized until renal mass is reduced to 25% or less in dogs, this suggests a reduction of dietary phosphorus intake to 25% of the normal, or less, in all dogs with azotemic chronic renal failure. Unfortunately, "normal" dietary phosphorus intake is difficult to define. In the absence of clear guidelines for dietary phosphorus intake, the goal should be to reduce dietary phosphorus intake until normophosphatemia ensues. The U.S. National Research Council (NRC) suggests a minimum daily requirement of 89 mg P/kg BW/day. On the basis of this recommendation, low phosphorus diets could be arbitrarily defined as 45–90 mg P/kg BW/day and very low phosphorus diets as less than 45 mg P/kg BW/day. It should be emphasized that these levels are relative to the usual or traditional (normal) amounts of phosphorus in canine diets. Other animals, such as human beings, frequently ingest diets with markedly less phosphorus. For example, the average human being ingests approximately 15 mg of dietary P/kg/day and people with renal failure being fed a phosphorus restricted diet may ingest as little as 6.5 mg P/kg/day.[10] Studies of dogs with induced renal failure indicate that although it is difficult to precisely predict the appropriate level of dietary phosphorus intake necessary to eliminate hyperphosphatemia, ingestion of diets containing the NRC minimum for phosphorus intake will induce hyperphosphatemia in all azotemic dogs.[15,36] Obviously, dietary phosphorus levels greater than this will worsen hyperphosphatemia and have been shown to enhance the decline to end-stage renal disease in dogs.[15,37]

A variety of factors besides GFR and total dietary phosphorus intake play a role in the appropriate level of dietary phosphorus intake. For example, inorganic salts of phosphorus are more readily absorbed than the relatively unavailable organic sources of phosphorus.[35] In addition, diets with high vegetable matter content will contain a large amount of phytin, which will bind phosphorus and make it less available for intestinal absorption. A high dietary calcium content will act similarly.

Phosphorus binding agents: In animals where available diets do not allow the establishment of normophosphatemia, intestinal phosphorus binders can be added. There are three basic types of phosphorus binders: salts of aluminum, magnesium or calcium. Each has distinct advantages and disadvantages. Magnesium-containing phosphorus binders pose considerable risk for the development of hypermagnesemia in animals with renal failure and are not often used because of this. The addition of 30–90 mg of AlOH/kg BW/day to meals can be used initially. Alternatively, a variety of calcium salts such as calcium carbonate, calcium acetate, calcium lactate and calcium gluconate are commonly employed in uremic people and can be readily obtained. When administered with food, the calcium serves to bind dietary phosphorus. When administered between meals, there is less phosphorus binding and the absorption of calcium will be enhanced.

Calcium carbonate has been employed in uremic human beings at dosages from 50–300 mg/kg body weight. Substantially higher doses may be required in dogs, although a reasonable starting dose is 100 mg/kg/day with meals. Calcium salts should be used with caution, as they may produce hypercalcemia in susceptible animals and accordingly, they should not be coadministered with active vitamin D metabolites. Calcium acetate may be somewhat less prone to produce hypercalcemia than the carbonate salt.[22] Some calcium salts are alkalinizing and thus have the usually desirable side-effect of reducing the extent of metabolic acidosis. The patient's acid–base status and plasma calcium and phosphorus concentrations should be carefully monitored during oral therapy with calcium salts.

Orally administered calcium can be bound (e.g. to intestinal phosphorus) or alternatively, may be absorbed. When administered between meals, the absorption of calcium will be enhanced. Although this minimizes the effectiveness of calcium as a phosphorus binder, it will allow the calcium to be absorbed and thus may suppress the secretion of parathyroid hormone, thereby reducing the extent of renal secondary hyperparathyroidism. Other means to lessen the severity of renal secondary hyperparathyroidism include dietary phosphorus restriction or the administration of calcitriol[21] or cimetidine.[56] There is conflicting evidence regarding the efficacy of cimetidine as a therapy for renal secondary hyperparathyroidism.[30] The rationale for reversing the extent of renal secondary hyperparathyroidism is based upon the hypothesis that parathyroid hormone is a uremic toxin, leading to glucose intolerance, encephalopathy and osteodystrophy, as well as a variety of other clinical abnormalities.[66–68,93] In those animals with uremic osteodystrophy or glucose intolerance, measures designed to lower the extent of renal secondary hyperparathyroidism are warranted. Hyperphosphatemic animals should be treated with dietary phosphorus restriction with the addition of intestinal phosphorus binding agents, if required to achieve normophosphatemia. The dietary calcium:phosphate ratio should be greater than 1.0 and studies of rodents with nephrocalcinosis indicate that a higher ratio may be beneficial.[51] Dietary supplementation with calcium, however, is contraindicated in hyper-

calcemic dogs and oral supplementation with calcium salts is appropriate only in hypocalcemic or normocalcemic animals.

Calcitriol: The most promising new treatment for renal secondary hyperparathyroidism is the use of the active vitamin D metabolite, calcitriol. This agent, administered orally between meals at an initial dose of 6.5 ng/kg/day, may reduce the extent of hyperparathyroidism in affected dogs. The most frequently encountered side-effect is the development of hypercalcemia. Calcitriol should not be used in hyperphosphatemic animals because it may increase serum calcium and phosphorus, enhancing soft tissue mineralization. Therapy aimed at normalizing plasma phosphorus concentration, including dietary phosphorus restriction and oral administration of calcium-free intestinal phosphorus binders, should be successfully employed prior to the institution of calcitriol therapy. Calcitriol therapy should be based upon reliable measurements of plasma parathyroid hormone (PTH) and ionized calcium concentrations. Dose adjustments based on frequent measurements of PTH are necessary. Another method to monitor the extent of renal secondary hyperparathyroidism is to measure fractional clearance (excretion) of phosphate, FE of P (%)

$$100 \times \frac{\text{urine [P]} \times \text{plasma [creatinine]}}{\text{urine [creatinine]} \times \text{plasma [P]}}.$$

This index of phosphate balance should be interpreted with caution. In addition to plasma PTH levels, FE of P is altered by dietary intake of phosphate, plasma phosphate concentration and changes in renal function (GFR). If a dog is placed on a low phosphorus diet or intestinal phosphate binders, then this clearance should decrease and the FE of P can be used in this manner as an index of owner compliance.

Dietary Protein

Dietary protein restriction can be employed in dogs with chronic renal failure for two reasons: to reduce the extent of uremia and to slow the rate of progression of renal disease. In uremic animals, the former rationale for dietary protein restriction is critical. It is important to emphasize that, at this time, there is no clear evidence that high dietary protein intake results in progression of renal disease in dogs. There is reason, however, to be concerned about some side-effects of protein intake. These include ingestion of high levels of dietary phosphorus, the generation of metabolic acidosis and the imposition of an oxidant or hypermetabolic stress on renal tubular cells. The first is best treated by selection of low phosphorus diets, which may or may not be reduced in protein content. Dietary protein and phosphorus restriction are not synonymous and, within limits, each can be separately controlled in a diet. (Unfortunately, most available dietary preparations modify these two nutrients simultaneously or preferentially lower dietary protein content.) Treatment of metabolic acidosis does not require dietary protein restriction and is best managed with dietary alkalinization or the selection of diets utilizing protein that is of vegetable source or low in sulfur-containing amino acids. The role of oxidant or hypermetabolic stress in progressive renal disease in dogs remains to be established. Unless further evidence regarding the role of protein in progressive renal disease in dogs emerges, dietary protein restriction in dogs with renal disease that are either mildly azotemic or nonazotemic cannot be based upon scientific rationale at this time.

In uremic animals, guanidinosuccinic acid and other by-products of protein metabolism are toxic, so uremic animals should be fed a low protein diet. The level of dietary protein restriction is ideally based upon knowledge of the minimum daily requirement of protein. Although this is unclear, uremic dogs should be fed a protein source that is of high biological value and given 2.0–3.5 g protein/kg BW/day. If this does not alleviate clinical signs, further restriction of dietary protein intake may be attempted, provided that a high biological value protein source is used. Because protein malnutrition leads to protein catabolism and generation of uremic toxins,

further restriction may actually be counter-productive unless the animals are supplemented with amino acids or their keto analogues.[71,97] This form of therapy has not been carefully evaluated in dogs with renal disease.

Dietary Sodium and Systemic Hypertension

Dogs with spontaneous renal disease frequently have systemic hypertension.[28,99] Elevated systemic arterial blood pressure can produce a variety of end-organ lesions, most commonly affecting the kidneys or eyes.[62,83,86] Although the presence of mild to moderate systemic hypertension may lead to progressive renal injury in rats and people, this link remains to be established in dogs. Indeed, studies in rodents suggest that local hypertension within the glomerulus is the critical determinant of progressive renal disease.[5,11,52] These studies indicate that maneuvers that control systemic arterial blood pressure without lowering GCP do not protect the kidney.[5]

In dogs with preexistent renal disease, however, the presence of systemic hypertension could be interpreted as a risk factor for progressive loss of renal function and effective management of systemic hypertension is appropriate. Newer techniques make measurement of mean arterial pressure possible in dogs.[28,39,47,53,76,99] In order to identify systemic hypertension in dogs, methods that have been carefully validated must be employed. There are several obstacles that must be overcome to allow this to be done. To directly measure pressure within the systemic arterial system requires cannulation of a systemic artery and methods that rely upon this technique are referred to as direct techniques: direct methods measure the actual systemic arterial blood pressure. Other techniques that employ the Doppler principle or oscillometry are referred to as indirect techniques because these techniques do not actually measure blood pressure but rather some correlate thereof. Consequently, each indirect method must be tested for reliability and validated for use in dogs. Some instruments that indirectly measure blood pressure have been evaluated in dogs.[28,39,47,53,76,99] At this time, these instruments have not been compared and no single unit can be recommended.

There are two additional factors that present substantial problems. First, there is an artifactual elevation of blood pressure that occurs as a result of the apprehension of the dog being evaluated. The degree of artifactual elevation may be minimized by obtaining blood pressure measurements in a calm, quiet environment or relying upon measurements obtained at home by trained pet owners. Multiple measurements should be obtained each time blood pressure is assessed and a diagnosis of systemic hypertension established only if blood pressure values consistently exceed normal values. Using direct blood pressure measurements in conscious dogs, systemic hypertension in dogs has been defined as sustained systolic pressure in excess of 180 mmHg or diastolic pressure in excess of 95 mmHg.[99] This same study identified normal values for dogs of systolic as 148 ± 16 mmHg, diastolic as 87 ± 8 mmHg and mean as 102 ± 9 mmHg. Because they measure a correlate of blood pressure, each indirect method for blood pressure measurement will yield slightly (or perhaps markedly) different normal values and care should be taken in the interpretation of blood pressure measurements obtained with these methods. In those dogs in which blood pressure is abnormal, other supportive data should be sought. In particular, a careful retinal examination or repeat measurements should be relied upon prior to establishing a diagnosis of systemic hypertension. Another problem with the measurement of blood pressure in animals is the minute-to-minute variation that occurs within an individual animal with substantial variation throughout the day.[46] Consequently, a single blood pressure measurement may not accurately reflect the true (or average) blood pressure present in that animal throughout the day.

Any efforts designed to control blood pressure must be based upon reliable measurements. As more information regarding validation of these measurement techniques in dogs becomes available, it will be possible for veterinarians to reliably measure blood pressure in dogs.

Although carefully controlled studies have yet to be performed to confirm the effectiveness of dietary sodium restriction in dogs with spontaneous systemic hypertension, experimental studies have established a relationship between the level of dietary sodium intake and the level of blood pressure within systemic arteries. In addition, a link between obesity and systemic hypertension has been clearly established in dogs.[85] Consequently, the first steps in the control of blood pressure should be dietary sodium restriction and normalization of body weight. Both of these should be accomplished gradually. In particular, obese animals that are azotemic should not be subjected to severe energy restriction, which might induce a catabolic state thereby worsening azotemia. Sequential determinations of serum concentrations of urea nitrogen and creatinine are indicated during dietary restriction of energy and/or sodium in an azotemic animal.

Experimental studies demonstrate that reduction of sodium intake can lower arterial pressure in dogs with renal insufficiency.[24,59,61,85] However, these studies have employed diets far in excess of normal dietary sodium intake for pet dogs. A recent study indicated that a low sodium diet might have an opposite effect.[48] Clearly, the effects of other nutrients, besides sodium, on blood pressure need to be carefully evaluated in dogs. Although many pharmacological agents are available for the therapy of systemic hypertension, the evidence available indicates that systemic hypertension in dogs with renal disease is salt-sensitive. Commercially available diets provide 100–300 mg Na/kg body BW/day in the dog and dietary sodium restriction to 15–50 mg Na/kg BW/day should be the first step of therapy in hypertensive animals. Although not thoroughly tested, greater reductions in dietary sodium intake that approach the U.S. NRC[74] minimum daily requirement of 11 mg Na/kg may further blood pressure reduction. However, animals with renal disease require several days to adjust to changes in dietary sodium intake[31,89] and any change in sodium intake should be accomplished slowly, over 7–14 days. A rapid reduction in sodium intake in a dog with chronic renal failure may result in extracellular fluid volume contraction and hypotension.

If dietary sodium restriction is ineffective, the second line of therapy in animals documented to have systemic hypertension is the use of pharmacological agents. Several classes of compounds have been utilized in human beings with systemic hypertension. In dogs, very little information is available on the use of these agents. One study compared the effects of a calcium channel antagonist with that of an angiotensin-converting enzyme inhibitor on systemic hypertension, renal hemodynamics and progression of renal disease in diabetic Beagles.[18] The angiotensin-converting enzyme inhibitor was more effective than the calcium antagonist in lowering the extent of glomerular hypertension and proteinuria in diabetic Beagles. On the basis of this study, there is reason to believe that angiotensin-converting enzyme inhibitors are superior for single agent therapy in animals with renal disease. However, other classes of antihypertensive agents, such as antiadrenergic agents (e.g. prazosin or propranolol) or diuretics (e.g. thiazides) have also been recommended for the treatment of dogs with hypertension and renal disease.[86]

Dietary Lipid

As noted above, modification of dietary lipid intake can lower plasma lipid concentrations and alter renal eicosanoid metabolism. There is reason to believe that both of these effects might benefit dogs with chronic renal failure by slowing the rate of progressive renal injury. At least two approaches to lowering plasma cholesterol concentration have been attempted in humans and rats: lowering dietary intake of saturated fat and cholesterol and pharmacological agents aimed at inhibiting endogenous production of cholesterol. Little is known of the effective-

ness of either in dogs with renal disease. It is plausible that a diet that is high in polyunsaturated fatty acids (fish or vegetable oil as the principal source of dietary fat) may reduce plasma cholesterol in dogs with induced renal disease. Alternatively, the selective addition of a polyunsaturated fat such as fish or vegetable oil to the diet of affected dogs may have untoward effects, such as those observed in some studies of rodents with experimental renal disease.[8,88]

Trials evaluating the consequences of therapies aimed at modifying renal eicosanoid metabolism have not been reported in dogs. There are at least two potential approaches. First, the use of nonsteroidal antiinflammatory agents in animals with renal disease will block the production of eicosanoids by cyclooxygenase. This may lower the extent of glomerular hypertension, thereby also lowering GFR. This has the obvious potential side-effect of interfering with renal autoregulation and gastrointestinal complications from nonsteroidal antiinflammatory drug toxicity. The second method for modifying renal eicosanoid metabolism is to feed a diet rich in n3 fatty acids (e.g. fish oil) to lower renal eicosanoid production and to shift the production from the usual arachidonic acid metabolites to a different series of eicosanoids. (Alternatively, dietary supplementation with plant oils will provide an abundant source of linoleic acid, the precursor of arachidonic acid and thus might actually enhance renal production of eicosanoids.)

Although hyperlipidemia is present in many dogs with renal dysfunction, adverse effects of hypercholesterolemia in these animals remain to be established. Consequently, the use of aggressive measures aimed at lowering plasma cholesterol concentrations or modifying renal eicosanoid metabolism in affected dogs seems unwarranted at this time.

Metabolic Acidosis

Animals with renal failure should have their acid–base status assessed regularly. Unless they have concurrent respiratory disease, measurement of plasma bicarbonate or total carbon dioxide concentration is adequate. An effort should be made to prevent the development of metabolic acidosis because it may be associated with enhanced renal ammoniagenesis and progression of renal disease. Although several factors contribute to the genesis of acidosis in these patients, the defect present is an inability of the kidney to excrete the daily acid load. Some dietary proteins, particularly those of animal origin or those high in sulfur-containing amino acids, are acidifying. In contrast, vegetable proteins do not generally have this effect. As noted above, high dietary phosphorus intake or inadequate circulating levels of activated vitamin D may contribute to the genesis of renal secondary hyperparathyroidism, which suppresses renal bicarbonate reabsorption.[6] Dietary phosphorus restriction should be employed for reasons already cited; however, the use of this maneuver alone should not be expected to produce much change in the acid–base status of the patient.[15] The management of acidosis in dogs with renal failure can be accomplished by the addition of alkalinizing agents to the diet, dosed to effect. Because sodium bicarbonate will increase intake of sodium and thus may potentially elevate systemic arterial pressure, alkalinizing salts of potassium (e.g. 35 mg of potassium citrate/kg BW tid–bid) or calcium (e.g. 100 mg/kg/day of calcium carbonate or calcium acetate) should be employed. The dose of these agents should be titrated to effect, based upon serial measurements of plasma total CO_2 or bicarbonate.

Dietary Potassium

Dogs with polyuric renal failure frequently exhibit hypokalemia. Consequently, the potassium content of diets for dogs with renal failure should exceed the U.S. NRC recommended minimum of 89 mg potassium/kg BW/day,[74] unless hyperkalemia is documented. In hypokalemic animals, additional supplementation of diets with potassium (1–6 mEq/kg BW/day) can be utilized.

In hyperkalemic animals, appropriate medical therapy should be instituted, particularly in those with bradycardia or electrocardiographic abnormalities. Emergency measures in markedly hyperkalemic dogs (serum potassium in excess of 8.0 mEq/l) may include the combined administration of insulin and glucose (0.5 U regular insulin/kg IV bolus followed by 2.0 g dextrose/kg IV bolus; alternatively, a solution of 10 U regular insulin/100 ml 20% dextrose may be infused IV at a rate of 50 ml/kg/hr). The parenteral administration of bicarbonate (1–3 mEq/kg IV over 30 min, given only if the dog is acidemic) is an alternative therapy to shift potassium intracellularly in hyperkalemic dogs. The goal of this therapy is a serum potassium below 7.5 mEq/l. Although hyperkalemia is generally observed only terminally in dogs with renal failure, dogs are occasionally observed with hyperkalemia and moderate renal dysfunction (e.g. amphotericin B nephrotoxicity).[87] In these patients, it is appropriate to restrict dietary potassium content. Because bicarbonaturia will enhance distal tubular secretion of potassium, the diet may be supplemented with an alkalinizing agent, administered to effect, with a goal of achieving a normal potassium without causing alkalemia. Alternatively, oral potassium exchange resins, such as sodium polystyrene sulfonate (0.5 g/kg BW tid–bid PO) can be used to alleviate mild to moderate hyperkalemia that is not immediately life-threatening.[101] However, this agent is an ion exchange resin that supplies sodium in exchange for potassium and thus may exacerbate systemic hypertension. In addition, it may interfere with alkalinizing agents or cause constipation and therefore this agent should be used with caution.

Dietary Energy Intake

Malnutrition is a common problem in dogs with renal failure. The catabolism of body protein stores for energy results in the generation of uremic toxins. Inadequate intake of protein and/or energy following a renal insult will reduce the renal compensatory response and may prevent the regeneration of adequate renal tissue to support life. Adequate dietary intake is particularly important in dogs in which the functional reserve of other organs may be limited by concurrent or related disease processes. Malnutrition may be further compounded by vomiting, malabsorption, maldigestion and disorders of protein and carbohydrate metabolism. Further, dogs with renal disease may have increased energy needs.

With the advent of a variety of special diets, the recent emphasis in veterinary medicine has been to minimize the intake of nutrients such as protein, sodium and phosphorus. However, the maintenance of adequate energy intake should take precedence. The non-uremic dog should initially ingest approximately 75 kcal/kg BW/day. Adjustments in energy intake must depend upon reliable, serial measurements of BW. In the edematous animal, changes in body fluid content will dramatically affect BW and an assessment of the adequacy of calorie intake should be accomplished by a careful physical evaluation of lean body mass.

Calorie supplements can be added to the diet when weight is not maintained. For uremic animals, low protein sources of fat include vegetable oils (9 kcal/g), margarine (7 kcal/g) and cream (4 kcal/ml). Carbohydrate sources of energy, such as honey, jelly or sugar contain about 3.5 kcal/ml.

As noted above, there is currently no clear rationale for restricting dietary protein intake in nonazotemic animals. If specially formulated low protein diets are employed in affected animals, then diets with high biological value protein should be used. Although uremic human beings are frequently placed on a low protein/low phosphorus diet, it is important to emphasize that the degree of restriction (≤0.6 g protein/kg/day)[10] is markedly greater than that commonly utilized in veterinary medicine (2.0–3.0 g protein/kg/day) and would probably not be tolerated by dogs.

Nephrotic Syndrome

Patients with severe proteinuria may develop the hallmarks of the nephrotic syndrome: hypo-albuminemia, edema and hyperlipidemia. Dietary sodium restriction should be employed as it may lessen the extent of edema and hypoalbuminemia. The rationale for increasing dietary protein intake to compensate for urinary protein loss in nephrotic animals has been challenged. In humans and rats with marked proteinuria, the degree of proteinuria seems to parallel dietary protein intake.[57,75] Consequently, gradual reduction of dietary protein intake is recommended as it may reduce the extent of the proteinuria. As dietary protein intake is modified, these patients should be carefully monitored by sequential measurements of the urinary protein:creatinine ratio (or 24-hr urinary protein excretion) and plasma albumin concentration. If dietary protein restriction results in a persistent worsening of hypoalbuminemia it is warranted to gradually increase dietary protein content. In uremic, proteinuric patients, protein restriction will be required. These animals may develop a deficiency of antithrombin III[81,82,98] that, coupled with the presence of hypoalbuminemia,[98] predisposes them to the development of thromboses. Appropriate therapy in dogs with antithrombin II deficiency may, therefore, include antithrombotic agents.[82]

References

1. Abel, R. M., Abbott, W. M. and Fischer, J. E. (1973) Intravenous essential L-amino acids and hypertonic destrose in patients with acute renal failure. Effects on serum potassium, phosphate, and magnesium. *American Journal of Surgery*, **123**, 632–638.
2. Abel, R. M., Beck, C. H. and Abbott, W. M. *et al.* (1973) Improved survival from acute renal failure after treatment with intravenous essential amino acids and glucose. *New England Journal of Medicine*, **288**, 695–699.
3. Alfrey, A. C., Karlinsky, M. and Haut, L. (1980) Protective effect of phosphate restriction on renal function. In *Phosphate and Minerals in Health and Disease*. Eds. S. G. Massry and H. Jahn. pp. 209–218. New York: Plenum Press.
4. Allen, T. A., Jaenke, R. S. and Fettman, M. J. (1987) A technique for estimating progression of chronic renal failure in the dog. *Journal of the*

American Veterinary Medical Association, **190**, 866–868.

5. Anderson, S., Rennke, H. G. and Brenner, B. M. (1986) Therapeutic advantage of converting enzyme inhibitors in arresting progressive renal disease associated with systemic hypertension in the rat. *Journal of Clinical Investigation,* **77**, 1993–2000.

6. Arruda, J. A., Nascimento, L., Westenfelder, C. and Kurtzman, N. A. (1977) Effect of parathyroid hormone in urinary acidification. *American Journal of Physiology,* **232**, F429–F439.

7. Bacia, J. J., Spyridakis, L. K., Barsanti, J. A. and Brown, S., (1986) Ibuprofen toxicosis in a dog. *Journal of the American Veterinary Medical Association,* **189**, 918–919.

8. Barcelli, U. O., Weiss, M. and Pollak, V. E. (1982) Effects of dietary prostaglandin precursor on the progression of experimentally induced chronic renal failure. *Journal of Laboratory and Clinical Medicine,* **100**, 786–797.

9. Barsanti, J. A. and Finco, D. R. (1985) Dietary management of chronic renal failure in dogs. *Journal of the American Animal Hospital Association,* **21**, 371–376.

10. Barsotti, G. *et al.* (1981) Effects on renal function of a low-nitrogen diet supplemented with essential amino acids and ketoanalogues and of hemodialysis and free protein supply in patients with chronic renal failure. *Nephron,* **27**, 113–117.

11. Brenner, B. M., Meyer, T. W. and Hostetter, T. H. (1982) Dietary protein intake and the progressive nature of renal disease: the role of hemodynamically mediated glomerular injury in the pathogenesis of progressive glomerular sclerosis in aging, renal ablation and intrinsic renal disease. *New England Journal of Medicine,* **307**, 652–659.

12. Brown, S. A., Barsanti, J. A. and Crowell, W. A. (1985) Gentamicin-associated acute renal failure in the dog. *Journal American Veterinary Medical Association,* **186**, 686–690.

13. Brown, S. A., Finco, D. R., Crowell, W. A., Choat, D. C. and Navar, L. G. (1990) Single-nephron adaptations to partial renal ablation in the dog. *American Journal of Physiology,* **258**, F495–F503.

14. Brown, S. A. and Finco, D. R. (1990) The chronic course of renal function following 15/16 nephrectomy in dogs. *Proceedings of the Eighth Annual Veterinary Medical Forum.* p. 1127. Washington DC: American College of Veterinary Internal Medicine.

15. Brown, S. A., Crowell, W. A., Barsanti, J. A., White, J. V. and Finco, D. R. (1991) Beneficial effects of dietary mineral restriction in dogs with marked reduction of functional renal mass. *Journal of the American Society of Nephrologists,* **1**, 1169–1179.

16. Brown, S. A., Finco, D. R. and Crowell, W. A., *et al.* (1991) Dietary protein intake and the adaptations to partial nephrectomy in dogs. *Journal of Nutrition,* **121**, S125–S127.

17. Brown, S. A. (1993) Unpublished observations.

18. Brown, S. A., Walton, C. L., Crawford, P. and Bakris, G. (1993) Long-term effects of anti-hypertensive regimens on renal hemodynamics and proteinuria in diabetic dogs. *Kidney International* **143**, 1210–1218.

19. Bourgoignie, J. J., Gavellas, G., Martinez *et al.* (1987) Glomerular function and morphology after renal mass reduction in dogs. *Laboratory and Clinical Medicine,* **109**, 380–388.

20. Bovee, K. C., Kronfeld, D. S. and Ramberg, C., *et al.* (1979) Long-term measurement of renal function in partially nephrectomized dogs fed 56, 27, or 19% protein. *Investigative Urology,* **16**, 378–384.

21. Chew, D. J. and Nagode, L. A. (1992) Calcitriol in the treatment of chronic renal failure. In *Current Veterinary Therapy XI.* Ed. R. W. Kirk. pp. 857–860. Philadelphia: W.B. Saunders.

22. Chew, D. J., DiBartola, S. P., Nagode, L. A. and Starkey, R. J. (1992) Phosphorus restriction in the treatment of chronic renal failure. In *Current Veterinary Therapy XI.* Ed. R. W. Kirk. pp. 853–857. Philadelphia: W.B. Saunders.

23. Clapp, M. J. L., Wade, J. D. and Samuels, D. M. (1982) Control of nephrocalcinosis by manipulating the calcium:phosphorus ratio in commercial rodent diets. *Laboratory Animals,* **16**, 130–132.

24. Coleman, T. G. and Guyton, A. C. (1969) Hypertension caused by salt loading. *Circulation Research,* **25**, 153–160.

25. Cooper, J. E. (1976) An unexplained high incidence of calcified lesions in dog kidneys. *Veterinary Record,* **98**, 220.

26. Coulter, D. B. and Keith, J. C. (1984) Blood pressure obtained by indirect measurement in conscious dogs. *Journal of the American Veterinary Medical Association,* **184**, 1375–1378.

27. Cousins, F. B. and Geary, C. P. M. (1966) A sex-determined renal calcification in rats. *Nature,* **211**, 980–981.

28. Cowgill, L. D. and Kallett, A. J. (1983) Recognition and management of hypertension in the dog. In *Current Veterinary Therapy Vol. VIII.* Ed. R. W. Kirk. pp. 1025–1028. Philadelphia: W.B. Saunders.

29. Cowgill, L. D. (1992) Application of recombinant human erythropoietin in dogs and cats. In *Current Veterinary Therapy XI.* Eds. R. W. Kirk and J. D. Bonagura. pp. 484–487. Philadelphia: W.B. Saunders.

30. Cunningham, J., Segre, G. V., Slatopolsky, E.

and Avioli, L. V. (1984) Effect of histamine H2–receptor blockade on parathyroid status in normal and uraemic man. *Nephron, 38*, 17–21.

31. Danovitch, G. M., Bourgoignie, J. J. and Briocker, N. S. (1977) Reversibility of the salt losing tendency of chronic renal failure. *New England Journal of Medicine, 269*, 14–19.

32. Deen, W. M., Maddox, D. A. and Robertson, C. R. *et al.* (1974) Dynamics of glomerular ultrafiltration in the rat. VII. Response to reduced renal mass. *American Journal of Physiology, 227*, 556–562.

33. DiBartola, S. P., Buffington, C. A., Chew, D. J., McLoughlin, M. A. and Sparks, R. A. (1993) Development of chronic renal disease in cats fed a commercial diet. *Journal of the American Veterinary Medical Association, 202*, 744–751.

34. Feinstein, E. I. *et al.* (1981) Clinical and metabolic responses to parenteral nutrition in acute renal failure. *Medicine, 60*, 124–137.

35. Finco, D. R., Barsanti, J. A. and Brown, S. A. (1989) Influence of dietary source of phosphorus and urinary excretion of phosphorus and other minerals by male cats. *American Journal of Veterinary Research, 50*, 263–266.

36. Finco, D. R., Brown, S. A. and Crowell, W. A. *et al.* (1992) Effects of phosphorus/calcium-restrictive 32% protein diets in dogs with chronic renal failure. *American Journal of Veterinary Research, 53*, 157–163.

37. Finco, D. R., Brown, S. A., Crowell, W. A., Duncan, R. J., Barsanti, J. A. and Bennett, S. E. (1992) Effects of dietary phosphorus and protein in dogs with chronic renal failure. *American Journal of Veterinary Research, 53*, 2264–2271.

38. French, S. W., Yamanaka, W. and Ostred, R. (1967) Dietary induced glomerulosclerosis in the guinea pig. *Archives of Pathology, 83*, 204–210.

39. Geddes, L. A., Combs, W. and Denton, W. *et al.* (1980) Indirect mean arterial pressure in the anesthetized dog. *American Journal of Physiology, 238*, H664–H666.

40. Giger, U. (1991) Serum erythropoietin concentrations in polycythemic and anemic dogs. *Proceedings of the Ninth Annual Scientific Meeting*, pp. 143–145. New Orleans, LA: American College of Veterinary Internal Medicine.

41. Gimenez, L. F., Walker, W. G., Tew, W. P. and Hermann, J. A. (1982) Prevention of phosphate-induced progression of uremia in rats by 3–phosphocitric acid. *Kidney International, 22*, 36–41.

42. Goedegebuure, S. A. and Hazewinkel, H. A. W. (1986) Morphological findings in young dogs chronically fed a diet containing excess calcium. *Veterinary Pathology, 23*, 594–605.

43. Goulding, A. and Malthus, R. S. (1969) Effect of dietary magnesium on the development of nephrocalcinosis in rats. *Journal of Nutrition, 97*, 353–358.

44. Grauer, G. F. and Thrall, M. A. H. (1986) Ethylene glycol poisoning. In *Current Veterinary Therapy IX*. Ed. R. W. Kirk. pp. 206–212. Philadelphia: W.B. Saunders.

45. Grone, H., Walli, A. and Grone, E. *et al.* (1989) Induction of glomerulosclerosis by dietary lipids. *Laboratory Investigation, 60*, 433–446.

46. Hall, J. (1992) Personal communication.

47. Hamlin, R. L., Kittleson, M. D. and Rice, D. *et al.* (1982) Noninvasive measurement of systemic arterial pressure in dogs by automatic sphygmomanometry. *American Journal of Veterinary Research, 43*, 1271–1273.

48. Hansen, B., DiBartola, S. P., Chew, D. J., Brownie, C. and Nagode, L. (1992) Clinical and metabolic findings in dogs with chronic renal failure fed two diets. *American Journal of Veterinary Research, 53*, 326–334.

49. Harris, D. C. H., Chan, L. and Schrier, R. W. (1988) Remnant kidney hypermetabolism and progression of chronic renal failure. *American Journal of Physiology, 23*, F267–F276.

50. Haut, L. L., Alfrey, A. C., Guggenheim, S., Buddington, B. and Shcrier, N. (1980) Renal toxicity of phosphate in rats. *Kidney International, 17*, 722–731.

51. Hoek, A. C., Lemmens, A. G., Mullinik, J. W. M. A. and Beynen, A. C. (1988) Influence of dietary calcium:phosphorus ratio on mineral excretion and nephrocalcinosis in female rats. *Journal of Nutrition, 118*, 1210–1216.

52. Hostetter, T. H., Olson, J. L. and Rennke, H. G. *et al.* (1981) Hyperfiltration in remnant nephrons: a potentially adverse response to renal ablation. *American Journal of Physiology, 241*, F85–F92.

53. Hunter, J. S., McGrath, C. J. and Thatcher, C. D. *et al.* (1990) Adaptation of human oscillometric blood pressure monitors for use in dogs. *American Journal of Veterinary Research, 51*, 1439–1442.

54. Ibels, L. S., Alfrey, A. C., Haut, L. and Huffer, W. E. (1978) Preservation of renal function in experimental renal disease by dietary restriction of phosphate. *New England Journal of Medicine, 128*, 122–126.

55. International symposium on renal adaptation to nephron loss. (1978) *Yale Journal of Biological Medicine, 51*, 235–430.

56. Jacob, A. L., Canterbury, J. M. and Gavellas, G. *et al.* (1981) Reversal of secondary hyperparathyroidism by cimetidine in chronically uremic dogs. *Journal of Clinical Investigation, 67*, 1753–1760.

57. Kaysen, G. A. *et al.* (1986) Effects of dietary protein intake on albumin homeostasis in nephrotic patients. *Kidney International,* **29**, 572–577.

58. Klahr, S. (1989) The modification of diet in renal disease study. *New England Journal of Medicine,* **320**, 864–866.

59. Langston, J. B., Guyton, A. C. and Douglas, B. H. *et al.* (1963) Effect of changes in salt intake on arterial pressure and renal function in partially nephrectomized dogs. *Circulation Research,* **12**, 508–513.

60. Leonard, C. D., Luke, R. G. and Siegel, R. R. (1975) Parenteral essential amino acids in acute renal failure. *Urology,* **6**, 154–159.

61. Liard, J. F. (1981) Regional blood flows in salt loading hypertension in the dog. *American Journal of Physiology,* **240**, H361–H367.

62. Littman, M. P., Robertson, J. L. and Bovee, K. C. (1988) Spontaneous systemic hypertension in dogs: five cases (1981–1983). *Journal of the American Veterinary Medical Association,* **193**, 486–494.

63. Long, J. M. *et al.* (1974) Fat–carbohydrate interaction: Effects on nitrogen-sparing total intravenous feeding. *Surgery Forum,* **25**, 52–57.

64. MacDougall, D. F., Cook, T., Steward, A. P. and Cattell, V. (1986) Canine chronic renal disease: prevalence and types of glomerulonephritis in the dog. *Kidney International,* **29**, 1144–1151.

65. MacKay, E. M. and Oliver, J. (1935) Renal damage following the ingestion of a diet containing an excess of inorganic phosphate. *Journal of Experimental Medicine,* **61**, 319–332.

66. Massry, S. G. (1987) Parathyroid hormone: a uremic toxin. *Advances in Experimental and Medical Biology,* **223**, 1.

67. Massry, S. G. (1987) In *Uremic Toxins.* Eds. S. Ringoir, R. Vanholder and S. G. Massry. pp. 1–19. New York: Plenum.

68. Massry, S. G. (1989) Pathogenesis of uremic toxicity: parathyroid hormone as a uremic toxin. In *Textbook of Nephrology.* Eds. S. G. Massry and R. J. Glassock. p. 1126. Baltimore: Williams & Wilkins.

69. Mault, J. R. *et al.* (1983) Starvation: a major contributor to mortality in acute renal failure. *Transactions of the American Academy of Artificial and Internal Organs,* **29**, 39–49.

70. Maxie, M. G. (1985) The urinary system. In *Pathology of Domestic Animals.* Eds. K. V. F. Jubb, P. C. Kennedy and Palmer. pp. 343–411. New York: Academic Press.

71. Mitch, W. E., Abras, E. and Walser, M. (1982) Long-term effects of a new ketoacid-amino acid supplement in patients with chronic renal failure. *Kidney International,* **22**, 48–53.

72. Moorhead, J. F., Chan, M. K. and Varghese, Z. (1986) The role of abnormalities of lipid metabolism in the progression of renal disease. In *The Progressive Nature of Renal Disease.* Ed. W. E. Mitch. pp. 133–148. New York: Churchill Livingstone.

73. Nath, K. A., Hostetter, M. K. and Hostetter, T. H. (1985) Pathophysiology of chronic tubulo-interstitial disease in rats. *Journal of Clinical Investigation,* **76**, 667–675.

74. National Research Council. (1985) Nutrient requirements of dogs. Washington, DC: National Academy Press.

75. Neugarten, J., Feiner, H. D. and Schacht, R. G. *et al.* (1983) Amelioration of experimental glomerulonephritis by dietary protein restriction. *Kidney International,* **24**, 595–601.

76. Pettersen, J. C., Linartz, R. R. and Hamlin, R. L. *et al.* (1988) Noninvasive measurement of systemic arterial blood pressure in the conscious Beagle dog. *Fundamental and Applied Toxicology,* **10**, 89–97.

77. Polzin, D. J., Osborne, C. A. and Hayden, D. W. *et al.* (1983) Influence of reduced protein diets on morbidity, mortality and renal function in dogs with induced chronic renal failure. *American Journal of Veterinary Research,* **45**, 506–517.

78. Polzin, D. J., Osborne, C. A., Adams, L. G. and Lulich, J. F. Medical management of feline chronic renal failure. In *Current Veterinary Therapy XI.* Ed. R. W. Kirk. pp. 848–853. Philadelphia: W. B. Saunders.

79. Polzin, D. J., Leininger, J. R. and Osborne, C. A. *et al.* (1988) Development of renal lesions in dogs after 11/12 reduction of renal mass. *Laboratory Investigation,* **58**, 172–183.

80. Purkeson, M. L., Hoffsten, P. E. and Klahr, S. (1976) Pathogenesis of the glomerulopathy associated with renal infarction in rats. *Kidney International,* **9**, 407–417.

81. Rasedee, A., Feldman, B. F. and Washabau, R. (1986) Naturally occurring canine nephrotic syndrome is a potentially hypercoagulable state. *Acta Veterinaria Scandinavica,* **27**, 369–377.

82. Relford, R. L. and Green, R. A. (1992) Coagulation disorders in glomerular diseases. In *Current Veterinary Therapy XI.* Eds. R. W. Kirk and J. D. Bonagura. pp. 827–829. Philadelphia: W.B. Saunders.

83. Remillard, R. L., Ross, J. N. and Eddy, J. B. (1991) Variance of indirect blood pressure measurements and prevalence of hypertension in dogs. *American Journal of Veterinary Research,* **52**, 561–565.

84. Robertson, J. L., Goldschmidt, M. S. and Kronfeld, D. S. *et al.* (1986) Long-term responses to high dietary protein intake in dogs with 75% nephrectomy. *Kidney International,* **29**, 511–519.

85. Rocchini, A. P., Moorehead, C. P., Wentz, E. and Bondie, D. (1989) Pathogenesis of weight related changes in blood pressure in dogs. *Hypertension*, **13**, 922–928.

86. Ross, L. A. (1989) Hypertensive diseases. In *Textbook of Veterinary Internal Medicine*. Ed. S. J. Ettinger. pp. 2047–2056. Philadelphia: W. B. Saunders

87. Rubin, S. I. (1986) Nephrotoxicity of amphotericin B. In *Current Veterinary Therapy IX*. Ed. R. W. Kirk. pp. 1142–1146. Philadelphia: W. B. Saunders.

88. Scharschmidt, L. A., Gibbons, N. B., McGarry, L. *et al.* (1987) Effects of dietary fish oil on renal insufficiency in rats with subtotal nephrectomy. *Kidney International*, **32**, 700–709.

89. Schmidt, R. W., Bourgoignie, J. J. and Bricker, N. S. (1974) On the adaptation in sodium excretion in chronic uremia: the effects of proportional reduction' of sodium intake. *Journal of Clinical Investigation*, **53**, 1736–1741.

90. Shigematsu, T., Caverzasio, J. and Bonjour, J. (1993) Parathyroid removal prevents the progression of chronic renal failure induced by high protein diet. *Kidney International*, **44**, 173–181.

91. Slatopolsky, E. *et al.* (1971) On the pathogenesis of hyperparathyroidism in chronic experimental renal insufficiency in the dog. *Journal of Clinical Investigation*, **50**, 492–498.

92. Slatopolsky, E. (1972) On the prevention of secondary hyperparathyroidism in experimental chronic renal disease using'proportional reduction' of dietary phosphorus intake. *Kidney International*, **2**, 147–151.

93. Slatopolsky, E. and Bricker, N. S. (1973) The role of phosphorus restriction in the prevention of secondary hyperparathyroidism in chronic renal disease. *Kidney International*, **4**, 141–145.

94. Toback, F. G. (1977) Amino acid enhancement of renal regeneration after acute tubular necrosis. *Kidney International*, **12**, 193–198.

95. Toback, F. G. *et al.* (1983) Amino acid administration enhances renal protein metabolism after acute tubular necrosis. *Nephron*, **33**, 238–243.

96. VanCamp, I., Ritskes–Hoitinga, J. and Lemmesn, A. G. *et al.* (1990) Diet-induced nephrocalcinosis and urinary excretion of albumin in female rats. *Laboratory Animal*, **24**, 137–141.

97. Walser, M. (1986) Ketoacid therapy and the progression of renal disease. In *The Progressive Nature of Renal Disease*. Ed. W. E. Mitch. pp. 231–244. New York: Churchill Livingstone

98. Walter, E., Deppermann, D., Andrassy, K. *et al.* (1981) Platelet hyperaggregability as a consequence of the nephrotic syndrome. *Thrombosis Research*, **23**, 473–479.

99. Weiser, M. G., Spangler, W. L. and Griggle, D. H. (1977) Blood pressure measurements in the dog. *Journal of the American Veterinary Medical Association*, **171**, 364–368.

100. White, J. V., Finco, D. R., Crowell, W. A., Brown, S. A. and Hirakawa, D. A. (1991) Effect of dietary protein on functional, morphologic and histologic changes of the kidney during compensatory hypertrophy. *American Journal of Veterinary Research*, **52**, 1357–1365.

101. Willard, M. D. (1986) Treatment of hyperkalemia. In *Current Veterinary Therapy IX*. Ed. R. W. Kirk. pp. 94–101. Philadelphia: W. B. Saunders

102. Wilmore, D. W. and Dudrick, S. J. (1969) Treatment of acute renal failure with intravenous essential L-amino acids. *Archives of Surgery*, **99**, 669–673.

103. Woolfson, A. M. M., Heatley, R. V. and Allison, S. P. (1979) Insulin to inhibit protein catabolism after injury. *New England Journal of Medicine*, **300**, 14–17.

104. Yoshida, Y., Fogo, A. and Ichikawa, I. (1989) Glomerular hemodynamic changes vs. hypertrophy in experimental glomerular sclerosis. *Kidney International*, **35**, 654–660.

CHAPTER 19

Canine Lower Urinary Tract Disease

ASTRID E. HOPPE

Introduction

Clinical signs such as hematuria and dysuria are the most common reasons for the clinician to start an investigation of the lower urinary tract in the dog. Careful evaluation of the history and a physical examination are essential prerequisites for diagnostic accuracy. Laboratory, radiographic, ultrasonographic and biopsy procedures should be selected on the basis of a tentative diagnosis formulated from evaluation of the history and physical examination, with the purpose of confirming or eliminating diagnostic probabilities. In most instances specific diagnoses can be established by the use of reasoning and logic based on objective findings and an effective treatment, prognosis and prevention can be formulated.

Diseases of the Ureters

When the ureters are affected by disease processes the primary abnormality is often present in the urinary bladder or the kidneys and the signs of ureteral disease are often over-shadowed by signs related to the underlying abnormality.

Obstructive Ureteropathy

The ureters may be occluded by calculi, neoplasms and blood clots. The significance of the obstruction is usually the effect that it has upon the kidneys (hydronephrosis). Bilateral obstruction usually occurs as a result of disease processes at the trigone of the urinary bladder and may create a uremic crisis. A good prognosis is dependent on successful treatment of the primary cause and on recovery of sufficient renal function.

Vesicoureteral Reflux

Vesicoureteral reflux is regurgitation of urine from the bladder into the ureters. The ureters enter the bladder wall through the ureterovesical valves, which protect the kidneys from infected urine in the bladder. Primary vesicoureteral reflux (i.e. without urinary tract infection) has been reported in up to 50% of otherwise normal dogs less than 6 months-of-age.[18] It occurs more frequently in females than males and is more often bilateral than unilateral. It usually disappears as the animals get older and occurs in about 10% of adult dogs.[18]

Secondary vesicoureteral reflux may occur as a result of inflammation of the vesicoureteral junction and ureters, surgical damage to the trigone, ectopic ureters and neurogenic disease of the bladder. Reflux of infected urine to the pelvis and medulla may result in acute pyelitis or pyelonephritis. Manual compression of the urinary bladder to induce micturition may also induce vesicoureteral reflux[28] and should be avoided in patients suspected of having urinary tract infection.

Diagnosis is based on observing the reflux of contrast medium up the ureters by retrograde urethrocystography and the treatment of secondary cases involves removal of the underlying cause.

Ectopic Ureters

The most common cause of nonneurogenic urinary incontinence in young female dogs is an ectopic ureter[35,63,64] and affected dogs commonly have a history of persistent incontinence since birth or weaning. An ectopic ureter will generally empty into the urethra, the uterus or the vagina. Bilateral involvement occurs in approximately 25% of cases.[35,63]

Although ectopic ureters do occur in male dogs, the prevalence is approximately 20 times higher in females.[42,55,63] On the basis of epidemiological studies, a genetic predisposition in some breeds has been suspected (Siberian Husky, West Highland White Terrier, Fox Terrier and miniature and toy Poodles) and familial occurrence in Siberian Huskies and Labrador Retrievers has been found.[35,39,63]

Ectopic ureters are often associated with additional acquired and congenital anomalies of the urinary system. For instance, hypoplasia of the kidney, urinary bladder and urethra has been observed. Megaureter and hydronephrosis without evidence of obstruction of the urine outflow often involves the affected ureter and kidney. Because there is no functional sphincter of an ectopic ureter, reflux predisposes the associated kidney to ascending bacterial infections, and pyelonephritis in combination with an ectopic ureter may be observed.

The diagnosis may be established by the combination of a history of persistent urinary incontinence in a dog since birth, intravenous urography and cystoscopy. There is no effective medical treatment for patients with urinary incontinence due to an ectopic ureter and surgical repair is the treatment of choice.

Diseases of the Bladder and the Urethra

Urolithiasis

Increasing numbers of dogs have been found to have clinically significant uroliths. No precise epidemiological data are available, but the prevalence estimated by Hesse and Bruhl[29] was thought to be 1–3% and the incidence in the same investigation was 0.3–0.8%. Brown et al.[12] estimated that urolithiasis occurred in 0.4–2.8% of dogs. Despite improved methods of urolith removal and specific methods of after-care, mortality rates for dogs with urolithiasis have been quoted as being 15–20%.[15]

The incidence and composition of uroliths may be influenced by a variety of factors including: species, breed, sex, age, diet, anatomical abnormalities, urinary tract infection, medication and urine pH. Uroliths may be found throughout the urinary pathway and may occur in more than one site. However, the majority of uroliths found in dogs are in the lower urinary tract. Uroliths in the ureters are uncommon in dogs and usually originate in the renal pelvis. Urethral uroliths originate from the urinary bladder.

Urolith Diagnosis

Uroliths are usually suspected on the basis of typical findings obtained by history and physical examination. Urinalysis, quantitative urine culture and radiography are often required to confirm urolithiasis and to determine if uroliths are associated with predisposing disorders of the urinary tract. The four most common mineral types of uroliths found in dogs are magnesium ammonium phosphate (struvite), oxalate, cystine and ammonium

---

FACTORS THAT MAY AID ESTIMATION OF MINERAL COMPOSITION OF CANINE UROLITHS

- Breed and sex
- Quantitative analysis of uroliths from earlier urolith operations, or uroliths passed during micturition
- Radiographic density and physical characteristics of uroliths

Degree of radiopacity:	(a) Struvite	++ to ++++
	(b) Oxalate	++++
	(c) Cystine	+ to ++
	(d) Calcium phosphate	++++
	(e) Ammonium urate	0 to ++
	(f) Silica	++ to ++++

- Urine pH:

  (a) Struvite and calcium apatite uroliths — usually alkaline
  (b) Ammonium urate uroliths — acid or neutral
  (c) Cystine uroliths — acid
  (d) Calcium oxalate uroliths — variable
  (e) Silica uroliths — acid or neutral

- Identification of crystals in the urine sediment. However, only cystine crystals are pathognomonic for a disease (cystinuria) usually leading to urolith formation
- Type of bacteria, if any, isolated from urine:

  (a) Urease-producing bacteria (staphylococci and *Proteus* spp.) are typically associated with struvite uroliths
  (b) Urinary tract infection is often absent in patients with calcium oxalate, cystine, ammonium urate and silica uroliths

- Hypercalcemia may be associated with calcium-containing uroliths
- The cyanide–nitroprusside test should be performed if cystine uroliths are suspected[34]

FIG. 19.1 Factors that may aid estimation of the mineral composition of canine uroliths.

urate. Less common types of uroliths found are calcium phosphate, silica, carbonate, xanthine, drugs and drug metabolites.

A definitive diagnosis of the type of the urolith is dependent on analysis of the mineral composition of calculi. Although simple qualitative chemical analyses are commonly used, they should be avoided because both false-positive and false-negative results are common. Quantitative analyses performed by qualified laboratories are recommended.[67] Effective medical treatment of urolithiasis is dependent on a good estimation of urolith composition (Fig. 19.1).

Struvite Urolithiasis

Etiopathogenesis

The most common type of mineral found in uroliths of dogs is magnesium ammonium phosphate or struvite. Pure struvite uroliths are uncommon because most struvite calculi contain a small quantity of calcium phosphate and occasionally ammonium urate. Oversaturation of urine with magnesium ammonium phosphate is a prerequisite for struvite urolith formation, but several factors, including urinary tract infection, alkaline urine, diet and genetic predisposition may influence urolith formation. Of these, urinary tract infection appears to be the most significant in dogs.

Clinical and experimental studies of dogs have demonstrated a close relationship between formation of struvite uroliths and urinary tract infection with urease-producing bacteria.[16,26,41,66] The enzyme urease hydrolyzes urea to form ammonia and carbon dioxide that through a series of reactions results in a progressively alkaline environment (Fig. 19.2). Hyperammonuria, hyper-

$$NH_2 - CO - NH_2 + H_2O \underset{\text{urease}}{\rightleftharpoons} 2NH_3 + CO_2$$

Further hydrolysis yields:

$$NH_3 + H_2O \rightleftharpoons NH_4^+ + OH^-$$

$$CO_2 + H_2O \rightleftharpoons H_2CO_3 \rightleftharpoons H^+ + HCO_3^-$$

$$HCO_3^- \rightleftharpoons H^+ + CO_3^{2-}$$

FIG. 19.2 Bacterial urease hydrolyzes urea, leading to an increased urinary concentration of ammonium and carbonate and subsequent formation of struvite and carbonate apatite.

carbonaturia, as well as alkaluria are dependent on the quantity of urea in the urine, which in turn is dependent on the quantity of dietary protein intake.

In a small percentage of dogs with struvite urolithiasis, some investigators have found both urine and the inside of uroliths to be sterile.[66] Several observations suggest that dietary or metabolic factors may be involved in the formation of sterile struvite uroliths. This has been shown in pilot studies of clinical cases revealing dogs with frequently alkaline urine, but without identifiable bacteria and no detectable urease.[48] However, the pathogenesis of sterile struvite urolith formation is still unclear.

Urolith Characteristics

Struvite uroliths are white or pale yellow in color and are found predominantly in the bladder, from where they often pass into the urethra. They may be present singly or in large numbers, have rapid growth and sometimes become rather large in size.

Clinical experience from the Swedish University of Agricultural Sciences, as well as from other investigations, has shown that struvite uroliths can form within a month following urinary tract infection with urease producing staphylococci.[41] They also have a tendency to recur following surgical removal or medical dissolution.[14] Any breed can be affected and the uroliths are more often found in female than in male dogs. Puppies usually do not form uroliths. However, infection-induced struvite uroliths have been detected in puppies as young as 2½ weeks (Hoppe, unpublished data).

Treatment and Prevention

Management of struvite urolithiasis includes: relief of obstruction to urine outflow if necessary, elimination of existing uroliths, eradication of urinary tract infection and prevention of recurrence of uroliths.

Medical dissolution: The objectives and current recommendations of medical treatment of uroliths are:

1. Increase the solubility of crystalloids in urine, which could be accomplished by administration of urine acidifiers.

2. Eradicate or control urinary tract infection. Because of the quantity of urease produced by bacterial pathogens, it may be impossible to acidify urine. Therefore, sterilization of urine with appropriate antimicrobial agents is an important objective in decreasing the concentration of struvite crystals preventing further growth of uroliths or promote their dissolution.

3. Decrease the concentration of crystalloids in urine by stimulation of thirst, thereby increasing urine volume.

4. Reduce the quantity of calculogenic crystalloids in urine. Dietary change is an example of this method.

Dietary considerations: Calculolytic diets have been formulated to reduce urine concentration of urea (which is the substrate of urease), phosphorus and magnesium (Fig. 19.3). These diets contain a reduced quantity of high-quality protein, reduced quantities of phosphorus and magnesium and are supplemented with sodium chloride to stimulate thirst. The efficacy of the diets in inducing urolith dissolution has been confirmed by controlled experimental and clinical studies in dogs.[31,49,50]

Response to therapy should be evaluated every fourth week and dissolution or growth of the uroliths should be measured radiographically (Figs. 19.4–19.6). Antibiotic therapy should be maintained until urolith dissolution occurs, which could range from 1 to 6 months.[31] The time required to induce

COMPOSITION OF TWO CALCULOLYTIC DIETS		
Nutrient (%DM)	Diet 1	Diet 2
Protein	11.6	7.6
Fat	22.0	26.2
Carbohydrates	60.3	59.3
Fiber	2.3	2.4
Ash	3.8	4.5
Calcium	0.39	0.27
Phosphorus	0.36	0.13
Magnesium	0.086	0.017
Sodium	0.73	1.2
Caloric density	508 kcal/100 g (2100 kJ/100 g)	517 kcal/100 g (2137 kJ/100 g)
All nutrients, with the exception of caloric density, are expressed as a percentage of dry matter.		

FIG. 19.3 Composition of two calculolytic diets.

dissolution of sterile struvite is usually shorter than that required for infection-induced struvite.

Because calculolytic diets are restricted in protein and supplemented with sodium chloride they should not be given to patients with heart failure, nephrotic syndrome, hypertension or to growing dogs.[1,31]

Calcium Oxalate Urolithiasis

Etiopathogenesis and Prevalence

Calcium oxalate uroliths are the most common type of uroliths in humans in the West. Several studies have suggested that this urolith is less common in dogs and represents only between 3 and 10% of the uroliths found in

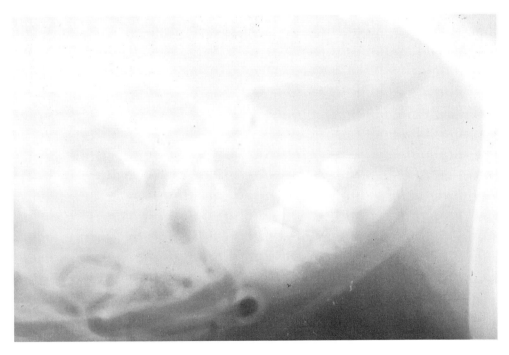

FIG. 19.4 Lateral survey radiograph of the abdomen of a 9-year-old female Cocker Spaniel illustrating multiple radiodense uroliths in the urinary bladder.

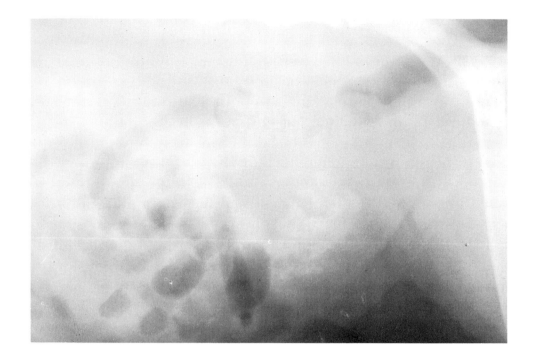

FIG. 19.5 Lateral survey abdominal radiograph of the dog shown in Fig. 19.4 obtained 8 weeks after initiation of dietary therapy to induce urolith dissolution. Progressive reduction in size and density of the uroliths is seen.

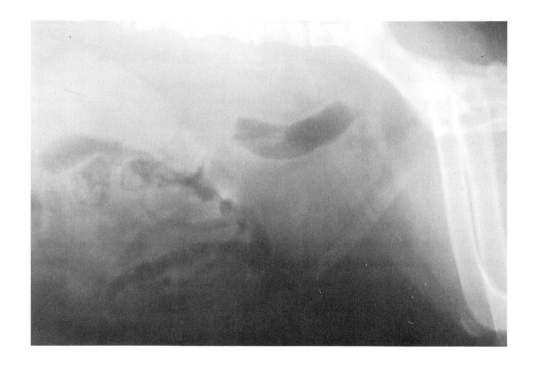

FIG. 19.6 Lateral survey abdominal radiograph of the dog shown in Figs 19.4 and 19.5 obtained 12 weeks after initiation of dietary therapy to induce urolith dissolution. Uroliths are no longer detected.

this species.[8,29,46] Recent investigations, however, have shown that uroliths composed primarily of calcium oxalate accounted for as many as 25% of canine uroliths submitted for analyses at the University of Minnesota Urolith Center.[47] This is also in agreement with the author's clinical experience in Sweden, where calcium oxalate, next to struvite, is the most common urolith found in dogs.

Factors incriminated in the etiopathogenesis include: hypercalciuria and hyperoxaluria, of which hypercalciuria probably is the most important predisposing factor in dogs as well as in humans. Most of the information concerning calcium oxalate uroliths, however, comes from extrapolation of data from humans: caution is therefore necessary when using these data in dogs.

Normocalcemic hypercalciuria is thought to be the most common finding in humans, but has also been encountered in dogs with calcium oxalate uroliths and results from either intestinal hyperabsorption of calcium or decreased renal tubular reabsorption of calcium. In contrast, hypercalcemic hypercalciuria results from an increased glomerular filtration of calcium, but this is a relatively infrequent cause of calcium-containing uroliths in dogs. Potential causes of hypercalcemia such as primary hyperparathyroidism, pseudo-hyperparathyroidism and malignant lymphoma are rare in dogs.

Urolith Characteristics and Detection

Calcium oxalate crystals form in urine as calcium oxalate monohydrate or calcium oxalate dihydrate. Urinalysis may reveal the presence or absence of these crystals. Determination of serum calcium, ionized calcium and parathyroid hormone concentrations may help to elucidate the underlying mechanism of the disease. Calcium oxalate uroliths are more commonly found in males (approximately 70%) than females and mostly in older dogs.[51] They may be detected anywhere in the urinary tract, are very radio-dense and may vary in size from fractions of a millimetre to several centimetres. Recurrence is common: up to 50% of dogs have shown recurrence within 3

years of surgical removal of uroliths in some investigations.[47]

Medical Treatment and Prevention

Attempts to dissolve calcium oxalate uroliths have been disappointing and current treatment is surgical removal followed by preventive strategies (Fig. 19.7). In some dogs, however, calcium oxalate uroliths are clinically silent and there may not be a need for surgical intervention. In general, medical treatment should have the initial goal of reducing the urine concentration of calculogenic substances.

Dietary considerations: Although there is agreement that excessive consumption of calcium and oxalate should be avoided, it is not advisable to restrict dietary calcium in calcium oxalate urolith formers unless they have absorptive hypercalciuria. Even then, only moderate restriction is recommended to prevent negative calcium balances in the body.[57] Moderate dietary restriction of sodium is recommended, for those with active urolith formation because consumption of high levels of sodium increases the renal excretion of calcium.[47,57] Reduction in dietary phosphorus is known to increase hypercalciuria due to the activation of vitamin D and the subsequent increased intestinal calcium absorption.[1,57] Therefore this dietary restriction should be avoided, although it may contribute to calcium oxalate urolith formation.

At present, a diet moderately restricted in protein, calcium, oxalate and sodium may be considered to prevent recurrence of calcium oxalate uroliths in dogs.[51]

Thiazide diuretics: Although the use of thiazides in humans has become common, the role of these substances in calcium oxalate urolith prevention is far from clear. The limited clinical experience in dogs makes this preventative measure unsafe.

Citrates: Citrates are calcium oxalate crystal inhibitors because they have the ability to form salts with calcium that are more soluble than calcium oxalate. However, there have been no studies examining the efficacy of citrate in dogs with calcium oxalate urolithiasis.

RECOMMENDATIONS FOR CALCIUM OXALATE UROLITH PREVENTION

Hypercalcemia hypercalciuria

• If primary hyperparathyroidism is confirmed, surgical removal of the parathyroid glands should be performed
• Increase urine volume (add water to food, but avoid exessive dietary sodium supplements)

Normocalcemia and high recurrence rate of calcium oxalate uroliths

• Increase urine volume (add water to food, but avoid exessive dietary sodium supplements)
• Consider change to diet that does not contain excessive calcium, oxalate or protein

FIG. 19.7 Recommendations for calcium oxalate urolith prevention.

Cystine Urolithiasis

Etiopathogenesis and Biological Behavior

Cystinuria is an inborn metabolic disease characterized by excessive urinary excretion of cystine and the dibasic amino acids lysine, arginine and ornithine. In normal dogs, circulating cystine is freely filtered at the glomerulus and most of it is actively reabsorbed in the proximal tubules. Cystinuric dogs reabsorb a much smaller proportion of the amino acid from the glomerular filtrate[7] and some may even have net cystine excretion.[6] The solubility of cystine in urine is pH dependent: it is relatively insoluble in acidic urine, but becomes more soluble in alkaline urine. The exact mechanism of cystine urolith formation is unknown and not all cystinuric dogs form uroliths.

In a recent study, the urinary excretion of 20 amino acids was investigated in 24 stone-forming cystinuric dogs and 15 normal dogs.[33] Compared with normal dogs, most cystinuric dogs showed significantly increased excretion of the dibasic amino acids, cystathionine, glutamic acid, threonine and glutamine. A most significant finding in the same study was the great variation in urinary cystine excretion. Seven out of 24 cystinuric dogs showed normal cystine excretion when compared with normal dogs. This suggests that factors other than just excretion of cystine have to be considered as causes of the formation of cystine uroliths. For example, the cystinuric dogs in this study were found to have a lower diuresis than normal dogs. Therefore, the cystinuric dogs also produced urine with a higher cystine concentration, which increases the risk of cystine urolith formation.

The precise genetic mode of inheritance of canine cystinuria is unknown. However, both sex-linked and autosomal recessive patterns have been suggested.[9,10] Surprisingly, cystine uroliths are often not recognized until maturity and the average age of detection is approximately 3–5 years.[9,34] Uroliths commonly recur within 6–12 months, because cystinuria is an inherited defect (in some dogs within 4–8 weeks), unless prophylactic therapy is initiated. Apart from the occurrence of uroliths, cystinuric dogs have no other defects and normal renal function. The disease would have remained a physiological curiosity if cystine had not been the least soluble naturally occurring amino acid and thus potentially leads to the formation of cystine uroliths.

Urolith Characteristics, Prevalence and Diagnosis

Cystine uroliths account for 3.5–27% of the uroliths in the dog, probably depending on the breeds of dog encountered in specific surveys.[9] Many dog breeds have been reported to develop cystine uroliths. In Sweden and Germany, cystine urolithiasis is particularly a problem in the Dachshund and it accounts for approximately 4 and 18.8%, respectively, of uroliths analyzed from the dog populations in those countries.[24,30] With rare exceptions, cystine uroliths have been reported only in male dogs.[13,30]

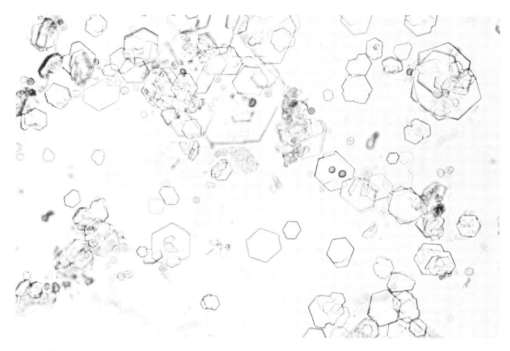

FIG. 19.8 Photomicrograph of cystine crystals formed in urine from a 6-year-old male Shih Tzu dog.

Detection of characteristic flat hexagonal cystine crystals provides strong support for a diagnosis of cystinuria (Fig. 19.8). However, far from all dogs with cystine uroliths have concomitant cystine crystalluria. If a sufficient quantity of cystine is present in urine (>10 mmol/mol creatinine), the cyanide–nitroprusside test result for cystine will be positive.[34] Although false positive reactions due to medications with sulfur-containing drugs have been reported,[56] the author's experience is that this test is a reliable and simple diagnostic procedure. For quantitative measurement of urinary cystine excretion, urine should be collected for 24 hr and quantified by, for instance, ion exchange chromatography.[33]

Treatment and Prevention

Current recommendations for dissolution of cystine uroliths and prevention encompass reduction in the urine concentration of cystine and increasing the solubility of cystine in urine. Therapeutic approaches may be divided into four categories: reduction and change of dietary protein intake, aimed at reducing cystine production and excretion; increase in diuresis; increase in cystine solubility; and conversion of cystine to a more soluble compound.

Dietary modification: Attempts have been made in humans to design diets low in methionine to decrease the excretion of cystine. The results of such diets have been mainly disappointing.[19] A protein-restricted diet, designed for dissolution of canine struvite uroliths (Fig. 19.3, Diet 1), was used in two dogs with cystine uroliths at the Swedish University of Agricultural Sciences. However, the uroliths did not dissolve.[34]

Increase in diuresis: Increase of water intake provides a progressive reduction in urinary cystine concentration and reduces the likelihood of precipitation.

Alteration of cystine solubility: Cystine solubility can be enhanced by inducing an alkaline pH, but the solubility does not increase significantly until the pH is above 7.5.[19] Administration of bicarbonate and citrate, for example, has been advocated for improving solubility, but the author's clinical experience, as well as that of others, suggests that not much practical benefit is achieved.[9,34,61]

Conversion of cystine to a more soluble

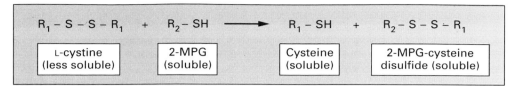

FIG. 19.9 Disulfide exchange reaction.

compound: Chemical modification of the cystine molecule into a more soluble form with D-penicillamine or 2-mercaptopropionylglycine (2-MPG) has been suggested.[40] With this treatment, decreased cystine excretion into the urine is expected and the likelihood of urolith formation is diminished (Fig. 19.9). Although D-penicillamine is effective in prevention of the formation and sometimes in the dissolution of cystine uroliths, this treatment is accompanied by frequent complications that limit its use. In dogs, the most prominent side-effect is vomiting.[9] Another property of penicillamine is its chelation of metals. In a study of 11 normal Beagles given D-penicillamine orally and intravenously, significantly increased excretion of calcium, copper, zinc, chromium, cobalt, iron and magnesium was found.[32]

2-MPG is chemically related to D-penicillamine, but it has a higher oxidation–reduction potential and may therefore be even more effective in a disulfide exchange reaction.[40] A clinical study covering 1–6 years was performed during oral treatment of 25 cystinuric dogs with 2-MPG.[34] The drug was effective in urolith dissolution when given at a dose of approximately 40 mg/kg body

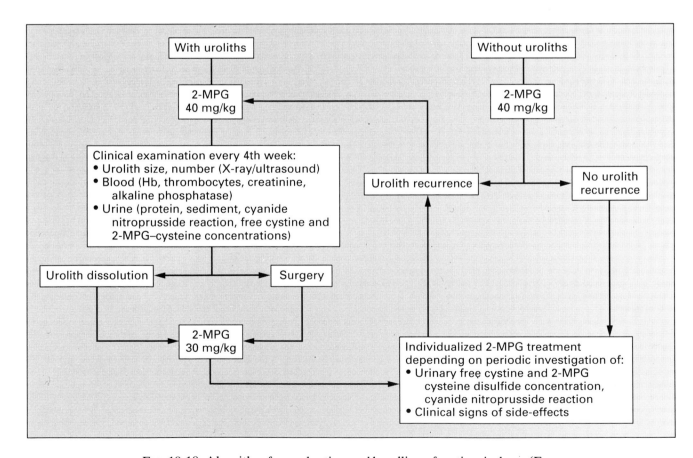

FIG. 19.10 Algorithm for evaluation and handling of cystinuric dogs. (From Hoppe *et al.*[34])

weight (BW). Thus, in 15 dogs with uroliths found in the bladder on 17 occasions, complete urolith dissolution was achieved on nine occasions (53%). During prophylactic therapy with 2-MPG, a dose of 30 mg/kg BW is recommended. Most dogs tolerate 2-MPG well. However, transient side-reactions, like thrombocytopenia, may occur.[34] Figure 19.10 summarizes the author's recommendations with regard to evaluation and handling of cystinuric dogs with and without uroliths.

Urate Urolithiasis

Etiopathogenesis and Prevalence

Urate uroliths are relatively uncommon in dogs, comprising approximately 2–8% of the uroliths analyzed.[8,15,29] Ammonium acid urate (also known as ammonium urate) is the most common salt of urate uroliths found in dogs.[62] Uroliths composed of sodium acid urate and uric acid may occur but are uncommon.

Various breeds of dogs have been reported to produce urate urolithiasis. Although urate uroliths are commonly encountered in Dalmatian dogs, approximately 30–60% of all canine urate uroliths have been found in other breeds.[8,52] Most urate uroliths occur in males (70%) and are most frequently detected in dogs aged 3–6 years.[52] Following surgical removal of urate uroliths, the recurrence rate has been found to be as high as 33–50% in both Dalmatians and other breeds.[14,21,65]

In non-Dalmatian dogs, almost all urate formed from degradation of purine nucleotides is metabolized by hepatic uricase to allantoin, which is excreted by the kidneys and is very soluble. In Dalmatians, only 30–40% of the uric acid is converted to allantoin.[58] The defect of the uric acid metabolism in Dalmatians is thought to be due to impaired transport of urate across the hepatocyte cell membrane.[25] Also, intestinal uptake of hypoxanthine and urate is delayed and renal reabsorption of urate in the proximal tubules is reduced in Dalmatian dogs.[11,59] The definitive cause of urate urolith formation in Dalmatian dogs, however, remains unknown. Because all Dalmatians excrete increased quantities of

urate in their urine and only a few form urate stones, it leads to a predisposition, rather than being a primary cause, of urolith formation.

Regardless of cause, severe hepatic dysfunction may predispose dogs to urate urolithiasis, especially ammonium urate uroliths. A high incidence of ammonium urate uroliths has been observed in dogs with portal vascular anomalies.[27] Hepatic dysfunction in these dogs is associated with reduced hepatic conversion of uric acid to allantoin and ammonia to urea.

Urolith Characteristics

Urate uroliths usually have a smooth surface, and are most often located in the bladder in Dalmatians, whereas both kidneys and the bladder are often involved in dogs with portal vascular anomalies.[27,58] Ammonium acid urate uroliths are poorly radiopaque and could be difficult to detect on plain radiographs. Ultrasonography is ideal for detection of radiolucent bladder uroliths.

Treatment and Prevention

Recommendations for medical dissolution of canine ammonium acid urate uroliths include: calculolytic diets, administration of xanthine oxidase inhibitors (allopurinol), alkalinization of urine and if necessary, eradication or control of urinary tract infections (Fig. 19.11).

Calculolytic diets: The aim of dietary modification is to reduce urine concentration of ammonium and urate. A protein-restricted diet designed for moderate to severe renal insufficiency has been tested and seems to be superior to the diets designed for dissolution of struvite uroliths.[54]

Xanthine oxidase inhibitors: Allopurinol binds to and inhibits the action of xanthine oxidase (Fig. 19.12), thereby decreasing the production of uric acid by inhibiting the conversion of hypoxanthine to xanthine and xanthine to uric acid. The result is a reduction in the serum and urine uric acid concentrations within a couple of days.[23] The recommended dose for dissolution of ammonium

---

RECOMMENDATIONS FOR MEDICAL DISSOLUTION OF AMMONIUM URATE UROLITHS

- Feed a calculolytic diet
- Initiate therapy with allopurinol (30 mg/kg/day, divided bid or tid)
- If necessary, supplement the diet with sodium bicarbonate to achieve a urine pH of 7.0
- If necessary, eradicate or control urinary tract infection with appropriate antimicrobial agents
- Assess the urinary tract by radiography or ultrasonography every 4th week to determine persistence of uroliths and progress of dissolution
- Evaluate serial urinalyses; pH, specific gravity, and sediment for urate crystals
- Continue the full dissolution protocol until 1 month following disappearance of uroliths detected by radiography or ultrasonography

FIG. 19.11 Recommendations for medical dissolution of ammonium urate uroliths.

---

acid urate uroliths in dogs is 30 mg/kg BW/day divided into two or three subdoses.[53] If urate crystalluria or hyperuricuria persists after uroliths are dissolved, allopurinol may be given at a dose of 10–20 mg/kg per day.

Alkalinization: Administration of alkaline agents, such as sodium bicarbonate or potassium citrate, appears to prevent renal tubular production of ammonia and subsequent acid metabolites, possibly because ammonium ions appear to precipitate urates in dog urine. The aim of treatment with urine alkalinizers is to maintain a urine pH of approximately 7.0. Dosage of urine alkalinizers should be individualized for each patient.

Cystitis and Urethritis

Etiopathogenesis

Cystitis is inflammation of the urinary bladder and the most common cause is bacterial infection. Bacteria usually reach the bladder by ascending migration from the urethra sometimes derived from infection in the genital tract (i.e. the prostate and the uterus). Inflammation of the urethra may arise in association with urinary tract infections (UTI) or with trauma caused by for instance catheterization, excessive licking, masturbation, obstruction with calculi or surgery.

The predominant uropathogens are the Gram-negative fecal flora accounting for approximately 75% of UTI in dogs.[43] The most commonly isolated pathogens in dogs are *Escherichia coli*, *Staphylococcus* spp., *Proteus* spp., *Pseudomonas* spp., *Streptococcus* spp. and *Klebsiella* spp. Most investigations have identified *E. coli* as the most common isolate.[4,43]

One of the most important predisposing factors for infection is the retention of urine

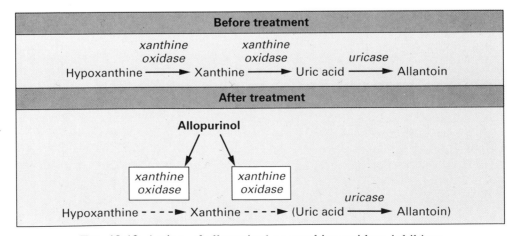

FIG. 19.12 Action of allopurinol as xanthine oxidase inhibitor.

GUIDELINES FOR ANTIMICROBIAL SELECTION AGAINST COMMON UROPATHOGENS		
Microbe	**Antimicrobial recommended**	**Approximate dose**
Staphylococcus, Proteus and *Streptococcus* *Escherichia coli*	Ampicillin Amoxicillin Trimethoprim/sulfa	10–20 mg/kg, qid PO 22 mg/kg, bid PO 15 mg(combined)/kg, bid PO
	Enrofloxacin Nitrofurantoin	5 mg/kg, sid PO 4 mg/kg, tid PO
Pseudomonas	Enrofloxacin Tetracycline	5 mg/kg, sid PO 20 mg/kg, tid PO
Klebsiella	Enrofloxacin Trimethoprim/sulfa	5 mg/kg, sid PO 15 mg(combine)/kg, bid PO
Enterobacter	Trimethoprim/sulfa	15 mg(combined)/kg, bid PO
	Enrofloxacin	5 mg/kg, sid PO

FIG. 19.13 Guidelines for antimicrobial selection for common uropathogens.

in the bladder. This may be caused by: urethral obstruction due to calculi or neoplasms; neurological derangement of micturition caused by spinal cord disease, pelvic nerve disease or chronic distension of the bladder; or acquired or congenital defects of the bladder wall including diverticula or neoplasia. One predisposing iatrogenic cause for UTI is catheterization, especially when using indwelling catheters.[5]

Diagnosis

Diagnostic tests include radiography, ultrasonography and prostatic fluid examination in addition to history, physical examination and urinalysis. Hemogram and clinical chemistries are usually normal in simple lower UTI.

Clinical signs: Clinical signs often include hematuria, dysuria, stranguria, pollakuria and urinary incontinence. However, these signs are not specific for cystitis or UTI because various diseases like urolithiasis, neoplasia and trauma could cause similar signs. It is important to note that a number of cases of significant bacteriuria are asymptomatic and the only way to detect such infections is to perform complete urinalysis with urine culture. The bladder may be painful on abdominal palpation, but signs of systemic illness, such as fever and vomiting, are uncommon with lower UTI.

Urinalysis: A complete urinalysis is the most rapid method for identification of UTI. The urine may be cloudy and hemorrhagic, with a bad odor. Alkalinuria could be a normal finding, but may also result from infection with urea-splitting bacteria. Hematuria and pyuria are supportive of urinary tract inflammation, especially in the presence of proteinuria, which is due to leakage of plasma proteins across the inflamed bladder wall into the urine. Definitive diagnosis of UTI is based on isolation of organisms from a properly collected urine specimen, if possible by cystocentesis. In most instances, a single strain of bacteria is found, but two or more strains of organisms may be isolated.

Treatment

Besides recognition and correction of impairments of host defenses, treatment of UTI requires the administration of appropriate antimicrobial therapy (Fig. 19.13). Owners should be carefully instructed on the importance of correct administration of medication. Acute, uncomplicated bacterial cystitis should be treated with antibiotics for 10–14 days. To ensure that treatment has been effective, a urine sample for culture should be collected 5 days after therapy is concluded.

Complicated infections: Complicated infections are generally associated with prostatic infection (male dogs), abnormalities in the bladder or the urethra, renal infection or failure, urolithiasis, diabetes mellitus, hyperadrenocorticism or prolonged steroid administration. A complete urinalysis and urine culture should always be performed to determine the causative agent. Antibiotic sensitivity testing is desirable, especially if previous antibiotic therapy has been given. Therapy in male dogs with prostatitis should be maintained for at least 3 weeks, with careful follow-up. Successful therapy of other complicated infections is dependent on eradication of the underlying complicating factors. In general, therapy of complicated infections often requires weeks or months, depending on the underlying cause. Urinalysis and urine culture should be performed during therapy and 1–2 weeks after its conclusion.

Recurrent infections: Recurrent infections are common in some dogs. If the same agent is found as in the previous UTI, the occurrence of a chronic infection (e.g. chronic pyelonephritis, chronic cystitis and chronic prostatitis) of the urinary tract should be assumed and localized. If a new agent is found, reinfection has taken place. A diagnostic plan with a careful search for underlying complicating factors should be performed, including quantitative culture and careful clinical and radiographic investigations. If infections recur frequently, a low daily dose (30–50% of the daily dose at night) of antibiotics can be given. Drugs usually recommended with recurrent Gram-negative infections are trimethoprim/sulfa, cephalexin and nitrofurantoin. For recurrent Gram-positive infections ampicillin is recommended.[44,45] However, long-term administration of antibiotics is not risk free. The experience of the author, as well as of others, is that various side-effects, such as keratoconjunctivitis sicca and anemia, have to be considered during treatment with trimethoprim/sulfa combinations.

Manipulation of urine pH by alkalinization (sodium bicarbonate) or acidification (ammonium chloride) may enhance the activity of antimicrobials.[2] For instance, the activity of chloramphenicol and erythromycin may be enhanced at an alkaline pH, and acidification of urine may increase the activity of nitrofurantoin, tetracycline, penicillin, ampicillin and carbenicillin.[2,22]

Urinary Incontinence

Urinary incontinence may be defined as the loss of voluntary control of micturition that results in frequent or constant involuntary passage of urine. It occurs as the result of several different disease mechanisms and should be considered a sign of urinary bladder and urethral dysfunction and not as a diagnosis.

Etiopathogenesis

On the basis of pathogenesis, it is clinically useful to classify diseases associated with urinary incontinence as neurogenic incontinence or nonneurogenic incontinence.

Neurogenic Incontinence

Paralytic bladder: A paralytic bladder may be caused by diseases that damage the nerve supply to the urinary bladder and urethra and include: fractures and luxations of vertebrae, ruptured intervertebral discs, myelitis, pachymeningitis or osteomyelitis, neoplasia of the spinal cord or surrounding structures and congenital anomalies of the spinal cord or the vertebrae.

The affected dog cannot micturate normally and therefore the urinary bladder becomes overdistended with urine and when intravesical urine pressure exceeds the urethral resistance, dribbling of urine occurs.

Cord bladder: Cord bladders are caused by lesions situated between the brain and the sacral spinal cord and include trauma and destructive lesions of the spinal cord. Initially there is a temporary paralysis of the bladder similar to that seen in cases with paralytic urinary bladder. If the spinal reflex center in the sacral cord is not damaged, it may periodically stimulate the bladder to contract without the patient's awareness.

Reflex dyssynergia: Reflex dyssynergia is caused by partial spinal cord or cauda equina lesions, resulting in failure of adequate urethral relaxation when bladder contraction occurs. Dogs with reflex dyssynergia have the sensation and desire to void. Attempts to micturate are often associated with straining and are typically characterized by small interrupted spurts of urine that subsequently stop completely. Large residual volumes of urine in the bladder are common and overdistension of the bladder may occur, resulting in the loss of detrusor contractility.

Nonneurogenic Incontinence

In patients with nonneurogenic urinary incontinence, the nerve supply to the urinary bladder and the urethra are normal, resulting in normal micturition.

Sphincter mechanism incompetence: Urinary incontinence in dogs with sphincter mechanism incompetence is frequently associated with reduced urethral resistance to urine flow during the storage phase of micturition.[36,60] This urethral incompetence occurs most commonly in the female and is often idiopathic. Incontinence may be continuous, but usually occurs when the animal is sleeping.

Estrogens, as well as testosterone, enhance the sensitivity of the α-adrenergic receptors in the smooth muscle of the bladder neck and urethra to sympathetic stimuli. In bitches, a lack of circulating estrogens appears to be important in the development of urinary incontinence. As treatment with the appropriate hormone often produces improvement, the condition has been termed *hormone-responsive incontinence*. In one study, half of the juvenile bitches with this disorder showed an improvement around the time of their first estrus and others improved with estrogen therapy.[37]

Caudal displacement of a portion of the bladder into the pelvic canal, referred to as *pelvic bladder*, has been cited as a cause of urinary incontinence.[20] In association with an intrapelvic bladder there is generally a short urethra, which causes the bladder to lie within the pelvic cavity. If the bladder neck is intrapelvic an increase in intraabdominal pressure is transmitted to the bladder body alone, raising the intravesical pressure and if the urethral resistance in that individual is poor, leakage of urine may occur. However, most male and some female dogs with pelvic bladders do not have incontinence, so other factors may also be involved.

Anatomic abnormalities: The most common *congenital* abnormality causing urinary incontinence is ectopic ureters (see diseases of the ureters). *Acquired* abnormalities like chronic cystitis, chronic urethritis, neoplasia and urolithiasis reduce the filling capacity of the bladder and reduce sphincter closure.

Diagnosis

The results of historical and physical examination usually allow urinary incontinence to be placed into a neurogenic or nonneurogenic category. Subsequently, neurological examination, observation of the voiding pattern, urinalysis, urine culture and plain and contrast uroradiographic studies may be needed to determine the cause of the urinary incontinence. Palpation of the bladder before and after voiding may be helpful in determining residual urine volume. A large residual volume may be encountered in disorders of emptying and with functional or structural obstruction to urine outflow. Obstruction may be ruled out if catheterization can be performed without difficulty.

Treatment

The initial goal should be to eliminate the primary cause. However, permanent damage of the nerve supply to the urinary bladder and advanced generalized inflammation or neoplasia of the bladder wall or the urethra are usually unresponsive to therapy. In patients with overdistension of the urinary bladder, regular elimination of urine helps to prevent cystitis and atonicity of the bladder wall. Catheterization of a nonfunctional bladder should be avoided if possible, because bacterial infection is a frequent complication.

Sphincter mechanism incompetence: The aim of treatment of sphincter mechanism incompetence is to enhance the urethral sphincter closure. This can be achieved by medical therapy or surgical intervention. α-Adrenergic drugs (phenylpropanolamine, 1.5 mg/kg bid) or ephedrine (1–2 mg/kg bid) are the most appropriate for the treatment of sphincter incompetence. Resolution of incontinence has been found to occur in three out of four incontinent bitches.[3] Because of their effect on circulation, sympathomimetic drugs should not be given to dogs with hypertension or cardiac arrhythmias. Alternatively, estrogens can be used because they appear to increase the sensitivity of α-adrenergic receptors to catecholamines.[17] However, blood cells should be counted regularly for signs of bone marrow depression.

In bitches that do not respond satisfactorily to medical treatment, surgery is an alternative using the technique of retropubic urethropexy or colposuspension.[36] Long-term evaluation of 150 incontinent bitches treated with colposuspension showed satisfactory results, revealing complete cure in 53%, improvement in 37% and failure to respond in 9%.[38]

Neurogenic urinary incontinence: The main goal in medical treatment of neurogenic urinary incontinence is to stimulate urinary bladder contraction. Parasympathomimetic drugs like bethanechol (direct cholinergic, 5–25 mg, total dose, tid PO) or neostigmine (anticholinesterase, 5–15 mg, total dose, as required, PO) may be used to treat reflex dyssynergia and bladder atonia. Treatment should be avoided if there is urinary or intestinal obstruction and it should be used with care in cases of gastric ulcers, bronchial constriction or cardiac arrythmia.

To assist in bladder emptying, decreasing the urethral pressure could be of value. For relaxation of the smooth muscle of the urethra, α-adrenergic antagonists like phenoxybenzamine (0.25 mg/kg bid PO) have been used. This may be of value in treatment of reflex dyssynergia and urethral spasm due to inflammation or outlet resistance due to damage caused by, for instance, neoplasia. Dosages may need to be adjusted after a trial period.

References

1. Abdullahi, S. H., Osborne, C. A., Leininger, J. R., Fletcher, T. F. and Griffith, M. D. (1984) Evaluation of a calculolytic diet in female dogs with induced struvite urolithiasis. *American Journal of Veterinary Research,* **45**, 1508–1519.
2. Allen, T. A. (1986) Urinary tract infections. In *Nephrology and Urology.* Ed. E. B. Breitschwerdt. pp. 89–107. New York: Churchill Livingstone.
3. Arnold, S., Arnold, P., Hubler, M., Casal, M. and Rüsch, P. (1989) Incontinentia urinae bei der kastrierten Huendin: Haeufigkeit und Rassedisposition. *Schweizer Archiv fuer Tierheilkunde,* **131**, 259–263.
4. Barsanti, J. A. (1984) Genitourinary tract infections. In *Clinical Microbiology and Infectious Diseases of the Dog and Cat.* Ed. C. E. Greene. pp. 269–271. Philadelphia: W. B. Saunders.
5. Barsanti, J. A. (1985) Urinary tract infections due to indwelling bladder catheters in dogs and cats. *Journal of the American Veterinary Medical Association,* **187**, 384–388.
6. Bovee, K. C., Thier, S. O. and Segal, S. (1974) Renal clearance of amino acids in canine cystinuria. *Metabolism,* **1**, 51–58.
7. Bovee, K. C. (1984b) Urolithiasis. In *Canine Nephrology.* Ed. K. C. Bovee. pp. 335–341. Media, PA: Harwal Publications.
8. Bovee, K. C. and McGuire, T. (1984) Qualitative and quantitative analysis of uroliths in dogs. Definitive determination of chemical type. *Journal of the American Veterinary Medical Association,* **185**, 983–988.
9. Bovee, K. C. (1986) Canine cystine urolithiasis. In *Canine Urolithiasis II. Veterinary Clinics of North America.* Ed. C. A. Osborne. pp. 211–216, Philadelphia: W. B. Saunders.
10. Brand, E., Cahill, G. F. and Kassell, B. (1940) Canine cystinuria. Family history of two cystinuric Irish terriers and cystine determinations in dog urine. *Journal of Biological Chemistry,* **133**, 431–436.
11. Briggs, D. M. and Harley, E. H. (1896) The fate of administered purines in the Dalmatian coach hound. *Journal of Comparative Pathology,* **96**, 267–278.
12. Brown, N. O., Parks, J. L. and Green, R. W. (1977) Canine urolithiasis: retrospective analysis of 438 cases. *Journal of the American Veterinary Medical Association,* **170**, 414–418.

13. Brown, N. O., Parks, J. L. and Green, R. W. (1977) Canine urolithiasis: retrospective analysis of 438 cases. *Journal of the American Veterinary Medical Association,* **170**, 414–418.

14. Brown, N. O., Parks, J. L. and Green, R. W. (1977) Recurrence of canine urolithiasis. *Journal of the American Veterinary Medical Association,* **170**, 419–422.

15. Clark, W. T. (1974) The distribution of canine urinary calculi and their recurrence following treatment. *Journal of Small Animal Practice,* **15**, 437–444.

16. Clark, W. T. (1974) Staphylococcal infection of the urinary tract and its relation to urolithiasis in dogs. *The Veterinary Record,* **95**, 204–206.

17. Creed, K. E. (1983) Effect of hormones on urethral sensitivity to phenylephrine in normal and incontinent dogs. *Research in Veterinary Science,* **34**, 177–181.

18. Christie, B. A. (1971) Incidence and etiology of vesicoureteral reflux in apparently normal dogs. *Investigative Urology,* **10**, 184.

19. Dent, C. E. and Senior, B. (1955) Studies on the treatment of cystinuria. *British Journal of Urology,* **27**, 317–332.

20. Dibartola, S. P. and Adams, W. W. M. (1983) Urinary incontinence associated with malposition of the urinary bladder. In *Current Veterinary Therapy VIII.* Ed. R. W. Kirk. pp. 1089–1092. Philadelphia: W. B. Saunders.

21. Finco, D. R., Rosin, E. and Johnson, K. (1970) Canine urolithiasis: a review of 133 clinical and 23 necropsy cases. *Journal of the American Veterinary Medical Association,* **157**, 1225–1228.

22. Finco, D. R. and Barsanti, J. A. (1983) Urinary acidifiers. In *Current Veterinary Therapy VIII.* Ed. R. W. Kirk. pp. 1095–1097. Philadelphia: W. B. Saunders.

23. Foreman, J. W. (1984) Renal handling of urate and other organic acids. In *Canine Nephrology.* Ed. K. C. Bovee. pp. 135–151. Media, PA: Harwal Publications.

24. Frank, A. (1992) Department of Chemistry, National Veterinary Institute, Uppsala, Sweden. Personal comunication.

25. Giesecke, D. and Tiemeyer, W. (1984) Defect of uric acid uptake in Dalmatian dog liver. *Experimentia,* **40**, 1415–1416.

26. Goulden, B. E. (1968) Clinical observations on the role of urinary infection in the aetiology of canine urolithiasis. *The Veterinary Record,* **83**, 509–514.

27. Hardy, R. M. and Klausner, J. S. (1983) Urate calculi associated with portal vascular anomalies. In *Current Veterinary Therapy VIII.* Ed. R. W. Kirk. pp. 1073–1076. Philadelphia: W. B. Saunders.

28. Harrison, L. Cass, A., Cox, C. and Boyce, W. (1974) Role of bladder infection in the etiology of vesicoureteral reflux in dogs. *Investigative Urology,* **12**, 123–124.

29. Hesse, A. and Brühl, M. (1988) Comparative aspects of urolithiasis in man and animals (dogs, cats). *Proceedings of the Third Annual Symposium, Intercongress, WSAVA, ESVNU.* Barcelona. pp. 183–194.

30. Hoffmann, J., Nährig, M. and Hesse, A. (1992) Cystinuria and cystine urolithiasis. *Proceedings of the First Annual Meeting, ISVNU, Intercongress, WSAVA.* Rome.

31. Hoppe, A. E., Bellström, P. O., Gustavsson, E. I., Jerre, S. P. and Sevelius, E. (1987) Dietary management of urolithiasis in the dog. *Tijdschrift voor diergeneeskunde,* 111S–116S.

32. Hoppe, A., Denneberg, T., Frank, A., Kågedal, B and Petersson, L. R. (1993) Urinary excretion of metals during treatment with D-penicillamine and 2-mercapto-propionylglycine in normal and cystinuric dogs. *Journal of Pharmacology and Therapeutics,* **16**, 93–102.

33. Hoppe, A., Denneberg, T., Jeppsson, J.-O. and Kågedal, B. (1993) Urinary excretion of amino acids in normal and cystinuric dogs. *British Veterinary Journal,* **149**, 253–268.

34. Hoppe, A., Denneberg, T., Jeppsson, J.-O. and Kågedal, B. (1993) Canine cystinuria: an extended study on the effects of 2-mercaptopropionylglycine on cystine urolithiasis and urinary cystine excretion. *British Veterinary Journal,* **149**, 235–251.

35. Holt, P. E., Gribbs, C. and Pearson, H. (1982) Canine ectopic ureter—a review of twenty-nine cases. *Journal of Small Animal Practice,* **23**, 195–208.

36. Holt, P. E. (1985) Urinary incontinence in the bitch due to sphincter mechanism incompetence: surgical treatment. *Journal of Small Animal Practice,* **26**, 237–246.

37. Holt, P. E. (1985b) Urinary incontinence in the bitch due to sphincter mechanism incompetence: prevalence in referred dogs and retrospective analysis of sixty cases. *Journal of Small Animal Practice,* **26**, 181–190.

38. Holt, P. E. (1990) Long-term evaluation of colposuspension in the treatment of urinary incontinence due to incompetence of the urethral sphincter mechanism in the bitch. *The Veterinary Record,* **127**, 537–542.

39. Johnston, G. R., Osborne, C. A., Wilson, J. W. and Yano, B. L. (1977) Familial ureteral ectopia in the dog. *Journal of the American Animal Hospital Association,* **13**, 168–170.

40. Kallistratos, G., Fenner, O. and Berg, U. (1973) Cystinurie und L-cysteinbildung beim Hund. *Experimentia,* **29**, Fasc 7, 791.

41. Klausner, J. S., Osborne, C. A., O'Leary, T. P., Muscoplat, C. M. and Griffith, D. P. (1980) Experimental induction of struvite uroliths in miniature Schnauzer and Beagle dogs. *Investigative Urolology*, **18**, 127–132.

42. Lennox, J. S. (1978) A case report of unilateral ectopic ureter in a male Siberian husky. *Journal of the American Animal Hospital Association*, **14**, 331–336.

43. Ling, G. V., Biberstein, E. L. and Hirsh, D. C. (1979) Bacterial pathogens associated with urinary tract infections. *Veterinary Clinics of North America*, **9**, 617–630.

44. Ling, G. V. (1979) Treatment of urinary tract infections. *Veterinary Clinics of North America*, **9**, 795.

45. Ling, G. V. (1984) Therapeutic strategies involving antimicrobial treatment of the canine urinary tract. *Journal of the American Veterinary Medical Association*, **185**, 1162–1164.

46. Ling, G. V. and Rudby, A. L. (1986) Canine uroliths: analysis of data derived from 813 specimens. *Veterinary Clinics of North America*, **16**, 303–316.

47. Lulich, J. P., Osborne, C. A., Felice, L. J., Polzin, D. J., Thumchai, R. and Sanderson, S. (1992) Calcium oxalate urolithiasis. *Proceedings from the Sixteenth Annual Waltham/OSU Symposium for the Treatment of Small Animal Diseases*. pp. 69–74.

48. Osborne, C. A., Klausner, J. S., Polzin, D. J. and Griffith, D. P. (1986) Etiopathogenesis of canine struvite urolithiasis. In *Canine Urolithiasis I. The Veterinary Clinics of North America*. Ed. C. A. Osborne. pp. 79–80. Philadelphia: W. B. Saunders.

49. Osborne, C. A., Polzin, D. J., Kruger, J. M., Abdullahi, S. U., Leininger, J. R. and Griffith, D. P. (1986) Medical dissolution of canine struvite uroliths. In *Canine Urolithiasis II*. Ed. C. A. Osborne. *Veterinary Clinics of North America*. pp. 353–372. Philadelphia: W. B. Saunders.

50. Osborne, C. A., Polzin, D. J. and Kruger, J. M. (1986) Medical dissolution and prevention of canine struvite uroliths. In *Current Veterinary Therapy IX*. Ed. R. W. Kirk. pp. 1177–1187. Philadelphia: W. B. Saunders.

51. Osborne, C. A., Poffenbarger, E. M., Klausner, J. S., Johnston, S. D. and Griffith, D. P. (1986) Etiopathogenesis, clinical manifestations and management of canine calcium oxalate urolithiasis. *Veterinary Clinics of North America*, **16**, 133–170.

52. Osborne, C. A., Clinton, C. W., Bamman, L. K., Moran, H. C., Coston, B. R. and Frost, A. P. (1986) Prevalence of canine uroliths: Minnesota Urolith Center. *Veterinary Clinics of North America*, **16**, 27–44.

53. Osborne, C. A., Kruger, J. M., Johnston, G. R. and Polzin, D. J. (1986) Dissolution of canine ammonium urate uroliths. *Veterinary Clinics of North America*, **16**, 375–388.

54. Osborne, C. A., Kruger, J. M., Johnston, G. R., O'Brien, T. D., Polzin, D. J., Lulich, J. P., Davenport, M. P., Clinton, C. W. and Hoppe, A (1987) Medical management of canine uroliths with special emphasis on dietary modification. *Companion Animal Practice*, **1**, 72–75.

55. Owen, R. R. (1973) Canine ureteral ectopia: a review. 2. Incidence, diagnosis and treatment. *Journal of Small Animal Practice*, **14**, 419–427.

56. Pahira, J. J. (1987) Management of the patient with cystinuria. *Urology Clinics of North America*, **14**, 339–346.

57. Pak, C. Y. C. (1984) Dietary management of idiopathic calcium urolithiasis. *Journal of Urology*, **133**, 123–125.

58. Porter, P. (1963) Urinary calculi in the dog. II. Urate stones and purine metabolism. *Journal of Comparative Pathology*, **73**, 119–125.

59. Roch–Ramel, F., Wong, N. L. M. and Dirks, J. H. (1976) Renal excretion of urate in mongrel and Dalmatian dogs: a micropuncture study. *American Journal of Physiology*, **231**, 326–331.

60. Rosin, A. E. and Barsanti, J. A. (1981) Diagnosis of urinary incontinence in dogs: role of the urethral pressure profile. *Journal of the American Veterinary Medical Association*, **178**, 814–822.

61. Segal, S. and Thier, S. (1989) Cystinuria. In *The Metabolic Basis of Inherited Disease II*. Eds. C. R. Scriver, A. L. Beaudet, W. S. Sly and D. Valle. pp. 2479–2496. 6th edn. Donald S. Fredrickson, Library of Congress, Wyngaarden.

62. Senior, D. F. (1992) Urate urolithiasis. *Proceedings from the Sixteenth Annual Waltham/OSU Symposium for the Treatment of Small Animal Diseases*. pp. 59–67.

63. Smith, C. W., Stowater, J. L. and Kneller, S. K. (1981) Ectopic ureter in the dog: a review of cases. *Journal of the American Animal Hospital Association*, **17**, 245–248.

64. Tanger, C. H. (1981) A review of ectopic ureters and methods of surgical correction. *Southwestern Veterinarians*, **34**, 113–117.

65. Weaver, A. D. (1970) Canine urolithiasis, chemical composition and outcome of 100 cases. *Journal of Small Animal Practice*, **11**, 93–107.

66. Weaver, A. D. and Pillinger, R. (1975) Relationship of bacterial infection in urine and calculi to canine urolithiasis. *The Veterinary Record*, **97**, 48–50.

67. Weissman, M., Klein, B. and Berkowitz, J. (1959) Clinical applications of infrared spectroscopy. Analysis of renal tract calculi. *Analytical Chemistry*, **31**, 1334–1338.

CHAPTER 20

Cardiovascular Disease

REBECCA L. STEPIEN and MATTHEW W. MILLER

Introduction

Cardiovascular disease and congestive heart failure are common medical conditions in the dog and cat. The discovery of taurine deficiency in 1987 as a reversible cause of feline dilated cardiomyopathy heightened awareness of the importance of dietary management in the prevention and therapy of cardiovascular disease.[81] More recently, myocardial L-carnitine deficiency has been implicated in the pathogenesis of myocardial disease in Boxer dogs[45] and subnormal plasma carnitine and taurine levels were demonstrated in American Cocker Spaniels with dilated cardiomyopathy.[49]

Little has been published concerning the effects of dietary alteration on companion animals with cardiac disease. Nutritional decisions are often based on physiological extrapolation and empirical recommendations. The purpose of this chapter is briefly to review the incidence of cardiovascular disease, current concepts of cardiovascular physiology and the diagnosis and management of feline and canine cardiovascular disease and to make dietary recommendations for dogs and cats with cardiovascular disease and congestive heart failure.

Incidence and Importance of Cardiovascular Disease in Dogs and Cats

A recent review has updated the veterinary database regarding the incidence of canine and feline cardiovascular disease.[7] One study of 5000 dogs listed the prevalence of canine heart disease at 11% with another 9% of the dogs classified as possibly having heart disease.[17] According to Buchanan, the incidence of heart disease has changed little, but the number of recognized cases of congenital cardiovascular lesions has increased. This increase may be due to better diagnostic techniques, increased disease incidence or regional differences in dog populations. Currently available information on frequency of congenital cardiovascular diseases is summarized in Fig. 20.1.

Compared to congenital heart disease, the current prevalence of acquired cardiovascular disease in the dog and cat is less well documented. Chronic valvular disease and dilated cardiomyopathy are the most commonly diagnosed acquired cardiac conditions in the dog; myocardial disease is most common in the cat.[25,27] Reported prevalence of vascular abnormalities is dependent in some cases on geographical location of the

353

FREQUENCY OF CONGENITAL CARDIOVASCULAR DISEASE		
Anomaly	**Dog (%)**	**Cat (%)**
Patent ductus arteriosus	31.7	11
Pulmonic stenosis	18.3	3
Aortic stenosis	22.1	6
Persistent right aortic arch (dog), miscellaneous vascular anomalies (cat)	4.5	8
Ventricular septal defect	6.6	15
Atrial septal defect	0.7	4
Tetralogy of fallot	2.7	6
Tricuspid dysplasia (dog), tricuspid or mitral dysplasia (cat)	2.4	17
Atrioventricular canal defect (cat)		4
Endocardial fibroelastosis (cat)		11

FIG. 20.1 The frequency of congenital cardiovascular disease in the dog (D) and the cat (C). (Modified from Buchanan)[7]

study (e.g. high prevalence of pulmonary vascular disease in heartworm endemic areas) or on the type of study (reported clinical prevalence of pulmonary thromboembolic disease is low; reported prevalence at postmortem is substantially higher).[55] Primary systemic vascular disease is uncommon in the general population, but the diagnosis of systemic hypertension secondary to chronic renal disease (dogs and cats) or hyperthyroidism (cats) is increasing in frequency, perhaps due to increased clinical awareness of these conditions and the introduction of improved methods of detection.[53,92] In cats, the most common acquired vascular disease diagnosed is aortic thromboembolic disease secondary to cardiomyopathy.[7]

The most frequently encountered problems requiring dietary modification are fluid retentive states associated with chronic congestive heart failure, systemic hypertension (primary or secondary) and specific nutritional deficiency-related myocardial disease (taurine and carnitine). To date, most veterinary studies have investigated nutritional deficiencies, where supplementation of the missing nutrient has completely or partially reversed clinical signs of the cardiac abnormality.[44,49,81]

Taurine

Taurine (2-aminoethanesulfonic acid) is an essential amino acid for cats. Although the exact role of taurine in "normal" metabolism

is not completely understood, it is clear that taurine depletion can result in a variety of clinical abnormalities in cats, including central retinal degeneration (FCRD),[38] platelet function abnormalities[39] and myocardial failure.[81]

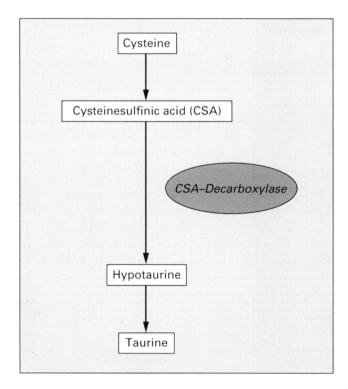

FIG. 20.2 Taurine biosynthesis: The rate limiting step in taurine synthesis is the decarboxylation of cysteine-sulfinic acid by cysteinesulfinic acid decarboxylase, an enzyme with low activity in the cat.

Exogenous taurine is mandatory in feline diets because low activity of the enzyme cysteinesulfinic acid decarboxylase results in low endogenous taurine synthesis (Fig. 20.2). In addition, unlike many other mammals who may conjugate bile acids with glycine when taurine supplies are low, conjugation of bile acids with taurine is obligatory in cats. This results in tissue taurine depletion when dietary taurine supplies are inadequate.

A completely carnivorous diet supplies abundant dietary taurine, but cereal and grain-based diets may provide only marginal or inadequate levels of taurine for cats.[39] Unfortunately, merely supplementing taurine to cats fed taurine-deficient diets is not universally successful in providing enough taurine to reverse taurine-deficient cardiac abnormalities, or maintain adequate tissue concentrations.

Taurine is either more available or better retained by cats in a dry diet format; availability of taurine in dry and canned diets is variable. Dr.y diets containing 1000–1200 mg taurine/kg dry weight of diet and wet diets containing 2000–2500 mg taurine/kg dry weight in the diet will support a normal plasma taurine balance in cats.[80]

Feline taurine-deficient dilated cardiomyopathy (TDDCM) is likely to occur as a clinical problem only after profound and prolonged taurine depletion. In spite of increased awareness by veterinarians, pet-food companies and pet owners, some pet and research animals may be receiving diets inadequate in taurine, especially if the diets provided are composed primarily of cereals or are based on a single type of processed food. TDDCM should be suspected in any case of feline dilated cardiomyopathy and a plasma taurine sample obtained prior to dietary supplementation.

Although ancillary testing leads to diagnosis of dilated cardiomyopathy, diagnosis of TDDCM depends on documentation of low plasma taurine concentrations. A plasma taurine concentration of <20 µmol/l in cats with dilated cardiomyopathy is consistent with TDDCM.[80] Normal feline plasma taurine concentrations range from approximately 50–120 µmol/l[39] but fasting plasma concentrations of >60 µmol/l indicate adequate dietary taurine in normal cats.[80] Plasma for taurine analysis is drawn into a preheparinized syringe and transferred to a plastic or silicone-coated glass tube, spun, separated immediately and frozen.[80] Whole blood taurine is thought to provide a better assessment of the taurine status and to overcome the problem of false high readings due to hemolysis of plasma. Whole blood taurine values >200 µmol/l are recognized as normal values.

Extensive reviews of treatment of TDDCM and related congestive heart failure are available.[26,27,80] After initial stabilization of congestive signs, therapy is directed to repletion of diminished tissue taurine stores. Oral taurine supplementation should begin concurrently with standard congestive heart failure therapy in cases of suspected TDDCM (initial dose: 250–500 mg PO bid). Echocardiographic parameters often improve after 3–6 weeks of supplementation and clinical signs of heart failure often resolve by 6–12 weeks, allowing discontinuation of diuretics, vasodilators and other specific therapy for congestive heart failure (Fig. 20.3). When echocardiographic improvement is seen, taurine supplementation may be reduced to 250 mg sid but the optimal total length of supplementation is unknown. Once the cat has been eating a diet with adequate taurine content for 12–16 weeks (see above), oral taurine supplementation may be discontinued. Adequate plasma taurine concentrations should be verified 4–6 weeks after discontinuing supplementation.[80]

Carnitine

L-Carnitine (β-hydroxy-γ-*N*-trimethylammonium) is a low molecular weight compound that is synthesized endogenously from lysine and methionine and required for transport of free fatty acids into the mitochondria of cardiac muscle. Fatty acids are esterified to L-carnitine and carried across the mitochondrial membrane where β-oxidation takes place, leading to generation of ATP. Inside the mitochondria, carnitine "scavenges" toxic acyl groups and other potentially toxic mitochondrial metabolites and transports them

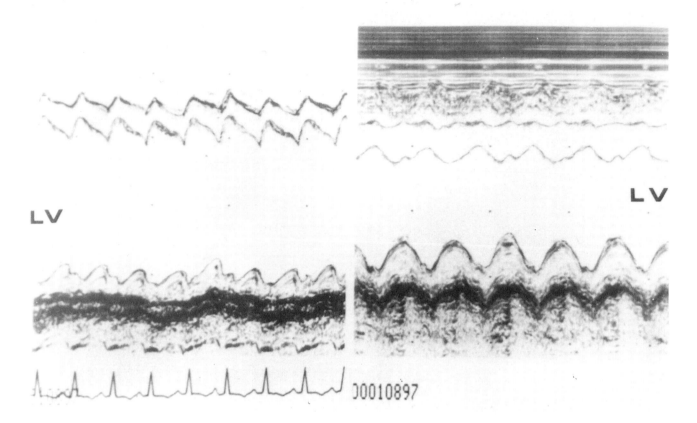

FIG. 20.3 M-mode echocardiographic tracing from a cat with taurine-deficient dilated cardiomyopathy. Left: Prior to therapy, left ventricular diameter (LV) is increased and both left ventricular wall and septal motion are decreased. Right: After successful dietary management, LV size is decreased and left ventricular wall and septal motion are improved. (Courtesy of Dr. Clarke Atkins)

out of the mitochondria as carnitine esters (Fig. 20.4).[40] Carnitine deficiency has been linked with primary myocardial disease in humans, turkeys and hamsters. Low myocardial carnitine concentrations were reported to be associated with dilated cardiomyopathy in a family of Boxer dogs, in unrelated Doberman Pinschers[44,45] and in several American Cocker Spaniels.[49] In one study, approximately 50–90% of unrelated dogs with DCM had myocardial carnitine deficiency.[40] Some dogs with DCM and low myocardial carnitine concentrations who have been supplemented with oral L-carnitine have had notable improvement in clinical signs[40] or have lived longer than supplemented dogs with normal myocardial carnitine concentrations.[42]

Carnitine deficiency may result from decreased synthesis or intake, intestinal mal-absorption, increased renal loss, increased esterification of free carnitine or transport defects.[40] The lack of correlation between plasma and myocardial carnitine concentrations suggests that a transport defect may be the basis for myocardial carnitine deficiency in dogs.[40] It has been proposed that low renal reabsorption of carnitine in dogs may make dogs particularly sensitive to carnitine deficiency.[34]

Dietary carnitine is found in greatest abundance in red meat and dairy products.[34] In a survey of 100 apparently healthy dogs fed a variety of nonmeat-based commercial diets, dogs had plasma carnitine concentrations 50% lower than those receiving meat-based diets and 50% lower than concentrations reported for humans and rats.[45] Although these findings suggest that nonmeat-based

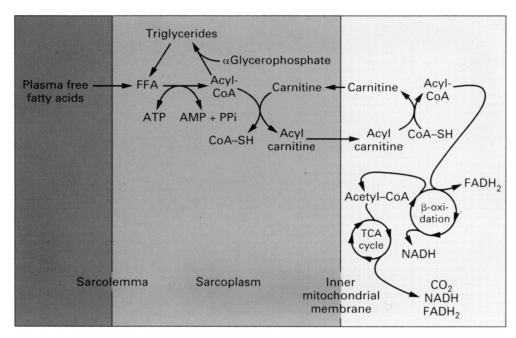

FIG. 20.4 Metabolism of long-chain fatty acids in the myocardium. Following entry into the cell, the fatty acid must first be activated to form an acyl-CoA. The acyl-CoA can then be transported into the mitochondrion, via a carnitine-dependent shuttle, to undergo β-oxidation to acetyl-CoA for entry into the tricarboxylic acid cycle, or it can be esterified in the cytosol to form triglyceride. FFA = free fatty acids; acyl-CoA = acyl-coenzyme A; acetyl-CoA = acetyl-coenzyme A; TCA = tricarboxylic acid; CoA-SH = reduced coenzyme A.

commercial diets may not provide adequate L-carnitine for some animals, no information is presently available regarding the effects of long-term dietary L-carnitine deficiency on normal or myopathic canine myocardium.

Detailed descriptions of diagnosis and conventional therapy of canine dilated cardiomyopathy have been published.[26,27,42] Carnitine deficiency can be referred to as "plasma," "myocardial" or "systemic" (low concentrations in both plasma and myocardial samples), based on the sample source. Carnitine concentrations in the plasma and myocardium of healthy dogs[41] and dogs with DCM[44] have been published. Plasma carnitine concentrations have been reported to be a specific, but not sensitive, indicator of myocardial carnitine concentrations[40] and normal plasma L-carnitine concentration does not guarantee normal myocardial carnitine concentration.[42] Approximately 80% of dogs with DCM identified as having myocardial carnitine deficiency have normal or elevated plasma carnitine concentrations.[40,43] Endomyocardial biopsies are obtained percutaneously, blotted dry to remove excess blood, wrapped in foil and snap frozen in liquid nitrogen. One milliliter of blood is placed in heparinized tubes, separated immediately and frozen.[40]

If myocardial carnitine deficiency is identified in a dog with DCM, supplementation may be instituted. Oral supplementation of L-carnitine (available at health food stores), has been shown to raise the myocardial carnitine concentration to within the normal range in some (but not all) dogs with myocardial carnitine deficiency-associated DCM. There is a biological difference between the optical isomers of carnitine that renders D-carnitine and racemic mixtures of D- and L-carnitine ineffective for supplementation in the dog. L-carnitine supplementation (approximately 2 g mixed with food tid) appears to have few adverse side-effects. If improvement is seen, it usually manifests as increased appetite and activity after 1–4 weeks of supplementation;

improvements in echocardiographic parameters may occur after 2–3 months. In dogs that show echocardiographic improvement, depressed cardiac indices may not return to normal but there is usually improvement or resolution of clinical signs. It is recommended that carnitine be viewed as an adjunctive therapy and conventional cardiac therapy for DCM be administered concurrently with supplementation. If endomyocardial biopsy is not available, carnitine may be supplemented presumptively, as it appears to cause few adverse reactions.[44]

Hypertension

In human beings, primary or "essential" hypertension is the most commonly diagnosed hypertensive disease. Hypertension in dogs and cats is usually secondary to renal disease[48,53] or diagnosed in concert with other systemic disorders, such as hyperthyroidism[53,58] hyperadrenocorticism, diabetes mellitus, polycythemia or pheochromocytoma.[15,104] In cases of hypertension secondary to a systemic disease, e.g. feline hyperthyroidism, the underlying cause is treated.[53] In renal disease, the primary control of hypertension is important, as the underlying disease is often not correctable.

It has been estimated that between 50 and 93% of dogs with renal disease have systemic hypertension and up to 80% of dogs with glomerular disease are hypertensive, indicating an increased risk for dogs with this type of renal lesion.[14] In one study, mild to moderate hypertension was present in 61% of 28 cats with chronic renal failure and 87% of 39 hyperthyroid cats.[53] Sustained systemic hypertension may result in the development of secondary problems, including vascular disease, ventricular hypertrophy leading to decreased cardiac function and heart failure, retinal hemorrhage and/or detachment and progressive renal damage.[14] Diagnosis is aided by a high clinical index of suspicion based on the presence of a high risk disease and compatible clinical presentation. The diagnosis may be confirmed via invasive[14,83] or noninvasive[53,58,79,83] methods. All animals

with documented hypertension should receive a full metabolic diagnostic work-up, with appropriate testing to rule out curable causes of hypertension. Although the morbidity and mortality associated with sustained hypertension in pet animals is not established, the detrimental long-term effects of hypertension in humans suggest that treatment of hypertension is indicated when the underlying disease process cannot be corrected. Renal disease is an underlying cause of hypertension that is common and usually incurable and the following discussion will focus on the therapy of primary and renal-related hypertension.

The pathogenesis of renal-related hypertension is unclear, but the inability to excrete high levels of dietary sodium, increased adrenergic stimulation and activation of the renin–angiotensin–aldosterone system (RAAS) appears to play a role in the maintenance of elevated arterial pressure.[14,19,92] Therapy is, if possible, directed toward correcting the underlying disease process: reducing circulating volume, primarily through diuretics and dietary management; attenuating the effects of adrenergic stimulation; and pharmacological alteration of blood pressure. If at any stage of the therapeutic regimen hypertension is relieved to an acceptable level, therapy is maintained at that level. Dietary sodium restriction alone may not control hypertension adequately[60]; therapeutic medical strategies have been outlined elsewhere.[14,92,93]

Dietary Management of Hypertension

Most discussions of the dietary management of systemic hypertension are limited to recommendations regarding reductions in dietary sodium intake.[34,85,91,93] This is a reasonable recommendation in that increased extracellular fluid volume may be a contributing factor in cases of systemic hypertension and excessive dietary sodium intake has been incriminated in the pathogenesis of primary hypertension in human beings. Commercial dog foods can contain a range of levels of sodium chloride.[85] It has been shown in many animal models of hypertension, and in many humans with hypertension secondary to renal disease, that arterial blood pressure may return to normal with restriction of the dietary sodium chloride intake.[97] It is presumed that dogs and cats with hypertension associated with renal disease would benefit from moderate dietary sodium restriction. Sodium intakes between 0.1–0.3% of the diet have been recommended in dogs and 0.4% of the diet in cats.[15,85,93]

Results of recently performed controlled investigations on hypertensive humans[51] and a population-based intervention study[4] have confirmed earlier suggestions that dietary ω3 fatty acids exert antihypertensive effects. The mechanism responsible for this effect has not been completely elucidated. Prostaglandins and other eicosanoids have been shown to be involved in a variety of processes affecting blood pressure regulation and are made from polyunsaturated fatty acids.[11] It has been suggested that alterations in blood pressure during dietary supplementation with ω3 fatty acids are due to changes in the endogenous synthesis of vasoactive eicosanoids.[51] Other possible mechanisms proposed include effects of ω3 fatty acids on renal function, a reduction in vascular responsiveness to systemic vasoconstrictors and a lowering of blood viscosity.[63] It is possible that some or all of these mechanisms could be operative in patients with hypertension of different origins. Elucidation of the mechanism of hypertension may allow identification of patients that will be most likely to benefit from ω3 polyunsaturated fatty acid therapy.

It has been suggested that the antihypertensive effects of ω3 fatty acids might be due in part to beneficial alterations in the renal handling of electrolytes or in the release or response to renin and related hormones.[63] Conversely, ω3 fatty acids may protect renal function in some situations by lowering arterial pressure and reducing the progressive renal dysfunction seen with systemic hypertension. ω3 fatty acid supplements have been reported to have beneficial effects on blood pressures in patients on hemodialysis with little residual renal function.[33] Therefore, the effects of ω3 fatty acids are probably not limited to altered renal function, but must also involve vasoactive hormones or vascular reflexes. Ongoing veterinary clinical studies evaluating the effect of dietary lipids on renal disease in dogs may provide an insight into the potential antihypertensive effect of ω3 fatty acids in this species.[6]

Pathophysiology of Congestive Heart Failure

An understanding of the pathophysiology of congestive heart failure (CHF) is essential when considering dietary adjustment. New information has recently come to light regarding neurohormonal activation (NHA) and physiological renal sodium management in edematous states. It is likely that modification of preexisting ideas will continue in the forthcoming years; some new findings may directly influence the clinical treatment of CHF.

Neurohormonal Response to Low Cardiac Output

Low cardiac output (CO) and systemic blood pressure may result from many pathological processes ranging from acute hemorrhage to chronic myocardial failure. Decreases in CO and blood pressure are detected by pressure–stretch receptors in the carotid sinus and aortic arch. Diminished tonic inhibition of afferent glossopharyngeal pathways to the central nervous system (CNS) leads to activation of the sympathetic nervous system (SNS). Increased SNS activity results in constriction of peripheral arterioles, increased stroke volume through augmentation of venous return, cardiac contractility and relaxation and increased heart rate. These responses directly increase CO and maintain perfusion to vital organs. SNS activity also stimulates renin[54] and antidiuretic hormone (ADH) release.[94] Renin is a proteolytic enzyme secreted by renal juxtaglomerular cells in response to decreased renal perfusion pressure detected by pressure–volume–stretch receptors in the afferent renal arteriole. Renin release may also be stimulated directly by adrenergic discharge or by decreases in the amount of sodium delivered to the distal tubule.[54]

Renin cleaves the serum globulin substrate, angiotensinogen, to release angiotensin-I (AT-I) into the circulation. Angiotensin-converting enzyme is released from pulmonary endothelial cells and converts circulating AT-I into the vasoactive hormone angiotensin-II (AT-II). AT-II is a potent arterial vasoconstricting agent. Acting directly on the proximal renal tubular cells, AT-II promotes avid sodium (and therefore water) reabsorption and retention.[73] AT-II causes renal efferent arteriolar vasoconstriction with afferent arteriolar vasodilation, leading to support of a previously decreased glomerular filtration rate (GFR).[32]

Apart from the direct effects on the proximal tubule and peripheral vasculature, there is recent evidence that increased levels of AT-II may, by altering protein synthesis in myocardial cells, contribute to myocardial hypertrophy.[28] AT-II also stimulates secretion of aldosterone (ALDO) from the adrenal gland. ALDO acts primarily on the distal tubule and collecting ducts to promote tubular reabsorption of sodium and chloride and secretion of potassium and hydrogen ions.

ADH is released from the posterior pituitary gland in response to osmotic or non-osmotic stimuli. In situations of pure water deficit (adipsia, diabetes insipidus), increases in plasma osmolality lead to secretion of ADH. Increases in the permeability of cortical and medullary collecting tubules, caused by ADH, allow reabsorption of free water through osmotic forces into the interstitium. ADH increases thirst via CNS mechanisms. Nonosmotic stimuli to ADH secretion are important in hemodynamic disorders (perceived volume depletion) and may be dependent on SNS activity.[94] Vasoconstrictive effects of ADH[68] may be dependent on the degree

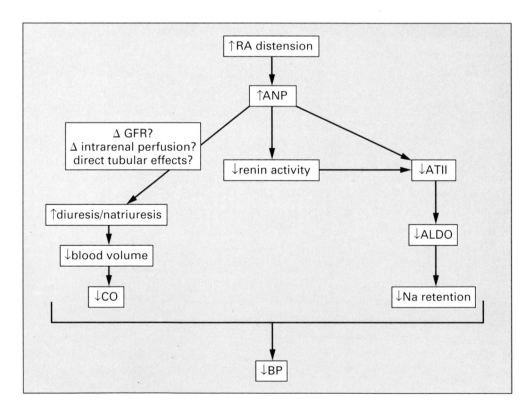

FIG. 20.5 Physiological effects of atrial natriuretic peptide (ANP) in normal dogs. See text for details. RA = right atrium; AT-II = angiotensin-II; GFR = glomerular filtration rate; ALDO = aldosterone; CO = cardiac output; BP = blood pressure.

of heart failure and activity of other neuro-hormonal mechanisms.[94]

Atrial peptides are endogenous peptides secreted from granules in the atrial myocardium in response to atrial distension. In healthy dogs, acute administration of atrial natriuretic peptide (ANP) has been shown to decrease arterial pressure as well as promote natriuresis[57,61] but these effects may not occur in chronic situations.[50] Natriuresis may result from changes in the GFR, intrarenal blood distribution or via direct or indirect tubular actions.[8,101] In healthy dogs and humans, ANP may exert an inhibitory effect on renin activity[88,95] and directly inhibit AT-II activity[65] resulting in decreased ALDO activity and decreased sodium retention. The antirenin effects of ANP may be dependent on the degree of activation of the RAAS.[50] The decrease in arterial blood pressure induced by ANP results from decreases in CO secondary to decreased total blood volume, a consequence of increased urine production.[57,67,88] In healthy individuals, ANP may play a role in daily sodium homeostasis by promoting mild diuresis in response to atrial distension thereby lowering arterial pressures (Fig. 20.5).[100]

Congestive Heart Failure

Development of clinical signs of congestive heart failure (edema, exercise intolerance) is the physical manifestation of the inability of the hemodynamic homeostatic mechanisms to compensate for chronic (and often progressive) decreases in CO and blood pressure. In the decompensated phase of CHF, myriad physiological processes are activated to maintain blood pressure and organ perfusion. SNS activation results in increases in heart rate and contractility, while peripheral arteriolar vasoconstriction and mild venoconstriction "centralize" blood volume to increase filling pressures and increase CO. Activation of the RAAS results in sodium and water retention, the volume of the vascular compartment is reduced due to AT-II-, ADH- and SNS-mediated peripheral

vasoconstriction and the effective circulating volume is increased. In mild CHF, the increase in effective circulating volume may increase cardiac filling pressures enough to augment CO. As arterial pressures increase, SNS stimulation is decreased via negative feedback, decreasing stimulation to the RAAS and to ADH release. When renal perfusion improves and SNS activity decreases, activation of the RAAS subsides, ADH production decreases[88] and a new stable (increased) plasma volume is maintained (Fig. 20.6).[21,103]

When activation of normal homeostatic mechanisms are insufficient to return CO and blood pressure to sufficient levels, or when reductions in baroreceptor sensitivity occur during CHF, sodium and water retention continues, as does increased SNS activity.[103] Hyperreninemia, hyperaldosteronemia and high circulating concentrations of norepinephrine have been documented in dogs with experimental[88,103] and naturally occurring CHF.[52,102] Norepinephrine- and AT-II-mediated increases in afterload cannot be overcome by a failing heart and symptoms of low CO develop. A low effective circulating volume is sensed by baroreceptors and ADH- and AT-II/ALDO-mediated chronic sodium and water retention and increased thirst in the face of impaired renal mechanisms of sodium and water excretion lead to congestive signs of edema (peripheral or pulmonary) and ascites. Sodium excretion is further reduced by a diminution of renal response to ANP.[88] The net effect of these changes is increasing congestive and low cardiac output signs and the inability of the normal feedback mechanisms to halt vasoconstriction and sodium retention.

Sodium Balance in Congestive Heart Failure

It has long been recognized that hyponatremia is associated with increased morbidity and mortality in human patients with CHF.[56] Total body sodium is related to intake, excretion and intracellular vs. extracellular distribution of sodium stores. Traditionally, low sodium diets have been recommended for

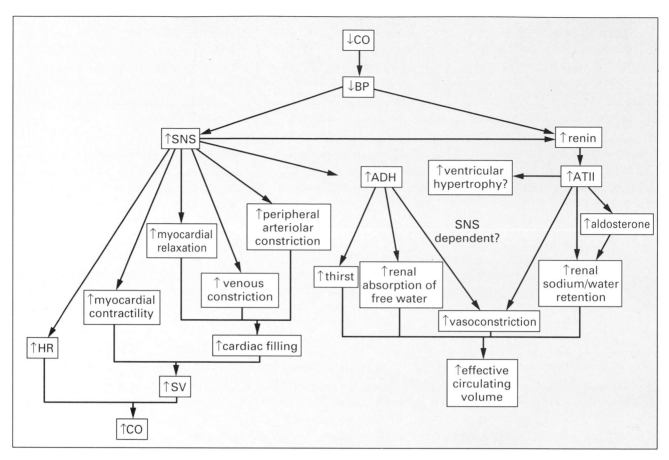

FIG. 20.6 Neurohormonal response to decreased cardiac output. See text for details. CO = cardiac output, BP = arterial blood pressure, SNS = sympathetic nervous system activity, ADH = anti-diuretic hormones, AT-II = angiotensin-II, HR = heart rate, SV = stroke volume.

veterinary patients in CHF to reduce sodium and water retention.[34,69,77,91] Because the application of drugs markedly affecting sodium and water excretion (loop diuretics and angiotensin converting enzyme inhibitors, ACEI) for the treatment of CHF, these recommendations for severe sodium restriction in humans have been modified.[96] Recent advances in the understanding of mechanisms by which sodium and water excretion are regulated are likely to affect future dietary sodium recommendations.

Hyponatremia in Congestive Heart Failure

Although the physiological processes underlying the development of hyponatremia in CHF states are not completely understood, it is probable that many mechanisms may play a contributory role. It is uncommon for hyponatremic patients with CHF to be sodium depleted[5,71]; RAAS-activation, ADH secretion, local renal factors (tubular flow kinetics) and diuretic use may all contribute to dilutional hyponatremia via retention of free water and increased thirst.[76] Clinically, hyponatremia results in neurological signs related to decreases in plasma osmolality and, when accompanied by dehydration, signs of low cardiac output.[90] Although hyponatremia is a complication of heart failure and heart failure therapy in some dogs, it does not occur in all cases, even when patients with similar disease severity are treated with similar medications.[5] Low serum sodium concentration has been correlated with increased plasma noradrenaline levels in humans with CHF[70] and in dogs with severe CHF due to dilated cardiomyopathy.[102] Hyponatremic animals in CHF anecdotally have a poorer prognosis.[5]

Low sodium diets tend to decrease plasma sodium in normal dogs, but the decreases may not be clinically significant in magnitude. In studies of normal dogs fed low sodium diets, the exact magnitude of "significant" decreases in plasma sodium that occurred was not specified[29,98]; other studies report no significant change in plasma sodium in normal dogs fed low sodium diets.[36,89] Chronic sodium restriction of normal dogs caused decreased cardiac output, stroke volume and plasma volume and increased heart rate and peripheral vascular resistance, changes that were fully reversed by a change to a sodium "replete" diet.[98] Plasma renin activity is elevated in dogs fed a low sodium diet[29,89] and maintenance of blood pressure in hyponatremic animals appears to be more dependent on RAAS activity than SNS simulation. In normal dogs, chronic sodium depletion resulted in blunted carotid baroreceptor reflexes in response to acute carotid sinus occlusion[89] and sodium depleted dogs show a greater elevation in blood pressure in response to renal artery constriction than sodium replete dogs, an elevation that was reversed by administration of ACEI.[15]

The consequences of sodium restriction in animals with CHF have not been fully explored. In dogs with experimentally induced CHF, significant changes in plasma volume, sodium excretion and voluntary water intake were not reflected in significant changes in plasma sodium, suggesting that plasma sodium concentration may be a relatively insensitive index of sodium/water balance in CHF.[103] Dogs with naturally occurring CHF, dogs with asymptomatic cardiac disease[35] and dogs with experimentally induced valvular disease and CHF[25] have an impaired ability to excrete dietary sodium as compared to normal dogs. Low sodium diets have therefore been recommended for veterinary CHF patients to decrease delivery of sodium to the distal tubule, thereby decreasing "facultative reabsorption" of sodium in the distal renal tubule.[34] Early studies of an experimental low sodium diet in unmedicated dogs with asymptomatic heart disease and in some cases with CHF demonstrated that low sodium intake was clinically beneficial in some dogs, but the number of dogs studied was low.[77] Some authors have concluded that sodium intake in dogs with CHF could be safely restricted to "any level required to relieve congestion."[69]

The effects of ACE inhibition on sodium balance in clinical heart failure in dogs and cats have not been reported in detail; recent developments in the understanding of the neurohormonal mechanisms involved in sodium and water balance in CHF and increased use of humorally active medications for the management of CHF may alter the universal application of severe sodium restriction for animals in CHF. The clinical experience of veterinarians indicates that animals in CHF receiving therapy not including ACEI benefit from decreased sodium intake via reduction of ascites and edema, but that hyponatremia, when it develops, is a severe complication and a therapeutic dilemma. It has been recommended that low sodium diets be instituted early to "delay" the onset of congestive signs, but studies to document this intended effect are lacking and decreases in cardiac output and activation of the RAAS resulting from a low sodium diet may theoretically contribute to eventual development of cardiac failure. When animals are in overt CHF, the decision to institute a low sodium diet (if the diet is accepted by the pet) seems rational and has proven beneficial in some documented cases[77] and has been recommended.[34,46,85,91] When a low sodium diet is instituted along with high doses of loop diuretics, the potential exists for exacerbation of the tendency to develop hyponatremia. In one study of hyponatremic humans with severe CHF stabilized with sodium restriction and diuretic and digitalis therapy, serum sodium concentrations were increased by the administration of ACEI.[76] The effects of a low sodium diet in combination with these medications in dogs with congestive heart failure have not been sufficiently examined, but may differ in the short- and long-term from normal dogs and dogs with compensated cardiac disease.

Potassium and Magnesium Balance in CHF

Electrolyte disorders other than those associated with sodium homeostasis are also important in heart disease. The use of potent diuretics, most notably the loop diuretics, may be associated with hypokalemia and hypomagnesemia. Hypokalemia represents the most common electrolyte abnormality encountered during therapy.[105] Reduced potassium intake associated with anorexia combined with renal potassium loss due to diuretic use, hyperaldosteronism and persistent chloride depletion may lead to the development of severe hypokalemia. High sodium diets increase the delivery of sodium chloride to the distal tubular potassium secreting sites, but very low salt diets may stimulate aldosterone-induced potassium secretion.

The need for therapy of hypokalemia in patients with CHF has been well documented.[75] Patients with CHF are frequently at risk for malignant ventricular arrhythmias and many of these patients receive cardiac glycosides. The enhanced automaticity of cardiac tissue in response to toxic levels of digoxin is increased by hypokalemia,[47] and the efficacy of the class I antiarrhythmics (e.g. lidocaine, procainamide, quinidine) is diminished by hypokalemia.[107]

Although hypokalemia is commonly associated with therapy for CHF, it should be noted that concomitant administration of diuretics or prostaglandin synthetase inhibitors, beta-blockers and angiotensin-converting enzyme inhibitors may result in a rise in serum potassium concentrations.[10] Empirical therapy with oral potassium supplementations should be avoided and decisions regarding potassium supplementation should always be based on serum electrolyte concentrations.

The problem of magnesium depletion has received growing attention as a potential major contributor to cardiovascular morbidity in human patients with CHF.[31,84] Unlike urinary excretion of calcium, which is enhanced only by loop diuretics, urinary magnesium wasting occurs with both thiazide and loop diuretic use.[20] Magnesium deficiency is most commonly detected in human patients with poor dietary magnesium intake or increased renal magnesium wasting due to diuretic use as well as to lengthy exposure to other drugs that exacerbate renal magnesium loss: most commonly ethyl alcohol, but also cis-platinum, amphotericin-B and certain amino-glycoside antibiotics including gentamycin and tobramycin. It is interesting to note that digitalis preparations, the toxicity of which can be exacerbated by hypomagnesemia,[47] also potentiate renal magnesium wasting.[30]

Serum magnesium concentrations often do not correlate well with other measures for determining magnesium homeostasis, probably because magnesium is largely bound to intracellular buffers or to bone, with only approximately 1% in the extracellular space.[86] Skeletal and cardiac muscle biopsies or measurements of free or total magnesium in circulating mononuclear cells are more reliable than serum magnesium concentrations[18] but these measures are not readily available. Serum magnesium concentrations are quite variable in dogs with CHF.[12,23] Veterinary clinical studies indicate that serum magnesium may not reflect total body magnesium homeostasis in dogs but serial measurements of serum magnesium in a given patient may reflect changes in total body magnesium homeostasis. Some ventricular arrhythmias may respond favorably to magnesium supplementation even in the face of normal serum magnesium concentrations. For the clinician, a high index of suspicion (a cachectic CHF patient receiving digoxin and chronic loop diuretic therapy) coupled with a low normal serum magnesium concentration, would be sufficient to warrant consideration of magnesium-replacement therapy.

Cardiac Cachexia

His countenance was pale, his pulse quick and feeble, his body greatly emaciated except for his belly, which was very large . . . (Withering, 1785.)[106]

A loss of total body fat and lean body mass, most notably skeletal muscle, has been associated with severe chronic congestive heart failure (CCHF) and referred to as cardiac cachexia. Cardiac cachexia, as defined by anthropometry and low serum albumin, is present in 35–50% of human patients with severe signs of CCHF.[9] Although a unified concept of the overall pathogenesis of cardiac cachexia has not been established, much information is available to help explain its occurrence.[1,82] Aspects of the pathogenesis can be categorized into decreased energy intake, reduced energy assimilation, increased energy utilization and iatrogenic factors.

Inappetence and anorexia are major causes of weight loss in patients with CCHF and cause inadequate intake of energy and vitamins.[37] Although reductions in nutrient intake have been shown, the causes for these reductions are not understood. Mild abdominal discomfort associated with ascitic distention is common in CCHF and may act as a deterrent to eating, especially if eating intensifies the discomfort. Compression of the stomach by ascitic fluid and an enlarged liver may lead to a feeling of satiety or fullness. Reduced gastric emptying associated with altered gastric motility has also been documented in CCHF.[3] Finally, diets made unpalatable by severe sodium and protein restriction may result in decreased energy intake.

Malabsorption of nutrients is well documented in patients with CHF. Using ^{131}I-triolein absorption tests investigators have documented significant steatorrhea in 56% of human patients with gross evidence of right heart failure; studies in a small number of these patients using an ^{131}I-oleic acid meal suggest that up to 40% of the patients with malassimilation and steatorrhea may suffer from a primary digestive defect.[2] Congestion of the intestinal mucosa may contribute to the malabsorption noted, whereas edema of the pancreas may cause functional impairment of pancreatic cells or blockage of acini leading to maldigestion. Other investigators have documented protein-losing enteropathy from secondary intestinal lymphangiectasia associated with tricuspid regurgitation.[99] Lymph-angiectasia resulting from systemic venous hypertension and lymphatic hypertension not only causes gastrointestinal protein loss, but is also associated with the loss of lymphocyte-rich lymph and resultant immunologic impairment.[99]

The nutritional status of patients with cardiac cachexia appears to be more influenced by systemic venous congestion than by cardiac output reduction. Most studies have noted that patients with malnutrition and cardiac cachexia have higher right atrial pressures and more severe tricuspid regurgitation than those cardiac patients with a normal nutritional status.[9,99] The importance of increased right atrial pressure is further supported by the finding of protein-losing enteropathy in patients with constrictive pericardial disease[78] and congenital lesions of the right atrium[66] in which left ventricular systolic function may be normal.

Abnormal energy expenditure accelerates the wasting process even when severe cardiac disease limits the activity level. An absolute or relative degree of hypermetabolism may result from increased myocardial oxygen consumption, hyperpnea, increased work of breathing and occasionally the presence of low grade fever.[13,62,74] SNS activation associated with CCHF[52,102] may play a contributing role in the pathogenesis of hypermetabolism associated with cardiac cachexia.

Use of cardiac glycosides, ACEI or diuretics may result in electrolyte imbalances. These imbalances may affect appetite and gastrointestinal (GI) motility or cause direct toxicity (e.g. digitalis glycosides). Rigidly enforced adherence to an appropriate but unpalatable diet may further decrease intake. Individual drugs may have detrimental effects on digestion; digitalis glycosides have been shown to have an inhibitory effect on sugar and amino acid transport in the small bowel.[16]

Recent investigation into the pathogenesis of cardiac cachexia has involved evaluation of the role of tumor necrosis factor (TNF). TNF (also referred to as cachectin) is a cytokine produced primarily by monocytes that has been shown to have

SUMMARY OF RELATIONSHIP BETWEEN NUTRIENTS AND CARDIOVASCULAR DISEASE

Nutrient	Associated condition	Dietary recommendations
Sodium	Sodium and fluid retention Systemic hypertension Congestive heart failure	Restrict sodium intake to between 0.1–0.3% of the diet in dogs and 0.4% in the cat*
Potassium	Hypokalemia: may predispose to arrhythmias and digitalis toxicity; decreases efficacy of Class I antiarrhythmics Hyperkalemia: usually of minimal clinical significance unless severe; most commonly associated with use of ACEI and/or potassium-sparing diuretics	Oral potassium supplementation: 3–5 mmol/kg/day Withdraw any K^+ supplements; consider switching from a potassium-sparing diuretic to a loop or thiazide diuretic
Magnesium	Hypomagnesemia: most commonly associated with thiazide or loop diuretic use; can predispose to digitalis intoxication, which in turn may hasten renal loss of magnesium[†]	Maintenance: 10 mg/kg/day $MgSO_4$ (based on magnesium content of product). Initial therapy with 20 mg/kg/day may be beneficial to replete animals with documented hypomagnesemia
Taurine	*Feline*: taurine deficient dilated cardiomyopathy (normal plasma taurine: 50–120 nmol/l) *Canine*: taurine deficiency has been documented in some American Cocker Spaniels with DCM (normal plasma taurine = 50 nmol/l)	Provide diet with adequate taurine 1000–2000 mg/kg dry wt (dry diet), 2000–2500 mg/kg dry diet wt (wet diet) Supplement 250–500 mg bid PO 500 mg bid PO
Carnitine	Low plasma and myocardial carnitine levels have been associated with DCM in dogs (normal plasma carnitine: 18.4–31.4 mmol/ml)	2 g of L-carnitine tid PO

*This degree of dietary sodium restriction may not be necessary when using ACEI and/or loop diuretics (e.g. furosemide).

†Free or total magnesium in circulating mononuclear cells or muscle cells (skeletal or cardiac) is probably a more accurate reflection of magnesium homeostasis than is serum magnesium concentration.

FIG. 20.7 A summary of nutrient associated conditions and the relevant dietary recommendations.

cytotoxic activity against certain tumor cells. TNF has also been shown to inhibit the activity of lipoprotein lipase, an enzyme involved in the hydrolysis of chylomicron-associated triglycerides to free fatty acids and glycerol.[72]

Clinical studies have documented elevated circulating concentrations of TNF in cachectic patients with chronic heart failure.[59,64] TNF concentrations were highest in patients who had high plasma renin activity, low serum sodium concentration and poor renal function. Renin and prostaglandins (especially prostaglandin E_2) are released in CCHF to regulate glomerular filtration and systemic vascular resistances. It is interesting to note that concentrations of prostaglandin E_2 similar to those measured in patients with hypona-tremic heart failure have been shown to stimulate TNF production from monocytes *in vitro*.[22,87]

Cardiac cachexia is debilitating, eventually leading to severe skeletal muscle wasting and a poor prognosis. Assuring adequate energy intake, combined with steps to normalize hemodynamic parameters, may lessen the severity of the clinical condition. In the future, possible therapeutic intervention might include attempts to alter circulating TNF concentrations in patients with elevated levels of this cytokine. Both oxpentifylline and n3 polyunsaturated fatty acids have been shown to suppress concentrations of TNF in humans.[24] Passive immunization against TNF might also be of benefit in patients with CHF and progressive wasting.

Conclusions

Historically, dietary recommendations regarding therapy of CHF have centered on restrictions of dietary sodium intake. Although reductions in dietary sodium intake are important, they are not the sole dietary manipulation that is appropriate in CHF. Supplementation of missing or inadequate nutrients (taurine and carnitine) or alteration in dietary constituents including supplementation of ω3 fatty acids represent an important potential adjunct to conventional therapy for cardiovascular disease. Prospective clinical trials evaluating the efficacy of dietary management in select cardiovascular diseases may provide information that will promote optimal therapy for veterinary patients with CHF (Fig. 20.7).

References

1. Ansari, A. (1987) Syndromes of cardiac cachexia and the cachetic heart: current perspective. *Progress in Cardiovascular Disease,* **30,** 45–60.
2. Berkowitz, D., Croll, M. N. and Likoff, W. (1963) Malabsorption as a complication of congestive heart failure. *American Journal of Cardiology,* **11,** 43–47.
3. Bologna, A. and Costadoni, A. (1938) Osservazioni sulla funzonione gastropancreatica nei cardiopazienti. *Archio italiano di malattie dell'apparato digerente.,* **7,** 215–254.
4. Bonaa, K. H., Bjerve, K. S. and Straume, B. (1990) Effect of eicosapentanoic and docosahexaenoic acids on blood pressure in hypertension. *New England Journal of Medicine,* **322,** 795–801.
5. Bonagura, J. D. and Lehmkuhl, L. B. (1992) Fluid and diuretic therapy in heart failure. In *Fluid Therapy in Small Animal Practice.* Ed. S. P. DiBartola. pp. 529–553. Philadelphia: W. B. Saunders.
6. Brown, S. A. (1992) Role of dietary lipids in renal disease in the dog. *Proceedings of the Tenth Annual ACVIM Forum.* pp. 568–569. Madison, WI: Omnipress. (Abstr.)
7. Buchanan, J. W. (1992) Causes and prevalence of cardiovascular disease. In *Kirk's Current Veterinary Therapy XI: Small Animal Practice.* Eds. R. W. Kirk and J. D. Bonagura. pp. 647–655. Philadelphia: W. B. Saunders.
8. Burnett, J. C., Opgenorth, T. J. and Granger, J. P. (1986) The renal action of atrial natriuretic peptide during control of glomerular filtration. *Kidney International,* **30,** 16–19.
9. Carr, J. G., Stevenson, L. W., Walden, J. A. and Heber, D. (1989) Prevalence and hemodynamic correlates of malnutrition in severe congestive heart failure secondary to ischemic or idiopathic dilated cardiomyopathy. *American Journal of Cardiology,* **63,** 709–712.
10. Chakko, S. C., Frutchey, J. and Gheorghiade, M. (1989) Life-threatening hyperkalemia in severe heart failure. *American Heart Journal,* **117,** 1083–1090.
11. Cinotti, G. A. and Pugliese, F. (1989) Prostaglandins and hypertension. *American Journal of Hypotension,* **2,** 10s–15s.
12. Cobb, M. A. and Michell, A. R. (1992) Plasma electrolyte concentrations in dogs receiving diuretic therapy for cardiac failure. *Journal of Small Animal Practice,* **33,** 526–529.
13. Cohn, A. E. and Steele, J. M. (1934) Unexplained fever in heart failure. *Journal of Clinical Investigations,* **13,** 853–868.
14. Cowgill, L. D. and Kallett, A. J. (1983) Recognition and management of hypertension in the dog. In *Current Veterinary Therapy VIII: Small Animal Practice.* Ed. R. W. Kirk. pp. 1025–1028. Philadelphia: W. B. Saunders.
15. Cowgill, L. D. and Kallett, A. J. (1986) Systemic hypertension. In *Current Veterinary Therapy IX: Small Animal Practice.* Ed. R. W. Kirk. pp. 360–364. Philadelphia: W. B. Saunders.
16. Csky, T. Z., Hetzog, H. J. and Fernald, J. W. (1961) Effects of digitalis on active intestinal sugar transport. *American Journal of Physiology,* **200,** 459–462.
17. Detweiler, D. K. and Patterson, D. F. (1965) Prevalence and types of cardiovascular disease in dogs. *Annals of the New York Academy of Science,* **127,** 481.
18. Dorup, I., Skajaa, K., Clausen, T. and Kjeldsen, K. (1988) Reduced concentrations of potassium, magnesium and sodium pumps in human skeletal muscle during treatment with diuretics. *British Medical Journal,* **296,** 455–467.
19. Dukes, J. (1992) Hypertension: a review of the mechanisms, manifestations and management. *Journal of Small Animal Practice,* **33,** 119–129.
20. Dyckner, T., Wester, P. O. and Widman, L. (1988) Amiloride prevents thiazide induced intracellular potassium and magnesium losses. *Acta Medica Scandinavica,* **224,** 25–32.
21. Dzau, V. J., Colucci, W. S., Hollenberg, N. K. and Williams, G. H. (1981) Relation of the renin–angiotensin–aldosterone system to clinical state in congestive heart failure. *Circulation,* **63,** 645–651.

22. Dzau, V. J., Packer, M., Lilly, L. S., Swartz, S. L., Hollenberg, N. K. and Williams, G. H. (1984) Prostaglandins in severe congestive heart failure: Relation to activation of the renin–angiotensin system and hyponatremia. *New England Journal of Medicine*, **310**, 347–352.

23. Edwards, N. J. (1991) Magnesium and congestive heart failure. *Proceedings of the Ninth Annual ACVIM Forum*. pp. 679–680. Madison: Omnipress.

24. Endres, S., Ghorbani, R., Kelley, V. E., Geoerilis, K., Lonnemann, G., Van Der Meer, J. W. M., Cannon, J. G., Rogers, T. S., Klempner, M. S., Weber, P. C., Schaefer, E. J., Wolff, S. M. and Dinarello, C. A. (1989) The effect of dietary supplementation with *n*-3 polyunsaturated fatty acids on the synthesis of interleukin-1 and tumor necrosis factor by mononuclear cells. *New England Journal of Medicine*, **320**, 265–271.

25. Ettinger, S. J. (1989) Valvular heart disease. In *Textbook of Veterinary Internal Medicine*. Ed. S. J. Ettinger. pp. 1031–1050. Philadelphia: W. B. Saunders.

26. Fox, P. R. (1989) Myocardial diseases. In *Textbook of Veterinary Internal Medicine*. Ed. S. J. Ettinger. pp. 1097–1131. Philadelphia: W. B. Saunders.

27. Fox, P. R. (1988) Feline myocardial disease. In *Canine and Feline Cardiology*. Ed. P. R. Fox. pp. 435–466. New York: Churchill Livingstone.

28. Francis, G. S. (1990) Neuroendocrine activity in congestive heart failure. *American Journal of Cardiology*, **66**, 33D–39D.

29. Fray, J. C. S., Johnson, M. D. and Barger, A. C. (1977) Renin release and pressor response to renal arterial hypotension: effect of dietary sodium. *American Journal of Physiology*, **233**, H191–H195.

30. Gottlieb, S. S. (1989) Importance of magnesium in congestive heart failure. *American Journal of Cardiology*, **63**, 39G–46G.

31. Gottlieb, S. S., Baruch, L. and Kukin, M. L. (1990) Prognostic importance of the serum magnesium concentration in patients with congestive heart failure. *Journal of the American College of Cardiologists*, **16**, 827–836.

32. Gottlieb, S. S. and Weir, M. R. (1990) Renal effects of angiotensin converting enzyme inhibition in congestive heart failure. *American Journal of Cardiology*, **66**, 14D–21D.

33. Hamazaki, T., Nakazawa, R. and Tateno, S. (1984) Effects of fish oil rich in eicosapentanoic acid on serum lipid in hyperlipidemic hemodialysis patients. *Kidney International*, **26**, 81–84.

34. Hamlin, R. L. and Buffington, C. A. T. (1989) Nutrition and the heart. *Veterinary Clinics of North American: Small Animal Practice*, **19**, 527–538.

35. Hamlin, R. L., Smith, C. R. and Ross, J. N. (1967) Detection and quantification of subclinical heart failure in dogs. *Journal of the American Veterinary Medical Association*, **150**, 1513–1515.

36. Hamlin, R. L., Smith, R. C., Smith, C. R. and Powers, T. E. (1964) Effects of a controlled electrolyte diet, low in sodium, on healthy dogs. *Veterinary Medicine/Small Animal Clinician*, **59**, 748–751.

37. Harrison, J. V. (1945) Diet therapy in congestive heart failure. *Journal of the American Dieticians Association*, **21** (Abstr.) 86–95.

38. Hayes, K. C., Carey, R. E. and Schmidt, S. Y. (1975) Retinal degeneration associated with taurine deficiency in the cat. *Science*, **88**, 949–951.

39. Hayes, K. C. and Trautwein, E. A. (1989) Taurine deficiency syndrome in cats. *Veterinary Clinics of North America: Small Animal Practice*, **19**, 403–413.

40. Keene, B. W. (1992) L-carnitine deficiency in canine dilated cardiomyopathy. In *Current Veterinary Therapy XI: Small Animal Practice*. Ed. R. W. Kirk. pp. 780–783. Philadelphia: W. B. Saunders.

41. Keene, B. W., Atkins, C. E., Kittleson, M. D., Rush, J. E. and Shug, A. L. (1988) Frequency of myocardial carnitine deficiency associated with spontaneous canine dilated cardiomyopathy. *Proceedings of the Sixth Annual ACVIM Forum*. p. 787 (Abstr.).

42. Keene, B. W., Kittleson, M. D., Rush, J. E., Pion, P. D., Atkins, C. E., DeLellis, L. D., Meurs, K. M. and Shug, A. L. (1989) Myocardial carnitine deficiency associated with dilated cardiomyopathy in Doberman Pinschers. *Proceedings of the Seventh Annual ACVIM Forum*. p. 126 (Abstr.).

43. Keene, B. W., Kittleson, M. E., Atkins, C. E., Rush, J. E., Eicker, S. W., Pion, P. and Regitz, V. (1990) Modified transvenous endomyocardial biopsy technique in dogs. *American Journal of Veterinary Research*, **51**, 1769–1772.

44. Keene, B. W., Mier, H. C., Meurs, K. M., Scmidt, M. J. and Shug, A. L. (1991) Dietary L-carnitine deficiency and plasma L-carnitine concentration in dogs. *Proceedings of the Ninth Annual ACVIM Forum*. p. 890 (Abstr.).

45. Keene, B. W., Panciera, D. P., Atkins, C. E., Regitz, V., Schmidt, M. J. and Shug, A. L. (1991) Myocardial L-carnitine deficiency in a family of dogs with dilated cardiomyopathy. *Journal of the American Veterinary Medical Association*, **198**, 647–650.

46. Keene, B. W. and Rush, J. E. (1989) Therapy of heart failure. In *Textbook of Veterinary Internal Medicine*. Ed. S. J. Ettinger. pp. 939–975. Philadelphia: W. B. Saunders.

47. Kelly, R. A. (1990) Cardiac glycosides and congestive heart failure. *American Journal of Cardiology,* **65,** 10e–16e.
48. Kittleson, M. D. and Oliver, N. B. (1983) Measurement of systemic arterial blood pressure. *Veterinary Clinics of North American: Small Animal Practice,* **13,** 321–336.
49. Kittleson, M. D., Pion, P. D. and DeLellis, L. A. (1991) Dilated cardiomyopathy in American Cocker Spaniels—taurine deficiency and preliminary results of response to supplementation. *Proceedings of the Ninth Annual ACVIM Forum.* p. 879 (Abstr.).
50. Kivlighn, S. D., Lowmeier, T. E., Yang, H. M. and Shin, Y. (1990) Chronic effects of a physiologic dose of ANP on arterial pressure and renin release. *American Journal of Physiology,* **258,** H1491–H1497.
51. Knapp, H. and Fitzgerald, G. A. (1989) The antihypertensive effects of fish oil: a controlled study of polyunsaturated fatty acid supplements in essential hypertension. *New England Journal of Medicine,* **320,** 1037–1043.
52. Knowlen, G. G., Kittleson, M. D., Nachreiner, R. F. and Eyster, G. E. (1983) Comparison of plasma aldosterone concentration among clinical status groups of dogs with chronic heart failure. *Journal of the American Veterinary Medical Association,* **183,** 991–996.
53. Kobayashi, D. L., Peterson, M. E., Graves, T. K., Lesser, M. and Nichols, C. E. (1990) Hypertension in cats with chronic renal failure or hyperthyroidism. *Journal of Veterinary Internal Medicine,* **4,** 58–62.
54. Laragh, J. H. (1985) Atrial natriuretic hormone, the renin–aldosterone axis and blood pressure–electrolyte homeostasis. *New England Journal of Medicine,* **313,** 1330–1340.
55. LaRue, M. J. and Murtaugh, R. J. (1990) Pulmonary thromboembolism in dogs: 47 cases (1986–1987). *Journal of the American Veterinary Medical Association,* **197,** 1368–1372.
56. Leaf, A. (1962) The clinical and physiologic significance of the serum sodium concentration. *New England Journal of Medicine,* **267(24),** 24–30, 77–83.
57. Lee, R. W. and Goldman, S. (1989) Mechanism for decrease in cardiac output with atrial natriuretic peptide in dogs. *American Journal of Physiology,* **256,** H760–H765.
58. Lesser, M., Fox, P. R. and Bond, B. R. (1992) Assessment of hypertension in 40 cats with left ventricular hypertrophy by Doppler-shift sphygmomanometry. *Journal of Small Animal Practice,* **33,** 55–58.
59. Levine, B., Kalman, J., Mayer, L., Fillit, H. M. and Packer, M. (1990) Elevated circulating levels of tumor necrosis factor in severe chronic heart failure. *New England Journal of Medicine,* **323,** 236–241.
60. Littman, M. P., Roertson, J. L. and Bovee, K. C. (1988) Spontaneous systemic hypertension in dogs: five cases (1981–1983). *Journal of the American Veterinary Medical Association,* **193,** 486–493.
61. Maher, E., Cernacek, P. and Levy, M. (1989) Heterogenous renal responses to atrial natriuretic factor I. Chronic caval dogs. *American Journal of Physiology,* **257,** R1057–R1067.
62. McIlroy, M. B. (1959) Dyspnea and the work of breathing in diseases of the heart and lungs. *Progress in Cardiovascular Disease,* **1,** 284–297.
63. McMillan, D. E. (1989) Antihypertensive effects of fish oil. *New England Journal of Medicine,* **321,** 1610–1616.
64. McMurray, J., Abdullah, I., Dargie, H. J. and Shapiro, D. (1991) Increased concentrations of tumor necrosis factor in "cachectic" patients with severe chronic heart failure. *British Heart Journal,* **66,** 356–358.
65. Metzler, C. H. and Ramsay, D. J. (1989) Physiologic doses of atrial peptide inhibit angiotensin II-stimulated aldosterone secretion. *American Journal of Physiology,* **256,** R1155–R1159.
66. Miller, M. W., Bonagura, J. D., DiBartola, S. P. and Fossum, T. W. (1989) Budd–Chiari-like syndrome in two dogs. *Journal of American Animal Hospital Association,* **25,** 277–283.
67. Mizelle, H. L., Hildebrandt, D. A., Gaillard, C. A., Brands, M. W., Montani, J. P., Smith, M. J. and Hall, J. E. (1990) Atrial natriuretic peptide induces sustained natriuresis in conscious dogs. *American Journal of Physiology,* **258,** R1445–R1452.
68. Mohring, J., Glanzer, K. and Maciel, J. A. (1980) Greatly enhanced pressor response to antidiuretic hormone in patients with impaired cardiovascular reflexes due to idiopathic orthostatic hypotension. *Journal of Cardiovascular Pharmacology,* **2,** 367–376.
69. Morris, M. L., Patton, R. L. and Teeter, S. M. (1976) Low sodium diet in heart disease: how low is low? *Veterinary Medicine/Small Animal Clinician,* **71,** 1225–1227.
70. Nicod, P., Biollax, J., Waeber, B., Goy, J. J., Poliker, R., Schlapfer, J., Schaller, M. D., Turini, G. A., Nussberger, J., Hofbauer, K. G. and Brunner, H. R. (1986) Hormonal, global and regional haemodynamic responses to a vascular antagonist of vasopressin in patients with congestive heart failure with and without hyponatraemia. *British Heart Journal,* **56,** 433–439.
71. Oh, M. S. and Carrol, H. J. (1992) Disorders of

sodium metabolism: hypernatraemia and hyponatraemia. *Critical Care Medicine,* **20,** 94–103.

72. Oliff, A. (1988) The role of tumor necrosis factor (cachetin) in cachexia. *Cell,* **54,** 141–142.

73. Olsen, M. E., Hall, J. E., Montani, J. P., Guyton, A. C., Langform, H. G. and Cornell, J. E. (1985) Mechanisms of angiotensin II natriuresis and antinatriuresis. *American Journal of Physiology,* **249,** F299–F307.

74. Olson, R. E. (1959) Myocardial metabolism in congestive heart failure. *Journal of Chronic Disease,* **9,** 442–464.

75. Packer, M. (1990) Potential role of potassium as a determinant of morbidity and mortality in patients with systemic hypertension and congestive heart failure. *American Journal of Cardiology,* **65,** 45E–46E.

76. Packer, M., Medina, N. and Yushak, M. (1984) Correction of dilutional hyponatraemia in severe chronic heart failure by converting-enzyme inhibition. *Annals of Internal Medicine,* **100,** 782–789.

77. Pensinger, R. R. (1964) Dietary control of sodium intake in spontaneous congestive heart failure in dogs. *Veterinary Medicine/Small Animal Clinician,* **59,** 752–757, 784.

78. Peterson, V. P. and Hastrup, J. (1963) Protein-losing enteropathy in chronic constrictive pericarditis. *Acta Medica Scandinavica,* **173** (Abstr.) 401–410.

79. Pettersen, J. C., Linartz, R. R., Hamlin, R. L. and Stoll, R. E. (1988) Noninvasive measurement of systemic arterial blood pressure in the conscious Beagle dog. *Fundamental Applied Toxicology,* **10,** 89–97.

80. Pion, P. D., Kittleson, M. D. and Rogers, Q. R. (1989) Cardiomyopathy in the cat and its relation to taurine deficiency. In *Current Veterinary Therapy X: Small Animal Practice.* Ed. R. W. Kirk. pp. 251–262. Philadelphia: W. B. Saunders.

81. Pion, P. D., Kittleson, M. D., Rogers, Q. R. and Morris, J. G. (1987) Myocardial failure in cats associated with low plasma taurine: a reversible cardiomyopathy. *Science,* **237,** 764–768.

82. Pittman, J. G. and Cohen, P. (1964) The pathogenesis of cardiac cachexia. *New England Journal of Medicine,* **271,** 403–408.

83. Podell, M. (1992) Use of blood pressure monitors. In *Current Veterinary Therapy XI: Small Animal Practice.* Eds. R. W. Kirk and J. D. Bonagura. pp. 834–837. Philadelphia: W. B. Saunders.

84. Ralston, M. A., Murnane, M. R., Unverferth, D. V. and Leier, C. V. (1990) Serum and tissue magnesium concentrations in patients with heart failure and serious ventricular arrhythmias. *Annals of Internal Medicine,* **113,** 841–849.

85. Ralston, S. L. and Fox, P. R. (1988) Dietary management, nutrition and the heart. In *Canine and Feline Cardiology.* Ed. P. R. Fox. pp. 219–228. New York: Churchill Livingston.

86. Reinhart, R. A. (1988) Magnesium metabolism. A review with special reference to the relationship between intracellular content and serum levels. *Archives of Internal Medicine,* **148,** 2415–2428.

87. Renz, H., Gong, J. H., Nain, M. and Gemsa, D. (1988) Release of tumor necrosis factor from macrophages: enhancement and suppression are dose dependently regulated by prostaglandin E2 and cyclic nucleotides. *Journal of Immunology,* **141,** 2388–2393.

88. Riegger, G. A. J. and Leibau, G. (1982) The renin-angiotensin-aldosterone system, antidiuretic hormone and sympathetic nerve activity in an experimental model of congestive heart failure in the dog. *Clinical Science,* **62,** 465–469.

89. Rocchini, A. P., Cant, J. R. and Barger, A. C. (1977) Carotid sinus reflex in dogs with low- to high-sodium intake. *American Journal of Physiology,* **322,** H196–H202.

90. Rose, B. D. (1989) In *Clinical Physiology of Acid–Base and Electrolyte Disorders.* 3rd edn. pp. 617–618. New York: McGraw–Hill.

91. Ross, J. N. (1987) Heart failure. In *Small Animal Clinical Nutrition III.* Eds. L. D. Lewis, M. L. Morris and M. S. Hand. 3rd edn. pp. 11.1–11.38. Topeka, KS: Mark Morris Associates.

92. Ross, L. A. (1989) Hypertensive disease. In *Textbook of Veterinary Internal Medicine.* Ed. S. J. Ettinger. pp. 2047–2056. Philadelphia: W. B. Saunders.

93. Ross, L. A. and Labato, M. A. (1989) Use of drugs to control hypertension in renal failure. In *Current Veterinary Therapy X: Small Animal Practice.* Ed. R. W. Kirk. pp. 1201–1204. Philadelphia: W. B. Saunders.

94. Schrier, R. W. (1988) Pathogenesis of sodium and water retention in high-output and low-output cardiac failure, nephrotic syndrome, cirrhosis and pregnancy (first of two parts). *New England Journal of Medicine,* **319,** 1065–1072.

95. Shenker, Y., Sider, R. S., Ostafin, E. A. and Grekin, R. J. (1985) Plasma levels of immunoreactive ANF in healthy subjects and patients with edema. *Journal of Clinical Investigation,* **76,** 1684–1687.

96. Spann, J. F. and Hurst, J. W. (1990) The recognition and management of heart failure. In *The Heart, Arteries and Veins.* Ed. J. W. Hurst. 7th edn. pp. 418–441. New York: McGraw–Hill.

97. Stamler, R., Stamler, J. and Grimm, R. (1987) Nutritional therapy for high blood pressure: final

report of a four year randomized controlled trial of the Hypertension Control Program. *Journal of the American Medical Association,* **257,** 1484–1496.

98. Stephens, G. A., Davis, J. O., Freeman, R. H., DeForrest, J. M., Seymour, A. A., Rowe, B. P. and Williams, G. M. (1980) Hemodynamic, fluid and electrolyte changes during sodium depletion in conscious dogs. *Proceedings of the Society of Experimental Biology and Medicine,* **163,** 416–420.

99. Strober, W., Cohen, L. S., Waldmann, T. A. and Braunwald, E. (1968) Tricuspid regurgitation: a newly recognized cause of protein-losing enteropathy, lymphocytopenia and immunologic deficiency. *American Journal of Medicine,* **44,** 842–850.

100. Verburg, K. M., Freeman, R. H., Davis, J. O., Villarreal, D. and Vari, R. C. (1986) Control of atrial natriuretic factor release in conscious dogs. *American Journal of Physiology,* **251,** R947–R956.

101. Villarreal, D., Freeman, R. H. and Brands, M. W. (1990) ANF and postprandial control of sodium excretion in dogs with compensated heart failure. *American Journal of Physiology,* **258,** R232–R239.

102. Ware, W. A., Lund, D. D., Subieta, A. R. and Schmid, P. G. (1990) Sympathetic activation in dogs with congestive heart failure caused by chronic mitral valve disease and dilated cardiomyopathy. *Journal of the American Veterinary Medical Association,* **197,** 1475–1481.

103. Watkins, L., Burton, J. A., Haber, E., Cant, J. R., Smith, F. W. and Barger, A. C. (1986) The renin–angiotensin–aldosterone system in congestive heart failure in conscious dogs. *Journal of Clinical Investigation,* **57,** 1606–1617.

104. Wheeler, S. L. (1986) Canine pheochromocytoma. In *Current Veterinary Therapy IX: Small Animal Practice.* Ed. R. W. Kirk. pp. 977–981. Philadelphia: W. B. Saunders.

105. White, R. J., Chamberlain, D. A., Hamer, J., McAlister, J. and Hawkins, L. A. (1969) Potassium depletion in severe heart disease. *British Medical Journal,* **2,** 606–610.

106. Withering, W. (1785) *An Account of the Foxglove and Some of its Medicinal Uses.* Birmingham, UK: Swinney.

107. Woosley, R. L. (1987) Pharmacokinetics and pharmacodynamics of antiarrhythmic agents in patients with congestive heart failure. *American Heart Journal,* **114,** 1280–1291.

CHAPTER 21

Endocrine Disorders

IAN E. MASKELL and PETER A. GRAHAM

Introduction

Objective study of dietary management of endocrine disease has largely been restricted to diabetes mellitus. There are, therefore, very few well supported dietary recommendations for other endocrine diseases.

Hyperthyroidism

Hyperthyroidism has become the most common endocrine disorder of the cat and, in nearly all cases, it arises from functional adenomatous hyperplasia of the thyroid gland. Affected cats are usually >6 years old and display a gamut of clinical signs including weight loss, polyphagia, polydipsia, hyperactivity, poor hair coat, diarrhea and cardiac dysrhythmias with or without the presence of palpable goiter.[82,83] Diagnosis is confirmed by an increased serum thyroxine concentration that is often accompanied by increased concentrations of alanine aminotransferase and akaline phosphatase.[83] Triiodothyronine suppression tests may be useful in equivocal cases. It is important to identify any concurrent illness, such as chronic renal failure, that may affect the choice of therapy. Treatment is by surgical thyroidectomy,[3] chronic carbimazole or methimazole therapy[74,82] or radioactive iodine therapy. Dietary recommendations are limited to provision of sufficient caloric intake to restore lost weight and consideration of specific recommendations for geriatric cats or any concurrent illness.

Hypothyroidism

Hypothyroidism is often referred to as the most common endocrine disease of the dog: it is usually the result of lymphocytic thyroiditis or thyroid atrophy. Middle-aged dogs are most often affected and they can present with a number of clinical signs including: alopecia, hyperpigmentation, lethargy, mental disinterest, cold intolerance and bradycardia. Diagnosis is based on a low serum thyroxine concentration.[12,13] The possibility of low thyroxine concentrations in nonthyroidal illness, however, means that a thyrotropin stimulation test is considered necessary for definitive diagnosis. Treatment is by oral levothyroxine therapy.[12,13] Once medical therapy is undertaken, recommendations for dietary management are limited to caloric restriction for those dogs that are overweight.

Hyperadrenocorticism

Spontaneous hyperadrenocorticism (Cushing's disease) occurs in aged dogs and cats and results from pituitary-dependent adrenocortical hyperplasia or adrenocortical neoplasia. Affected animals classically present with polyuria, polydipsia, polyphagia, muscle wasting, weak abdominal muscles, thin skin, hair loss and recurrent infections.[15,73] Diagnosis is by ACTH stimulation or dexamethasone suppression tests. Elevated plasma concentrations of alkaline phosphatase, alanine aminotransferase and glucose are commonly found as are neutrophilia, lymphopenia and eosinopenia. Treatment is medical, by o,p'-DDD or ketoconazole, or surgical, by adrenalectomy.[15,73] The hypercortisolemia of hyperadrenocorticism causes insulin resistance and therefore glucose intolerance in affected animals. These animals therefore can be considered as prediabetic and managed under a similiar dietary strategy to diabetic animals (see below).

Hypoadrenocorticism

Hypoadrenocorticism results from immune mediated destruction of the adrenal cortex (Addison's disease) or excessive o,p'- DDD administration. Clinical signs may include: lethargy, depression, anorexia, vomiting, diarrhea, polyuria, polydipsia, weakness, dehydration and bradycardia.[24] Diagnosis of this condition is by ACTH stimulation test. Other laboratory findings include eosinophilia and lymphocytosis (relative to other unwell animals), prerenal azotemia, hyperkalemia, hyponatremia and a Na:K ratio <27:1.[24] Treatment is by oral fludrocortisone or injections of desoxycortisone pivalate. The addition of salt to food has been recommended but appears to be unnecessary when medical therapy is adequate.

Hyperparathyroidism/Hypoparathyroidism

The management of disorders of calcium homeostasis is dealt with in Chapter 22.

Diabetes Mellitus

Diabetes mellitus is a clinically and etiologically heterogeneous group of hyperglycemic disorders. Hyperglycemia results from an absolute or relative deficiency of the hormone insulin concomitant with an absolute or relative excess of glucagon. It is a relatively common endocrine disease of humans and dogs, affecting 0.25–2% of the American human population, depending on the type of diabetes[85] and between 0.0005–0.5% of the canine population.[14,47] Diabetes mellitus is less common in cats and affects only 0.00066–0.25% of the population.[14,71] Insulin is produced in the pancreatic islets of Langerhans and is secreted into the blood in response to a rise in the plasma concentration of glucose or amino acids. Insulin normally suppresses mobilization of peripheral fat and protein and their subsequent conversion to glucose (gluconeogenesis) while promoting the storage of fat, protein and carbohydrate. Hyperglycemia results when there is insufficient insulin because gluconeogenesis can proceed uncontrolled and because insulin is required to transport glucose into cells from plasma. When plasma glucose concentration is greater than 10–12 mmol/l, the renal threshold is exceeded and osmotic diuresis occurs. The energy deficit created in peripheral tissues by the reduced uptake of glucose is met by the mobilization of fat stores and their metabolism in the liver to produce ketones that can be used as an alternative energy source. Ketogenesis is suppressed by insulin in the normal animal. Mobilization of fat to the liver also causes fatty infiltration and hepatic swelling that in turn is associated with elevations in the plasma concentrations of alkaline phosphatase, alanine aminotransferase and occasionally γ-glutamyl transferase.

Dietary advice for diabetic dogs draws heavily on the human experience because both are omnivorous and have similar nutritional needs. The dog, however, tends toward being carnivorous, and has a slightly higher requirement for dietary protein. Cats are obligate carnivores, and although recom-

FACTORS THAT CAUSE OR PROMOTE DIABETES MELLITUS IN DOGS AND CATS		
Failure of:		
Insulin production	**Insulin transit**	**Target tissue sensitivity**

Autoimmunity
Pancreatitis
Insular amyloid
Islet cell hypoplasia
Chemical toxicity
Pancreatic neoplasia
Pancreatectomy
	Insulin antibodies	Obesity
		Hormonal antagonism
		    Glucagon: glucagonoma, infection, uremia
		    Growth hormone: acromegaly, progesterone/progestagens
		    Glucocorticoids: endogenous/exogenous
		    Catecholamines: pheochomocytoma, stress
		    Progesterone/progestagens: endogenous/exogenous?
		Autoimmunity?

FIG. 21.1 Factors that cause or promote diabetes mellitus in dogs and cats.

mendations for feline diabetics are often similar to those for humans and dogs, there is no scientific evidence to support this. Cats are more adept at handling dietary protein and fat than carbohydrate. The following discussion presents current recommendations for nutritional management of the canine and feline diabetic, with particular emphasis on the role of different forms of dietary carbohydrate. The specific nutritional requirements of the cat are discussed and the likelihood that diabetic cats have very different requirements than omnivores is proposed.

Canine Diabetes Mellitus

Etiology and Classification

Diabetes mellitus can be caused or exacerbated by one or more of the following factors: decreased insulin production, target tissue insensitivity or insulin transport failure (Fig. 21.1). For example, in the dog, failure of insulin production could be the result of immune-mediated islet cell destruction or chronic pancreatitis. Peripheral insulin insensitivity can result from obesity, hypercortisolemia of hyperadrenocorticism or other hormonal antagonism. Failure of insulin transport, e.g. by the formation of insulin antibodies, is possible but rarely occurs as a recognizable clinical syndrome in dogs.[23,25,34] Dogs that are, or are about to become, diabetic primarily as a result of insulin insensitivity initially exhibit compensatory elevations in insulin production.[21,41,51,52] After a sustained period of increased output, which may last months, insulin secreting B-cells may become exhausted,[21,41,51,52] creating a diabetic that has a combination of failed insulin production and target tissue insensitivity. In such cases, if the primary cause of insulin resistance can be identified and eliminated early, the dog need not be diabetic for the rest of its life.[4,11,21]

The best example of a "cured" diabetic dog is the bitch with metestrus diabetes who is spayed before B-cell exhaustion occurs.[4,21] Metestrus diabetes is associated with progesterone excess and consequent induction of elevated concentrations of growth hormone.

In human medicine, diabetes mellitus has been broadly classified as being idiopathic (insulin dependent diabetes mellitus (IDDM) and noninsulin dependent diabetes mellitus (NIDDM) or secondary in origin (Fig. 21.2).[90] Secondary diabetes mellitus can result from a

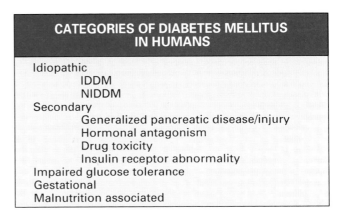

FIG. 21.2 Categories of diabetes mellitus in humans.

number of causes such as generalized pancreatic disease or injury, effects of antagonistic hormones or drug toxicities. The etiopathologically descriptive terms Type I and Type II have been used synonymously with the clinically descriptive terms IDDM and NIDDM by many in human medicine. Some authors restrict the use of the term Type I diabetes mellitus to refer solely to diabetes mellitus resulting from immune-mediated islet cell cytotoxicity and the term Type II diabetes mellitus to refer to those diabetics who do not have immune-mediated islet cell toxicity nor any other known cause of diabetes mellitus. This has been the cause of confusion in applying classifications of human diabetes mellitus to the veterinary field. Nearly all canine diabetics require insulin therapy and are therefore insulin dependent. This does not mean, however, that all diabetic dogs fit the human classification of IDDM/Type I, particularly because most canine diabetes mellitus appears to be secondary in origin *and* insulin dependent.[14,21] The authors recommend that canine diabetics should be referred to by their state of insulin dependency and by their underlying cause when it is known or can be surmised.

Veterinarians should remember that a state of insulin dependence in animals can arise from a number of causes acting singly or in combination (Fig. 21.1), each of which may have differing prognoses or management requirements. The relative prevalence of canine diabetes mellitus resulting from each cause is difficult to ascertain. Dixon and Sandford found as many as four out of eight diabetic dogs to have evidence of pancreatic inflammation at postmortem examination.[19] It has since been suggested that pancreatitis is a relatively unimportant cause of canine diabetes mellitus.[14] Hormonal antagonism, especially by progesterone induced growth hormone secretion, progestagens or cortisol, is thought to be a more common etiology.[2,23] Indeed, metestrus associated diabetes mellitus or underlying hyperadrenocorticism has been confirmed in approximately 5 and 13% of a group of 80 diabetic dogs, respectively, at Glasgow University (Graham, P. A. and

Nash, A. S., unpublished observations). Islet cell autoimmunity has also been recorded in some diabetic dogs[31,78] and a condition resembling human Type II diabetes has been reported in about 10% of diabetic dogs[52] in which insulin replacement therapy was necessary. It is likely that autoimmunity, pancreatic inflammation, hormonal antagonism and Type II disease all have a role in the development of canine diabetes mellitus. There are no reported studies that accurately define their relative importance.

Diagnosis

Diabetes mellitus should be suspected in dogs presented for investigation of polyuria and polydipsia. When polydipsia has not been noted, polyphagia, weight loss, exercise intolerance, ketotic breath, recurrent urinary tract infection or cataract formation should lead to a high index of suspicion for the presence of the disease (Fig. 21.3). Differential diagnoses include those diseases associated with polyuria and polydipsia (Fig. 21.4).

Diagnosis is largely based on persistent hyperglycemia with glucosuria, although some pitfalls need to be avoided. Positive urine glucose results can be found in animals that do not have diabetes mellitus, e.g.:

- *Stress* can be associated with marked hyperglycemia and moderate glucosuria.
- *Renal disease.* Some renal diseases result in a leak of glucose into urine despite normal plasma glucose concentrations. Primary

---

PRESENTING SIGNS OF DIABETES MELLITUS

Polyuria
Polydipsia
Polyphagia
Weight loss
Exercise intolerance
Ketotic breath
Recurrent urinary tract infection
Cataracts

FIG. 21.3 Presenting signs of diabetes mellitus.

DIFFERENTIAL DIAGNOSIS OF POLYURIA /POLYDIPSIA
Diabetes mellitus
Chronic renal failure
Pyometra/toxemia
Hyperadrenocorticism
Hyperthyroidism
Diabetes insipidus
(central, nephrogenic)
Hypercalcemia
Acromegaly
Primary renal glucosuria
Psychogenic polydipsia
Hypoadrenocorticism
Hypokalemia
Hepatopathy
Gastroenteropathy

FIG. 21.4 Differential diagnosis of polyuria/polydipsia.

renal glucosuria, Fanconi syndrome, is a rare disease in which there is failure of glucose resorption in the renal tubule.

- *Interfering substances.* Salicylates and vitamin C are capable of seriously affecting urine test strip and tablet results.

Hyperglycemia can be present in animals that do not have diabetes mellitus, e.g.:

- *Stress* will increase plasma glucose by causing release of cortisol and catecholamines that promote gluconeogenesis and antagonize the effects of insulin. This is most obvious in animals that become stressed in a pre-blood sample struggle or have severe nondiabetic illness.
- α_2 *agonist sedatives.* Xylazine and medetomidine can elevate plasma glucose and cause glucosuria. Therefore, if animals are sedated to take blood samples, these agents should be avoided.
- *Intravenous fluid therapy.* Intravenous administration of dextrose saline can cause significant hyperglycemia.
- *Glucocorticoid* administration will elevate plasma glucose.

Intravenous and oral glucose tolerance tests are often quoted but rarely necessary to diagnose diabetes mellitus. If an animal requires a glucose tolerance test for an initial diagnosis of diabetes mellitus it is unlikely that it is going to be a suitable candidate for insulin replacement therapy. Measurement of post-glucose administration serum insulin concentrations may help determine which animals have insulin producing capacity. If only post-glucose blood glucose concentrations are measured then the test is more useful to prove animals *do not* have diabetes mellitus, e.g. when diabetics have apparently self-cured.

Therapy of Canine Diabetes Mellitus

The management of uncomplicated canine diabetes mellitus depends on the combination of appropriate insulin replacement therapy and appropriate diet. Recommendations for the management of ketoacidotic diabetes mellitus have been extensively reviewed.[14,23,58,91]

Insulin Replacement Therapy

There are many insulin preparations currently available with different times of onset, peak activity and duration of activity.[16,23,28,60,72] Idealized pharmacokinetics of the common types are shown in Fig. 21.5.[16,23,28,29,60] Awareness of the pharmacokinetic properties of the particular preparation selected to treat a diabetic dog will allow a suitable dietary timetable to be chosen. Times of onset and peak/duration of activity vary between dogs and will generally be much less than suggested by the human based literature.

Lente (mixed insulin zinc suspensions) and Biphasic (e.g. 30/70 insulins) are mixtures of two insulin types and generally have two peaks of activity (Fig. 21.5).[23,29,59,84] In some dogs, adequate glycemic control can be achieved with a single daily injection of Isophane, Lente or 30/70 insulin in combination with one meal fed at the time of injection and another fed 6–8 hr later (Isophane and 30/70 preparations) or just before the second peak of insulin activity (Lente preparations).[14,23,39,54,60,61]

Dividing the volume of food evenly between the two meals works well with the Lente insulins

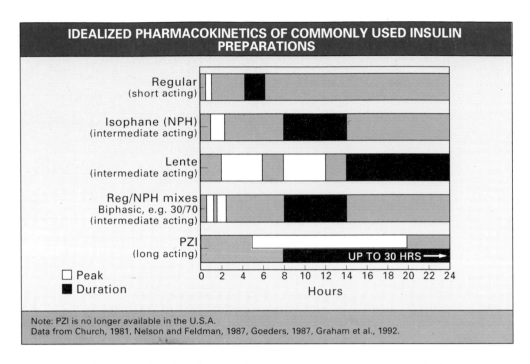

FIG. 21.5 Idealized pharmacokinetics of commonly used insulin preparations.

but increasing the proportion fed in the first meal, when injected insulin is more active, may be valuable when using an Isophane or 30/70 insulins.[28] Generally, increasing the number of daily meals reduces postprandial hyperglycemia and improves glycemic control provided meals are fed while the injected insulin is still active. In many dogs, however, the duration of activity of the intermediate type insulins is shorter than expected, and maybe only 12 or 14 hr.[16,28,61] These dogs are best managed with twice daily injections of an intermediate type insulin. In such circumstances feeding should be either at injection times (2 meals/day) or additionally at times of peak insulin activity (3–4 meals/day). Again, dividing the daily ration into as many meals as the owners daily routine allows will improve control. Longer acting preparations, for example, Protamine zinc insulin (PZI) and Ultralente can be used once daily in dogs with rapid insulin metabolism. PZI, however, is no longer available in the U.S.A. There are currently four insulin preparations available with British Veterinary Product Licenses for use in the dog and cat.

The onset, peak and duration of an insulin preparation in any particular dog can be determined by generating a 24 hr plasma glucose curve. These curves assist in the choice of preparation based on the time of nadir glucose concentration and duration of action and will also be useful when deciding on feeding times and proportions. Large and continuous rises in plasma glucose concentrations following meals suggest that the effect of insulin is waning. The time of nadir plasma glucose indicates peak insulin effect. If measurement of plasma glucose concentration is performed only once daily for the purposes of stabilization or periodic monitoring, it is best done around its estimated nadir if hypoglycemia and insulin induced hyperglycemia are to be avoided.

To generate a 24 hr profile, it is usually recommended that analyses be performed every 2 hr. This can be expensive, especially when overnight staff costs are considered. Costs can often be lowered by reducing the number of samples and concentrating on those periods of the day when insulin is most effective or when meals are being fed. The use of reagent strips and a portable glucometer to measure blood glucose can help reduce costs further.

An alternative approach to assess the efficacy of an insulin preparation is to rely on nadir plasma glucose at first (taken at an estimated time of peak insulin effect, e.g. just prior to the second meal for a single daily intermediate acting insulin regimen) and then after 2–3 weeks measure glycosylated hemoglobin, plasma concentrations of alkaline phosphatase, alanine aminotransferase or glycosylated proteins (fructosamine).[76] If these indicators of glycemic control remain elevated despite adequate nadir plasma glucose concentrations, then the regimen should be altered and the monitoring repeated. This approach will probably be more expensive if it has to be repeated more than once.

Where repeated biochemical analyses are unavailable or prohibitively expensive, careful clinical monitoring becomes very important. Initially, single daily injections of an intermediate-acting insulin and daily monitoring of urine glucose and ketone concentrations are useful in the early stages of stabilization. Clues to the effectiveness of this "first choice" regimen will become apparent within a few weeks. Treatment failure will manifest itself in the form of nocturnal polydipsia (if morning insulin injection) or failure to regain lost weight in the face of adequate (3.5–7 mmol/l) nadir plasma glucose concentrations. When either occurs a different approach must be implemented.

Stable diabetic dogs that are starved or fed alternative diets as part of the management of transient vomiting or diarrhea must continue to receive some insulin replacement therapy. This is because of insulin's inhibitory action on ketogenesis. If a diabetic is deprived of both food and insulin then ketogenesis proceeds unabated and the dog becomes even more unwell. Generally, administering one-quarter to one-half of the usual insulin requirement is safe and will inhibit ketone production. It is not necessary to control plasma glucose concentrations finely during the recovery period. This recommendation also applies to dogs that are anorexic as the result of other illness. This situation arises commonly in diabetics because of compromised immunological function and an increased risk of infection.

Dietary Management

Rationale of contemporary dietary management: Before exogenous insulin became widely available human diabetics were advised to eat diets low in carbohydrate that were consequently high in fat. In 1931 human diabetics at London Hospital were managed on diets containing 15, 68 and 17% energy from carbohydrate, fat and protein, respectively.[39] Arteriosclerosis and high blood cholesterol, prevalent in diabetics, became associated with this high fat recommendation. This outmoded low carbohydrate–high fat philosophy was based on the premise that because diabetics cannot regulate blood sugar levels, dietary sugar should be restricted. Fat is the alternative energy source that diabetics can utilize well, and in the short term (i.e. minutes/hours) is the substrate that the body itself would naturally mobilize in the absence of adequate carbohydrate. In the mid- to long-term (days/months), however, continual transport and metabolism of dietary (or stored) fat may manifest in clinical complications, e.g. ketoacidosis or hepatic lipidosis. Old ideas die hard, and even in the 1980s many veterinary clinics were still using diets unnecessarily low in carbohydrate and high in fat or protein.

Modern dietary management of diabetics relies on controlling the proportion and physical form of the energy-giving nutrients in the diet. As early as 1935 a high carbohydrate–low energy diet was found to accomplish good diabetic control,[75] although the concept was not fully embraced by the medical profession for several decades. The current recommendation is a diet high in *complex* carbohydrates (starch and dietary fiber, 50–55% of energy), no simple sugars (e.g. sucrose), restricted in fat (<20% of energy) and moderate in protein (14–30% of energy). Carbohydrate is the primary component of the regimen and heavily influences the overall profile of the diet. The rationale for the

NUTRITIONAL RECOMMENDATIONS FOR HUMAN DIABETICS COMPARED WITH THE CURRENT AVERAGE U.K. DIET, THE 1931 DIABETIC DIET AND CANINE RECOMMENDATIONS				
	Approx. content of usual U.K. diet	Recommendations for 1990s diabetics	Typical 1930s profile of diabetic diet	Canine recommendations
Carbohydrate (% energy)	45	50–55	15	50–55
Fat (% energy)	40	30–35	68	<20
Protein (% energy)	12–15	10–15	17	14–30
Dietary fiber (g/day)	19–25	>30g	—	?

Note: For humans this data assumes energy values as follows;
1g dietary fat: 9 kcal (37 kJ)
1g dietary protein: 4 kcal (17 kJ)
1g dietary carbohydrate: 3.75 kcal (16 kJ)
Average energy expenditure of a male human = 2510 kcal (10500 kJ)

FIG. 21.6 Nutritional recommendations for human diabetics compared with the current average U.K. diet, the 1931 diabetic diet and canine recommendations.

carbohydrate recommendation and significance of the different types will be discussed in some detail. Figure 21.6 compares 1990 nutritional recommendations for human diabetics with the current average U.K. diet and the original low carbohydrate–high fat recommendation.[7]

Carbohydrate: the key to diabetic dietary management: Carbohydrate is the preferential substrate for energy production in omnivorous mammals. Diabetics experience considerable difficulty regulating blood sugar levels. Even with appropriate insulin therapy, there may be wide fluctuations in diurnal blood glucose concentrations making delivery to metabolizing tissue erratic. Using appropriate dietary carbohydrate the gut can act as a reservoir, slowly releasing monosaccharides into the blood over an extended period. If correctly managed, this effect can complement the action of exogenous insulin. There is also evidence that elevated dietary carbohydrate may improve glucose metabolism by enhancing tissue sensitivity to insulin, partly by increasing the number and binding affinity of cellular insulin receptors.[8,9,81]

In addition fiber has been associated with increased insulin sensitivity in the liver and peripheral tissues[89] and alterations in gastrointestinal endocrine secretions (e.g.

glucagon) implicated in control of nutrient metabolism.[57]

The two major groups of dietary carbohydrate, simple sugars and complex carbohydrates (starches and dietary fiber) are discussed in Chapter 7.

Complex carbohydrates: Starch and dietary fiber have become increasingly important in the management of diabetes mellitus. It becomes apparent that although a general understanding of the concept of dietary fiber is important, specific recommendations are required. Dietary fiber can be classified as soluble or insoluble. Soluble fibers tend to be more viscous and physiologically active in the gastrointestinal tract. The viscosity of soluble fibers depends on the source and guar gum, for example is considerably more viscous than pectin.

Guar is the most widely investigated "viscous" fiber, although most studies have been related to humans. It reduces postprandial increases in plasma glucose and insulin concentrations in normal and diabetic humans.[27,36] These effects have been maintained for months in human diabetics. Studies in dogs have revealed a similar pattern. Blaxter *et al.* found that 20 g guar granules (approximately 8 g/400 kcal) sprinkled daily on the diet of both normal and diabetic dogs reduced hyperglycemia for at least 4 hr after feeding.[5] Oversupplementation with guar can cause loose feces and severe flatulence. Guar also has been shown to lower plasma cholesterol in diabetic, normal and hyperlipidemic humans. This effect is thought to be mediated through interactions with bile acid circulation because fat absorption is not affected.[33]

The effects of insoluble/nonviscous fibers seem to be less dependent on their dietary source, probably because their effects are primarily mechanical. In humans, high levels of dietary fiber (DF, 65–95 g/day; 12–17 g/400 kcal) are considered less effective than more "physiological" levels of around 35–45 g/day DF (6–8 g/400 kcal) that are associated with significant improvements in both mean blood glucose levels and 24 hr profiles.[42] Foods with higher fiber levels are often less palatable, which may partially explain this effect, but it is of interest that moderate levels are at least as effective. Blaxter *et al.* showed that 20 g wheat bran (approximately 8 g/400 kcal) sprinkled daily on the diet of both normal and diabetic dogs reduced hyperglycemia for at least 4 hr after feeding, although the effect was less pronounced than that of guar.[5] Dogs seem to be able to tolerate diets with higher insoluble fiber levels than humans and there is evidence that levels of up to 20 g/400 kcal may be beneficial for the canine diabetic.[30]

An evaluation of different forms of dietary fiber in diets for diabetic dogs has been reported by Nelson and colleagues (1991). They showed that a diet high in insoluble (15% cellulose in dry matter, DM), or soluble fiber (15% pectin in DM), containing more than 50% digestible carbohydrate in the DM significantly reduced postprandial plasma glucose, concentration and fluctuations of 24 hr mean plasma glucose and 24 hr urine glucose excretion in six dogs with alloxan-induced IDDM after 8 weeks of feeding. The results suggested no difference between the soluble and insoluble dietary fiber contrary to previous reports that soluble fiber is normally more effective in improving glycemic control.[35,37,48] There have, however, been other reports that both types are equally effective.[86,87] Even though the diets in this study contained equal amounts of cellulose or pectin, the former contained 28 g DF/400 kcal and the latter 22 g DF/400 kcal. Furthermore, the daily caloric intake was lower for dogs on the pectin-based diet. These findings suggest that similar glycemic control was achieved on a daily intake of less (64%) dietary fiber from the pectin diet compared to the cellulose diet. The levels of dietary fiber in these diets was higher than in other studies, but did not adversely effect palatability.

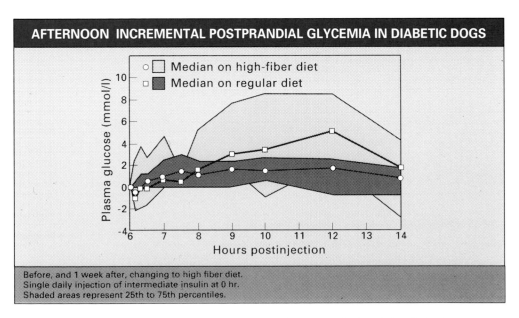

FIG. 21.7 Afternoon incremental postprandial glycemia in eight diabetic dogs.

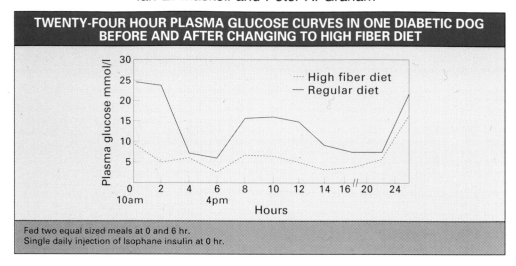

FIG. 21.8 Twenty-four hour plasma glucose curves in one diabetic dog
before and after changing to high fiber diet.

In another study involving eight naturally occurring canine diabetics, postprandial plasma glucose was monitored before and 1 week after switching from a regular to a high starch diet containing insoluble (18 g/400 kcal) and soluble (4 g/400 kcal) dietary fiber derived from a combination of pea hulls (mostly insoluble) and guar gum. This high fiber regime significantly reduced the degree of fluctuation of plasma glucose and resulted in a smoother, less erratic postprandial glycemic curve (Fig. 21.7).[30] Figure 21.8

shows how the 24 hr plasma glucose curve of one of the diabetic dogs changed after the high fiber diet was introduced.

A companion study,[50] involving nondiabetic Labradors investigated effects on postprandial glycemia and insulinemia of three variants of a high starch canned diet (one with no dietary fiber supplement NF, one with a powdered fiber pea hulls and guar gum supplement added at feeding NF + F and one containing the same fiber supplement added before processing, HF). The two high fiber

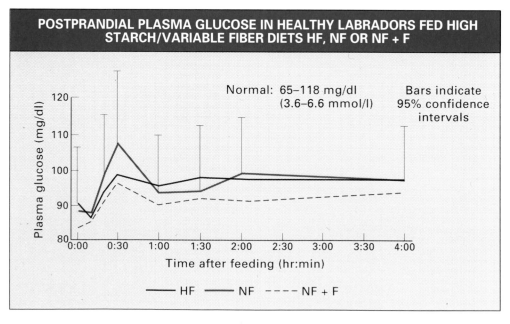

FIG. 21.9 Postprandial plasma glucose in healthy labradors fed high starch/
variable fiber diets HF, NF or NF + F (see text for details).

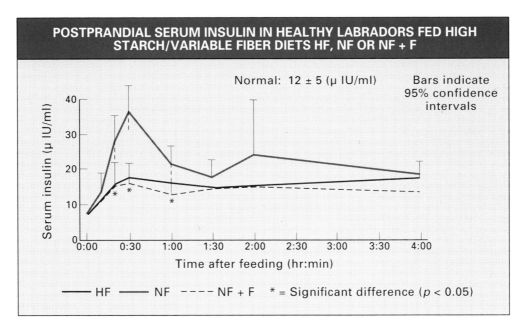

FIG. 21.10 Postprandial serum insulin in healthy Labradors fed high starch/variable fiber diets HF, NF or NF + F.

diets had similar profiles to the diet described in the previous study.

There were no significant differences in postprandial plasma glucose curves between the diets (see Figs. 21.9 and 21.10). Postprandial serum insulin was, however, significantly elevated at 20, 30 and 60 min after feeding NF when compared to NF + F or HF. These findings indicate that a high starch/high fiber regimen reduces insulin requirements for plasma glucose homeostasis and that the process of canning does not depress the physiological effects of dietary fiber. If these findings are transposed to diabetic dogs, the inference is that less exogenous insulin would be required to maintain good glycemic control. The findings with the eight naturally occurring diabetics corroborated this observation.[30]

In conclusion, it appears that to gain optimal effects from dietary fiber it is necessary to provide a blend of both viscous and insoluble sources. Viscous fibers tend to be more potent, therefore lower levels are required.

Diets high in complex carbohydrates: There are three approaches to formulating diets high in complex carbohydrates: using foods naturally high in fiber (e.g. cereal based

diets); supplementing regular petfood with fiber-rich supplements (e.g. guar, bran); or incorporating commercially available fiber enriched products. There are pros and cons with each approach.

Fiber occurring naturally in a foodstuff is likely to be more physiologically effective, particularly in slowing the rate of digestion, by virtue of being in intimate contact with the matrix of the food. Formulating canine diets high in naturally occurring fiber is not easy, achieving nutritional balance is tricky and palatability is often poor. Supplementing low fiber diets is simple but the fiber is less intimately associated with the matrix of the diet and results can be unpredictable. It is also possible for the "picky" eater to preferentially select the nonfiber component of the diet. Furthermore, there is the danger that supplementation of a nutritionally balanced commercial diet may result in an inadvertent dilution of certain dietary nutrients. Commercially prepared products overcome many of these difficulties, fiber-rich ingredients are included at levels to match the nutrient profile of the diet and through processing the fiber becomes homogeneous with the dietary matrix.

Dietary fat and protein: Because carbohydrate contributes much of the dietary energy, fat is only required to provide the shortfall and acts as a delivery route for fat-soluble vitamins. Fat does improve palatability, especially for dogs, but restriction is recommended because of the tendency for diabetics to develop hyperlipidemia and lipid-related complications, such as ketoacidosis, fatty liver and possibly pancreatitis.

Fat restriction, in the presence of high dietary carbohydrate does not yield a low calorie diet, but one of moderate energy level. Obesity tends to exacerbate diabetes and therefore obese diabetics should have measures for weight reduction built into their diabetic regimen (see below).

Protein is an essential dietary requirement to maintain body function. Insulin is involved in tissue uptake and anabolism of amino acids, therefore, although diabetics have a protein requirement, excess is undesirable. Consequently dietary protein content should be moderated and ideally of a high biological value.

Appropriate diet and dietary priority: Obesity in humans and dogs has been associated with decreased insulin sensitivity[41,51,52,78]; improved sensitivity is often achieved with weight loss. Many diabetic dogs are obese when first presented and this has led to the common "blanket" recommendation to feed calorie restricted diets to all canine diabetics. Such a practice is good for obese dogs but ill advised for the large proportion of diabetic dogs that are considerably underweight when presented. Such dogs may have to undergo an initial stabilization period with a diet that is calorie dense, followed by a diet that is moderately calorie restricted once weight has been regained.

Although complex carbohydrate diets are an excellent adjunct to management when diabetes mellitus is the primary clinical presentation, coexisting disease, such as chronic renal failure, hepatic failure, congestive cardiac failure, maldigestion or malabsorption, may require primary consideration in terms of dietary strategy.

Exercise

Exercise is the third component of a triangular relationship with diet and insulin which ultimately controls plasma glucose in the treated diabetic dog. Although sometimes considered as an enemy to diabetic stability, exercise may actually be a useful tool in the management of hyperglycemia. Provided that injected insulin is still active, exercise can be used to blunt postprandial hyperglycemia. The exercise routine should be reasonably constant from day-to-day and should be avoided around times of peak insulin action and when meals are due.

Feline Diabetes Mellitus

Etiology and Classification of Feline Diabetes Mellitus

Both IDDM and NIDDM have been diagnosed in cats.[69] IDDM is the most common form of feline diabetes and is characterized by hypoinsulinemia and little or no response to insulin secretagogues.[26,44,69] Clinical recognition of NIDDM is more common in the cat than the dog and accounts for approximately 20% of diabetic cats.[60,63] Clinical features of human NIDDM evident in cats with NIDDM include mild clinical signs, lack of ketoacidosis and a close relationship with obesity. The transient and sometimes reversible nature of diabetes mellitus in cats further clouds this issue. Causes also remain unclear although there does appear to be a relationship with age, obesity and sex. Pancreatic islet amyloidosis is found in 65% of diabetic cats and is more severe and extensive than in non-diabetics.[95] Islet amyloidosis alone, however, does not appear to be sufficient to cause diabetes when other diabetogenic factors are absent.[69] Other factors predisposing to feline diabetes include viral infections, immune-mediated disease and causes of insulin resistance including endogenous cortisol or growth hormone excess, administration of progestins or corticosteroids, obesity or systemic illness.[60]

Obesity

Obesity is considered an important etiological factor in feline diabetes mellitus, and glucose tolerance and insulin sensitivity are impaired in obese cats.[58] However, the significance remains unclear. Panciera *et al.* classified cats above 6.8 kg, as obese and found only 3.8% of diabetic cats in their study were obese.[71] In another study of 13 diabetic cats,[55] 11 were classified as obese, but only one weighed more than 6.8 kg. Thus assessing obesity in cats must be clarified.

Age

Older cats are most susceptible to diabetes (90% are over the age of 5 years).[71,79] In a cohort of 333 cats a relationship between diabetes mellitus and age was clear, 73.5% of the cats being more than 7 years-old. A similar phenomenon has been observed in humans, where an age related reduction in B-cell function has been associated with the onset of diabetes.[10] It is also possible to speculate on other age related factors that may induce insulin resistance. Renal disease, hyperthyroidism, acromegaly, obesity, hyperadrenocorticism, hepatic and pancreatic disease are known to lead to insulin resistance in humans and develop most often in older cats.[93]

Sex

An interesting aspect of feline diabetes is the predominance among male cats.[55,71] In other species intact females are recognized as being the more susceptible group.[14,21] Incidence is particularly high among neutered male cats and it is speculated that this may be related to the use of progestagens such as megestrol acetate to combat urine spraying and aggressive behavior. Although it is difficult to explain this aspect of feline diabetes mellitus it is suggestive that the feline disease may have other peculiarities.

Drugs

An incidence of diabetes mellitus in cats of 1 in 400,[71] indicates a similar frequency to that observed in dogs,[49] and is higher than previous estimates.[38,53,79] The authors accounted for this in terms of possible increased disease recognition, incidence or use of diabetogenic hormones such as megestrol acetate and corticosteroids.

Breed

Despite previous claims of susceptibility among Siamese cats,[79] most observations do not support the theory of a congenital element in the etiology of feline diabetes mellitus.[71]

Diagnosis

The presenting signs of diabetes mellitus in cats are similar to those in dogs (Fig. 21.3). Unlike diabetic dogs, however, diabetic cats are resistant to the formation of cataracts[14] but may develop a hind-limb neuropathy that results in a plantigrade posture.[14,23] Differential diagnoses of feline diabetes mellitus include those diseases associated with polyuria, polydipsia and weight loss (Fig. 21.4), the most important of which are hyperthyroidism, chronic renal failure and intestinal malabsorption. Polyuria and polydipsia may not be noticed by owners of cats that spend much of their time outdoors. In this situation, poor coat condition, recurrent infection and weight loss may be the only clues to the presence of diabetes mellitus.

Diagnosis of feline diabetes mellitus is notoriously more difficult than its canine counterpart. This is because of great susceptibility to stress hyperglycemia. The initiating stress may be severe illness or the struggle involved in obtaining a blood or urine sample. Therefore, feline blood glucose concentrations need to be interpreted with caution. Blood glucose concentrations of up to 15 mmol/l (270 mg/dl) may be present in stressed nondiabetic cats and so to make a diagnosis of feline diabetes mellitus it is often necessary to take multiple fasting samples. Urine samples obtained at home from gravel or plastic bead litter trays are more reliable than those obtained under stressful conditions. The presence of ketonuria with glucosuria confirms a diagnosis of diabetes mellitus. Glycosylated protein (fructosamine) concentrations may also be useful where diagnosis proves difficult.[76]

Therapy

The two main forms of feline diabetes mellitus, IDDM and NIDDM, each require a different therapeutic approach. Differentiating between the two types in the clinical situation is not always easy. IDDM tends to occur in thin or normal weight aged male cats. Cats that are ketotic on presentation are insulin dependent. The initial management of such cats has been extensively reviewed.[14,23,58,91] Cats metabolize insulin more quickly than dogs[14,88] and uncomplicated insulin-dependent diabetic cats usually require at least twice daily injection with an intermediate acting insulin preparation.[88] This protocol, however, may cause wide fluctuations in blood glucose concentration and once or twice daily injections of a long acting preparation such as PZI or ultralente insulin may be preferable (PZI is no longer available in the U.S.A.). Meals should be fed at injection times to cats receiving twice daily insulin injections and 6–10 hr apart to those on once daily injection regimens. Blood glucose concentration should be checked sequentially so that a blood glucose curve can be generated. Interpretation of such a curve will help identify times of peak effect and duration of insulin action. Adjustments in insulin dose can be made on the basis of nadir blood glucose concentrations once a satisfactory blood glucose curve has been generated and the best time to sample identified. Most cats will stabilize at an insulin dose of 3–5 IU/cat. Like the dog, cats requiring doses of greater than 2 IU/kg should be investigated for poor insulin storage on administration, or for underlying diseases associated with insulin resistance such as infection, hyperadrenocorticism or acromegaly.[14,23,34,68,93,94] To ensure accurate dosing, cats should be administered insulin from a 0.5 ml insulin syringe or the preparation should be diluted 1:10 with water or saline for injection immediately prior to injection. Veterinary licensed 40 IU/ml insulin preparations and syringes are available that may allow for safer dosing of cats.

NIDDM can occur in any age of cat and is often associated with obesity and/or progestagen therapy. It can sometimes be managed with weight loss or withdrawal of the diabetogenic therapy alone but, many NIDDM cats will require supportive insulin replacement therapy in addition to weight control.[58] This will be especially apparent in cats that are stressed or have an additional, perhaps temporary, illness. Oral hypoglycemic therapy may be of value in managing this particular group of diabetic cats. The primary effect of these drugs is to stimulate insulin secretion by the pancreas. In healthy cats glipizide has been shown to stimulate insulin secretion within 10 min of oral administration. In feline diabetics glipizide (5 mg/cat bid) is effective in improving glycemic control (i.e. blood glucose <200 mg/dl) in certain cats when used in conjunction with dietary therapy and obesity correction.[62,66] Vomiting, hypoglycemia and raized plasma alanine aminotransferase concentrations are reported as side-effects.[62,66] Glipizide therapy is maintained until hypo- or euglycemia occur, based on serial blood glucose estimations, when it is discontinued or reduced.[62,66] Because these drugs do not increase insulin synthesis, they are of less use in the management of IDDM but they do improve the efficiency of insulin in NIDDM and thereby can be an alternative approach to management.

Unfortunately, it is not always easy to decide which diabetic cats are insulin dependent and which are not. Basal insulin concentrations and insulin response to secretagogue administration are useful in only a few cases.[62] Cats with ketoacidosis, severe clinical signs or which are underweight must receive insulin-replacement therapy as part of their initial management.[63] Those with only mild clinical signs that are obese and generally healthy may respond best to an initial trial period of weight loss ± oral hypoglycemic therapy.[63,68] These cats should be monitored regularly to assess the response to therapy and requirement for insulin replacement.

Cats are resistant to cataract formation and so fine glycemic control may be less important than in dogs as long as polyuria and polydipsia are absent and hepatopathy is minimized.

Nutritional Peculiarities and Dietary Recommendations

There are a number of features specific to healthy and diabetic cats that suggest that dietary strategies applicable to the dog and human are less appropriate.

Recommendations for management of feline diabetics have tended to concentrate on insulin requirements,[55] with cursory remarks that diet should maintain constant nutrient delivery and be offered at the same time daily to match insulin activity. Where more specific dietary advice has been made there has been a tendency to treat cats the same as dogs and recommend a high fiber, high starch diet.[63,65] This strategy is not only without clinical support but does not take into account some well established observations of peculiarities that predispose the cat to being poorly adapted to diets high in carbohydrate.

Feline Carbohydrate Metabolism

Hexokinase and glucokinase work together in the liver of most mammals to control the early stages of glucose metabolism. The feline liver exhibits normal hexokinase activity, but glucokinase activity is virtually absent.[1] Hexokinase reaches optimal activity at relatively low glucose concentrations. Glucokinase is implicated with "mopping up"

high levels of glucose, is optimal at high glucose concentrations and converts glucose to glycogen for storage in the liver.[46] The implication is that the cats normal metabolism is unaccustomed to high levels of glucose. Another interesting observation is that in humans, insulin stimulates the synthesis of glucokinase and hence this enzyme is deficient in human diabetics; in this regard the human diabetic and the normal cat are similar.

Cats exhibit an inability to metabolize high intakes of dietary sucrose (i.e. the disaccharide of glucose and fructose), which manifests as fructosuria.[20] The mechanism is unclear but the preference of hexokinase for glucose over fructose may lead to preferential metabolism of glucose because fructose exceeds cats' limited metabolic capacity. Although sucrose is uncommon in the feline diet the mechanism highlights a fundamental difference in simple carbohydrate metabolism.

In contrast with this inability to efficiently utilize dietary carbohydrate, cats fed high protein–low carbohydrate diets adequately maintain blood glucose concentrations and continue to do so even when deprived of food for over 72 hr.[43] This suggests that the normal cat is better equipped to maintain its essential glucose requirements from gluconeogenic precursors, i.e. amino acids, rather than dietary carbohydrate.

The normal feline pancreas releases insulin in a biphasic pattern in response to a continuous glucose or amino acid stimulus. Relative response to these stimuli is, however, different from that observed in most other species.[18] In comparison with the rat, B-cells of the feline pancreas are stimulated less by glucose. In addition, arginine has been shown to be a more potent secretagogue relative to glucose in the cat than in dogs or rats.[17,18,22,40] Furthermore, insulin response continues for longer in response to amino acids than glucose. Curry *et al.* also showed that glucose has much less impact on glucagon homeostasis than amino acids.[18] In summary amino acids are more important modulators of pancreatic hormone release in cats than omnivores.

Previous Dietary Reports

Reports of the management of diabetic cats are limited to those that make no reference to diet or case studies where few details are given. No studies have investigated the relative merits of different diets in the management of diabetic cats nor are there any reports of the effects of dietary fiber and complex carbohydrates on glycemic control in this species.

Moise and Reimers reported considerable success through individualization of insulin therapy in 13 diabetic cats. However they made only passing reference to diet (commercial canned cat food) during initial treatment.[55] It is assumed that all cats were maintained on regular cat food that did not apparently interfere with long-term management.

Laflamme et al. reported the use of a two phase dietary strategy in the management of a feline diabetic with concurrent renal insufficiency.[45] During rehabilitation a liquid diet (protein = 20% metabolizable energy, ME, fat = 45% ME, carbohydrate = 25% ME) containing appreciable amounts of highly digestible carbohydrate was utilized effectively. Over a 6 day period the diet was fed enterally until the cat began to accept normal canned food. The authors postulated that the liquid diet reversed the cats' anorexia and helped combat the secondary hepatic lipidosis. The potential adverse effects of the soluble carbohydrates were attenuated by small frequent feedings in conjunction with appropriate insulin therapy. Once the cat had been stabilized, a protein and phosphorus restricted diet also high in fat (approximately 60% of ME and 20% each from protein and carbohydrate) was implemented for long-term management. The authors reported that the cat was well stabilized as long as the insulin dosage was correctly adjusted and the same type and volume of diet was eaten consistently.

Nelson reported the use of a high fiber–low calorie diet in the management of an obese (8.2 kg), 8-year-old feline diabetic.[59] Obesity and insulin requirements declined simultaneously until after 6 weeks when the cat weighed 6.4 kg and no longer required insulin treatment. Whilst the author was clear that the cat had NIDDM and that weight reduction was the key to management, this is one of the only reports of successful management of feline diabetes mellitus with a high fiber diet, and it is easy to make the incorrect assumption that the fiber was the key to managing the diabetes. It is interesting to speculate how the cat would have progressed had weight loss been achieved with a low fiber calorie restricted diet.

A further detailed case report indicated that a cat with IDDM was controlled very well, long term, on a regular canned diet (no details given) in conjunction with appropriate insulin therapy (1 U PZI/kg; cat weighed 5.9 kg).[80] Half the daily food allowance was given in the morning before insulin was administered, the rest 8 hr later when insulin activity was judged to be optimal.

Dietary Recommendations

These case studies illustrate that a diversity of diets can be used successfully, and make it clear that little conclusive evidence exists to support any particular dietary strategy for managing feline diabetics. Dietary recommendations for cats are commonly translated from canine or human recommendations. Because of the connection, in small animal medicine, between fiber and weight reduction in obese patients (i.e. it is often assumed that calorie restriction can only be achieved by inclusion of dietary fiber) there is reason for confusion as to the exact role that dietary fiber may play in the management of diabetics.

It is clear, however, that the cat digests, absorbs and metabolizes carbohydrate differently from dogs and humans, the tendency being to *tolerate* carbohydrate but utilize protein and fat more efficiently. The primary rationale behind high starch diets in diabetic management is to reduce the intake of dietary fat, because fat metabolism is compromised in omnivorous diabetics. It is clear, however, that the normal cat metabolizes fat more efficiently and it appears that the diabetic cat is similar, particularly once insulin therapy has been initiated. The value of dietary fiber in the management of cats with diabetes mellitus is unproven, in fact very little work has investigated how normal cats cope with dietary fiber. It may be that the addition of dietary fiber to a normal feline diet will moderate the rate of small intestinal absorption. The subsequent moderation of nutrient delivery may be useful in optimizing insulin activity. However, because fat tends to slow the rate of digestion, fiber may add no incremental benefit to such a diet.

Conclusions

In conclusion, there is currently no evidence to support the use of a particular diet for the management of feline diabetes mellitus except that semimoist foods, high in simple sugars, are contraindicated. It seems prudent therefore to recommend that diabetic cats be fed on a normal diet (most energy from fat and protein, low in carbohydrate) in conjunction with appropriate insulin therapy when required. In cats receiving insulin-replacement therapy, it is important that the type and quantity of diet and meal times are as consistent as possible.

Regardless of the type of diet used, initial nutritional strategy should be individualized to meet the specific needs of thin and obese diabetic cats. Weight gain is a priority in underweight diabetics, and because reestablishment of glycemic control is necessary for weight gain the feeding of a moderately energy dense diet in conjunction with adequate insulin therapy is recommended. Once normal weight has been achieved, calorie intake and nutrient profile can be reassessed with insulin dose to optimize glycemic control. Conversely, weight reduction is the priority for obese diabetic cats in order to improve insulin sensitivity. In this case, caloric intake ought to be restricted, ideally by feeding a palatable low calorie product that may or may not be high in dietary fiber (see Chapter 7). The use of standard commercial diets helps to simplify management, and maintains the emphasis on calorie intake and correct insulin dose.

Feline diabetes is commonly associated with coexisting disease in other organ systems. Where diabetes occurs in conjunction with a condition for which a dietary strategy is well established, it seems prudent to advocate the strategy relevant to the other condition while attempting to control blood glucose concentrations with exogenous insulin, oral hypoglycemics or weight loss as appropriate.

References

1. Ballard, F. J. (1965) Glucose utilisation in mammalian liver. *Comparative Biochemistry and Physiology*, **14**, 437–443.
2. Bauer, J. (1988) Feline lipid metabolism and hepatic lipidosis. *Proceedings of the Twelfth Kal Kan Symposium*. pp. 75–78.
3. Birchard, S. J., Peterson, M. E. and Jacobson, A. (1984) Surgical treatment of hyperthyroidism: Results of 85 cases. *Journal of the American Animal Hospital Association*, **20**, 705–709.
4. Blaxter, A. (1990) Effects of ovariohysterectomy in nine entire bitches with diabetes mellitus. *BSAVA Congress Proceedings*, p. 190.
5. Blaxter A. C., Cripps P. J. and Gruffydd–Jones T. J. (1990) Dietary fibre and post prandial hyperglycaemia in normal and diabetic dogs. *Journal of Small Practice*, **31**, 229–233.
6. Bonagura, J. D. (Ed.) *Current Veterinary Therapy XI Small Animal Practice*. pp. 330–334. Philadelphia: W. B. Saunders.
7. British Diabetic Association (1992) Dietary Recommendations for people with diabetes: an update for the 1990s. *Diabetic Medicine*, **9**, 189–202.
8. Brunzell, J. D., Lerner, R. L. and Hazzard, W. R. (1971) Improved glucose tolerance with high carbohydrate feeding in mild diabetes. *New England Journal of Medicine*, **284**, 531–544.
9. Brunzell, J. D., Lerner, R. L. and Porte, D. (1974) Effect of fat free high carbohydrate diet on diabetic subjects with fasting hyperglycaemia. *Diabetes*, **23**, 138–142.
10. Cahill, G. F. (1988) Beta-cell deficiency, insulin resistance or both? *New England Journal of Medicine*, **318**, 1268–1270.
11. Campbell, K. L. and Latimer, K. S. (1984) Transient diabetes mellitus associated with prednisone therapy in a dog. *Journal of the American Veterinary Medical Association*, **185**, 299–301.
12. Chastain, C. B. (1992) Unusual manifestations of hypothyroidism in dogs. In *Current Veterinary Therapy XI Small Animal Practice*. Eds. R. W. Kirk and J. D. Bonagura. pp. 330–334. Philadelphia: W. B. Saunders.

13. Chastain, C. B. and Ganjam, V. K. (1986) Hypothyroidism and thyroiditis. In *Clinical Endocrinology of Companion Animals*. pp. 135–156. Philadelphia: Lea and Febiger.

14. Chastain, C. B. and Ganjam, V. K. (1986) Diabetes mellitus. In *Clinical Endocrinology of Companion Animals*. pp. 257–302. Philadelphia: Lea and Febiger.

15. Chastain, C. B. and Ganjam, V. K. (1986) Spontaneous hyperadrenocorticism. In *Clinical Endocrinology of Companion Animals*. pp. 363–395. Philadelphia: Lea and Febiger.

16. Church, D. B. (1981) The blood glucose response to three prolonged duration insulins in canine diabetes mellitus. *Journal of Small Animal Practice*, **22**, 301–310.

17. Curry, D. L. and Bennett, L. L. (1973) Dynamics of insulin release by perfused rat pancreas: effects of hypophysectomy, growth hormone, adrenocorticotrophic hormone and hydrocortisone. *Endocrinology*, **93**, 602–609.

18. Curry, D. L., Morris, J. G., Rogers, Q. R. and Stern, J. S. (1982) Dynamics of insulin and glucagon secretion by the isolated cat pancreas. *Comparative Biochemistry and Physiology*, **72A**, 333–338.

19. Dixon, J. B. and Sandford, J. (1962) Pathological features of spontaneous canine diabetes mellitus. *Journal of Comparative Pathology*, **72**, 153–164.

20. Drochner, W. and Muller–Schlosser, S. (1980) Digestibility and tolerance of various sugars in cats. In *Nutrition of the Dog and Cat*. Ed. R. S. Anderson. pp. 101–111. Oxford: Pergamon.

21. Eigenmann, J. E. (1989) Pituitary–hypothalamic diseases. In *Textbook of Veterinary Internal Medicine*. Ed. S. J. Ettinger. 3rd edn. pp. 1579–1609. Philadelphia: W. B. Saunders.

22. Fajans, S. S., Christensen, H. N., Floyd, J. C. and Pek, S. (1974) Stimulation of insulin and glucagon release in the dog by a non-metabolisable arginine analogue. *Endocrinology*, **94**, 230–233.

23. Feldman, E. and Nelson, R. (1987) Diabetes mellitus. In *Canine and Feline Endocrinology and Reproduction*. pp. 229–273. Philadelphia: W. B. Saunders.

24. Feldman, E. C. and Nelson, R. W. (1992) Desoxycortisone pivalate (DOCP) treatment of canine and feline hypoadrenocorticism. In *Current Veterinary Therapy XI Small Animal Practice*. Eds. R. W. Kirk and J. D. Bonagura. pp. 353–355. Philadelphia: W. B. Saunders.

25. Feldman E. C., Nelson R. W. and Karam J. H. (1983) Reduced antigenicity of pork insulin in dogs with spontaneous insulin-dependent diabetes mellitus (IDDM). *Diabetes, 32*, 153A.

26. Finn, J. P., Martin, C. L. and Manns, J. G. (1970) Feline pancreatic islet cell hyalinosis associated with diabetes mellitus and lowered serum-insulin concentration. *Journal of Small Practice, 11*, 607.

27. Fuessl, H. S., Williams, G., Adrian, T. E. and Bloom, S. R. (1987) Guar sprinkled on food: effect on glycaemic control, plasma lipids and gut hormones in non-insulin dependent diabetes mellitus. *Diabetic Medicine, 4*, 463–468.

28. Goeders, L. A., Esposito, L. and Peterson, M. E. (1987) Absorption kinetics of regular and isophane insulin in the normal dog. *Domestic Animal Endocrinology, 4*, 43–50.

29. Graham, P. A., Nash, A. S. and McKellar, Q. A. (1992) The pharmacokinetics of a highly purified porcine insulin zinc suspension (IZS-P) in dogs with naturally-occurring diabetes mellitus. *BSAVA Congress Proceedings*, p. 167.

30. Graham, P. A., Maskell, I. E. and Nash, A. S. (1993) The effects of feeding a commercially produced high fibre diet on post prandial hyperglycaemia in naturally occurring diabetic dogs. *BSAVA Congress Proceedings*, p. 195.

31. Haines, D. M. and Penhale, W. J. (1985) Autoantibodies to pancreatic islet in canine diabetes mellitus. *Veterinary Immunology and Immunopathology*, **8**, 149.

32. Hoenig, M. and Ferguson, D. C. (1989) Diabetes mellitus in the dog and cat. *Companion Animal Practice, 19*, 12–16.

33. Ide, T., Horii, M., Yamamoto, T. and Kawashima, K. (1990) Contrasting effects of water-soluble dietary fibres on bile acid conjugation and taurine metabolism in the rat. *Lipids*, **25(6)**, 335–340.

34. Ihle, S. L. and Nelson, R. W. (1991) Insulin resistance and diabetes mellitus. *The Compendium of Continuing Education: Small Animal*, **13**, 197–205.

35. Jenkins, D. J. A. (1980) Dietary fibre and carbohydrate metabolism. In *Medical Aspects of Dietary Fibre*. Eds. G. A. Pilar and R. M. Kay. pp. 175–191. New York: Plenum.

36. Jenkins, D. J. A., Goff, D. V., Leeds, A. R. Alberti, R. G., Wolever, T. M. S, Gassuld, M. A. and Hockaday, T. D. R. (1976) Unabsorbable carbohydrates and diabetes: decreased post prandial hyperglycaemia. *Lancet, 2*, 170–174.

37. Jenkins D. J., Wolvever, T. S. and Bacon, S, (1980) Diabetic diets: high carbohydrate combined with high fibre. *American Journal of Clinical Nutrition*, **33**, 1729–1733.

38. Johnson, K. H., Hayden, D. W. and O'Brien T. D. (1986) Spontaneous diabetes mellitus—islet amyloid

complex in adult cats. *American Journal of Pathology*, **125**, 416–419.

39. Joslin, E. P. (1928). *The Treatment of Diabetes Mellitus*. 4th edn. Philadelphia: Lea and Febiger.
40. Kanazawa, Y., Kuzuya, T., Ide, T. and Kosaka, K. (1966) Plasma insulin responses to glucose in femoral, hepatic and pancreatic veins in dogs. *American Journal of Physiology*, **2**, 442–448.
41. Kaneko, J. J., Mattheeuws, D., Rottiers, R. P. and Vermeulen, A. (1977) Glucose tolerance and insulin response in diabetes mellitus of dogs. *Journal of Small Animal Practice*, **18**, 85–94.
42. Karlstrom, B., Vessby, B., Asp, N.-G., Boberg, M., Lithell, H. and Berne, C. (1987) Effects of leguminous seeds in NIDDM patients. *Diabetes Research*, **5**, 199–205.
43. Kettlehut, I. C., Foss, M. C. and Migliorini, R. H. (1978) Glucose homeostasis in a carnivorous animal (cat) and in rats fed a high-protein diet. *American Journal of Physiology*, **239**, R115–R121.
44. Kirk, C. A., Feldman, E. C. and Nelson, R. W. (1993) Diagnosis of naturally acquired Type I and Type II diabetes mellitus in cats. *American Journal of Veterinary Research*, **54**, 463–467.
45. Laflamme *et al.* (1993) Enteral nutrition as an adjunct to the management of a complicated case of diabetes mellitus in a cat. *Journal of the American Animal Hospital Association*, **29**, 64–266.
46. Lehninger, A. L. (1982) Glycolysis: A central pathway of glucose catabolism. In: *Principles of Biochemistry*. pp. 397–434. New York: Worth.
47. Ling, G. V., Lowenstine, L. J., Pulley, L. T. and Kaneko, J. J. (1977) Diabetes mellitus in dogs: a review of initial evaluation, immediate and long-term management, and outcome. *Journal of American Veterinary Medical Association*, **170**, 521–530.
48. Manshire, A., Henry, C. and Hartog, M. (1982) Unrefined dietary carbohydrate and dietary fibre in treatment of diabetes mellitus. *Journal of Human Nutrition*, **35**, 99–101.
49. Marmor , M., Willeberg, P. and Glickman, L. T. (1982) Epizootiologic patterns of diabetes mellitus in dogs. *American Journal of Veterinary Research*, **43**, 465–470.
50. Maskell, I. E. (1993) The effect of supplementing a high starch diet with processed and unprocessed dietary fibre on postprandial glycaemia and insulinaemia in healthy Labrador Retrievers. *BSAVA Congress Proceedings*, p. 210.
51. Mattheeuws, D., Rottiers, R., Baeyens, D. and Vermeulen, A. (1984) Glucose tolerance and insulin response in obese dogs. *Journal of the American Animal Hospital Association*, **20**, 287–293.
52. Mattheeuws, D., Rottiers, R., Kaneko, J. J. and Vermeulen, A. (1984) Diabetes mellitus in dogs: relationship of obesity to glucose tolerance and insulin response. *American Journal of Veterinary Research*, **45**, 98–103.
53. Meier, H. (1960) Diabetes mellitus in animals. *Diabetes*, **6**, 485–489.
54. Milne, E. M. (1987) Diabetes mellitus: an update. *Journal of Small Animal Practice*, **28**, 727–736.
55. Moise, N. S. and Reimers T. J. (1983) Insulin therapy in cats with diabetes mellitus. *Journal of the American Veterinary Medical Association*, **182**, 158–164.
56. Mooney, C. T., Thoday, K. L. and Doxey, D. L. (1992) Carbimazole therapy of feline hyperthyroidism. *Journal of Small Animal Practice*, **33**, 228–235.
57. Morgan, L. M., Goulder, T. J. and Tsioladis, D. (1979) The effect of unabsorbable carbohydrate on gut hormones: modification of postprandial GIP secretion by guar. *Diabetologia*, **17**, 85–89.
58. Nelson, R. W. (1989) Disorders of the endocrine pancreas. In *Textbook of Veterinary Internal Medicine*. Ed. S. J. Ettinger. 3rd edn. pp. 1677–1720. Philadelphia: W. B. Saunders.
59. Nelson, R. W. (1989) The role of fiber in managing diabetes mellitus. *Veterinary Medicine*, **84**, 1156–1160.
60. Nelson, R. W. (1992) Disorders of the endocrine pancreas. In *Essentials of Small Animal Internal Medicine*. Eds. R. W. Nelson and C. G. Couto. pp. 561–586. Missouri: Mosby Year Book.
61. Nelson, R. W. and Feldman, E. C. (1992) Insulins: Characteristics and indications in diabetic dogs and cats. *Proceedings of the ACVIM*, pp. 354–356.
62. Nelson, R. W. and Feldman, E. C. (1992) Noninsulin-dependent diabetes mellitus in the cat. *Proceedings of the ACVIM*, pp. 351–353.
63. Nelson, R. W. and Feldman, E. C. (1992) Treatment of feline diabetes mellitus In *Current Veterinary Therapy XI Small Animal Practice*. Eds. R. W. Kirk and J. D. Bonagura. pp. 364–367. Philadelphia: W. B. Saunders.
64. Nelson R. W., Himsel, C. A., Feldman, E. C. and Bottoms, G. D. (1990) Glucose tolerance and insulin response in normal weight and obese cats. *American Journal of Veterinary Research*, **51(9)**, 1357.
65. Nelson, R. W., Ihle, F. L. Lewis, L. D., Salisbury, S. K., Miller, T., Bergdall, V. and Bottoms, G. D. (1991) Effects of dietary fiber supplementation on glycemic control in dogs with alloxan-induced diabetes mellitus. *American Journal of Veterinary Research*, **52(12)**, 2060–2066.

66. Nichols, R. (1992) Recognising and treating canine and feline diabetes mellitus. *Veterinary Medicine,* **87(3),** 211–222.

67. Nichols, R. (1992) Management of the difficult to regulate diabetics. *Proceedings of the ACVIM,* pp. 314–316.

68. Norsworthy, G. D. (1993) The difficulties in regulating diabetic cats. *Veterinary Medicine,* **88,** 342–348.

69. O'Brien T. D., Hayden, D. W. and Johnson, K. H. (1981) High dose intravenous glucose tolerance test and serum insulin and glucagon levels in diabetic and non diabetic cats: relationships to insular amyloidosis. *Veterinary Pathology,* **22,** 250–261.

70. O'Brien T. D., Hayden, D. W., Johnson, K. H. and Fletcher, T. S. (1986) Immunohistochemical morphometry of pancreatic endocrine cells in diabetic, normoglycaemic glucose-intolerant and normal cats. *Journal of Comparative Pathology,* **96,** 357–369.

71. Panciera, D. L., Thomas, C. B., Eicker, S. W. and Atkins, C. E. (1990) Epizootiologic patterns of diabetes mellitus in cats: 333 cases (1980–1986). *Journal of the American Veterinary Medical Association,* **187,** 1504–1508.

72. Peterson, M. E. (1992) Insulin and insulin syringes. In *Current Veterinary Therapy XI Small Animal Practice.* Eds. R. W. Kirk and J. D. Bonagura. pp. 356–358. Philadelphia: W. B. Saunders.

73. Peterson, M. E. and Randolph, J. F. (1989) Endocrine diseases. In *The Cat: Diseases and Clinical Management.* Ed. R. G. Sherding. pp. 1095–1163. New York: Churchill Livingstone.

74. Peterson, M. E., Kintzer, P. P. and Hurvitz, A. I. (1988) Methimazole treatment of 262 cats with hyperthyroidism. *Journal of Veterinary Internal Medicine,* **2,** 150–157.

75. Rabinowich, I. M. (1935) Effects of high carbohydrate-low calorie diet upon carbohydrate tolerance in diabetes mellitus. *Canine Medical Association Journal,* **33,** 136–144.

76. Reusch, C. E. Liehs, M. R., Hoyer, M. and Vochezer, R. (1993) Fructosamine: a new parameter for diagnosis and metabolic control in diabetic dogs and cats. *Journal of Veterinary Internal Medicine,* **7,** 177–182.

77. Sai, P., Debray–Sachs, M., Jondet, A., Gepts, W. and Assaw, R. (1984) Anti-beta-cell immunity in insulinopenic diabetic dogs. *Diabetes,* **33,** 135–140.

78. Salans L. B., Knittle, J. R. and Hirsch, J. (1983) Obesity, glucose intolerance and diabetes mellitus. In *Diabetes Mellitus: Theory and Practice.* Eds. M. Ellenburg and H. Rifkin. 3rd edn. pp. 469–479. New York: Medical Examination.

79. Shaer, M. A. (1977) Clinical survey of thirty cats with diabetes mellitus. *Journal of the American Animal Hospital Association,* **13,** 23–27.

80. Smerdon, T. (1990) A case of diabetes mellitus in the cat. *Bulletin of the Feline Advisory Bureau,* **27(4),** 70; continued in **28(1),** 51.

81. Sun, J. V., Tepperman, H. M. and Tepperman, J. A. (1977) Comparison of insulin binding by liver plasma membranes of rats fed a high glucose diet or a high fat diet. *Journal of Lipid Research,* **18,** 533–539.

82. Thoday K. L. and Mooney, C. T. (1992) Medical management of feline hyperthyroidism. In *Current Veterinary Therapy XI Small Animal Practice.* Eds. R. W. Kirk and J. D. Bonagura. pp. 338–345. Philadelphia: W. B. Saunders.

83. Thoday, K. L. and Mooney, C. T. (1992) Historical, clinical and laboratory features of 126 hyperthyroid cats. *Veterinary Record,* **131,** 257–264.

84. Tunbridge, W. M. G. and Home, P. D. (1991) Treatment of diabetes mellitus. In *Diabetes and Endocrinology in Clinical Practice. Clinical Practice Series.* Eds. W. M. G. Tunbridge and P. D. Home. pp. 92–133. London: Edward Arnold.

85. Unger, R. H. and Foster, D. W. (1985) Diabetes mellitus. In *William's Textbook of Endocrinology.* Eds. J. D. Wilson and D. W. Foster. pp. 1018–1080. Philadelphia: W. B. Saunders.

86. Vaaler, S. (1986) Diabetic control is improved by guar gum and wheat bran supplementation. *Diabetic Medicine,* **3,** 230–233.

87. Villaume, C., Beck, B. and Gariot, P. (1984) Long term evaluation of the effect of bran ingestion on meal-induced glucose and insulin responses in healthy men. *American Journal of Clinical Nutrition,* **40,** 1023–1026.

88. Wallace, M. S., Peterson, M. E. and Nichols, C. E. (1990) Absorption kinetics of regular, isophane and protamine zinc insulin in normal cats. *Domestic Animal Endocrinology,* **7,** 509–516.

89. Weinstock, R. S. and Levine, R. A. (1988) The role of dietary fibre in the management of diabetes mellitus. *Nutrition,* **4,** 187–193.

90. World Health Organisation (1985) Diabetes Mellitus Technical Report Series 727.

91. Wheeler, S. L. (1988) Emergency management of the diabetic patient. *Seminars in Veterinary Medicine and Surgery: Small Animal,* **3,** 265–273.

92. Wolfe, A. M. (1988) Management of the diabetic cat. *Proceedings of the Twelfth Kal Kan Symposium,* pp. 91–95.

93. Wolfsheimer, K. J. (1989) Insulin resistant diabetes mellitus. In *Current Veterinary Therapy X*. Ed. R. W. Kirk. pp. 1012–1019. Philadelphia: W. B. Saunders.

94. Wolfsheimer, K. J. (1990) Problems in diabetes mellitus management. *Problems in Veterinary Medicine,* **2,** 591–601.

95. Yano, B. L., Hayden, D. W. and Johnson, K. H. (1981) Feline insular amyloid: association with diabetes mellitus. *Veterinary Pathology,* **18,** 621–627.

CHAPTER 22

Skeletal Disease

HERMAN A. W. HAZEWINKEL

Introduction

Many nontraumatic orthopedic disorders may be related to food intake. A deficiency, but more frequently an excess, of one or more nutrients may play an important role in a variety of skeletal diseases.

In this chapter, the relationship between the most relevant skeletal diseases and prolonged deviant food intake will be discussed. In most cases, optimal food intake prevents the described skeletal diseases, or at least diminishes their severity. For a few, dietary correction alone is sufficient to restore skeletal integrity. In some, however, permanent effects on important organ systems obstruct any therapy. In most of the disorders, dietary correction should precede surgical treatment.

In addition, nutritional management as a therapeutic factor is considered for a number of orthopedic conditions not caused by faulty diet.

Bone Composition

Bone is a specialized form of connective tissue with a complex chemical and physical composition. Apart from its cellular fraction and the water phase (10%), it is composed of organic matrix and a mineral phase. The cellular fraction includes osteoblasts (organic matrix-forming cells) and osteoclasts (calcified matrix-resorbing cells). The organic matrix, which is about 20% of the bone volume, is composed of 90% collagen fibers with a high content of hydroxyproline and 10% aminopolysaccharides, noncollagen proteins and a small quantity of lipids. The mineral phase encompasses about 70% of the bone volume, mainly in the form of hydroxyapatite crystals and amorphous calcium phosphate, as well as small quantities of other elements. Of the total body calcium and phosphorus, 99% and 80%, respectively, are present in the skeleton.

Skeletal Growth

Skeletal growth includes cartilage maturation as part of endochondral ossification, osteoid formation and mineralization, as well as the bone modelling influenced by skeletal homeostasis and calcium homeostasis. The latter is under the influence of parathyroid hormone (PTH), calcitonin and vitamin D metabolites.

Young dogs and cats should have sufficient

food during the growth phase to meet their needs, which can be 2.5–3.6 times the adult maintenance value.[43,59] A food of good quality and palatability and shown to be nutritionally adequate for growth should be provided. Optimal nutrition in an otherwise satisfactory environment permits optimal growth as ordained by the genotype.

Although zinc, copper and other elements may influence the development of skeletal abnormalities,[42] they appear to be of minor significance in small animal orthopedics. Vitamins, including the fat-soluble vitamins A and D, have both direct and indirect influences on skeletal mineralization.

Skeletal Disease Caused by Nutritional Errors

Undernutrition During Growth

Slight underfeeding, with respect to the energy intake, may slow the growth of puppies but will not influence the adult size of the dog. After a period of inhibited growth due to malnutrition or illness of short duration, the animal will grow at a greater rate than average for its age (catch-up growth). Influences on growth factors other than nutrition must be considered when judging an animal that is small for its age, including the extent of variation in growth patterns of normal healthy animals.

Young animals have a great need for calcium to mineralize newly formed cartilage and osteoid. Depending on the dietary regimen and hormonal status, 225–900 mg/kg body weight (BW)/day is deposited in the skeleton, of which 100–225 mg/kg/day should be absorbed from the intestine.[22,30,42] Low calcium absorption, either from a low or poorly available calcium content of the food, can cause a decrease in the circulating calcium concentration, resulting in *nutritional secondary hyperparathyroidism*. This situation can be created by feeding an all-meat diet.

This condition results in massive osteoclasia, causing the skeleton to weaken to such an extent that it can withstand neither body weight nor muscle forces. The result is the development of skeletal deformities, which include bowing of the long bones and calcaneus, compression fractures in cancellous bone (metaphyseal and epiphyseal areas), deformation of the pelvis and vertebrae and greenstick fractures of long bones. No changes occur at the growth plates.

Vitamin D metabolites stimulate calcium and phosphate absorption, increase bone cell activity and influence endochondral ossification and calcium excretion. Vitamin D is absorbed by the intestine, hydroxylated in the liver to 25-OH vitamin D, and then further hydroxylated in the kidney to $24,25\text{-}OH_2$ vitamin D or to the most active metabolite, $1,25\text{-}OH_2$ vitamin D. Unlike other species, puppies do not synthesize sufficient quantities of vitamin D (by UV irradiation) to meet all their requirements. A dietary deficiency of vitamin D may therefore result in *hypovitaminosis D (rickets)*, although clinical cases in cats and dogs are exceedingly rare.[27]

Undernutrition in Adults

Permanent bone turnover occurs in the adult dog and cat and consists of bone resorption as well as new bone formation. In the adult dog, calcium deposition and resorption are equal in magnitude: approximately 4–8 mg/kg BW/day. Daily losses of calcium by endogenous fecal and urinary excretion (i.e. 10–30 and 1–7 mg/kg BW/day, respectively) can easily be compensated for by a balanced diet.[42]

Natural menopause and other causes of estrogen deficiency in women are characterized by bone loss, leading to pathological fractures of the vertebrae, proximal femur and wrist. This explains the extensive concern for nutrition in elderly women. Although ovariectomy is a common practice in female dogs and cats, and some osteoporotic changes are noticeable 17–36 weeks after surgery,[12,41] this practice does not result in a problem of practical significance in companion animal orthopedics.

Overnutrition During Growth

In order to avoid the classic skeletal diseases that are due to calcium deficiency, phosphate excess or vitamin D deficiency, some owners tend to oversupplement the diet with calcium, with or without a proportional addition of phosphate.[32] A high calcium content increases the circulating calcium concentration and eventually increases calcitonin secretion and decreases PTH secretion. A chronic hypercalcitoninemic state causes decreased activity of osteoclasts, which are of utmost importance for skeletal modelling during growth.

In young animals, unlike adults, excessive energy intake does not cause a substantial increase in fat deposition, but rather, a more rapid rate of growth. From different studies on young dogs, it can be concluded that, provided there is an adequate protein and essential fatty acid supply, it does not appear to matter for growing dogs if the proportion of energy comes from carbohydrate, fat or protein. If the diet supplies sufficient amounts of specific nutrients, the amount of energy will regulate the rate of growth within the animal's genetic possibilities.[18]

A variety of studies in large and giant breed dogs have shown that puppies should not be fed with the aim of giving the maximum possible rate of weight gain that the dog is capable of, even though the period of growth will be reduced. Overnutrition resulting in excess body weight and thus overloading of the juvenile skeleton and its supportive system may contribute to the development of a variety of multifactorial diseases that have been described in *ad libitum* fed dogs, including *osteochondrosis*,[29,37] *hip dysplasia*[29,33] and fissures and fragmentation of the coronoid process.[18] There are a number of manifestations of the osteochondrosis syndrome, including *osteochondritis dissecans,* retained ulnar cartilage core (and the *radius curvus* syndrome) and fragmentation of the coronoid process.[54] Decreased skeletal remodelling, often seen in conjunction with osteochondrosis, is a feature of two further disease entities, the *wobbler syndrome* and *enostosis*.[25,29]

Because a high protein diet (30% on a dry matter basis) did not increase the frequency or severity of skeletal abnormalities in giant breed dogs (as compared with control dogs when fed isoenergetic diets),[46] it may be concluded that it is excess BW during rapid growth, rather than high protein content of the diet, that can be deleterious for skeletal development.

An additional problem that dogs may encounter during the period of rapid growth (chiefly between 1½–3 months-of-age) is abnormal angulation of the carpal joint (hyperextension or hyperflexion), probably due to a discrepancy between the increase in BW and the carpal support (Fig. 22.1). Enforced bandaging for 10 days will allow the carpus to gain strength and restore the normal posture of the dog.

Skeletal abnormalities as listed above are seldom diagnosed in dogs of small breeds or (with the exception of hip dysplasia) in cats. To date, there are no indications that patella luxations in dogs or cats, or avascular femoral head necrosis (Legg–Calvé–Perthes disease) in dogs are related to nutrition.

However, owners of all young dogs and cats should be aware of the severe clinical effects that may result from oversupplementation of the diet with the fat-soluble vitamins A and D (which may accumulate within the body). *Hypervitaminosis A* develops, particularly in cats, after the prolonged intake of foods rich in vitamin A (such as raw liver, fish liver oils), resulting in severe skeletal abnormalities. Although young pets need approximately twice the amount (200 IU/kg BW) of vitamin A required by adults, supplementation of a commercial petfood with a vitamin additive can result in giving 100 times the normal requirement.[32]

Massive intakes of vitamin D (or even more so of its metabolites) can cause hypercalcemia together with hyperphosphatemia, anorexia, polyuria, vomiting, muscle weakness and lameness. The circulating high levels of calcium and phosphate are due to increased bone resorption, increased absorption from the gastrointestinal tract and, eventually, to tubular mineralization. Vitamin D intoxication in

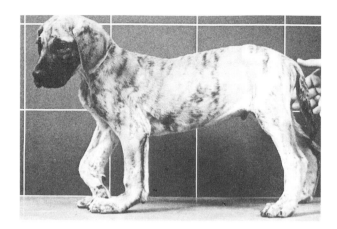
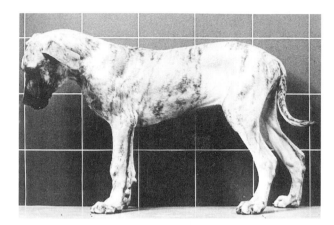

FIG. 22.1 Great Dane pup, age 8 weeks, with hyperflexion and hyperextension of the carpal joints (left), and after bandaging for 10 days (right).

dogs and cats is characterized by mineralization of the soft tissues, including blood vessels, alveoli and renal tubules, together with pathological changes in the gastrointestinal tract and the heart.[19,32,52] Many petfoods contain more than 10 times the NRC minimum recommended levels of vitamin D.[32] Extra supplementation with vitamin D can cause increased calcium and phosphate absorption, with deleterious effects on skeletal development and, probably, also on kidney function.[32]

Overnutrition in Adults

Feeding adult animals a diet based on some variation of the dietary requirements prescribed for growing animals does not usually result in serious problems. The most important clinical problems in relation to orthopedics in adult companion animals are excessive energy intake, excessive phosphorus intake together with decreased renal function and excessive vitamin A intake, especially in cats.

In most countries of the Western world, the major nutrition-related clinical problems in orthopedics are due to an excessive energy intake that, in the normal active companion animal, will be stored as body fat. In a study of dogs visiting veterinary practices in the U.K.,[11] between 28 and 44% of the dogs were recorded as overtly obese. About 25% of the obese dogs had orthopedic problems, whereas 10% of all dogs seen in a variety of veterinary

practices had orthopedic problems.[11] From this study, it is not clear if the high body weight preceded the orthopedic problems, which included osteoarthrosis, herniated intervertebral disks and cruciate ligament ruptures. It may also be possible that existing abnormalities are manifested when the skeleton is overloaded; the clinical complaints of arthrosis, for example, will be aggravated when the leg is overloaded, due to overuse or excess body weight.

No dietary constituents have been implicated as the cause of these conditions, other than the mechanical effect of increased BW that might cause a shear or tear of the stabilizing structures and an overload of the articular cartilage. A weight control program (Chapter 10), as well as restricted exercise and medication when indicated, may play an important role in the conservative treatment.

As with young animals, oversupplementation with vitamins A and D can be detrimental to adult dogs and cats. Hypervitaminosis A is most frequently seen in adult cats, and may result in ankyloses of vertebral, shoulder and elbow joints, in particular. The effects of vitamin D intoxication in adults are the same as those described for young animals.

The calcium and phosphate content of commercial petfoods can significantly exceed the recommended minimum levels, although this may not have a deleterious effect on healthy adults.[32] However, dogs and cats with severe renal dysfunction cannot efficiently excrete phosphate, which accumulates and results in a lowering of circulating calcium concentrations. To compensate for this, more parathyroid hormone is secreted *(renal secondary hyperparathyroidism)*, and this leads to osteodystrophy, particularly of the mandible and maxilla (rubber jaw). However, the systemic effects of renal failure are of greater clinical significance.

Nutritional Therapy in Skeletal Disease

Other than dietary correction for specific nutritional errors, a number of skeletal conditions may benefit from dietary manipulation.

Any obese animal affected by arthrosis is likely to benefit from a weight reduction program.

Fractures

When properly treated in a healthy animal, traumatic fractures heal via the process of primary or secondary bone healing. The latter has much in common with endochondral ossification and thus needs the same nutritional support. Optimal conditions can be reached with a balanced commercial petfood containing optimal levels of calcium, phosphate, vitamin A and vitamin D. Excessively high doses of these nutrients have been shown to retard bone healing.[8] Pain due to trauma or surgery causes distress, which depletes reserves of protein and diminishes immune competence.[36] In addition, the dietary requirements of ascorbic acid, and probably of other nutrients, are increased. In surgical patients, the fasting period before and after anesthesia may be detrimental. Therefore, a palatable food formulated to meet the needs of young, growing animals must be considered. Even obese patients should be kept in a positive energy balance, although this demands special management of fracture treatment and post-operative mobilization.

Pathological fractures (including compression fractures in cancellous bone and greenstick fractures in cortical bone) arise due to

poor mineralization of the skeleton. This may be caused by low calcium intake, high phosphate intake or low vitamin D intake. Where spontaneous fractures, especially of the mandibles, occur in old dogs, severe renal dysfunction should be suspected. Dietary correction and, therefore, normalized skeletal mineralization must precede fracture treatment or surgical correction.

Renal Osteodystrophy: Renal Secondary Hyperparathyroidism, Rubber Jaw

In dogs and cats with severe loss of kidney function, phosphate will accumulate and thereby decrease circulating calcium concentrations. This may occur in combination with decreased hydroxylation of 25-OH vitamin D in the failing kidney, causing diminished calcium absorption. As a result, the parathyroid glands will secrete more PTH (renal secondary hyperparathyroidism), causing an

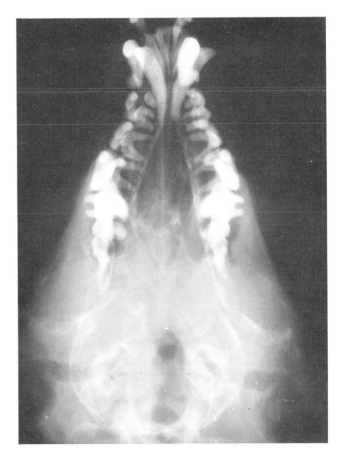

FIG. 22.2 Renal secondary hyperparathyroidism causes decalcification of mandible and maxilla.

increase in osteoclastic activity to normalize circulating calcium concentrations, with the eventual development of osteodystrophy and loosening of teeth in the mandible and maxilla (Fig. 22.2)

Renal osteodystrophy is characterized microscopically by increased osteoclast activity and inadequate mineralization of normal osteoid. High phosphate intake will, therefore, aggravate renal failure by: promoting mineral depositions in soft tissues, not only the kidney, but also periarticular tissue, tendon sheaths and foot pads; and decreasing circulating calcium concentrations with the subsequent production of hyperparathyroidism.[40]

It is generally believed that in clinical cases of renal failure, a high protein intake may cause even more renal damage and more accumulation of uremic toxins, reflected by rising urea and creatinine concentrations. Using the plasma creatinine concentration as a parameter for kidney function, dietary protein (of high biological value) can be reduced to 4–9 g/1000 kJ, together with a gradual decrease in dietary phosphate to 6–8 mg/1000 kJ (Chapters 16 and 18).

Skeletal Tumors

Diet is a factor in carcinogenesis; dietary deficiencies or excesses do not cause cancer but can modify it. Vitamin A, or its synthetic forms, has been demonstrated to inhibit tumor growth during the promotional phase, whereas vitamin A deficiency has been related to the development of odontomas. Not only vitamin A, but also vitamins C, D and E are claimed to be protective in the initiation phase of a variety of tumors.[72] Reports on selenium are conflicting; it can inhibit development of certain tumors, although it is called carcinogenic by other investigators.[62]

One of the known contributory causes of osteosarcoma is long standing nutritional osteodystrophy related to a high liver diet as has been identified in cats. It has been suggested that bone disorders with prolonged periods of excessive cellular stimuli are prone to neoplastic change.[67]

Specific Syndromes

Osteochondrosis

Osteochondrosis is a disturbance of endochondral ossification, characterized by an abnormal maturation of chondrocytes and thus a delay in cartilage mineralization. The condition is widespread among a variety of species, including dogs.[54,63] In dogs, osteochondrosis is seen primarily in certain large and giant breeds (i.e. over 25 kg adult BW), and is more common in males and rapidly growing females.[54,63] The breeds with the highest risk include the Great Dane, Labrador and Golden Retrievers, Newfoundland and Rottweiler.[44,63,69]

Where the disturbance in endochondral ossification occurs in articular cartilage, osteochondritis dissecans (OCD) may be a consequence. In OCD, part of the articular cartilage becomes detached and may be fragmented, mineralized or even ossified. This occurs together with inflammation of both the joint and the endochondral bone in the area of the cartilage lesion (Fig. 22.3).

The disturbance in endochondral ossification can also occur in growth plate cartilage, with irregular growth plates, enlarged cartilage cores and decreased growth in length as a consequence. Delay in ossification of the secondary ossification centers may be a further manifestation of the disturbance in endochondral ossification.

Osteochondrosis is a multifactorial disease in which the genetic makeup and nutrition play significant roles. The disease frequently affects a variety of breeds, with each breed having a prevalence for the specific location of osteochondrosis. OCD lesions are seen most commonly in the shoulder and stifle joints of Great Danes, the shoulder, elbow and hock joints of Labrador and Golden Retrievers, the elbow joints of Newfoundlands and the shoulder and hock joints of Rottweilers.[63] Animals affected at multiple sites are commonly seen. In one study, 66% of the dogs with OCD had more than one joint affected, in 5% three joints were affected.[63]

Although histologically osteochondrosis

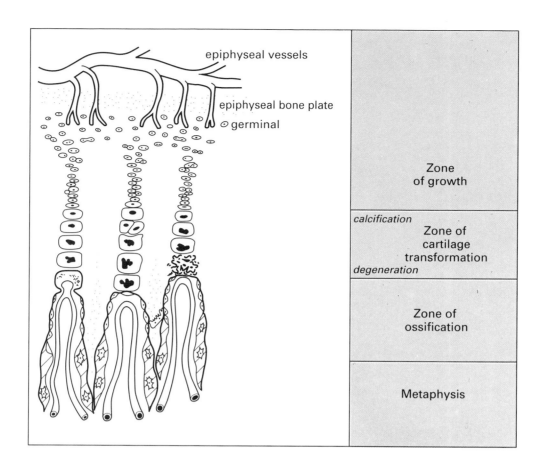

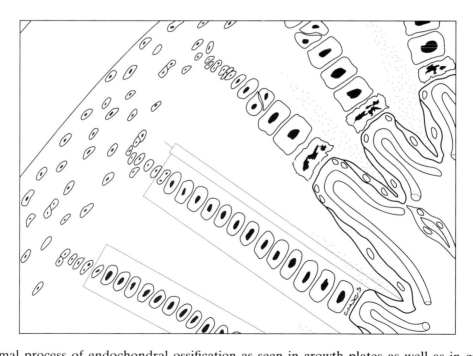

FIG. 22.3 Normal process of endochondral ossification as seen in growth plates as well as in growing articular cartilage (left). Osteochondrosis: due to a disturbance of cartilage-cell maturation, mineralization of the intercellular substance is delayed (right). Cartilage becomes thickened and vulnerable to microtrauma. Thickened cartilage due to disturbed chondrocyte maturation is known as osteochondrosis, whereas detached cartilage is called osteochondritis dissecans (OCD).

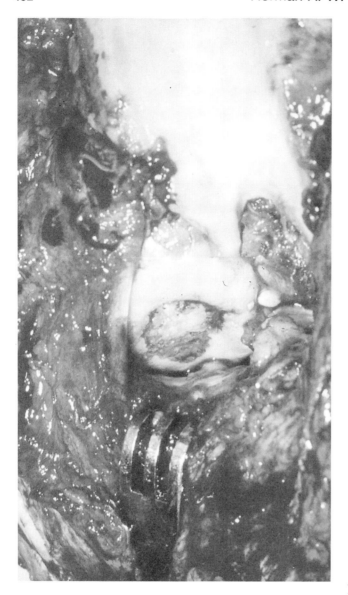

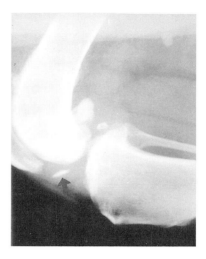

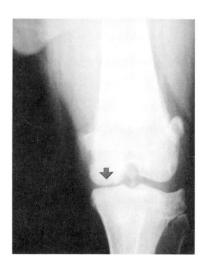

FIG. 22.4 Osteochondrotic lesion (arrow) in stifle joint of Great Dane, 8 months-of-age, as seen at surgery (left) and on radiographs (right).

may also be diagnosed in nonweight-bearing growth plates (such as that of the rib), microtrauma may play a significant role in causing the fissures in affected cartilage. This may be concluded from the fact that OCD is mostly seen on convex, weight-bearing areas (Fig. 22.4).

Disturbances in growth plates that lead to clinical manifestation of osteochondrosis (including bilateral radius curvus syndrome or rear leg exotorsion) are seen mainly in giant breed dogs. Detachment of the anconeal process of the ulna or the supraglenoid process of the scapula can also be seen in this respect.

Pathophysiology

In undisturbed endochondral ossification, the intercellular material between the columns of cartilage cells mineralizes when a certain stage of maturity is reached. Chondrocytes are then sealed off from their nutritional resources (i.e. synovia in the case of articular cartilage and epiphyseal vessels in the case of growth plate cartilage), which results in their death and disintegration. Endothelial cells of capillaries then invade the noncalcified horizontal septa between the chondrocytes, and the lacuna of the disintegrated chondrocyte. Osteoblasts align themselves along the partially

NUTRITIONAL RESEARCH IN GROWING GREAT DANES					
	Hedhammar[29]	Meyer[43]	Lavelle[37]	Nap[46]	Hazewinkel[25]
Skeletal changes	++ In *ad lib* group	+ In *ad lib* group	+ In *ad lib* group	– In neither group	++ In high Ca intake group
Food:					
Protein	36.0	39.5–48.1	29.6	21	21
Fat	13.7	7.7–13.4	14.4	9.9	9.9
Ca	2.05	0.82–3.0*	2.3	1.0	3.3
P	1.44	0.39–1.6*	1.6	0.9	0.9
ME in kcal/100 g DM (kJ/100g DM)	501 (2094)	407–438 (1700–1830)*	431 (1802)	359 (1500)	402 (1680)
Variables	*Ad lib or* 66% of the *ad* *lib* amount	*Ad lib* (with rice) *or* 70–80% *without rice	*Ad lib or* 60% of the *ad* *lib* amount	Protein content of 12.9% *or* 21% or 30% on a dry matter basis, but all isoenergetical	Calcium 1.1% *or* 3.3% on a dry matter basis. Both *ad* *lib*, but 1.1% group eat more

* In addition to other contents.

FIG. 22.5 Nutritional research in growing Great Danes.

resorbed cartilaginous cores and deposit osteoid that becomes bone when mineralized. Primary spongiosa (mineralized cartilage surrounded by mineralized osteoid) can be eroded by osteoclasts to become cancellous bone (more oriented osteoid fibers without a cartilaginous core) or, when present in the metaphyseal area of growing bone, it may be eroded locally to become a medullary cavity.

In osteochondrosis, maturation of the cartilage cells is disturbed and therefore mineralization of the intercellular substance, death and disintegration of chondrocytes, capillary ingrowth, osteoblast introduction, as well as bone formation, do not occur or are delayed. This causes elongated cartilage columns in articular cartilage as well as in growth plate cartilage (Fig. 22.3).

Because osteochondrosis is so frequently seen in large breed dogs, a variety of studies have been performed in different institutes to elucidate the role of nutrition in the manifestation of osteochondrosis.

The original study was performed by Hedhammar *et al.*[29] on 12 pairs of Great Danes raised on food rich in protein, calcium, phosphorus and energy. More frequently, skeletal diseases, including osteochondrosis and delayed skeletal modelling, were observed in the dogs fed this formula *ad libitum*,

whereas the restricted-fed dogs (which received two-thirds of the amount of the *ad libitum* group) showed less severe signs. This opened a whole series of investigations performed by others (Fig. 22.5).

Overload, either by overnutrition using a basal food enriched with rice or simply by simulating high BW through the use of sand belts in the scapular region, did aggravate the signs of osteochondrosis.[43] In a controlled study in Great Danes with high food intake of more balanced commercial foods, the *ad libitum* fed dogs revealed more frequent osteochondrosis of the shoulder compared to the dogs fed 60% of the *ad libitum* amount.[37] In a recently published study performed in Great Danes raised on food high only in protein, no differences in either occurrence or severity of osteochondrosis occurred when compared with the normal or low protein fed dogs.[46] In studies in Great Danes where the food only differed in its calcium content (with or without a constant ratio to phosphorus), progressively more severe disturbances of osteochondrosis were seen in the proximal humerus as well as in growth plates of long bones and of nonweight-bearing areas, such as the ribs.[25]

This leads to the practical conclusion that chronic intake of too much of a balanced food[37] or of a food enriched with calcium[25] with or without other constituents[29,43] will cause an increase in frequency and severity of signs of osteochondrosis in young dogs of large breeds (Fig. 22.5). Minor changes of endochondral ossification, without clinical significance, have also been demonstrated in miniature poodles raized on a food with high calcium content.[45]

In young dogs, calcium is absorbed in the intestine by means of both uncontrolled passive diffusion and active, controlled absorption. It has been demonstrated that Great Danes raized on a food according to the NRC[48] recommendations (1.1% Ca) absorbed 45–60% of the ingested amount of calcium, whereas dogs with triple that amount of calcium in their diet, absorbed 23–43%. Thus, the dogs fed the high calcium-containing diet, absorbed considerably higher amounts of calcium.

Intake of food, and especially of calcium, causes the release of gastrointestinal hormones, some of which will cause calcitonin (CT) release from the thyroid glands. An increased CT concentration prevents calcium release from the skeleton by influencing the bone-resorbing osteoclasts. The absorbed calcium will, therefore, be routed to the skeleton without influencing the concentration of calcium in the extracellular fluid at each meal. In the fasting periods and periods between meals, this loosely deposited calcium is freed and used for a variety of life-saving processes. The series of studies in Great Danes have demonstrated that daily intake of a calcium rich diet leads to hyperplasia of CT producing cells, reduced osteoclast activity and disturbed endochondral ossification.[25,51] Although it is not fully understood if calcium plays a direct role in disturbing chondrocyte maturation, or if it is mediated by CT or a relative deficiency of other minerals at a cellular level, there is little doubt that high calcium intake has a deleterious effect on endochondral ossification, with osteochondrosis as a consequence.

Diagnosis

In dogs with osteochondrosis of articular cartilage, without detachment of cartilage, no specific clinical signs will be apparent. In those cases where the cartilage becomes detached (i.e. OCD), osteoarthrosis and inflammation of the subchondral bone will occur (Fig. 22.4). Osteochondritis dissecans is, therefore, associated with lameness, pain upon (hyper) extension and flexion of the affected joint, joint effusion and subchondral sclerosis bordering an indentation of the articular surface, which is visible on radiographs. In addition, a variety of diagnostic methods (including paracentesis, arthroscopy and other imaging techniques) can be employed.

Measurement of circulating concentrations of calcium or phosphorus will *not* provide a good insight into the dietary content or absorption rate of these elements, and does not give any indication to support the diagnosis of osteochondrosis. This is based on the fact that plasma calcium concentration is very well maintained within certain limits and phosphorus is related to calcium homeostasis, although its circulating level is affected more by intake. Alkaline phosphatase will be increased where there is a high rate of calcium deposition in the skeleton, but will be high in all young animals during the period of rapid growth. Calcium-regulating hormones, i.e. PTH, CT and vitamin D, can only be determined in highly specialized laboratories and will only provide an insight when measured repeatedly.

In summary, blood investigation will not be of great value to support the diagnosis in the acute phase of osteochondrosis development. With clinically suspected OCD, a thorough clinical and radiological investigation will suffice in most cases, although additional techniques may be required in other circumstances.

Where osteochondrosis occurs in the growth plates, no specific clinical signs may be expected when the retained cartilage core is small or temporary. In longitudinal radiological studies of the distal growth plate of the ulna of Great Danes, a slight flattening or indentation could be seen at the age of 5 months.[70] From clinical experience, it can be stated that most of these slight abnormalities will not cause clinical disturbances. When a severe flattening of the metaphyseal area

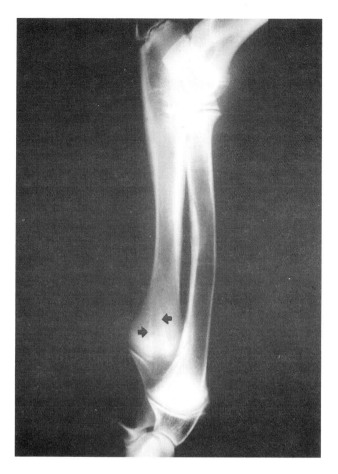 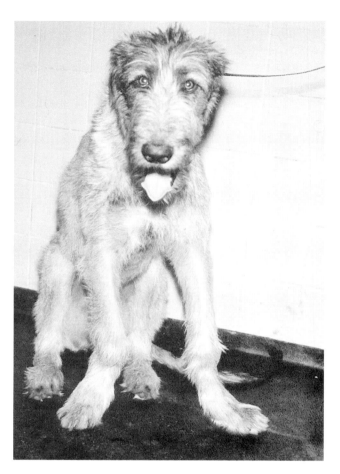

FIG. 22.6 Mild retained cartilage cores may disappear spontaneously (left) or develop into the radius curvus syndrome (right).

develops, or a deep cartilage core can be seen, an impaired growth in length of the radius and ulna can be expected (Fig. 22.6). The short ulna and curved radius, together with the valgus deformation of the feet, complete the radius curvus syndrome.

Prognosis

Osteochondrosis in articular cartilage will not develop into OCD in all cases. Based on controlled studies where both shoulder joints were radiographed, it can be concluded that, although 45–65% of the dogs had disturbed, radiologically detectable, abnormal contours of the humeral head, only 3–5% were clinically affected on both sides.[69] In cases where the cartilage flap becomes detached, surgical treatment can help to shorten the period of lameness and is likely to minimize secondary changes within the joint.

In the radius curvus syndrome, severe shortening of the ulna may cause irreversible, abnormal development of the carpus and/or detachment of the anconeal process. The latter is only possible when the elongated radius pushes the humerus against the incompletely ossified anconeal process.

Therapy

Dietary correction at an early stage may positively influence the spontaneous resolution of disturbed endochondral ossification.[70] Osteochondrosis in joint cartilage and growth plates could disappear, but dietary modification will not normalize cases of OCD in which there is detached cartilage, or where a more severe curvature of the radius exists.[54] Surgical correction will be indicated in most of these cases.

Dietary correction entails a decrease in intake of energy (protein, fat, carbohydrates), calcium and vitamins to the minimal requirements for dogs.[49] No pharmacological or other medications have been identified that may be used to support this nutritional management.

Decreased Skeletal Remodelling

Decreased skeletal remodelling may occur in two separate entities, canine wobbler syndrome and enostosis, which are seen either alone or in conjunction with osteochondrosis. As with osteochondrosis, the etiology may be multifactorial, but the influence of diet has been demonstrated by the results of studies in rapidly growing dogs of large breeds.[25,29] Canine wobbler syndrome is seen particularly in young Great Danes, Mastiffs and Irish Wolfhounds and is unrelated to the spondylolisthesis and consequent ligamentous hypertrophy as seen in the aged Doberman Pinscher. Enostosis, also named *panosteitis eosinophilica*, is seen in a variety of dog breeds at a young age, particularly in the German Shepherd dog.[39]

Pathophysiology

Skeletal growth occurs in two ways, growth in length and modelling in shape. The latter includes an adaptation to changes in body size, muscle-pull and BW. The load of hydroxyapatite crystals may cause a shift in electrons that can influence osteoblastic and osteoclastic activity. This and other still unexplained mechanisms may form the basis of Wolff's law, which states that bone is laid down where it is needed. However, the integrity of the skeleton is subordinate to calcium homeostasis, which includes the strict regulation of calcium concentration in the extracellular fluid, thus allowing a variety of biological processes, such as heart muscle contraction and the blood-clotting cascade, to be maintained.

A chronic excessive calcium intake will result in high calcium absorption in young dogs, especially of large breeds. Calcium is not significantly excreted in the urine or via the endogenous fecal pathway, but is mainly routed to the bone in these dogs. This process, as occurs in osteochondrosis, can be summarized as nutritional, hypercalcitoninism-induced, decreased osteoclastic activity. Thus high calcium intake diminishes bone remodelling: adaptation of the diameter of foramina to the proportional growth of spinal cord or blood vessels may be delayed. This can be the cause of certain forms of canine wobbler syndrome and enostosis.

Wobbler syndrome: In several studies, Great Danes fed a diet with a high calcium content (i.e. 2–3 times the recommended amount) displayed a delayed expansion of the cervical vertebral canal in proportion to the growth of the spinal cord. Compression of the spinal cord causes myelin degeneration of both the ascending and descending tracts, the extent of which is related to the severity of clinical and radiological signs[25,29] (Fig. 22.7).

Enostosis: In dogs fed high calcium-containing diets, a decreased endosteal osteoclastic resorption, together with an increase in new periosteal bone formation has been observed.[29] The nutrient canals and foramina of the cortex are often abnormal in shape; this may cause edema formation and, eventually, fibrosis in the medullary cavity. Edema may also extend throughout the cortex and underneath the periosteum, causing a loose periosteal attachment or excessive lamellar bone formation.

Diagnosis

Wobbler syndrome: Ataxia, uncoordinated hind limb gait, delayed correction reflexes and elicitation of a pain reaction on extension of the neck are all signs that can be seen in young dogs of large breeds in association with canine wobbler syndrome. These signs appear at the age of 6 months, unlike the unrelated ataxia in Doberman Pinschers, which appears at the age of 6 years. Although not pathognomonic for canine wobbler syndrome of young dogs, the presence of the crossed extensor

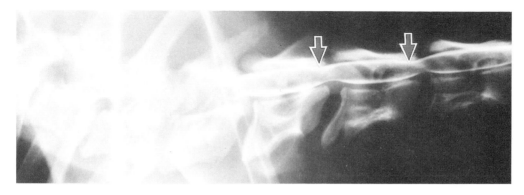

FIG. 22.7 Disproportionate widening of the spinal canal (arrows) causes compression of the spinal cord.[25]

reflex is of great help in making a clinical diagnosis. Due to the spinal degeneration, the *nucleus ruber* of the brain will not inhibit spinal reflexes; a positive crossed extensor reflex (i.e. the extension of the heterolateral leg when one leg is flexed in response to pinching of the interdigital skin, with the dog in lateral recumbency) may therefore be demonstrated in affected dogs. However, because the signs are not pathognomonic, plain and contrast radiological investigation is needed to confirm the diagnosis (Fig. 22.7).

Enostosis: Shifting lameness in dogs under 2 years-of-age is suggestive of this disease. This occurs due to the fact that all long bones are affected, but will vary in their degree of painfulness at any given time. A positive pain reaction upon deep palpation of bones, together with radioopaque areas in the medullary cavities, which arise close to the nutrient foramina, are conclusive.

Prognosis

Depending on the degree of myelin degeneration, normalization of locomotion can be anticipated in wobbler syndrome. Enostosis, although very painful during repetitive bouts, will always heal without long-term effects, before the dog is 2 years-old.

Therapy

Early dietary correction may halt the process of disproportionate remodelling of the skele-

ton. In wobbler syndrome, surgical decompression of the spinal cord may prevent further degeneration, although the clinical signs will not improve in most cases. Enostosis can be very painful: during these periods non-steroid antiinflammatory drugs (NSAIDs) can be prescribed. The owner should be aware that relapses may occur until the dog is 2 years-of-age.

Dietary correction includes a decrease in the amount of calcium and vitamins to the minimal requirements for dogs. On theoretical grounds, it may be beneficial to feed a calcium-deficient diet for a limited period to hasten osteoclast activity, but this has never been investigated.

Pathological Fractures Due to Low Calcium Intake: Nutritional Secondary Hyperparathyroidism, "Juvenile Osteodystrophy"

Pathological fractures, including folding of the cortical bones, compression of the cancellous bone spiculae and deformation of flat bones, in particular, may be encountered in companion animal practice. It is not only dogs and cats, but also cage birds and reptiles, that may be affected.

Under experimental conditions, pathological fractures may occur in dogs of small breeds when fed a diet of extremely low calcium content, whereas Great Danes may be similarly affected when the dietary calcium content is 50% of the recommended amount.[28,45,70]

INCIDENCE OF ABNORMALITIES OF THE ABAXIAL SKELETON IN CATS	
Hypertrophic osteopathy	1
Osteoporosis	2
Osteomyelitis	2
Leucoses	2
Osteopetrosis	3
Patella luxation	5
Hypervitaminosis A	8
Nutritional 2° hyperparathyroidism	21
Total	44

FIG. 22.8 Incidence of nontraumatic, nontumorous and extraarticular abnormalities of the abaxial skeleton in 44 referred cats.

In a survey carried out at Utrecht University, of 44 cats referred with different abnormalities of the abaxial skeleton (nontraumatic and nontumorous in origin, and extraarticular), the majority had pathological fractures due to low calcium intake (Fig. 22.8) Although 23% of all cats referred to the clinic were of oriental breeds, these breeds represented 38% of this group of 44 cats. Two-thirds of the group with nutritional secondary hyperparathyroidism (male:female = 3:4) were under 14 months-of-age (mean age 6 months), whereas the remainder were older (2–18 years, mean 6½ years) and were all castrated males.

Pathophysiology

Home-made meat-rich diets, especially those prepared from cardiac and skeletal muscle, are deficient not only in phosphorus, but also in calcium (Fig. 22.9). In addition, the disease may also be induced when the diet meets all other requirements but is deficient only in its calcium content, and cannot therefore support proper skeletal mineralization. In Great Danes the daily calcium accretion can be as high as 225–900 mg Ca/kg BW, of which 100–225 mg Ca/kg BW/day should be absorbed from the intestine. Poor availability of calcium due to complex formation with phytate or oxalate, high dietary phosphate content or inadequate vitamin D may produce the same symptoms.

A sufficient amount of calcium is present in home-made diets when the requirements of the NRC 1974[48] are followed for dogs (i.e 1.1% Ca on a dry matter basis), and the NRC 1986[50] requirements are met for cats (i.e. 0.8% Ca) with approximately the same content for phosphorus. Half these amounts can be fed safely to small and adult animals, but not to giant breeds or growing dogs.[28,45]

If the vitamin D content of the diet is low, this will cause other additional symptoms, which are described later. It should be noted that for the development of nutritional secondary hyperparathyroidism, the vitamin D content of the food can, and will, be adequate in most clinical cases.

Up to 95% of ingested calcium may be absorbed following the long-term feeding of a diet that is inadequate in calcium content. This increase in absorption efficiency is achieved by an increase in formation of the

ANALYSIS OF FOODS USED IN COMPANION ANIMAL NUTRITION						
Analysis (%)	Dry matter	Protein	Crude fat	Ca	P	Ca:P
Horse meat (lean)	25.5	21	2.9	0.03	0.18	0.2:1
Beef meat	26.3	22.7	2.3	0.03	0.18	0.2:1
Heart	24.8	16.7	6.4	0.01	0.20	0.05:1
Rumen	23.3	14.3	7.1	0.11	0.14	0.8:1
Liver	27.1	18.8	4.1	0.01	0.36	0.03:1
Kidneys	25.0	15	8	0.01	0.22	0.05:1
Poultry by-products	30.0	13	13	0.02	0.2	0.1:1
NRC recommendations (1974)	100	22	5.0	1.1	0.9	1.2:1

FIG. 22.9 Analysis of different foods used in companion animal nutrition (on dry matter basis) and NRC recommendations for dog food.

most active metabolite of vitamin D (1,25 di-hydroxyvitamin D). This metabolite is formed in the kidney under the influence of PTH. PTH is increasingly synthesized and secreted when there is a low calcium uptake and thus a reduction in the plasma calcium concentration. Both high 1,25-OH$_2$ vitamin D levels and hyperparathyroidism will increase the number, as well as the activity, of the bone resorbing osteoclasts. Osteoclasis will be augmented at the sites where osteoclasts are normally active in young growing bone, i.e. at the medullary aspect of cortical bone and at the periphery of the cancellous bone spiculae. The circulating calcium level is kept constant and is sufficient not to disturb significant processes in the body including blood clotting, muscle contraction and hormonal release, as well as mineralization of the newly formed cartilage of growth plates.

Diagnosis

Due to an efficient regulation mechanism, plasma calcium concentrations are kept constant. Plasma phosphate concentrations, always higher in young than in adult animals, are strongly influenced by the nutritional intake, so care must be exercized when interpreting plasma concentrations. Alkaline phosphatase, abundantly available in osteoblasts and liver cells, will be greatly increased whenever there is increased bone cell activity (including growth), whereas its hepatic origin can be related to purely hepatic related parameters, including the γ-glutamyl transferase (GGT) activity. Although it is possible to determine immunoreactive parathyroid hormone as well as vitamin D metabolites, it requires specialized laboratories, and a single measurement is of limited value.

The most practical and inexpensive diagnostic technique is radiological investigation of the long bones and axial skeleton.[58,70] Although it has been demonstrated that under standardized exposure, a mineral loss of at least 30% is required before the lesion is noticeable radiologically, the abnormal alignment due to greenstick and compression

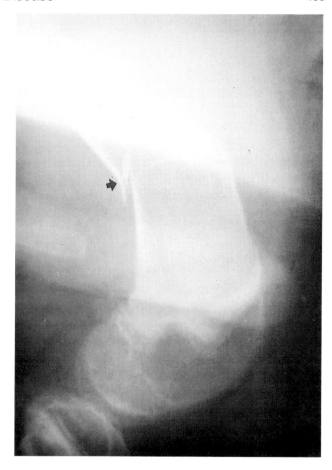

FIG. 22.10 Nutritional secondary hyperparathyroidism in a 5-month-old dog: radiograph showing thin cortex, greenstick fracture (arrow) and normal growth plate bordered by white metaphyseal area.

fractures, as well as bowing of bones due to the constant pull, is obvious. In addition, it should be noted that the growth plate has a normal width and that the metaphyseal area is usually more radioopaque than the rest of the bone (Fig. 22.10).

Prognosis

In the acute phase, compression of the vertebrae can cause compression of the spinal cord, especially in the lumbar area, which may result in posterior paralysis in severe cases. The constant muscle pull on pelvic bones, calcaneus, scapula and other prominences, can cause the weakened bone to become misshapen. In some locations this can be seen or palpated (Fig. 22.11). The long bones may be abnormally shaped and may require

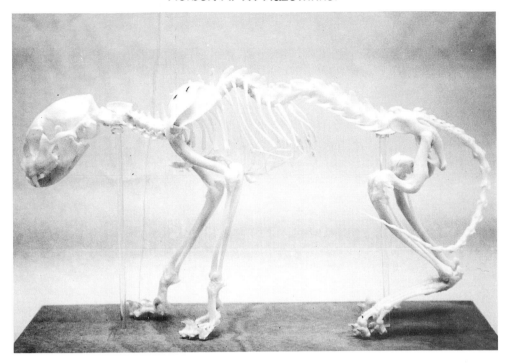

FIG. 22.11 Skeleton of a cat with nutritional secondary hyperparathyroidism. Note the alignment of the vertebrae and the shape of the pelvis and long bones.

surgical correction after mineralization is normalized, to permit normal use of the limb. Although the prognosis should be guarded, posterior paralysis may disappear 2 weeks after initiation of therapy. Abnormal alignment of the pelvic bones may cause repetitive bouts of constipation that will continue even after the normal mineralization status of the skeleton has been restored.

Therapy

Affected bones cannot withstand the load of a splint or cast and will form another greenstick fracture just proximal to its margin. In the acute phase, therapy is limited to good nursing and feeding a diet that fulfills the NRC requirements, without any injections of nutrients such as calcium or vitamin D. The amount of calcium needed exceeds the amount that can be safely injected by 1000-fold. As previously explained, it can be expected that the $1,25\text{-OH}_2$ vitamin D level is already extremely high in the patient, and vitamin D

injections will further aggravate the situation. Correction of the diet includes calculating the nutrient content of the food and supplying a diet that contains up to 1.1% Ca on a dry matter basis. Alternatively, the diet should be changed to a commercially available, complete and balanced dog or cat food, as appropriate. Adding calcium (as calcium carbonate or calcium lactate at a rate of 50 mg Ca/kg BW/day) will possibly accelerate osteoid mineralization, which will be noticeable within 3 weeks on radiographs (Fig. 22.12). Fracture treatment or corrective osteotomies must be postponed until the skeleton is firmly mineralized.

Rickets

Rickets (hypovitaminosis D) in young animals is seldom, if ever, seen in companion animal practice. However, because it is so often misdiagnosed by the owner or expected to play a significant role, it is pertinent to include a description of its pathophysiology, diagnosis and treatment.

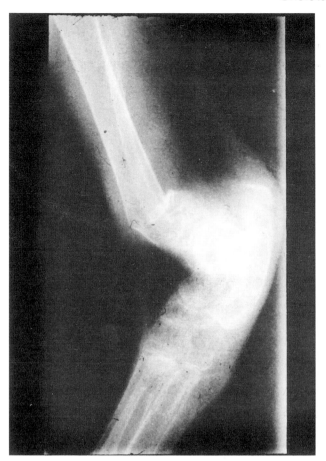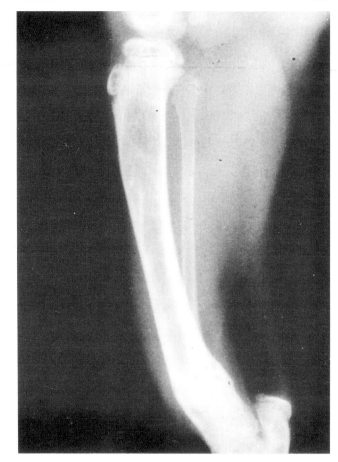

FIG. 22.12 Greenstick fracture of the distal tibia of a cat (left), mineralized
after 3 weeks of feeding a balanced cat food (right).

Pathophysiology

Vitamin D metabolites stimulate calcium
and phosphate absorption in the intestine and
reabsorption in the renal tubules, stimulate
osteoclasts, and are necessary for mineralization of newly formed osteoid and cartilage.
Vitamin D is absorbed in the intestine as one
of the fat-soluble vitamins, transported to,
and hydroxylated in, the liver, then further
hydroxylated in the kidney to $24,25\text{-OH}_2$
vitamin D or $1,25\text{-OH}_2$ vitamin D. It has been
demonstrated that dogs do not synthesize
vitamin D in their skin when radiated with
ultraviolet B light, unlike herbivores and
omnivores. Under experimental conditions,
young dogs developed all the signs of rickets
when fed a diet that contained calcium,
phosphorus and other constituents in amounts
according to NRC (1974)[48] guidelines, but
lacking only in its vitamin D content. Daily

exposure with ultraviolet B light did not
prevent or heal the hypovitaminosis D (Fig.
22.13).[27] Therefore, to meet their requirements, dogs rely on vitamin D in foods
including liver, fish, egg, milk (products) and
commercially available dog and cat foods.
Figure 22.14 lists the vitamin D_3 content of a
variety of foods used in diets for dogs and
cats. Synthesis of the most active metabolite
($1,25\text{-OH}_2$ vitamin D) takes place under the
influence of PTH, low plasma levels of
calcium and phosphorus, and during growth,
pregnancy and lactation.

Diagnosis

From the animal's history, it will be
apparent that the animal has been raized on
lean meat immediately after weaning. The
animal will exhibit bulging metaphyseal areas
of the radius–ulna and ribs.

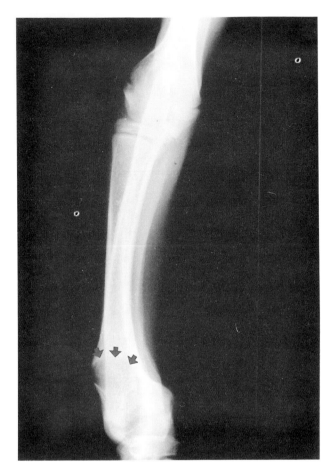

FIG. 22.13 Thin cortex, large diameter of medullary cavity and increased width of the growth plates (arrows) are typical for hypovitaminosis D (rickets).

Plasma concentration of calcium will be normal or low–normal whereas phosphorus concentration can be low in plasma and strongly elevated in urine (as high as 35 mmol/l/day). The latter can be explained by the hypocalemia-induced hyperparathyroidism,

VITAMIN D$_3$ CONTENT OF INGREDIENTS	
	Vitamin D$_3$ (IU/100 g DM)
Liver, calf, pork	20–100
Fish liver oil	5×10^3–6×10^6
Butter	10–120
Egg yolk	150–400
Milk	0.3–5
Dog food	50
Cat food	100

FIG. 22.14 Vitamin D$_3$ content of ingredients.

which will decrease the maximal tubular reabsorption of phosphorus. Measurement of vitamin D metabolites (which require highly specialized, expensive methods to determine) will provide the correct diagnosis of hypovitaminosis D. Levels of vitamin D, 25-OH vitamin D and 24,25-OH$_2$ vitamin D, in particular, will be low, whereas the 1,25-OH$_2$ vitamin D metabolite will be low–normal. It is easier and less expensive to take and interpret radiographs of the radius–ulna: typical findings include a thin cortex, larger diameter of the medullary cavity, bowed long bones and an increase in width of the growth plates (Fig. 22.13). In low calcium diets this increased width of growth plates will not occur, and greenstick fractures are more commonly seen. In adult animals, hypovitaminosis D is called osteomalacia; the bony changes are similar to rickets but less pronounced. Cartilaginous abnormalities will not occur in adults. Clinically, osteomalacia does not play a significant role in companion animal orthopedics.

In dogs and cats with severe kidney damage, an accumulation of plasma phosphate as well as a decreased renal synthesis of 1,25-OH$_2$ vitamin D can cause a parathyroid hormone-induced increase in bone resorption, together with a hypovitaminosis D-related decrease in bone mineralization. This renal osteodystrophy is part of the uremic syndrome, as discussed in Chapters 16 and 18.

Prognosis

When a diagnosis of hypovitaminosis D has been made, therapy should be introduced immediately. After 3 weeks, mineralization of the growth plates should be almost normal and there should be an improvement in mineralization of cortical and cancellous bone, as well as of callus around pathological fractures. Several weeks will be subsequently required before mineralization is complete. If mineralization has not improved within three weeks, the diagnosis should be reevaluated: collagen diseases, such as *osteogenesis imperfecta* or an inability to hydroxylate vitamin D metabolites should be considered.

Therapy

Commercially prepared food can be recommended as therapy because it contains adequate amounts of calcium and phosphorus, as well as vitamin D. In particular, commercial petfoods contain considerably more than the minimum recommendations for vitamin D (450 IU vitamin D/kg diet on a dry matter basis). Vitamin D injections carry the risk of oversupplementation, and should not be used to treat cases of rickets due to vitamin D deficiency, where dietary measures are also employed.[27] Corrective surgery should be postponed until mineralization of the skeleton is complete.

Hypervitaminosis A

In a survey of 44 feline patients with appendicular skeletal abnormalities (of non-tumorous and nontraumatic origin), referred to Utrecht University, hypervitaminosis A was diagnosed in eight cats (Fig. 22.15). All were neutered, domestic short haired cats, of mature age (3–13 years). Over a period of several years, four cats were fed a diet consisting of raw liver or fish, whereas the diets of the other four cats were, according to the owners, not suggestive of a high vitamin A intake.

Vitamin A intoxications are less frequently reported in dogs, either under experimental conditions[6] or as a consequence of feeding unbalanced, noncommercial diets, such as liver and/or cod liver oil.[10]

Pathophysiology

Vitamin A is one of the fat-soluble vitamins required to prevent a variety of abnormalities including reproductive failure, disorders of epithelial surfaces (bronchi, salivary glands, hair coat) and retinal degeneration. Vitamin A is also required for normal skeletal growth and development, and especially for osteoclastic activity.[21] Vitamin A_1 is present in a variety of foodstuffs originating from animals and saltwater fish, whereas vitamin A_2 (which is 30% less effective in mammals) is present in freshwater fish. In addition, β-carotene ($C_{40}H_{56}$), which is present in plants including corn, can be transformed into vitamin A ($C_{20}H_{29}OH$) with the aid of carotenase, as is present in the intestinal mucosa and liver cells of dogs (Fig. 22.16). Because cats lack this enzyme,[15] they rely totally on animal tissue or dietary supplements to provide a dietary source of vitamin A.

Dogs, unlike cats, are able to form retinyl esters that render vitamin A biologically inactive. They are also able to excrete 15–60% of the daily vitamin A intake as retinyl palmitate in urine. Although massive doses of vitamin A caused hypervitaminosis in kittens[60] and puppies[6] after several weeks, hypervitaminosis A is most likely to be seen

GENDER, AGE AND AFFECTED SITES OF CATS WITH HYPERVITAMINOSIS A					
		Localization of new bone formation			
Sex (all neutered)	Age (years)	Elbow joint	Stifle joint	Vertebrae	Additional findings
M	4	X			Tilted head
F	8	X			Horner's syndrome
F	3			X	Stridor nasalis
F	13	X		X	
M	12	X		X	Multiple thyroid adenoma
M	7			X	
M	11			X	
F	3	X	X	X	
M — male; F — female.					

FIG. 22.15 Gender, age and affected sites of eight European short haired cats with hypervitaminosis A.

VITAMIN A OR CAROTENE CONTENT OF INGREDIENTS		
	Vitamin A (IU/100 g DM)	Carotene (µg/100 g DM)
Ground beef	140	–
Raw egg	4000	–
Bovine liver	150,000	–
Cod liver oil	85,000	–
Corn	–	425
Dog food	500	–
Cat food	1000	–

FIG. 22.16 Vitamin A (IU) or carotene content of ingredients.

in companion animal practice as a disease of the mature cat, starting between 2 and 5 years-of-age (Fig. 22.15).[3,61]

Excessive amounts of vitamin A will reduce chondrocyte proliferation in the growth plates, depress periosteal osteoblastic activity and stimulate osteoclastic activity. This results in osteoporosis in addition to a variety of lipid infiltrations in parenchymal organs, including the liver.[17] New bone formation is seen particularly at the insertion sites of the tendons of the main muscles and at the origin of ligament and joint capsules. Periarticular tension forces in the osteoporotic bone may induce this new bone formation, which is similar to callus formation, eventually leading to ankylosis (Fig. 22.17).

Diagnosis

In dogs, a history of massive supply of cod liver oil or vitamin supplements will help to make the diagnosis. In cats, the dietary history is not always one of raw liver, fish or supplements: four out of eight of our patients and three out of 80 cats did not eat these foods, suggesting an individual predisposition.[55]

Kittens show reduced longitudinal growth and osteoporosis of the long bones, together with flaring of the metaphyseal regions.[7] Osteophyte formation and periosteal reaction are visible on radiographs but are less obvious than in adult cats.

In adult cats, a stiff neck and/or enlarged joints of front or rear legs, dull hair coat, hyper- or hyposensitivity of the skin, loss of nerve function (due to entrapment by newly formed bone) may be observed, as well as anorexia and weight loss. When there is bilateral involvement of the shoulder or elbow joints, with ankylosis of the affected joints, the cat may unload both front legs in a sitting position, often called the "marsupial position." When ankylosis of the cervical area occurs, the animal is unable to wash and groom itself.

A positive diagnosis can be made from radiographs of the cervical or thoracic vertebrae and the large joints of the legs: new bone formation without bone loss, leading to ankylosis of vertebrae (at the dorsal or ventral aspect) and of the joints, will be apparent (Fig. 22.17). To confirm the diagnosis, a liver biopsy will show fatty infiltration of the liver cells. Determination of retinol in plasma or liver can be conclusive. Liver retinol levels are most indicative because 20% of the hypervitaminosis A cats had normal plasma values in one study.[55]

Prognosis

Improvement may be observed, both clinically and even radiologically. Ankylotic joints will not regain their normal motion, but with a drastic change in diet, supportive analgesics when needed and time to heal, a dramatic improvement can be expected.

Therapy

The requirement for vitamin A in dogs varies with age, growth, illness (infections), pregnancy and lactation. It can be partially replaced by carotene: 1 µg carotene is equivalent to 0.5 IU vitamin A, approximately. In cats, the upper limit of vitamin A is reported

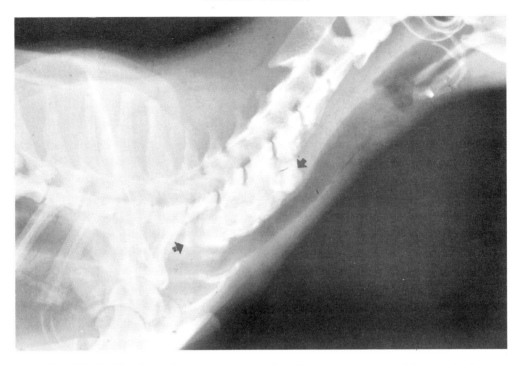

FIG. 22.17 New bone formation (arrows) at the ventral aspect of the cervical vertebrae of a 3-year-old cat, due to excessive vitamin A intake.

to be 100,000 IU/kg diet (dry matter), which is 10 times the requirement.[20] Cod liver oil or other supplements should be discontinued immediately. Vitamin A is thermolabile and will be destroyed above 70°C and therefore cooking of the liver will destroy most, although not all, of its vitamin A. Studies in larger groups of patients have demonstrated that the prognosis improves when the diet of the animal is changed completely, once the diagnosis is made.[55]

Because commercially available petfoods will be supplemented with vitamin A to the required amounts for cats, it is preferable to prescribe a balanced, home-made diet with a low vitamin A (and/or carotene for dogs) content.[10] Such a diet should not contain organ meats (including heart, kidneys, liver), whole eggs or milk (and milk products), but may include lean meat (veal, lamb, poultry), cottage cheese and vegetable oils. A low vitamin A diet, as prescribed by Donoghue,[10] is given as an example in Fig. 22.18.

This recipe is suitable for dogs and may be fed to cats for several weeks, provided that the liquid in which the chicken is cooked is also fed in order to provide adequate dietary levels of taurine.

Corticosteroids may relieve some nerve pain, but can hinder a decrease in plasma vitamin A concentration.[10] The other fat-soluble vitamins (i.e. D, E and K) diminish the toxic effects of vitamin A, possibly due to competitive absorption, but may be harmful to use as a therapy.[21] Therefore, a low dose of a nonsteroid antiinflammatory drug may be prescribed for animals showing signs of pain. A drastic change in a vitamin A deficient diet will allow the animal to deplete its vitamin A stores. Depending on the degree of intoxication, this will take up to 1 year. Improvement can be seen 2–4 weeks after therapy has been initiated.

A LOW VITAMIN A DIET		
Corn oil	9.1 g (2 tsp)	
Iodized salt	3.0 g (0.5 tsp)	
Bone meal	5.0 g (1 tsp)	as fed
Chicken, cooked	105 g	
Rice, cooked	61 g	

FIG. 22.18 A low vitamin A diet.

Hip Dysplasia

Hip dysplasia (HD) is a common developmental orthopedic disease in the canine species. This hereditary abnormality is most commonly seen in certain breeds (eg. St. Bernards, Rottweilers, Newfoundlands, Bernese Mountain dogs) but only infrequently in others (e.g. Afghan Hounds, Shetland Sheepdogs, Malamutes and Huskies).[9] A variety of studies report that dysplastic dogs are born with normal hips, but develop HD as a result of a disparity between the development of the bony part of the hip joint and its supporting soft tissues, ligaments, joint capsule and muscles.[1] This occurs during the first 6 months of life, during which time the tissues are soft and plastic with an elastic limit. Environmental factors, which are believed to influence the development of HD, have yet to be elucidated. However, research indicates that diet has, both quantitatively and qualitatively, a significant effect on the development of HD. Diet will not cure HD or alter the genetic status of the offspring in this respect, but it can influence the phenotypic expression of HD by optimizing the development of the hip joints of those animals potentially at risk. Diet can also play a role in conservative treatment in those dogs where HD has already developed.

Pathophysiology

In the canine hip joint, both the femoral head and the acetabulum are mainly cartilaginous at birth. Bone formation, and a change in position of the femoral head in relation to the femoral shaft, will take place via endochondral ossification and osteoclastic activity, respectively. In HD, joint laxity results in an incongruent hip joint where the dorsomedial part of the femoral head and the acetabular rim are in contact, while supporting almost half the body weight when walking. This causes microfractures and deformation of the acetabular rim, erosions of the cartilage and deformation of the subchondral bone.[13] The associated pathological changes are joint effusion, stretching and thickening of the joint capsule and the round ligament, as well as osteophyte formation.

Clinically, the dog may show pain at different stages of development of HD. In the immature dog, stretching of the joint capsule and microfractures of the cartilage will elicit pain, whereas in the mature dog (relative) overuse of the arthrotic joint will result in general signs of arthrosis. Such signs include pain upon rising, warming-out during exercise (initial stiffness that improves with walking), decreased range of movement and worsening of signs after rest following heavy exercise.[24]

Dietary factors that play a role in hip joint development and hip joint overload, both of clinical significance in canine HD, will be reviewed.

Overnutrition

It has been demonstrated that overnutrition in growing puppies causes a more rapid growth in bone length and a more rapid gain in BW, when compared with normal- or restricted-fed dogs.[2,29,33,34,37,43,47,58,68] Heavy BW will cause overloading of the cartilaginous skeleton, including the hip joints. This could be a significant factor, which may help to explain the increased frequency and severity of HD in overweight dogs.

The higher growth velocity in young dogs has been observed in bitch-fed puppies (vs. hand-reared puppies),[58] in dogs receiving *ad libitum* palatable commercial[2,34,37] energy enriched[29,33,43] food (vs. meal-restricted energy intake), in dogs fed unrestricted (vs. time-restricted food intake)[2] and in dogs fed high protein (vs. isoenergetic but lower protein) diets.[47] Not all these studies specifically researched hip joint development or status, but in those where this was the case, a detrimental effect of overnutrition was observed.[29,33,34,58] Dietary protein content may influence growth in body weight more than in body size, but has not been proven to play a role in the development of HD.[47,68]

A study, unique in its design and long-term follow-up has recently been published by Kealy *et al.*[34] in which 48 Labrador Retrievers, originating from seven different litters, were divided into two groups at the age of 8 weeks. Dogs of one group were allowed to eat a dry dog food *ad libitum*, whereas sex-matched littermates received 75% of the amount of food consumed by the littermates. The pairmates were both housed in the same indoor cage with an outdoor run to standardize other environmental influences, except for 15 min during feeding. Although others report in comparable studies that restricted-fed dogs approach the BW and size of *ad libitum* fed dogs in the second half of the first year of life,[2,29,37] the restricted-fed Retrievers in Kealy's study had an average BW of approximately 20 kg, being 78.3 ± 5.35% (mean ± SD) of that of the *ad libitum* fed pairmates, at the age of 30 weeks. The difference in mean BW of the two groups was even more apparent at the age of 2 years.

At 30 weeks-of-age, measurement of the Norberg angle on radiographs of all dogs revealed statistically significant differences ($p < 0.05$) in joint laxity. These differences were equally present at 2 years-of-age.[34] This long-term study with 24 pairs of littermates, raised under the same environmental conditions, demonstrates a considerably higher incidence of HD in the group of overfed dogs when compared with the group fed 25% less.[34] The severity of arthrosis (as reflected by crepitation upon palpation and the radiologically visible malformation and osteophyte formation of the affected hip joints, as well as of elbow and shoulder joints) in some of these dogs (personal observations) is frequently observed in family-owned dogs with severe lameness (Fig. 22.19).

The foregoing stresses the fact that HD can develop in young, overfed dogs, even under conditions of relatively restricted activity. This is probably due to overstressing the elasticity of the periarticular tissues and the resulting pathological cartilaginous and subchondral bone changes.

FIG. 22.19 Hip dysplasia is more frequent and more severe in overweight dogs.

Minerals and Vitamins

Calcium: In the growing skeleton a considerable amount of bone remodelling adapts skeletal configuration to a variety of changing forces acting on the bones. These forces include gravity, muscle forces, piezoelectrical forces and interrelated bone contacts. Some of these influences can be overruled by hormones acting on calcium homeostasis to prevent life-threatening changes in extracellular calcium concentration. Calcitonin secretion will increase due to high calcium intake and normalize plasma calcium concentration by decreasing osteoclastic activity.[23] In the growing animal, chronic high calcium intake will cause chronic hypercalcitoninism[25,29] and thus decrease osteoclastic activity and remodelling of the skeleton. A decreased remodelling of the proximal femur (i.e. a delayed antetorsion) was observed by Hedhammar *et al.*[29] in a group of Great Danes, a breed not particularly prone to HD, fed an *ad libitum* diet rich in calcium.[29] Others described delayed skeletal maturation in both Great Danes and in Poodles fed a dog food with a high calcium content (i.e. 3.3% Ca on a dry matter basis) when compared with controls fed a dog food according to NRC[48] guidelines.[45,70,71]

From these studies it can be concluded that an excessive dietary calcium intake decreases maturation of hip joint conformation as well as of the vulnerable cartilaginous template of the skeleton. This may coincide with overloading of the hip joint that is too immature in its development for the age and size of the dog and, therefore, may play a significant role in deformation of the hip joint at an early age.

Phosphorus: In the field of canine nutrition, there is now sufficient evidence to suggest that, within the range of nutrient levels normally encountered in practice, it is not the calcium to phosphorus ratio, but the absolute calcium amount in the daily ration that determines the occurrence of the skeletal abnormalities as described above.[28,45] A high dietary phosphorus content may bind more calcium in the intestine to form nonabsorbable complexes, but this is perhaps only the case with nonabsorbable phytates. A highly absorbable salt (as is present in bone meal) will cause the same skeletal effects as excessive calcium alone.[28]

Electrolytes: Electrolytes are present in body fluids including synovia. Differences in circulating cations (Na^+, K^+, Ca^{2+} and Mg^{2+}) and anions (Cl^-, $H_2PO_4^-$, HPO_4^- and the SO_4^{2-} as present in amino acids) influence the acid–base balance. The influence of electrolytes on the osmolality of body fluids as well as on the acid–base balance may play a role in the development of HD in young dogs.

Mean osmolality in synovial fluid from normal hip joints was significantly lower than that of synovia of hip joints of dysplastic Retrievers.[53] Whether this difference reflects the cause of joint laxity or the result of hyperperfusion of the joint capsule of the arthrotic joint needs to be elucidated.

In another study,[35] the dietary content of Na^+, K^+ and Cl^- ions differed (increased, increased and decreased, respectively) in the diets of three groups of dogs ($n = 177$) of five breeds (St. Bernard, German Shepherd, Coonhounds, English Pointers and Labrador Retrievers) originating from 27 litters. In these dogs, joint laxity was observed by the measurement of Norberg's angle on radiographs taken at 30 and 105 weeks-of-age. However, the acid–base balance and electrolyte content of body fluids were not noted.[35] It was found that the dogs on the dry dog food (moisture < 10%) with the low Na (0.32–0.43%), low K (0.39–0.70%) and high Cl (0.66–0.81%) content had a slight, but statistically significant, improvement in the Norberg angle when compared with the other groups. Only the Retrievers revealed a low Norberg angle of the hip joint, irrespective of the diet.[35] The clinical significance of these findings, the sensitivity and reproducibility of the radiographic procedure,[31,65] the influence of other electrolytes playing a role in the acid–base balance and osmolality[38] must all be further investigated before the optimum electrolyte content of the food can be established. The detrimental effects of prolonged dietary-induced acidosis on skeletal mineral content,[5] however, imply that further studies in this area would be valuable.

Vitamin D: Although not proven in research, vitamin D could play a role in the development of HD. Vitamin D metabolites will increase calcium absorption from the intestine and increase calcium resorption in the kidneys.[23] A vitamin D overdose, with the effect of increasing calcium concentration in the extracellular fluid, may mimic the decreased skeletal remodelling as well as the delayed ossification as previously mentioned, due to high calcium intake alone.[25,29]

Vitamin C: Conflicting results have been reported on the effect of dietary vitamin C.[4] Vitamin C overdose has been proven to increase calcium retention in dogs,[66] especially by decreasing calcium resorption from the skeleton. This can be either by altered bone-cell activity or due to physicochemical reactions. In either case it may coincide with a decreased osteoclastic activity and therefore should be prevented.

Thiamin: Humans, forced to live under extreme undernourished conditions, may develop thiamin (vitamin B_1) deficiency. The development of arthrotic changes in the hip joints of these patients have the same appearance as HD in man and dogs. Because thiamin deficiency causes gastrointestinal and neurological signs, fatal when not diagnosed at an early stage, this deficiency does not play a role in HD development in dogs.[56]

Diagnosis

A diagnosis of HD is made on the basis of the history and the clinical signs, including stiffness on rising, "bunny hopping," pain and lameness of the rear limbs and pain reaction or crepitation upon manipulation of the hip joints. Laxity of the hip joint can be tested by abduction of the proximal femur, preferably in a non-weight-bearing position:

1. using one hand as a fulcrum, medial to the proximal femur with the dog in lateral recumbency, and medial pressure of the stifle joint (Ortolani sign);

2. adducting the stifle with the dog in dorsal recumbency, with the femur perpendicular to the table top (Barden sign).[24]

Subluxation of the hip joint can be diagnosed by putting medially directed force on the greater trochanter. Radiographs in ventrodorsal (Fig. 22.20), lateral, flexed ("frog") views, as well as more specific views related to the acetabular rim[64] or the joint laxity[65] can be conclusive in diagnosing joint laxity, incongruency, subchondral sclerosis and osteophyte formation.

Prognosis

HD can cause great discomfort, depending on a variety of factors, including the severity of the disease, lifestyle, pain threshold and BW of the dog. A large variety of treatments, nonsurgical as well as surgical, are known, each with its own financial and prognostic consequences.

At least a temporary improvement in up to

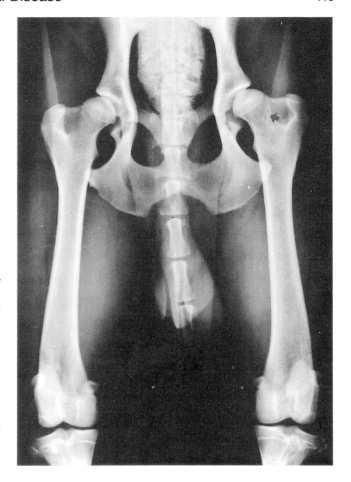

FIG. 22.20 Hip dysplasia with subluxation of the femoral heads and osteophytes (arrow), diagnosed on a ventrodorsal radiograph.

70% of all cases is claimed for most procedures, according to the reports of various orthopedic centers that were presented at the World Small Animal Veterinary Association meeting in Vienna, 1991. Long-term results and objective data on locomotion are lacking. An individual approach to each case should be followed, based on the clinical signs, economic and nursing possibilities, techniques available and reasonable lifestyle demands.

Prevention and Therapy

Overnutrition should be prevented by providing a limited amount of a well-balanced, home-made or commercial dog food. For both home preparations and commercial dog foods, special attention should be paid to the energy content and the amount of calcium.

The energy intake should be adapted to the individual needs of the animal, determined by breed, age, BW and activity. Control of body weight is discussed more fully in Chapter 10.

Rest *per se* can improve the clinical signs of HD in young and adult family-owned dogs, as was observed by force plate measurements before and after a period of 3 months cage rest.[24] In the nonsurgical treatment of HD, both dietary measures and activity restriction should be employed.

Supplementation of balanced home-made or commercially available, balanced dog foods, especially of the leading brands, should be prevented. Commercial dog foods differ in their palatability, digestibility and quality of ingredients rather than in the quality of calcium, phosphorus and vitamins. Because most petfood manufacturers guarantee that their products meet, at least, the NRC minimum recommendations,[48] adding minerals or vitamins to these foods will provide the dog with excesses. These excesses can be detrimental to hip joint development (especially in young, growing dogs of large breeds) and should not, therefore, be advocated by the veterinary surgeon, or discontinued immediately when brought to his or her attention.

Because the intake of the absolute amount of calcium per day has more influence on the occurrence of skeletal disturbances than the calcium ratio to other nutrients,[28] the absolute amount of intake of a balanced food needs to be limited also. Recommendations for an optimal calcium content of the food, according to the information currently available,[28,47,48] are that diets for growing dogs of large breeds differ considerably from those of small breeds: a calcium content of 0.7 g/1000 kJ is sufficient for undisturbed skeletal growth for large breed dogs, whereas a third of this (i.e. 0.23 g/1000 kJ) was sufficient for growing miniature poodles. In both groups of dogs 0.5 g P/1000 kJ metabolizable energy can be given, without being too concerned about the ratio of calcium to phosphorus, vitamins and other nutrients. For protein and vitamin D, this author considers the optimal content to be 15 g and 30 IU/1000 kJ, respectively.

In selected cases, development of the hip joint in young, growing dogs can be surgically optimized. Surgery in older and more arthrotic hip joints can also be performed when indicated, but will not be discussed here. To date, there are no proven benefits for herbs, microelements or ointments in the treatment of canine HD.

Information should be conveyed to the owner on the important effects of the quality and quantity of the daily ration, especially with regard to the effects of excesses in minerals and energy on prevention and alleviation of HD in both young and adult dogs.

Acknowledgements—The author would like to thank the Department of Veterinary Radiology, Utrecht University (Head: Professor K.J. Dik) for providing radiographs; Mr. B. Janssen for the drawings; and Ms. Fiona Taylor for her valuable contributions in the preparation of this chapter.

References

1. Alexander, J. W. (1992) The pathogenesis of canine hip dysplasia. *Veterinary Clinics of North America: Small Animal Practice*, **22**, 503–511.
2. Alexander, J. E. and Wood, L. L. H. (1990) Comparative growth study. In *Iams Update*, No. 1003, Dayton, Ohio.
3. Barnack, J. (1988) Untersuchung über die Häufigkeit des Verkommens von Skeletveränderungen bei Katzen. Diss. Berlin.
4. Belfield, W. O. (1976) Chronic subclinical scurvy and canine hip dysplasia. *Veterinary Medicine/Small Animal Clinician*, **74**, 1399–1401.
5. Ching, S. V., Fettman, M. J., Hamar, D. W. *et al.* (1989) The effect of chronic dietary acidification using ammonium chloride on acid–base and mineral metabolism in the adult cat. *Journal of Nutrition*, **119**, 902–915.
6. Cho, O. Y., Frey, R. A., Guffy, M. M. and Heipold, H. W. (1975) Hypervitaminosis A in the dog. *American Journal of Veterinary Research*, **36**, 1597–1603.
7. Clark, L., Seawright, A. A. and Gartner, R. J. W. (1970) Longbone abnormalities in kittens following vitamin A administration. *Journal of Comparative Pathology*, **80**, 113–117.

8. Copp, D. H. and Groenberg, D. M. (1945) Studies on bone fracture healing: 1. Effect of vitamins A and D. *Journal of Nutrition,* **29**, 261.

9. Corley, E. A. (1992) Role of the orthopedic foundation for animals in the control of canine hip dysplasia. *Veterinary Clinics of North America: Small Animal Practice,* **22**, 579–593.

10. Donoghue, S., Szanto, J. and Kronfeld, D. S. (1987) Hypervitaminosis A in a dog: an example of hospital dietetic. In *Nutrition, Malnutrition and Dietetics in the Dog and Cat.* Ed. A. T. B. Edney. pp. 94–96. Waltham Centre Press.

11. Edney, A. T. B. and Smith, P. B. (1986) Study of obesity in dogs visiting veterinary practices in the United Kingdom. *Veterinary Record,* **118**, 391–396.

12. Ferguson, H. W. and Hartles, R. L. (1970) The combined effects of calcium deficiency and ovariectomy on the bones of young adult cats. *Calcified Tissue Research,* **4** (Suppl.), 140.

13. Fox, S. M., Burns, J. and Burt, J. (1987) Canine hip dysplasia. *Veterinary Medicine: Small Animal Clinic,* **82**, 683–693.

14. Fraser, D. R. (1980) Regulation of the metabolism of vitamin D. *Physiological Reviews,* **60**, 551–613.

15. Gershoff, S. N., Andrus, S. B., Hegsted, D. M. and Leutini, E. A. (1957) Vitamin A deficiency in cats. *Laboratory Investigation,* **6**, 227–234.

16. Gershoff, S. N., Legg, M. A. and Hegsted, D. M. (1958) Adaptation to different calcium intakes in dogs. *Journal of Nutrition,* **64**, 303–312.

17. Goedegebuure, S. A. and Hazewinkel, H. A. W. (1981) Nutrition and bone metabolism. *Tijdschrift voor Diergeneeskunde,* **106**, 234–242.

18. Grøndalen, J. and Hedhammer, Å. (1982) Nutrition of the rapidly growing dog with special reference to skeletal disease. In *Nutrition and Behaviour in Dogs and Cats.* Ed. R. S. Anderson. pp. 81–88. Oxford: Pergamon.

19. Gunther, R. *et al.* (1988) Toxicity of a vitamin D_3 rodenticide to dogs. *Journal of the American Veterinary Medical Association,* **193**, 211–214.

20. Halle, J., (1992) Nutrition of the cat. *Journal of Veterinary Nutrition,* **1**, 17–30.

21. Hayes, K. C. (1971) On the pathophysiology of vitamin A deficiency. *Nutritional Reviews,* **29**, 3–6.

22. Hazewinkel, H. A. W. (1989) Calcium metabolism and skeletal development in dogs. In *Nutrition of the Dog and Cat.* Eds. I. H. Burger and J. P. W. Rivers. pp. 293–302. Cambridge: Cambridge University Press.

23. Hazewinkel, H. A. W. (1989) Nutrition in relation to skeletal growth deformities. *Journal of Small Animal Practice,* **30**, 625–630.

24. Hazewinkel, H. A. W. (1992) Diagnosis and conservative treatment of hip dysplasia in young dogs. *Tijdschrift voor Diergeneeskunde,* **S117**, 33–34.

25. Hazewinkel, H. A. W., Goedegebuure, S. A., Poulos, P. W. *et al.* (1985) Influences of chronic calcium excess on the skeletal development of growing Great Danes. *Journal of the American Animal Hospital Association,* **21**, 337–391.

26. Hazewinkel, H. A. W., Hackeng, W. H. L., Bosch, R. *et al.* (1987) Influences of different calcium intakes on calcitropic hormones and skeletal development in young growing dogs. *Frontiers in Hormonal Research,* **17**, 221–232.

27. Hazewinkel, H. A. W., How, K. L., Bosch, R. *et al.* (1987) Inadequate photosynthesis of vitamin D in dogs. In *Nutrition, Malnutrition and Dietetics in the Dog and Cat.* Ed. A. T. B. Edney. pp. 66–68. Waltham Centre Press.

28. Hazewinkel, H. A. W., Van den Brom, W. E., Van't Klooster, A. Th. *et al.* (1991) Calcium metabolism in Great Dane dogs fed diets with various calcium and phosphorus levels. *Journal of Nutrition,* **112**, S99–S106.

29. Hedhammar, Å., Wu, F., Krook, L. *et al.* (1974) Overnutrition and skeletal disease: an experimental study in growing Great Dane dogs. *Cornell Veterinarian,* **64** (Suppl.), 1–159.

30. Hedhammar, Å., Krook, L., Schryver, H. and Kallfelz, F. (1980) Calcium balance in the dog. In *Nutrition of the Dog and Cat.* Ed. R. S. Anderson. pp. 119–127. Oxford: Pergamon.

31. Heyman, S. J., Smith, G. K and Cofone, M. A. (1993) Biomechanical study of the effect of coxofemoral positioning on passive hip joint laxity in dogs. *American Journal of Veterinary Research,* **54**, 210–215.

32. Kallfelz, F. A. and Dzanis, D. A. (1989) Overnutrition: an epidemic problem in pet animal practice. *Veterinary Clinics of North America. Small Animal Practice,* **3**, 433–436.

33. Kasstrøm, H. (1975) Nutrition, weight gain and development of hip dysplasia. *Acta Radiologica,* **344** (Suppl.), 135–179.

34. Kealy, R. D., Olsson, S. E., Monti, K. L. *et al.* (1992) Effects of limited food consumption on the incidence of hip dysplasia in growing dogs. *Journal of the American Veterinary Medical Association,* **201**, 857–863.

35. Kealy, R. D., Lawler, D. F., Monti, K. L. *et al.* (1993) Effects of dietary electrolyte balance on subluxation of the femoral heads in growing dogs. *American Journal of Veterinary Research,* **54**, 555–562.

36. Kronfeld, D. S. (1983) Obesity and weight reduction programs. *Proceedings of the Postgraduate Committee of Veterinary Sciences,* **63**, 308.

37. Lavelle, R. B. (1989). The effects of the overfeeding of a balanced complete commercial diet to a young group of Great Danes. In *Nutrition of the Dog and Cat.* Eds. I. H. Burger and J. P. W. Rivers. pp. 303–315. Cambridge: Cambridge University Press.

38. Lemann, J. and Lennon, E. J. (1972) Role of diet, gastrointestinal tract and bone in acid-base homeostasis. *Kidney International,* 1, 275–279.

39. Lenehan, T. M., Van Sickle, D. C. and Biery, D. N. (1985) Canine panosteitis. In *Textbook of Small Animal Orthopaedics.* Eds. C. D. Newton and D. M. Nunamaker. Chapter 49. pp. 591–596. London: Lippincott.

40. Lewis, L. D., Morris, M. L. and Hand, M. S. (1987) *Small Animal Clinical Nutrition III.* Topeka, KS: Mark Morris Associates.

41. Malluchc, H. H. *et al.* (1988) Dihydroxyvitamin D_3 corrects bone loss but suppresses bone remodelling in ovariohysterectomized beagle dogs. *Endocrinology,* 122, 1998.

42. Meyer, H. (1983) *Ernährung des Hundes.* Stuttgart: Ulmer.

43. Meyer, H. and Zentek, J. (1991) Energy requirements of growing Great Danes. *Journal of Nutrition,* 121, S35–S36.

44. Milton, J. L. (1983) Osteochondritis dissecans in the dog. *Veterinary Clinics of North America: Small Animal Practice,* 13, 117–134.

45. Nap, R. C. (1993) Nutritional influences on growth and skeletal development in the dog. Thesis, Utrecht University. pp. 1–144.

46. Nap, R. C., Hazewinkel, H. A. W., Voorhout, G. *et al.* (1991) Growth and skeletal development in Great Dane pups fed different levels of protein intake. *Journal of Nutrition,* 121, S107–113.

47. Nap, R. C., Hazewinkel, H. A. W., Voorhout, G. *et al.* (1993) The influence of the dietary protein content on growth in giant breed dogs. *Journal of Veterinary and Comparative Orthopedics and Traumatology,* 6, 1–8.

48. National Research Council (1974) *Nutrient Requirements of Dogs.* Washington DC: National Academy Press.

49. National Research Council (1985) *Nutrient Requirements of Dogs.* Washington DC: National Academy Press.

50. National Research Council (1986) *Nutrient Requirements of Cats.* Washington DC: National Academy Press.

51. Nunez, E. A., Hedhammar, Å., Wu, F. M. *et al.* (1974) Ultrastructure of the parafollicular (C-) cells and the parathyroid cells in growing dogs on a high calcium diet. *Laboratory Investigations,* 31, 96–108.

52. O'Donnell, J. A. and Hayes, K. C. (1987) Nutrition and nutritional disorders. In *Diseases of the Cat.* Ed. J. Holzworth. Philadelphia: W. B. Saunders.

53. Olsewski, J,M., Lust, G. L., Rendano, V. T. *et al.* (1983) Degenerative joint disease: multiple joint involvement in young and mature dogs. *American Journal of Veterinary Research,* 44, 1300–1308.

54. Olssen, S. E. (1982) Morphology and physiology of the growth plate cartilage under normal and pathologic conditions. In *Bone in Clinical Orthopaedics.* Ed. G. Sumner–Smith. pp. 159–196. Philadelphia: W. B. Saunders.

55. Pobisch, R. and Onderscheka, K. (1976) Die Vitamin A-hypervitaminose bei der Katz. *Wiener Tierärztliche Monatschrift,* 63, 334–343.

56. Read, D. H., Jolly, R. D. and Alley, M. R. (1977) Polioencephalomalacia of dogs with thiamine deficiency. *Veterinary Pathology,* 14, 103–112.

57. Riser, W. H., Cohen, D. and Linqvist, S. (1964) Influence of early rapid growth and weight gain on hip dysplasia in the German Shepherd dog. *Journal of the American Veterinary Medical Veterinary Association,* 145, 661–668.

58. Riser, W. H. and Shirer, J. F. (1964) Radiographic differential diagnosis of skeletal diseases of young dogs. *Journal of the American Veterinary Radiological Society,* 5, 15–27.

59. Rivers, J. P. W. and Burger, I. H. (1989) Nutrient requirements of the dog and cat. Appendix 2. In *Nutrition of the Dog and Cat.* Ed. I. H. Burger and J. P. W. Rivers. pp. 7–11. Cambridge: Cambridge University Press.

60. Seawright, A. A., English, P. B. and Gartner, R. J. W. (1965) Hypervitaminosis A and hyperostosis of the cat. *Nature,* 206, 1171–1172.

61. Sherding, R. C. (1989) *The Cat, Diseases and Clinical Management.* New York: Churchill Livingstone.

62. Silverman, J. (1981) Nutritional aspects of cancer prevention: an overview. *Journal of the American Veterinary Medical Association,* 172, 1404.

63. Slater, M. R., Scarlett, J. M, Kaderley, R. E. *et al.* (1991) Breed, gender and age as risk factors for canine osteochondritis dissecans. *Journal of Veterinary and Comparative Orthopedics and Traumatology,* 4, 100–106.

64. Slocum, B. and Slocum, T. D. (1992) Pelvic osteotomy for axial rotation of the acetabular segment in dogs with hip dysplasia. *Veterinary Clinics of North America: Small Animal Practice,* 22 (3), 5–682.

65. Smith, G. K., Biery, D. N. and Gregor, T. P. (1990) New concepts of coxofemoral stability and the development of a clinical stress-radiographic method for quantitating hip joint laxity in the dog. *Journal of the American Veterinary Medical Association,* 196, 59–70.

66. Teare, J. A., Krook, L. and Kallfelz, F. A. (1979) Ascorbic acid deficiency and hypertrophic osteodystrophy in the dog. *Cornell Veterinarian,* **69**, 384–401.
67. Thielsen, G. H. and Madewell, B. R. (1987) Tumours of the skeleton. In *Veterinary Cancer Medicine.* 2nd edn. Philadelphia: Lea and Febiger.
68. Tvedten, H. W., Carrig, C. B. and Flo, G. L. (1977) Incidence of hip dysplasia in Beagle dogs fed different amounts of protein and carbohydrate. *Journal of the American Animal Hospital Association,* **13**, S95–S98.
69. Van Bree, H. J. J. (1991) Positive shoulder arthrography in the dog: the application in osteochondrosis lesions compared with other diagnostic imaging techniques. Thesis, University of Utrecht. pp. 1–173.
70. Voorhout, G. and Hazewinkel, H. A. W. (1987) A radiographic study on the development of the antebrachium in Great Dane pups on different calcium intakes. *Veterinary Radiology,* **28**, 152–157.
71. Voorhout, G., Nap, R. C. and Hazewinkel, H. A. W. (1993) A radiographic study on the development of the antebrachium in Great Dane pups, raized under standardized conditions. *Veterinary Radiology and Ultrasound* (accepted for publication).
72. Willett, W. (1989) The search for the causes of breast and colon cancer. *Nature,* **338**, 389.

CHAPTER 23

Skin Disease

RICHARD G. HARVEY

Introduction

Dermatological signs of nutritional deficiency may result from inadequate provision in the diet, impaired absorption from the gut or defective transport or metabolism of a specific nutrient.[35] The vast majority of dogs and cats are fed commercially prepared, nutritionally balanced diets and thus the overt deficiency of any one ingredient, particularly a major component of the diet such as protein, is very rare.[36,39] One authority has even suggested that clinical signs suggesting the deficiency of a single vitamin or mineral should raise suspicion of a genetic enzyme defect rather than a deficient diet.[28] However, there are complex relationships between many nutrients, be it within the gut or within the cytosol, and subtle changes in the amount of one nutrient may have profound affects on the biological availability, activity or expression, of another. Zinc-related dermatoses are a paradigm for the complex interrelationships between diet and metabolism. Diets may be absolutely deficient in zinc or a relative deficiency of dietary zinc may be induced by dietary excess of calcium, iron or phytate.[58] The expression of the clinical signs of deficiency is further dictated by individual differences in metabolism such as suboptimal absorption of zinc or an inability to utilize an apparently adequate dietary provision.[24,58] An additional twist is added by the observation that many of the signs of zinc deficiency can be alleviated by the provision of supplementary n6 fatty acids.[5,9] Thus, the term "nutritionally responsive dermatosis" does not necessarily imply absolute dietary deficiency of the specific nutrient, although the converse is true.

Most components of the diet have some function in one or more aspects of cutaneous homeostasis and a few, for example vitamin C and copper, are associated with well-defined cutaneous syndromes. In small animal practice clinical signs of disease that respond to supplementation appear to be confined to three nutrients, the essential fatty acids, vitamin A and zinc.[35,39] In addition a number of other dietary components, for example biotin and vitamin E, have been investigated for their potential in the control of skin disease both in dogs and cats. Because the skin has only a limited range of response the cutaneous signs of nutritional deficiency are similar, whatever the component implicated. The typical signs of a nutritional dermatosis are scale, erythema and alopecia and a greasy skin, often accompanied by secondary bacterial infection. These changes reflect alterations in the process of keratinization,

THE APPROACH TO THE DERMATOLOGICAL CASE

Complete clinical and dietary history

The aim is to assess the animal's health and management and to obtain a knowledge of the nature of the dermatosis and its progression. The presence of systemic disease may be suspected at this stage and dietary idiosyncrasies spotted. The presence or otherwise of contagion should be assessed. The response to previous treatments should be considered. While the history is being taken the behavior of the animal can be observed; a lack of interest in the surroundings may suggest pyrexia, lethargy or hypothyroidism for example.

Complete physical examination

Most practitioners start at the rostral end and progress caudally. Examine the oral cavity, palpate the superficial lymph nodes and auscultate the heart. The abdomen should be palpated and, in males, the scrotum and testicles should be assessed. The rectal temperature should be taken.

Detailed examination of the skin and its appendages

The coat is examined for the presence of primary and secondary hairs and their texture, crimp and gloss assessed. The pattern and distribution of any alopecia is noted. Oiliness and odor are apparent during this examination. Larger ectoparasites may be observed at this point. The surface of the skin is examined, including the external ear canals and the interdigital regions. The presence of primary lesions (macules, papules, pustules etc.) are particulary relevant. The secondary lesions (scale, crust, excoriation) are less helpful in themselves although their distribution may be valuable in arriving at a diagnosis. Thus in scabies the distribution of the secondary lesions on the pinnal margins, elbows and hocks is a valuable diagnostic clue.

Laboratory and other tests to narrow the list of differential diagnoses

Carefully taken and microscopically examined skin scrapings are taken from almost every dermatological case. Other tests such as trichograms, sellotape strips, aspiration cytology, punch or excision biopsy, intradermal testing, basal and dynamic endocrine assay and trial therapy may be indicated. Routine hematolgy and biochemistry are less helpful although in certain cases they may suggest a diagnosis. Thus elevated liver enzymes may be found in hypothyroidism, hyperadrenocorticism and epidermal metabolic necrosis (syn. hepatocutaneous syndrome, necrolytic migratory erythema or diabetic dermatopathy).

A definite diagnosis is made, a prognosis given and treatment prescribed.

FIG. 23.1 The approach to the dermatological case.

the fundamental metabolic activity of the skin and the adnexae.

Approach to the Case

Although the incidence of nutritionally responsive dermatoses is low the clinician should be aware of the possibility that they will be encountered and all case histories should include inquiries into the diet of the animal. An animal fed a commercially prepared diet should not be at risk from an absolute dietary deficiency, but may be exposed to low dietary fat and relative deficiencies of zinc. Internal disease such as malabsorption, hepatoses and renal failure may result in symptoms compatible with dietary protein or essential fatty acid deficiency and the clinician should ensure that all animals receive a complete physical examination prior to a detailed dermatological work-up (Fig. 23.1). It is important to assess the presence or absence of pruritus. The most common causes of pruritus are flea-bite hypersensitivity and other ectoparasites, pyoderma and hypersensitivities (Fig. 23.2). Nonpruritic dermatoses

THE MOST COMMON CAUSES OF PRURITIC DERMATOSES

Ectoparasites	Fleas
	Cheyletiellosis
	Sarcoptic mange
	Pediculosis
	Trombiculidiasis
Bacterial infections	Pyotraumatic dermatitis
	Superficial folliculitis
Hypersensitivities	Flea-bite hypersensitivity
	Atopy
	Dietary hypersensitivity and intolerance
	Allergic or irritant contact dermatitis
Otitis externa	
Anal sacculitis	

FIG. 23.2 The most common causes of pruritic dermatoses.

PRINCIPAL CAUSES OF NONPRURITIC DERMATOSES	
Ectoparasites Endocrinopathies Dermatophytosis Deep pyoderma Nutritional deficiencies	Demodicosis Hypothyroidism Hyperadrenocorticism Sertoli cell tumor
Note that secondary changes accompanying these conditions will ofter result in pruritus.	

FIG. 23.3 The principal causes of nonpruritic dermatoses.

are less common. Many of the causes of a nonpruritic dermatosis result in secondary changes (and pyoderma) and this results in pruritus. Examples include demodicosis, endocrinopathies, dermatophytosis, deep pyoderma and nutritional dermatoses (Fig. 23.3).

Most of the cutaneous signs of dietary deficiency are related to abnormal keratinization, whatever the dietary component. Thus the differential diagnoses that should be borne in mind are those of diseases characterized by the production of scale and crust (Fig. 23.4). A diagnosis of a diet-related dermatosis will be suggested by consideration of history, clinical examination and the elimination of

DIFFERENTIAL DIAGNOSIS OF DISEASES CHARACTERIZED BY THE PRODUCTION OF SCALE AND CRUST	
Disorders of keratinization	Idiopathic defects in keratinization Icthyosis Vitamin A-responsive dermatosis Zinc-responsive dermatosis Epidermal dysplasia of West Highland White Terriers Granulomatous sebaceous adenitis Color dilute alopecia Lichenoid-psoriasiform dermatoses of English Springer Spaniels Schnauzer comedone syndrome Idiopathic nasodigital hyperkeratosis Canine and feline acne Idiopathic ear margin dermatosis

Environmental and
  managemental causes
	Low humidity
	Irritant or too-frequent shampooing
	Repeated clipping of the coat
	Feeding of inadequate levels of protein or essential
	   fatty acids

Infectious causes	Ectoparasitic infection Bacterial skin disease Dermatophytosis
Hypersensitivities	Flea-bite hypersensitivity Atopy Dietary hypersensitivity and intolerance Drug eruption Endocrine hypersensitivity Allergic or irritant contact dermatitis
Disorders of metabolism	Malabsorption Hepatic disease Renal disease
Endocrinopathies	Hypothyroidism Hyperadrenocorticism Sertoli cell tumor Ovarian imbalance
Autoimmune disorders	Pemphigus foliaceus Systemic lupus erythematosus
Neoplasia	Cutaneous lymphoma

FIG. 23.4 The differential diagnosis of diseases characterized by the production of scale and crust.

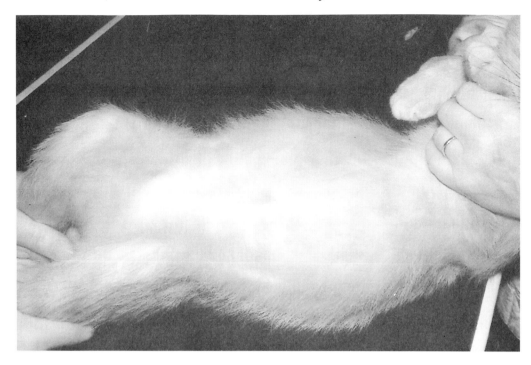

FIG. 23.5 Telogen defluxion. A cat with chronic renal failure (protein-losing nephropathy). Note the abdominal distension and alopecia on the ventral abdomen.

alternative differential diagnoses. Thus the elimination of infectious diseases (by examination of the animal and microscopic examination of skin scrapings) and a knowledge that a diet low in fat is being fed may suggest a diagnosis of essential fatty acid deficiency. In some cases, such as vitamin A responsive dermatosis for example, the diagnosis is made with the aid of histopathological examination of biopsy samples. In most cases, however, the diagnosis of a diet-related dermatosis is not made easily and may only be achieved after extensive investigations and trial therapies have eliminated alternative diagnoses.

If a deficient diet is identified then a nutritionally balanced diet should be substituted.[7] Oversupplementation of an inadequate diet may further imbalance the inadequate diet and cause additional problems. Supplements may be indicated during the transition phase or if clinical signs are severe, for example in zinc-responsive dermatosis. A number of nutritional supplements, for example vitamin A, the retinoids and essential fatty acids, may also be used for their therapeutic effects in situations where no evidence of deficiency or

metabolic defect is apparent. These indications will be discussed in the relevant sections on the parent nutrients.

Protein–Calorie Malnutrition

Pathophysiology

Epidermal turnover and hair growth have been estimated to require as much as 30% of a dog's daily protein intake.[7,28] Long-term protein deficiency is unlikely to be encountered in clinical practice but may be encountered in animals maintained for prolonged periods on very low protein diets or with chronic illness, such as chronic renal failure. There is hyperkeratosis, epidermal hyperpigmentation and, paradoxically, a loss of hair pigment.[39] The hair cycle enters telogen and the normal coat is gradually replaced by a thin, poor, lusterless coat and eventually alopecia may result (Fig. 23.5), *telogen defluxion*.[37] Short-term protein deficiency may occur in animals subject to suddenly increased maintenance requirements. Bitches in late gestation and at times of peak

FIG. 23.6 Anagen defluxion. Elderly Whippet with a history of acute nephritis some 8 weeks previously. Note the patchy alopecia resulting from fracture of weakened hair shafts as they emerge from the follicle.

lactation may be at risk, as are growing puppies and kittens, because owners may not realize that an increase in both quantity and quality of protein is necessary for animals at certain stages of their life. In humans the hair is a subtle indicator of nutritional deficiency and brittle, easily shed, lusterless hair accompanied by arrested anagen may result from the rapid onset of dietary protein deficiency.[11] Sudden cessation of anagen may occur in severe systemic disease, resulting in areas of weakness in the hair shaft. As the regrowing shaft emerges from the protection of the follicle, it breaks resulting in an uneven, rough, dry pelage (Fig. 23.6), *anagen defluxion*.[37] Note: the author has seen transient hair defects in orphan kittens subject to artificial rearing (Fig. 23.7).

Diagnosis

Consideration of the nutritional history will allow identification of the dietary cause. Physical examination and appropriate laboratory tests will confirm internal disease, if present. The clinical signs of scale and poor coat, in themselves, are not diagnostic although the thin, hyperpigmented skin and paradoxical loss of hair color will aid arrival at a diagnosis. Examination of the hair shafts and the preparation of a trichogram may allow identification of anagen and telogen hair bulbs and calculation of anagen/telogen ratios. However, the clinical relevance of any information gained is uncertain because of the profound effect of environment, particularly day length.[1,3] Histopathological examination of the skin would support a diagnosis but, again, changes are not specific.

Prognosis and Management

Uncomplicated protein–calorie deficiency is easily rectified by the institution of an appropriate level of protein in the diet. Chronic inanition as a consequence of internal disease carries a guarded to poor prognosis. Buffington recommends 16% of energy as high quality protein for maintenance of the adult dog whereas cats require at least 20%.[7] Canine and feline growth, gestation and lactation require 25–30% and 30–35% of daily energy as protein, respectively.[7,28]

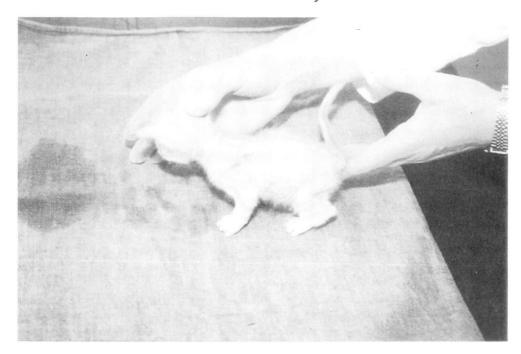

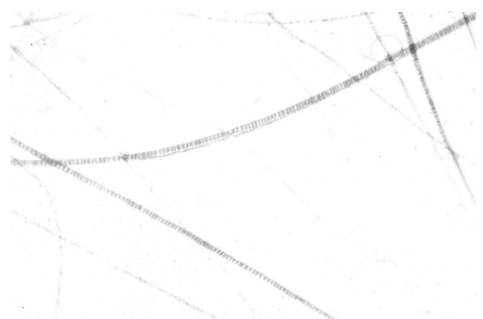

FIG. 23.7 Anagen defluxion (top) and hair shaft defects (bottom) in an orphaned kitten with alopecia consequent upon sudden cessation of hair growth.

Essential Fatty Acids

Pathophysiology

Essential fatty acids fulfill a number of structural roles in the body such as conferring fluidity on cell membranes.[55] The fluidity of cell membranes is of profound importance because it will affect the function of cell-bound receptors and transmembrane signal transducers that are present in all cell membranes, particularly those of the immune system.[57,60] Within the epidermis essential fatty acids fulfil two structural roles, binding water into the stratum corneum (and thus conferring suppleness and elasticity to the skin) and the

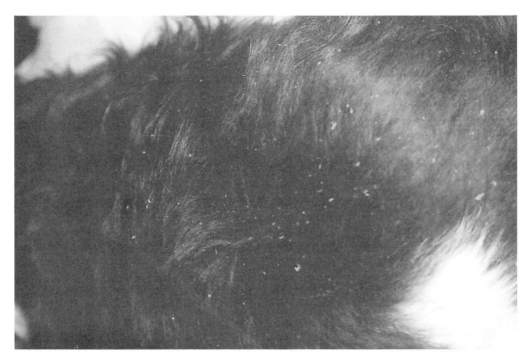

FIG. 23.8 Fine scale present on a Springer Spaniel fed a commercially prepared, dried, complete diet. The signs resolved after sunflower oil was added to the diet.

formation and maintenance of the epidermal permeability barrier.[22,29,63] In addition to structural functions the essential fatty acids also have biochemical roles as antioxidants and as precursors of eicosanoids, such as prostaglandins and leukotrienes.[20]

The essential fatty acids are all members of the n3 and n6 families and are derived from dietary α-linolenic and linoleic acid, respectively. In the dog these parent compounds are further metabolized by alternate desaturation and elongation into physiologically active derivatives that subserve some of the functions described above, particularly as sources of eicosanoids.[44] Cats lack δ6 desaturase and require linoleic and arachidonic acids in their diet.[14,31,32,34] Commercially prepared canned diets are adequately supplemented but linoleic acid content of the dried-type of diet may be inadequate if the food is stored for too long, particularly in a warm, damp environment.[26] Cutaneous signs compatible with essential fatty acid deficiency may accompany chronic internal disease such as malabsorption and liver disorders.

The early signs of essential fatty acid deficiency in the dog are a dull coat accompanied by a fine scale[28] (Fig. 23.8). Prolonged deficiency results in alopecia, a greasy skin particularly on the pinnae and feet, exfoliative dermatitis and secondary pyotraumatic dermatitis.[7,9] In the cat the earliest signs of deficiency are a dry, staring coat and profuse scale, later followed by varying degrees of erythema, alopecia and pruritus. Systemic signs such as poor wound healing, reduced immunocompetence and infertility may be seen in severe cases.[7,32,45] In the cat these signs do not become apparent until a diet deficient in essential fatty acids has been fed for many months.[32]

Diagnosis

The diagnosis of essential fatty acid deficiency may be supported by consideration of the clinical signs and nutritional history. The interpretation of the classical human assessments of essential fatty acid status such as the "triene ratio" (20:3n9/20:4n6), or indicators of desaturation (20:4n6/18:2n6) are open to misinterpretation and must be treated cautiously.[44]

(a)

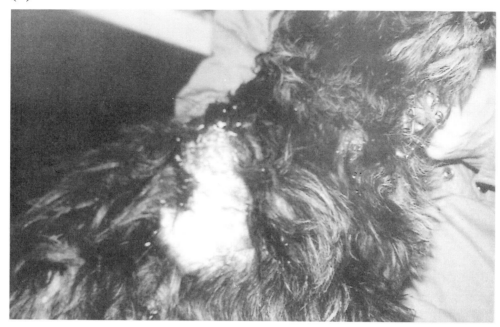

(b)

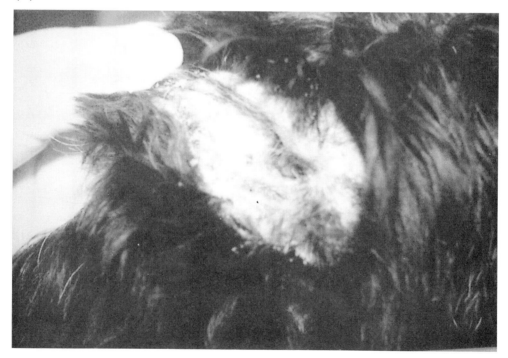

(c)

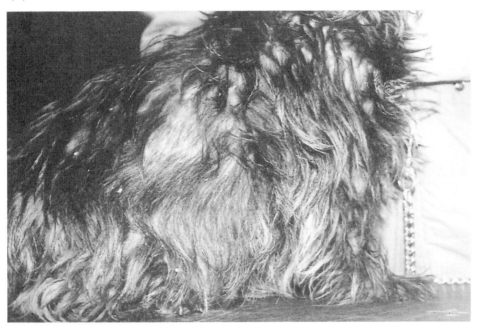

(d)

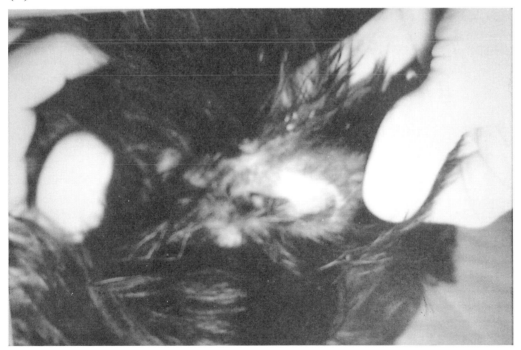

FIG. 23.9 Cairn Terrier with atopy. Note the scale, alopecia and unkempt coat on the trunk (a) and the scale and crusting on the pinna (b). After 8 weeks of supplementation with a source of linoleic and gamma linolenic acid (evening primrose oil) there was an improvement in coat condition, a resolution of the pinnal lesions and a regrowth of hair (c and d).

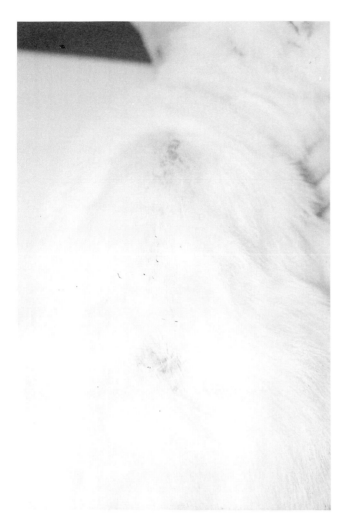
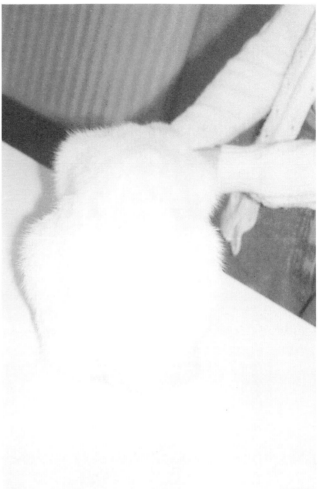

FIG. 23.10 Papulocrustous dermatitis in a domestic short hair (left). Note the resolution of lesions associated with dietary supplementation with n6 fatty acids (evening primrose oil) (right).

Prognosis and Management of Deficiency

Prognosis is good if nutritional deficiency is diagnosed. Dogs require 2% of calories as linoleic acid[7] whereas cats require at least 2% calories as linoleic acid and, in addition at least 0.04% as arachidonic acid.[7,34]

Therapeutics

Symptomatic Treatment of Defects in Keratinization

A percentage of dogs with dull coats and a fine scale will respond to dietary supplementation with fat.[35,39] The underlying cause in these instances is probably a relative nutritional deficiency and the traditional approach has been to administer animal fats to these animals. Recent advances in our understanding of lipid metabolism has resulted in greater stress being placed on replacing the animal fat with vegetable oil.[7,39] Recently, the use of sunflower oil in the management of idiopathic defects in keratinization has been reported.[8] Dogs were fed 1.5 ml/kg sunflower oil for 30 days and the improvement in the skin condition was accompanied by a decrease in serum and cutaneous 20:4n6 and by an increase in serum and cutaneous 18:2n6.

Management of Canine Atopy

It has been proposed that atopy in humans may be accompanied by a deficiency in $\delta6$ desaturase[33] and that may explain the reported efficacy of evening primrose oil in controlling pruritus associated with atopy.[67] Studies in the dog[30,38,48,49] have similarly documented that essential fatty acid supplements are useful in managing the pruritus and dermatological signs associated with canine atopy (Fig. 23.9). Whether there is an underlying enzyme defect in canine atopy is less clear. Scarff and Lloyd reported a double blind placebo-controlled study in which a significant reduction in erythema and an overall improvement in the clinical condition were reported.[48] Clinical experience of the author suggests that the improvement is dose-responsive and that, furthermore, there is often a considerable lag (2–6 weeks) between the beginning of supplementation and a beneficial effect being apparent to the owners. The author uses an initial dose of 127 mg evening primrose oil and 30.8 mg fish oil per 10 kg body weight (BW) sid (equivalent to 4 × 550 mg capsules evening primrose and fish oil/10 kg BW sid). Once remission is attained it is often possible to reduce the dose considerably. Although more expensive than systemic glucocorticoids the complete lack of side-effects in the vast majority of dogs makes it the medical treatment of choice in many instances. Some 30–40% of dogs can be sufficiently well controlled to obviate the requirement for systemic glucocorticoids.

Papulocrustous Dermatitis: "Miliary Dermatitis" in Cats

Miliary dermatitis is the common cutaneous manifestation of a number of dermatoses of which flea-bite hypersensitivity is the most common.[39] Other hypersensitivities, in particular atopy, to ectoparasites such as *Cheyletiella* spp. and dermatophytosis account for the majority of other cases. Management is best directed toward the underlying cause, where identified. Cases of atopy and a proportion of the cases of flea-bite sensitivity, however, will prove difficult to control without the use of systemic antipruritics such as glucocorticoids. In these cases supplementation with essential fatty acids may prove beneficial.[17] Recent work[18] has demonstrated that n6 fatty acids, or combinations of n6 and n3 fatty acids,

are more efficacious than n3 fatty acids alone in controlling the clinical signs (Fig. 23.10). The lack of side-effects makes this method of control attractive.

Other Feline Dermatoses

Feline acne is a common condition of cats associated with a defect in keratinization of the follicles. There may be associated bacterial and *Malassezia pachydermatis* infection. Some cases respond to essential fatty acid supplementation (Fig. 23.11) and others will respond to retinoid therapy (see below). Most cases are managed more conservatively with topical benzoyl peroxide or even benign neglect.

Vitamin A and Retinoids

Pathophysiology

Vitamin A (retinol) is formed in the gut by the reversible reduction of retinaldehyde.[2,46] In the dog, as in most mammals, dietary carotene is the source of the retinaldehyde whereas in the cat there is a dietary requirement for retinol because it cannot utilize carotenes.[7] Retinol is transported tightly bound to a binding protein and is stored in the liver.[2,39] Retinaldehyde is essential for the normal development and function of the retina and is an intermediate in the conversion of retinol to retinoic acid.[46] Retinoic acid interacts with the genome, via a nuclear receptor, to regulate cellular growth and development.[42,65] The interaction with the genome requires the presence of zinc and results in the expression of RNA and the synthesis of specific proteins. Retinoic acid is essential for a wide range of cell and tissue functions but its cutaneous role is particularly directed toward the proliferation and differentiation of the keratinocyte, where it regulates keratin gene expression.[6]

The cutaneous signs of vitamin A deficiency include hyperkeratosis of the epidermis, follicles and the sebaceous glands,[39] in addition to a poor coat and alopecia. There may be a papular eruption, a consequence of blocked,

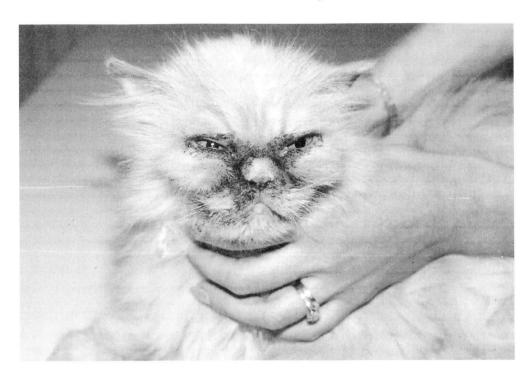

FIG. 23.11 Feline acne in a Persian cat. This is a severe case (top) with lesions apparent around the mouth, nares and eyes in addition to being present on the chin. Note the dramatic resolution (bottom) associated with dietary supplementation with n6 fatty acids (evening primrose oil).

hyperkeratotic sebaceous gland ducts.[39] There is increased susceptibility to bacterial infection.[7,39] Vitamin A deficiency is rare but is most likely to occur during growth, pregnancy or lactation[28] in animals on fat-restricted diets or those with poor fat absorption.[9] Vitamin A toxicity results in similar cutaneous signs to those seen in vitamin A deficiency.[9]

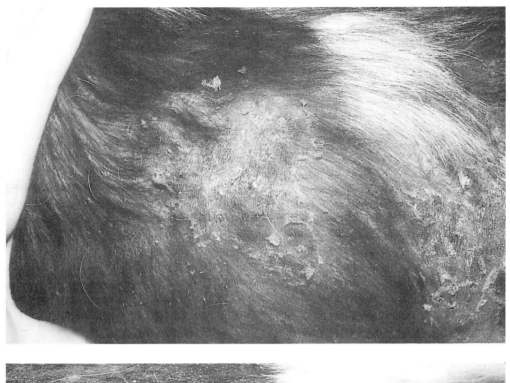

FIG. 23.12 Idiopathic seborrhea in a Cocker Spaniel. Note the alopecia, papules and scale (top). One month after treatment with 1 mg/kg sid etretinate there is a marked improvement in clinical condition (bottom) with hair regrowth and absence of papules. The associated otitis externa did not resolve.

Synthetic retinoids are derivatives of these natural retinoids and have been developed to exert only one of the biological actions of the natural group, thus minimizing the side-effects.[2,42] Retinoids are one of the most important additions to the human dermatologist's armamentarium, particularly with respect to the management of severe acne and other disorders of keratinization.[66] However, veterinary applications are, as yet, few.[27] Currently isotretinoin and etretinate are restricted to hospital pharmacies only. Furthermore, the cost of these products and their potential for causing toxicity to owner as well as to animal have limited their use to what are essentially clinical trials. As newer derivatives become available more veterinary applications may become apparent.

Diagnosis

Deficiency is unlikely in animals on commercially prepared diets and a nutritional history will suggest the diagnosis. A localized or generalized fine papular eruption may be noted. The papules are a consequence of blocked sebaceous ducts. Follicular and epithelial hyperkeratosis may be noted on histopathological examination of biopsy samples.[16,39] These findings may support the clinical suspicion of Vitamin A deficiency but are by no means pathognomonic; these are nonspecific signs.

Prognosis and Management of Deficiency

Dogs require 5000 IU/kg dry matter of diet and cats 35260 IU/kg dry matter of diet.[28] Great care should be taken to avoid overdosage and toxicity. Lewis recommends 3000 IU Vitamin A/lb BW (6540 IU/kg) as a once only dose.[28] This is equivalent to the liver's storage capacity and need not be repeated for at least 2 months. Muller *et al.* recommend a maximum of 400 IU/kg/day for 10 days.[39] Care must be taken in advising supplementation with vitamin A because toxicity can result if levels of four times the advised requirement are fed for prolonged periods.[28] Toxicity produces signs similar to those of a nutritional deficiency with pruritus, scale, erythema and alopecia being reported.[28]

Therapeutics

Vitamin A and Defects in Keratinization

Although vitamin A is closely involved with the process of epidermal differentiation, the

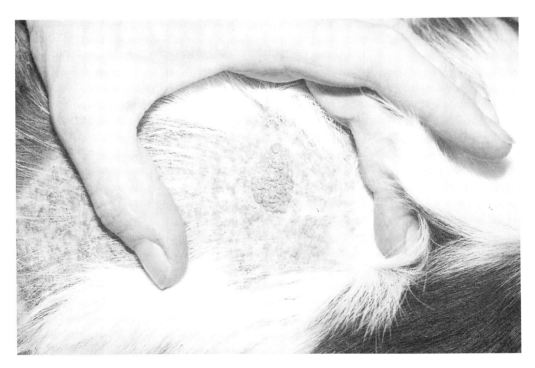

FIG. 23.13 Vitamin A responsive dermatosis. Predominantly found in Cocker Spaniels, the dermatosis is characterized by focal accumulations of tightly adherent crust, often in association with a more diffuse scale.

use of supplements in cases of generalized defects of keratinization (such as those occurring in the Cocker Spaniel and Basset Hound) has, in general, been disappointing.[13,27,35,39] Recently, a study was reported[41] in which Cocker Spaniels with idiopathic defects in keratinization (primary seborrhea) were treated with the synthetic retinoid etretinate. The dogs showed dramatic improvements with respect to the scale, crust and alopecia although the associated otitis externa did not respond. The author has used this product in a small number of cases (Fig. 23.12) with similar results. Etretinate is less useful in the management of keratinization defects in other breeds such as West Highland White Terriers and Basset Hounds.[41]

There is, however, one specific dermatosis that may respond to supplementation with vitamin A. Vitamin A responsive dermatosis is a rare condition almost entirely confined to Cocker Spaniels.[23,27,39] There is no evidence that affected animals have deficient diets. Signs are of a severe, and more or less generalized, defect in keratinization. There may be a greasy coat, loss of hair, mild to moderate pruritus and secondary pyoderma. The condition is characterized by focal areas of follicular plugging with plaque-like accumulations of keratin (Fig. 23.13). Histopathological examination of these areas reveals a markedly disproportionate follicular hyperkeratosis. The lesions slowly resolve with doses of 10,000 IU vitamin A daily.[23,27,39] Because this is in excess of the dietary requirement, care must be taken to eliminate other causes of the dermatosis before treatment is initiated.

Granulomatous sebaceous adenitis is an uncommon disease featuring a progressive scaling and alopecia. Many cases progress to a generalized, patchy dermatosis that is frequently refractory to treatment. Stewart *et al.* reported that isotretinoin was effective in the management of granulomatous sebaceous adenitis in two Hungarian Vizslas,[54] although this has not been the experience of clinicians using the drug in other breeds.

Other dermatoses characterized by defects in keratinization such as feline acne, Schnauzer comedo syndrome and congenital lamellar ichthyosis have been reported to respond to treatment with retinoids.[27]

Cutaneous Neoplasia

Multiple intracutaneous cornifying epitheliomata are benign neoplasms arising from the epithelial tissue. There are breed predispositions in the Norwegian Elkhound, Keeshond and Old English Sheepdog.[39] There is a single case report of a dog with multiple intracutaneous cornifying epitheliomata that resolved when the dog was treated with isotretinoin.[19] Although retinoids have shown some activity against precancerous epithelial lesions, they were ineffective in a trial investigating their affect in feline squamous cell carcinoma.[12]

Zinc Deficiency and Zinc-Responsive Dermatosis

Pathophysiology

Zinc is a component in numerous metalloenzymes that regulate lipid, protein and nucleic acid synthesis and metabolism.[4,35] It is essential for the interaction of vitamin A with the genome.[65] Zinc has been shown to exert an influence on most aspects of the immune response.[40] There is, however, no evidence that any of the clinical signs of dietary zinc deficiency can be related to dysfunction in any of the enzymes of which it is a component.[10] Furthermore, many of the cutaneous lesions of zinc deficiency can be reversed by the administration of essential fatty acids and this has led to the proposal that aberrations in fatty acid metabolism account for the majority of the signs associated with zinc deficiency.[5,9,10,21] Clinical signs of zinc deficiency may relate to absolute dietary zinc deficiency although this is considered rare in animals fed high quality, commercially prepared diets. More commonly there is a relative deficiency due to interaction with other dietary components. The absorption of zinc from the gut is inhibited by iron, copper and calcium that compete with zinc for absorption, and intestinal

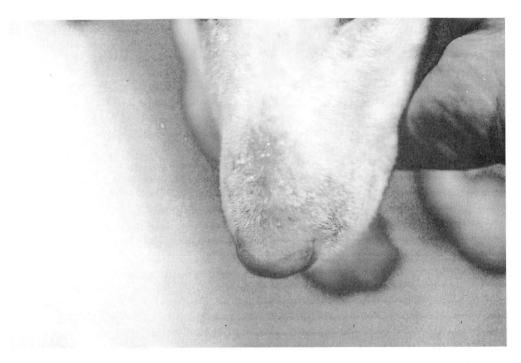

FIG. 23.14 Lethal acrodermatitis. Erythema, scale and crusting on the nasal
area of a Bull Terrier. Photo courtesy N. A. McEwan.

phytate and inorganic phosphate bind zinc and thus hinder absorption.[36,39,53,56] Factors such as age, sex and ambient temperature also influence the serum zinc levels.[36,39] Metabolic inability to utilize zinc may result in severe systemic signs of lethal acrodermatitis.[24] Naturally occurring zinc deficiency has not been reported in the cat.

In cases of experimental, absolute zinc deficiency or a metabolic inability to utilize zinc there may be systemic as well as cutaneous signs. These include growth retardation, emaciation, conjunctivitis and keratitis, dermatitis (Fig. 23.14), pododermatitis and thymic atrophy.[24,47] In less severe cases associated with relative dietary deficiency or interaction, the cutaneous signs predominate and appear over the pressure points and adjacent to the mucosae in a more or less symmetrical manner, around the mouth and periorbitally, peripinnal, perineal and on the limbs.[39,55] Areas of crusting are typically well demarcated with an erythematous border (Fig. 23.15). The coat is dull, harsh and occasional areas of achromotrichia may be noted. Lymphadenopathy is a common feature, particularly in younger animals.[53,58]

Two groups of animals appear to be at risk from zinc-responsive dermatosis and cutaneous lesions are similar in both groups.[39] The distinction between groups is not absolute and individuals may be encountered that have features of both syndromes. In the first group ("syndrome 1") certain individuals of many breeds of dog, but especially Siberian Huskies and Alaskan Malamutes, appear to be unable to absorb sufficient zinc from the intestine, even when fed a nutritionally balanced diet. The condition may be precipitated by adulthood or periods of stress.[7] The second type of zinc-responsive dermatosis ("syndrome 2") may occur in any breed of dog. Typically it is seen in rapidly growing animals fed inadequate diets or diets in which nutritional antagonism occurs, particularly diets oversupplemented with calcium or high in phytate.[39]

Diagnosis

The diagnosis of zinc deficiency is based on the case history, clinical examination, dietary history (particularly feeding soya or cereal-based diets), histopathological examination of cutaneous biopsy specimens and laboratory

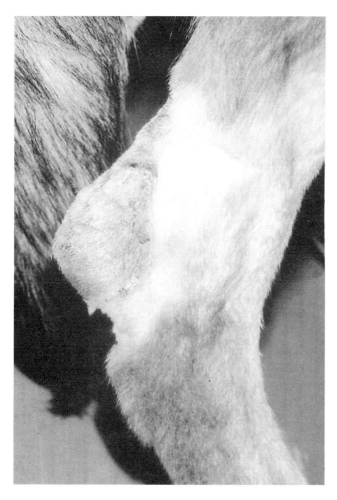

FIG. 23.15 Zinc-responsive dermatosis. Focal accumulation of scale and crust with accompanying erythema on the hock. Photo courtesy A. H. van den Broek.

analysis of tissue zinc concentration.[56] Histopathological confirmation of diagnosis based solely on the presence of parakeratotic hyperkeratosis is reported to be unreliable.[56] Confirmation of the deficiency by definitive laboratory testing is difficult. Zinc is a component of many materials used in collection and analytical equipment[56] and great care must be taken to avoid contamination. The diagnostic value of zinc concentration in leukocytes is minimal and that of zinc concentration in serum and hair poor, serving only to corroborate a diagnosis of zinc deficiency.[59]

Prognosis and Management of Deficiency

The prognosis for Bull Terriers with congenital acrodermatitis is very poor.[24] The prognosis for individuals exhibiting signs of zinc-responsive dermatosis is good. Zinc supplementation with oral zinc sulfate at 10 mg/kg, or zinc methionate at 1.7 mg/kg sid with food is the treatment of choice.[39]

Supplementation may be required for life in animals with an inability to absorb adequate zinc from normal diets, i.e. syndrome 1. However, if the animals are exhibiting signs compatible with syndrome 2 then supplementation may only be required until lesions have been resolved, assuming a nutritionally balanced diet is substituted for the inadequate one.

B Group Vitamins

B vitamins subserve numerous roles in many tissue as coenzymes and catalysts. Individual deficiency of B group vitamins is reported to be very rare in dogs and cats in clinical practice.[7,39] Signs of a nonpruritic, scaly alopecia have been reported.[28,39] Deficiency of water-soluble vitamins may occur in conditions associated with water loss such as enteritis or in cases where prolonged antibacterial therapy has reduced the synthetic ability of enteric flora.[28] Occasionally, animals on bizarre diets may exhibit signs of single B vitamin deficiency because dietary components inactivate specific vitamins, for example avidin in raw egg white will bind biotin; niacin in corn is bound and unavailable.[28] If B vitamin deficiency is suspected then dietary incompatibility should be investigated and therapy with the entire B vitamin group should be instituted with high levels being administered.[28]

Therapeutics

Biotin

Biotin is an intramitochondrial cofactor in the chain enlargement of fatty acids and possibly also the desaturation of EFA.[62] An early use of biotin supplementation was reported to be useful in the treatment of miliary dermatitis in the cat.[25] This has not been substantiated. More recently, the use of biotin in doses of 5 mg sid has been advocated for treatment of various dermatoses, although these were poorly controlled studies.[15,61]

Niacinamide

Niacinamide (the amide of niacin, syn. nicotinamide) has recently been reported to be useful in the management of canine autoimmune skin disease.[64] A dose of 500 mg tid in conjunction with 500 mg tetracycline tid was given to dogs weighing less than 10 kg. Those weighing more than 10 kg were given 250 mg of each drug tid. In 20 dogs with discoid lupus erythematosus treated as described, 40% had an excellent response.

Vitamin E Deficiency

Vitamin E is the generic name for a group of compounds, the tocopherols and tocoriols.[43] They function as antioxidants and free-radical scavengers, controlling tissue peroxidation. Naturally occurring vitamin E deficiency has not been reported in dogs.[50] "Megadoses" of vitamin E have been reported to be of benefit in the management of canine discoid lupus erythematous and acanthosis nigricans.[51,52] Reports from other clinicians are less favorable.

References

1. AlBadgered, F. A., Titkemeyer, C. W. and Lovell, J. E. (1977). Hair follicle cycle and shedding in male Beagle dogs. *American Journal of Veterinary Research*, **38**, 611–616.
2. Allen, J. G. and Bloxham, D. P. (1989). The pharmacology and pharmacokinetics of the retinoids. *Pharmacology and Therapeutics*, **40**, 1–27.
3. Baker, K. P. (1974). Hair growth and replacement in the cat. *British Veterinary Journal*, **130**, 327–334.
4. Banta, C. A. (1978). The role of zinc in canine and feline nutrition. *Nutrition of the Dog and Cat. Waltham Symposium Number 7*. pp. 317–327. Cambridge: Cambridge University Press.
5. Bettger, W. J., Reeves, P. G., Moscatelli, E. A., Reynolds, G. and O'Dell, B. L. (1979). Interaction of zinc and essential fatty acids in the rat. *Journal of Nutrition*, **109**, 480–488.
6. Blumenberg, M., Connolly, D. M. and Freedberg, I. M. (1992). Regulation of keratin gene expression: the role of the nuclear receptors for retinoic acid, thyroid hormone and vitamin D3. *Journal of Investigative Dermatology*, **98**, 42S–49S.
7. Buffington, C. A. (1987). Nutrition and the skin. *Proceedings of the Eleventh Annual Kal Kan Symposium for the Treatment of Small Animal Diseases*. pp. 11–16.
8. Campbell, K. L., Uhland, C. F. and Dorn, G. P. (1992). Effects of sunflower oil on serum and cutaneous fatty acid profiles in seborrhoeic dogs. *Veterinary Dermatology*, **3**, 29–35.
9. Cunnane, S. C., Manku, M. S., Horrobin, D. F. (1984). Accumulation of linoleic and γ-linolenic acids in tissue lipids of pyridoxine deficient rats *Journal of Nutrition*, **114**, 1754–1761.
10. Cunnane, S. C., Manku, M. S., Horrobin, D. F. and Davignon, J. (1981). Role of zinc in linoleic acid desaturation and prostaglandin synthesis. *Progress in Lipid Research*, **20**, 157–160.
11. Dawber, R. P. R., Ebling, F. J. G. and Wojnarowska, F. T. (1992). Disorders of the hair. In *Rook/Wilkinson/Ebling Textbook of Dermatology*, Eds. R. H. Champion, J. L. Burton and F. J. G. Ebling. 4th edn. pp. 2533–2638. Oxford: Blackwell Scientific Publications.
12. Evans, A. G., Madewell, B. R. and Stannard, A. A. (1985). A trial of 13cisretinoic acid for treatment of squamous cell carcinoma and preneoplastic lesions of the head in cats. *American Journal of Veterinary Research*, **46**, 2553–2557.
13. Fadok, V. A. (1986). Treatment of canine idiopathic seborrhoea with isotretinoin. *American Journal of Veterinary Research*, **47**, 1730–1733.
14. Frankel, T. and Rivers, J. P. W. (1978). The nutritional and metabolic impact of γ-linolenic acid (18:3n6) on cats deprived of animal lipid. *British Journal of Nutrition*, **39**, 227–231.
15. Fromageot, D. and Zaghroun, P. (1990). The potential role of biotin on canine dermatology. *Recueil de Médecine Vétérinaire*, **166**, 87–94.
16. Gross, T. L., Ihrke, P. J. and Walder, E. J. (1992). A macroscopic and microscopic evaluation of canine and feline skin disease. In *Veterinary Dermatology*. St Louis: Mosby Year Book.
17. Harvey, R. G. (1991). The use of essential fatty acid supplementation in the management of feline miliary dermatitis. *Veterinary Record*, **128**, 326–329.
18. Harvey, R. G. (1992). The effect of varying proportions of evening primrose oils and fish oil on cats with miliary dermatitis. *Veterinary Record* (in press).
19. Henfrey, J. I. (1991). Treatment of multiple intracutaneous cornifying epitheliomata using isotretinoin. *Journal of Small Animal Practice*, **32**, 363–365.

20. Horrobin, D. F. (1992). Nutritional and medical importance of gammalinolenic acid. *Progress in Lipid Research,* **31,** 163–194.

21. Huang, Y. S., Cunnane, S. C., Horrobin, D. F. and Davignon, J. (1982). Most biological effects of zinc deficiency corrected by γ-linolenic acid (18:3ω6) but not by linoleic acid (18:2ω6). *Atherosclerosis,* **41,** 193–207.

22. Imokawa, G. and Hattori, M. (1985). A possible function of structural lipids in the waterholding properties of the stratum corneum. *Journal of Investigative Dermatology,* **84,** 282–284.

23. Ihrke, P. J. and Goldschmidt, M. H. (1983). Vitamin A responsive dermatosis in the dog. *Journal of the American Animal Hospital Association,* **182,** 687–690.

24. Jezyk, P. F., Haskins, M. E., MacKaySmith, W. E. and Patterson, D. F. (1986). Lethal acrodermatitis in Bull Terriers. *Journal of the American Veterinary Medical Association,* **188,** 833–839.

25. Joshua, J. O. (1959). The use of biotin in certain diseases of the cat. *Veterinary Record,* **71,** 102.

26. Kelly, N. C. (1992). Practical evaluation and feeding of dog foods. *In Practice,* **14,** 8–15.

27. Kwochka, K. W. (1989). Retinoids in dermatology. In *Current Veterinary Therapy X.* Ed. R. W. Kirk. pp. 553–560. Philadelphia: W. B. Saunders.

28. Lewis, L. D. (1981). Cutaneous manifestations of nutritional imbalances. *Proceedings of the 48th Annual Meeting of The American Animal Hospital Association.* pp. 263–272.

29. Landman, L. (1986). Epidermal permeability barrier: transformation of lamellar granule disks into intercellular sheets by a membrane fusion process. *Journal of Investigative Dermatology,* **87,** 302–309.

30. Lloyd, D. H. and Thomsett, L. R. (1989). Essential fatty acid supplementation in the treatment of canine atopy. A preliminary study. *Veterinary Dermatology,* **1,** 14–44.

31. MacDonald, M. L., Rogers, Q. R. and Morris, J. G. (1983). Role of linoleate as an essential fatty acid for the cat independent of archidonate synthesis. *Journal of Nutrition,* **113,** 1422–1433.

32. MacDonald, M. L., Anderson, B. C., Rogers, Q. R., Buffington, C. A. and Morris, J. G. (1984). Essential fatty acid requirements of cats: pathology of essential fatty acid deficiency. *American Journal of Veterinary Research,* **45,** 1310–1317.

33. Manku, M. S., Horrobin, D. F., Morse, N. L., Wright, S. and Burton, J. L. (1984). Essential fatty acids in the plasma phospholipids of patients with atopic eczema. *British Journal of Dermatology,* **110,** 640–648.

34. McLean, J. G. and Monger, E. A. (1989). Factors determining the essential fatty acid requirements of the cat. In *Nutrition of the Dog and Cat. Waltham Symposium Number 7.* Eds. I. H. Burger and J. P. W. Rivers. pp. 329–342. Cambridge: Cambridge University Press.

35. Miller, S. J. (1989) Nutritional deficiency and the skin. *Journal of the American Academy of Dermatology,* **21,** 1–30.

36. Miller, W. H. (1989). Nutritional considerations in small animal dermatology. *Veterinary Clinics of North America: Small Animal Practice,* **19,** 497–511.

37. Miller, W. H. (1990). Symmetrical truncal hair loss in cats. *The Compendium on Continuing Education,* **12,** 461–470.

38. Miller, W. H., Griffen, C. E., Scott, D. W., Angarano, D. K. and Norton, A. L. (1989). Clinical trial of DVM Derm Caps in the treatment of allergic disease in dogs: a nonblinded study. *Journal of the American Animal Hospital Association,* **25,** 163–168.

39. Muller, G. H., Kirk, R. W. and Scott, D. W. (1989). *Small Animal Dermatology.* 4th edn. pp. 796–806. Philadelphia: W. B. Saunders.

40. Norris, D. A. (1985). Zinc and cutaneous inflammation. *Archives of Dermatology,* **121,** 985–989.

41. Power, H. T. and Ihrke, P. J. (1992). Use of etretinate for treatment of primary keratinisation disorders (idiopathic seborrhoea) in Cocker Spaniels, West Highland White Terriers and Basset Hounds. *Journal of the American Veterinary Medical Association,* **201,** 419–429.

42. Rees, J. (1992). The molecular biology of retinoic acid receptors: orphan from a good family seeks home. *British Journal of Dermatology,* **126,** 97–104.

43. Rice, D and Kennedy, S. (1988). Vitamin E: function and effects of deficiency. *British Veterinary Journal,* **144,** 482–496.

44. Rivers, J. P. W. and Frankel, T. L. (1978). Fat in the diet of cats and dogs. In *Nutrition of the Dog and Cat. Proceedings of the International Symposium on the Nutrition of the Dog and Cat, Hanover.* Ed. R. S. Anderson. pp. 67–99. Oxford: Pergamon.

45. Rivers, J. P. W and Frankel, T. L. (1981). Essential fatty acid deficiency. *British Medical Bulletin,* **37,** 59–64.

46. Ross, A. C. (1991). Vitamin A: current understanding of the mechanisms of action. *Nutrition Today,* **26,** 6–12.

47. Sanecki, R. K., Corbin, J. E. and Forbes, R. M. (1985). Extracutaneous histologic changes accompanying zinc deficiency in pups. *American Journal of Veterinary Research,* **10,** 2120–2123.

48. Scarff, D. H. and Lloyd, D. H. (1992). Double blind, placebo controlled, crossover study of evening

primrose oil in the treatment of canine atopy. *Veterinary Record*, **131**, 97–99.

49. Scott, D. W. and Buerger, R. G. (1988). Nonsteroidal anti inflammatory agents in the management of canine pruritus. *Journal of the American Animal Hospital Association*, **24**, 425–428.

50. Scott, D. W. and Sheffey, B. E. (1987). Dermatosis in dogs caused by vitamin E deficiency. *Companion Animal Practice*, **1**, 42–46.

51. Scott, D. W. and Walton, D. K. (1985). Clinical evaluation of oral vitamin E for the treatment of primary canine acanthosis nigricans. *Journal of the American Animal Hospital Association*, **21**, 345–350.

52. Scott, D. W., Walton, D. K., Manning, T. O., Smith, C. A. and Lewis, R. M. (1983). Canine lupus erythematosus II: discoid lupus erythematosus. *Journal of the American Animal Hospital Association*, **21**, 481–488.

53. Sousa, C. A., Stannard, A. A., Ihrke, P. J., Reinke, S. I. and Schmeitzel, L. P. (1988). Dermatosis associated with feeding generic dog food: 13 cases (1981–1982). *Journal of the American Veterinary Medical Association*, **192**, 676–680.

54. Stewart, L. J., White, S. D. and Carpenter, J. L. (1991). Isotretinoin in the treatment of sebaceous adenitis in two Vizslas. *Journal of the American Animal Hospital Association*, **27**, 65–71.

55. Stubbs, C. D. and Smith, A. D. (1990). Essential fatty acids in membrane: physical properties and function. *Biochemical Society Transactions*, **18**, 779–780.

56. Thoday, K. L. (1989). Diet-related zinc-responsive skin disease in dogs: a dying dermatosis? *Journal of Small Animal Practice*, **30**, 213–215.

57. Tiwari, R. K., Clandinin, M. T., Cinader, B. and Goh, Y. K. (1987). Effect of high polyunsaturated fat diets on the composition of B cell and T cell membrane lipids. *Nutrition Research*, **7**, 489–498.

58. van den Broek, A. H. M. and Thoday, K. L. (1986). Skin disease in dogs associated with zinc deficiency: a report of 5 cases. *Journal of Small Animal Practice*, **27**, 313–323.

59. van den Broek, A. H. M. and Stafford, W. L. (1988). Diagnostic value of zinc concentrations in serum, leucocytes and hair of dogs with zinc-responsive dermatosis. *Research in Veterinary Science*, **44**, 41–44.

60. Vajreswari, A. and Narayanareddy, K. (1992). Effect of dietary fats on some membrane bound enzyme activities, membrane lipid composition and fatty acid profiles of rat heart sarcolemma. *Lipids*, **27**, 339–343.

61. Volker, L. (1989). The influence of biotin on skin and hair changes in the dog. In *Nutrition of the Dog and Cat. Proceedings of the International Symposium on the Nutrition of the Dog and Cat, Hanover*. Ed. R. S. Anderson. pp. 173–179. Oxford: Pergamon.

62. Watkins, B. A. and Kratzer, F. H. (1987). Effects of dietary biotin and lineolate on polyunsaturated fatty acids in tissue phospholipids. *Poultry Science*, **66**, 2024–2031.

63. Wertz, P. W. and Downing, D. T. (1983). Ceramides of pig epidermis: structure determination. *Journal of Lipid Research*, **24**, 759–765.

64. White, S. D., Rosychuk, R. A. W., Reinke, S. I. and Paradis. M. (1992). Use of tetracycline and niacinamide for treatment of autoimmune skin disease in 31 dogs. *Journal of the American Veterinary Medical Association*, **200**, 1497–1500.

65. Wolf, G. (1990). Recent progress in vitamin A research: nuclear retinoic acid receptors and their interaction with gene elements. *Journal of Nutritional Biochemistry*, **1**, 284–289.

66. Wolverton, S. E. (1991). Retinoids. In *Systemic Drugs for Skin Diseases*. Eds. S. E. Wolverton and J. K. Wilkin. pp. 187–218. Philadelphia: W. B. Saunders.

67. Wright, S. and Burton, J. L. (1982). Oral evening primrose seed-oil improves atopic eczema. *Lancet* **ii**, 1120–1122.

APPENDIX

Home-Made Diets

SUSAN DONOGHUE and DAVID S. KRONFELD

Introduction

The better quality manufactured petfoods are complete and balanced for stages of the life cycle of healthy animals, matching national standards of composition or proven by feeding trials. Dietetic products are designed for the avoidance of conditions that lead to poor performance or disease, therapeutic products for the amelioration of existing disease. Despite this array of commercial products, home-made diets still have occasional use in clinical nutrition.

Home-made diets are fed to pets for a variety of reasons. Some owners prefer home-made diets because of a feeling, at times unfounded, that fresh ingredients are purer and safer. Some pets prefer home-made diets to commercial products, especially those low in protein, fat or salt. In the hospital, a home-made diet may be designed precisely for an individual patient. In remote locations, commercial products, especially those with special formulations, may be unavailable. When locale or economy dictate, plentiful local ingredients, such as fish in northern villages and corn meal in the subtropics, may be fed.

We designed a simple system of recipes in 1980; it has been tested and developed continuously. These recipes readily replace pet owners' formulations that are well-intentioned but often have an accumulation of deficiencies and excesses. The recipes here are designed for every useful purpose in dogs and cats. Readers should note, however, that the nutrient contents of most ingredients are variable, threatening the balance of the diet, and that wholesomeness (the absence of contaminating pathogens and toxins) is a constant challenge.

Diet Modifications for Performance and Disease

Canine Performance (for above-maintenance needs such as growth, reproduction, lactation, stress and after surgery or trauma): This diet is identical to the feline maintenance diet. Reverse the proportions of rice and meat, using 140 g meat and 70 g raw rice. This provides 31% protein, 41% fat and 28% carbohydrate (ME basis).

Canine and Feline Low Fat Diets (for weight control and for nonspecific gatrointestinal disorders, malabsorption, osmotic diarrhea, pancreatitis, lymphangiectasia): Start with the maintenance diet for dog or cat. Replace medium fat meat by very lean meat, lean

RECIPES OF HOME-MADE DIETS FOR MAINTENANCE*				
Ingredient	**Canine**		**Feline**	
	Volume	**Weight**	**Volume**	**Weight**
		grams		grams
Rice, raw long-grain, white	2/3 cup†	140	1/3 cup	70
Meat, lean	1/3 cup	70	2/3 cup	140
Liver	1/8 cup	30	1/8 cup	30
Bone meal	3 teaspoons	11	3 tsp	11
Corn oil	1 teaspoon	5	1 tsp	5
Iodized salt	1/2 teaspoon	2	1/2 tsp	2

The recipe for dogs at maintenance provides 17% protein (sufficient if protein is high quality), 31% fat and 53% carbohydrate [metabolizable energy (ME) basis]. The recipe for cats at maintenance provides 31% protein, 41% fat and 28% carbohydrate (ME basis).

COOKING: Place rice, salt, oil and bone meal in 1.5 cups boiling water. Cover and simmer for 10 min. Add meat and liver. Cover and continue to simmer for another 10 min. Cool.

STORAGE: Diet can be refrigerated for a few days or frozen.

FEEDING: The recipe for canine maintenance provides about 800 kcal (3350 kJ), enough for an average 10 kg dog for 1 day.

The recipe for feline maintenance also provides about 800 kcal (3350 kJ), enough food for an average 3.8 kg cat for 3 days.

If any of the recipes are increased or decreased, all proportions must be maintained as above. For example, if the recipe is doubled, all ingredients must be doubled.

*From: Kronfeld, D.S.: therapuetic diets for dogs and cats including a simple system of recipes. *Tijdschrift voor Diergeneeskunde* 111, 37S-46S, 1986.
†Standard 8 fl. oz. (approximately 240 ml) measuring cup.

poultry or lean fish. This reduces dietary fat by about half.

For further reductions in fat, replace corn oil with 2.5 g, or even 1.25 g, safflower oil. Commercial preparations of medium chain triglycerides may be used as well.

Low Fat, High Fiber Canine and Feline Diets (for weight reduction and specific fiber-responsive gastrointestinal disorders): Use 70 g each for rice and very lean meat, and add 40 g (2/3 cupful) wheat bran. One 8 oz cupful of wheat bran weighs only 60 g. It provides 190 kcal (800 kJ), 17% protein and 4% fat (dry matter basis). Increase water (for cooking) by 50%. This diet provides 600 kcal (2500 kJ) and 24% of energy from protein.

The lignin in bran may be irritative to some animals. α-Cellulose is purer.

Geriatric Canine and Feline Diets: Older dogs and cats have varying needs. Some will be overweight and others underweight, for example, and others will be chronically ill and quiet but still others will be as energetic as younger individuals.

Most older dogs and cats will thrive on their respective maintenance diets. Those that are underconditioned do well with fatty meats, and older underconditioned dogs enjoy the feline maintenance diet.

Feline and Canine Hypoallergenic Diets: Avoid suspected ingredients, especially protein sources such as meats, that the affected animal has been consuming. For example, beef may be replaced by lamb (traditionally), venison, rabbit, fish or dairy products.

Also consider replacing corn oil with sunflower or safflower oils, bone meal with dicalcium phosphate and liver with a commercial vitamin–trace mineral mixture produced for daily use by humans.

Consult tables of exchanges for further guidance.

Feline Diet for Lower Urinary Tract Disease: Use the feline maintenance diet, for its large contribution from meat promote acidic urine. In addition, replace bone meal with 3 g calcium carbonate or ground limestone; this lowers magnesium from 0.15 to 0.08%. Salt may be increased to 4 g to encourage water intake.

Chronic Kidney Disease Diets: Current recommendations for dietary management of chronic renal failure call for moderate protein and reduced dietary phosphorus.

Start most CRF dogs and cats on their respective maintenance diets. The next step is to replace bone meal with 3 g calcium carbonate, which lowers phosphorus from 1.1 to 0.3%. To lower protein, replace medium fatty meat with fatty meat.

Chronic Heart Failure Diets: Start with canine and feline maintenance diets. For moderate heart failure, reduce salt to 1 g to provide 0.25% sodium. For advanced heart failure, omit salt. This reduces sodium to 0.05%.

For patients treated with potassium-losing diuretics or suffering digitalis intoxication, use commercial ("Lite") salts containing half-potassium and half-sodium chloride. This change not only reduces sodium but increases potassium from 0.5 to 0.7%.

Canine Low Purine Diet: Start with the canine maintenance diet, and replace liver with a commercial vitamin–mineral daily supplement for humans. Replace meat with one egg mixed into 100 ml milk. Introduce diet slowly. If lactose intolerance persists, omit milk and use two eggs.

FOOD EXCHANGES: PROTEIN AND ENERGY SOURCES					
Food	Weight (g)	Energy (kcal)	Protein (% kcal)	Fat (% kcal)	Carbo (% kcal)
Beef	100	215	58	42	0
May be exchanged with:					
Pork, tenderloin	90	215	53	47	0
Goat, raw	130	214	47	53	0
Rabbit, stewed	100	216	56	44	0
Hare, raw	160	216	65	35	0
Horse, raw	180	212	64	33	3
Reindeer, forequarter	120	214	51	49	0
Reindeer, hindquarter	84	215	31	69	0
Chicken w/o skin	100	215	45	55	0
Herring, Atlantic	120	211	40	60	0
Menhaden	125	215	45	55	0
Sardines, Pacific, raw	134	214	50	50	0
Three hens' eggs	150	237	32	65	3
Cottage cheese, creamed	208	215	49	40	11
Yogurt, plain, whole milk	350	214	22	47	31
All meats, dairy products and eggs should be cooked or pasteurized prior to feeding.					

FOOD EXCHANGES: HIGH PROTEIN, LOW FAT SOURCES

Food	Weight (g)	Energy (kcal)	Protein (% kcal)	Fat (% kcal)	Carbo (% kcal)
Chicken (w/o skin) *May be exchanged with:*	100	177	64	36	0
Turkey, w/o skin	112	176	80	20	0
Guinea hen, w/o skin	160	176	78	22	0
Quail, w/o skin	130	174	68	32	0
Guinea pig, raw	184	176	84	16	0
Armadillo, raw	103	178	70	30	0
Cod, raw	227	177	96	4	0
Mussels, Atlantic and Pacific	180	171	64	22	14
Tuna, bluefin	122	177	73	27	0
Cottage cheese, low fat (1%)	226	164	71	13	16
Yogurt, plain, low fat milk	280	178	33	22	45

Liver should be used in small quantities only, as a source of vitamins and trace minerals. Never use liver from any species as a substitute for meat. Its high content of vitamin A will produce hypervitaminosis A in dogs and cats. All meats, dairy products and eggs should be cooked or pasteurized prior to feeding.

FOOD EXCHANGES: CARBOHYDRATE SOURCES

Food	Weight (g)	Energy (kcal)	Protein (% kcal)	Fat (% kcal)	Carbo (% kcal)
Rice (cooked, white, long-grain) *May be exchanged with:*	100	110	8	1	91
Rice, cooked, brown	93	110	9	5	86
Potato, white	145	110	11	1	88
Yam, cooked	105	110	9	2	89
Bread, wheat	45	110	15	15	70
Bread, corn	45	110	8	33	59
Bread, matzo	28	110	10	2	88
Noodles, cooked	94	110	15	10	75
Tortilla, corn	50	110	12	14	74

Ensure wholesomeness of all ingredients prior to feeding. Foods should be free of moulds, vermin feces and foreign objects.

FOOD EXCHANGES: OILS				
Food	**Weight (g)**	**Energy (kcal)**	**Fat (g)**	**PUFA (% fat)**
Corn oil	14	120	13.6	59
May be exchanged with:				
Cottonseed oil	14	120	13.6	52
Peanut oil	14	120	13.6	32
Safflower oil	14	120	13.6	74
Sesame oil	14	120	13.6	42
Soybean oil	14	120	13.6	58
Sunflower oil	14	120	13.6	65
Wheat germ oil	14	120	13.6	62
Less desirable:				
Almond oil	14	120	13.6	18
Coconut oil	14	120	13.6	1
Olive oil	14	120	13.6	8
Palm oil	14	120	13.6	10
Palm kernal oil	14	120	13.6	1

FOOD EXCHANGE: MACROMINERALS					
Food	**Calcium (%)**	**(g)**	**Phosphorus (%)**	**(g)**	**Other minerals**
Bone meal (11g)	24	2.6	12.6	1.4	Na, F
May be exchanged with:					
Phosphate, dicalcium, 12g	21	2.5	18	2.2	Na, F
Phosphate, curacao, 7g	36	2.5	14	1.0	Na, F
Phosphate, soft rock, 14g	17	2.4	9	1.3	Na, F
Phosphate, defluorinated, 8g	32	2.6	18	1.4	Na, F
Calcium carbonate, 6.5g	40	2.6	0	0	Na
Limestone, 6.8g	38	2.6	0	0	Na, F

List of Abbreviations

AAA	aromatic amino acids	BW	body weight
AAFCO	Association of American Feed Control Officials	CAH	chronic active hepatitis
ACE	angiotensin-converting enzyme	CBC	complete blood count
ACEI	angiotensin-converting enzyme inhibitors	CCHF	chronic congestive heart failure
ADF	acid-detergent fiber	CCK	cholecystokinin
ADH	anti-diuretic hormone	CF	crude fiber
AIHA	autoimmune hemolytic anemia	CHF	congestive heart failure
		CHO	carbohydrate
ALDO	aldosterone	CNS	central nervous system
ALP	alkaline phosphatase	CO	cardiac output
ALT	alanine aminotransferase	CoA	coenzyme A
ANP	atrial natriuretic peptide	CSA	cysteinesulfinic acid
ANS	autonomic nervous system	CSF	cerebrospinal fluid
AOAC	Association of Official Analytical Chemists	CT	calcitonin
		CTA	conditioned taste aversion
AST	aspartate aminotransferase	Da	daltons
AT-I	angiotensin-I	DALA	delta-amino levulinic acid
AT-II	angiotensin-II	DCM	dilated cardiomyopathy
ATP	adenosine triphosphate	DE	digestible energy
		DF	dietary fiber
BCAA	branched chain amino acids	DHEA	dehydroepiandrosterone
BHA	butylated hydroxyanisole	DIC	disseminated intravascular coagulation
BHT	butylated hydroxytoluene		
bid	twice daily	DM	dry matter
BP	arterial blood pressure	DSH	domestic short hair
BSA	body surface area	EC	European Community
BSP	bromosulfophthalein	EDTA	ethylenediamine tetraacetic acid
BT	bacterial translocation		

EFA	essential fatty acids	K_f	glomerular ultrafiltration coefficient
ELISA	enzyme-linked immunosorbent assay		
eod	every other day	LDL	low density lipoprotein
EPI	exocrine pancreatic insufficiency	LES	lower esophageal sphincter
		LI	large intestine
		LUTD	lower urinary tract disease
FA	fatty acids		
FCRD	feline central retinal degeneration	MCT	medium chain triglyceride
		MCV	mean corpuscular volume
FDA	Food and Drug Administration	ME	metabolizable energy
		2-MPG	2-mercaptopropionylglycine
FDP	fibrin degradation products	MRI	magnetic resonance imaging
FE of P	fractional excretion (clearance) of phosphorus		
		NCP	noncellulose polysaccharide
FeLV	feline leukemia virus	NDF	neutral detergent fiber
FFA	free fatty acids	NFE	nitrogen free extract
FIP	feline infectious peritonitis	NHA	neurohormonal activation
FIV	feline immunodeficiency virus	NIDDM	noninsulin dependent diabetes mellitus
FUS	feline urological syndrome	NRC	National Research Council
		NS	nutritional support
GABA	γ-aminobutyric acid	NSAID	nonsteriod antiinflammatory drug
GALT	gut associated lymphoid tissue		
GCP	glomerular capillary pressure	NSP	nonstarch polysaccharides
		oz	ounce
GDV	gastric dilatation–volvulus	OCD	osteochondritis dissecans
GFR	glomerular filtration rate		
GGT	γ-glutamyl transferase	PCM	protein–calorie malnutrition
GI	gastrointestinal	PCR	protein–calorie ratio
GIT	gastrointestinal tract	PCV	packed cell volume
GMC	giant migrating contraction	PDP	principal display panel
GTP	guanosine triphosphate	PEG	percutaneous endoscopic gastrostomy
Hb	hemoglobin	PG	propylene glycol
HD	hip dysplasia	PO	per os, orally
HDL	high density lipoprotein	PRSL	potential renal solute load
HE	hepatic encephalopathy	PS	portosystemic
IBD	inflammatory bowel disease	pSAP	negative logarithm of the struvite activity product
ICG	indocyanine green		
ICH	infectious canine hepatitis	PSS	portosystemic shunts
IDDM	insulin dependent diabetes mellitus	PT	prothrombin time
		PTH	parathyroid hormone
IL-1	interleukin 1	PTT	partial thromboplastin time
IM	intramuscular		
IP	intraperitoneal		
IV	intravenous	qid	every 6 hr/four times daily
KCS	keratoconjunctivitis sicca	RA	right atrium

RAAS	renin–angiotensin–aldosterone system	T_4	thyroxine
RAST	radioallergosorbent test	tblsp	tablespoon
RDA	recommended daily allowance	TDDCM	taurine-deficient dilated cardiomyopathy
RE	reticuloendothelial	TDF	total dietary fiber
RES	reticuloendothelial system	TEF	thermogenic effect of food
RMR	resting metabolic rate	TG	triglyceride
RQ	respiratory quotient	tid	every 8 hr/three times daily
RSS	relative supersaturation	TLI	trypsin-like immunoreactivity
		TNF	tumor necrosis factor
SCFA	short chain fatty acids	TPN	total parenteral nutrition
SG	specific gravity	tsp	teaspoon
SI	small intestine		
SIBO	small intestinal bacterial overgrowth	UTI	urinary tract infection
sid	once daily	VIP	vasoactive intestinal peptide
SNS	sympathetic nervous system	VLDL	very low density lipoprotein
SV	stroke volume	VMH	ventral medial hypothalamus
T_3	triiodothyronine	WCPN	WALTHAM Centre for Pet Nutrition

INDEX

Note: page numbers in *italic* denote figure citations.

455

β-Hydroxy-γ-*N*-trimethylammonium *see* Carnitine
Hyperadrenocorticism 374
  differentiation from obesity 138
Hyperammonemia 241
Hypercholesterolemia, and fats 114
Hyperfiltration theory, chronic renal failure, canine 318–20
Hyperglycemia
  in absence of diabetes 377
  stress 40
Hyperkalemia
  emergency measures 329
  sodium polystyrene sulphonate 329
Hyperlipidemia, lipid infusions 52
Hyperparathyroidism, secondary 3, 396, 399–400
  causes 399
  nutritional 396, 407–10
  phosphorus restriction 324–5
  renal 399
Hyperphosphatemia, management 316–17
Hypertension
  dietary recommendations 358–9
  management 358–9
  renal-related 358
  systemic, canine 321–2, 326–8
    and sodium 326–8
Hyperthermia, in poisoning 153
Hyperthyroidism 373
Hypervitaminosis A 155, 397, 398, 413–16
  skeletal disease 397, 413–16
Hypervitaminosis D, skeletal disease 155, 397–8
Hypoadrenocorticism 374
Hypoallergenic diets 446
Hypocalcemia 3
Hypokalemia
  canine renal failure 329
  carbohydrate excess 34, 54
  cardiovascular disease 364, 366
  feline renal failure 286–7
  treatment *279*
Hypomagnesemia 54
  cardiovascular disease 364, 366
  see also Magnesium deficiency
Hyponatremia, congestive heart failure 362–4
Hypophosphatemia, carbohydrate excess 34, 54
Hypothermia, in poisoning 153
Hypothyroidism 373–4
  differentiation from obesity *138*

Icterus *see* Jaundice
Idiopathic megacolon 237
Immune response
  food hypersensitivity 174–5
  normal 173–4

Immunosuppressive therapy, supplement 213
Inappetence 122–3, 125–6
  drugs *80–1*, 80
  hepatic disease 261
  and estrus 122
Inappropriate feeding practices 1–13
Incontinence, fecal, diphenoxylate 233
Incontinence, urinary
  canine 348–50
    diagnosis 349
    etiology and pathogenesis 348–9
    neurogenic 348, 350
    reflex dyssynergia 349
    sphincter mechanism 348, 350
Indospicine 158
  contaminated flesh 158
Infectious canine hepatitis (ICH) 249–50
Inflammatory bowel diseases
  cats 235–7
  dogs 211
Inflammatory mediators 175
Insulin replacement therapy
  cats 386
  dogs 377–9
  inappetence 81–2
  pharmacokinetics 378
Insulin resistance
  associated diseases 386
  and obesity 133–4
Interleukin-1, tumor cachexia 79
Intestinal lymphangiectasia 205
Intestinal phycomycosis 206–8
Intrinsic factor 112
Iodine
  deficiency 5–6
  excess 10
Irish Setters, gluten enteropathy 172, 177, 204–5
Iron
  deficiency 4
  excess 10
  in kidney failure 286
Isphagula, analysis *89*

Jaundice
  diagnostic approach 249
  differential diagnosis *248*
  poisoning 151, 157
Juvenile osteodystrophy 407–10

Keratinization defects
  etretinate 439
  vitamin A deficiency 438
Keratoconjunctivitis sicca 213–14